Discretization Methods in Structural Mechanics

International Union of Theoretical and Applied Mechanics

International Association for Computational Mechanics

G. Kuhn · H. Mang (Eds.)

Discretization Methods in Structural Mechanics

IUTAM/IACM Symposium Vienna/Austria, 5.–9.6.1989

Springer-Verlag
Berlin Heidelberg New York
London Paris Tokyo Hong Kong

Prof. Dr.-Ing. habil. Günther Kuhn
Universität Erlangen-Nürnberg
Am Pestalozziring 20
8520 Erlangen
FR Germany

Prof. Dr. techn. Herbert Mang, Ph. D.
Technische Universität Wien
Karlsplatz 13
1040 Wien
Austria

Library of Congress Cataloging-in-Publication Data
IUTAM/IACM Symposium (1989 : Vienna, Austria)
Discretization methods in structural mechanics : IUTAM/IACM Symposium, Vienna, Austria, 5.–9.6.1989 / G. Kuhn, H. Mang (eds.)
(IUTAM symposia)
At head of title: International Union of Theoretical and Applied Mechanics.

ISBN-13: 978-3-642-49375-1 e-ISBN-13: 978-3-642-49373-7

DOI: 10.1007/978-3-642-49373-7

1. Structural analysis (Engineering)--Data processing--Congresses.
2. Boundary element methods--Congresses.
3. Finite element method--Congresses.
I. Kuhn, G. (Günther)
II. Mang, H. (Herbert)
III. International Union of Theoretical and Applied Mechanics.
IV. Title. V. Series: IUTAM-Symposien.
TA647.I97 1989
624.1'71--dc20 90-9605

Softcover reprint of the hardcover 1st edition 1990

2161/3020 543210 – Printed on acid-free paper

Preface

The advent of the digital computer has given great impetus to the development of modern discretization methods in structural mechanics. The young history of the finite element method (FEM) reflects the dramatic increase of computing speed and storage capacity within a relatively short period of time. The history of the boundary element method (BEM) is still younger. Presently, intense scientific efforts aimed at extending the range of application of the BEM can be observed. More than 10 years ago, O.C. Zienkiewicz and his co-workers published the first papers on the coupling of FE and BE discretizations of subregions of solids for the purpose of exploiting the complementary advantages of the two discretization methods and reducing their disadvantages.

The FEM has revolutionized structural analysis in industry as well as academia. The BEM has a fair share in the continuation of this revolution. Both discretization methods have become a domain of vigorous, world-wide research activities. The rapid increase of the number of specialized journals and scientific meetings indicates the remarkable increase of research efforts in this important subdomain of computational mechanics.

Several discussions of this situation in the Committee for Discretization Methods in Solid Mechanics of the Society for Applied Mathematics and Mechanics (GAMM) resulted in the plan to submit a proposal to the General Assembly of the International Union of Theoretical and Applied Mechanics (IUTAM) to sponsor a pertinent IUTAM Symposium. The endorsement of this proposal by the aforementioned GAMM-Committee and the consent of the International Association for Computational Mechanics (IACM) to co-sponsor such a scientific activity paved the way to the IUTAM/IACM Symposium on Discretization Methods in Structural Mechanics in Vienna, Austria, from June 5 to 9, 1989.

The main reason for the rather extensive topic of this Symposium, as reflected by its title, was the merger, at the request of IUTAM, of the original proposal by Prof. H.A. Mang, emphasizing the application of discretization methods to nonlinear structural mechanics, with a proposal by a group of members of the GAMM Committee for Discretization Methods in Solid Mechanics, from the Federal Republic of Germany, around Prof. G. Kuhn, aimed at the application of the BEM to problems of structural mechanics. After some hesitation the undersigned organizers came to the conclusion that the advantages of this merger outweighed the disadvantages.

The organization of the scientific program was guided by the intention to present an unbiased view on the capacity of the FEM, the BEM, and of combinations of the two methods as timely tools for the numerical solution of challenging problems in the field of structural mechanics. The active support by the members of the Scientific Committee, which is gratefully acknowledged, facilitated the realization of this intention. Apart from the organizers, the Scientific Committee consisted of the following members:

J. Altenbach (Magdeburg/GDR)	J. H. Argyris (Stuttgart/FRG)	S. M. Atluri (Atlanta/USA)
N. Banichuk (Moscow/USSR)	R. H. Gallagher (Potsdam/USA)	P. Germain (Paris/France)
Feng Kang (Beijing/China)	M. Kleiber (Warsaw/Poland)	G. Maier (Milan/Italy)
T. H. H. Pian (Cambridge/USA)	A. Samuelsson (Göteborg/Sweden)	A. G. Ugodchicov (Gorki/USSR)
	Y. Yamamoto (Tokyo/Japan)	

81 scientists from 15 countries participated in the Symposium. The 42 oral presentations from leading experts in the area of computational mechanics gave a good impression of recent trends in discretization methods in structural mechanics. In some of these contributions fundamental mathematical and/or mechanical problems related to these methods were treated. Some other contributions were devoted to the application of discretization methods to structural analysis of beams, plates, shells, etc. A large segment of the presentations consisted of investigations of various mechanical phenomena, such as instability, vibrations, crack propagation, etc., by means of the FEM or the BEM.

The book contains written versions of 40 out of the 42 oral presentations. The manuscripts were to be submitted approximately two months after the Symposium, allowing the authors to consider comments from the discussion of their presentations.

The organizers are deeply indebted to the following institutions and corporations for their generous financial support or for their effective help in the preparation and organization of the Symposium:

International Union of Theoretical and Applied Mechanics (IUTAM)

International Association for Computational Mechanics (IACM)

Austrian Academy of Sciences

Federal Ministry of Science and Research of the Republic Austria

Daimler Benz AG, Stuttgart

Siemens AG Austria

Creditanstalt-Bankverein, Zweigstelle Rilkeplatz, Vienna

Institute for Applied Mechanics of the University of Erlangen-Nuremberg

Institute for Strength of Materials of the Technical University of Vienna

The organizers would like to express their gratitude to Prof. O. Hittmair, President of the Austrian Academy of Sciences, for his opening address on behalf of the Academy, which was hosting the Symposium in its historical building, and to Prof. F. Ziegler, Member of the Bureau of IUTAM, and Prof. R.H. Gallagher, President of Clarkson University and Member of the IACM Exectutive Council, for their opening addresses on behalf of IUTAM and IACM, respectively.

The organizers are also deeply indebted to their associates at the Institute for Applied Mechanics of the University of Erlangen-Nuremberg and the Institute for Strength of Materials of the Technical University of Vienna, respectively, for their help before, during and after the Symposium.

Finally, as the editors of this book, they would like to thank the editorial staff of Springer-Verlag for the advice and the efforts concerning publication of these proceedings.

G. Kuhn
Erlangen, FRG

H. A. Mang
Vienna, Austria

November 1989

List of Participants

Session No. 11	Abel	Prof., John F. School of Civil and Environmental Engineering Cornell University Hollister Hall Itaca, New York 14853, USA
Session No. 4	Altenbach	Prof. Dr., Johannes Technische Universität "Otto von Guericke" Magdeburg Sektion Maschinenbau Postschließfach 124, DDR-3010 Magdeburg
Session No. 4	Andelfinger	Dr., Ulrich Institut für Baustatik Universität Stuttgart Pfaffenwaldring 27, D-7000 Stuttgart 80
Session No. 5	Antes	Prof. Dr., Heinz Institut für Angewandte Mechanik Universität Braunschweig Beethovenstraße 51, Postfach 3329, D-3300 Braunschweig
Session No. 9	Atluri	Prof., Satya N. Center for Computational Mechanics Georgia Institute of Technology Atlanta, Georgia 30332-0356, USA
Session No. 1	Babuška	Prof., Ivo Institute for Physical Science and Technology The University of Maryland College Park, Maryland 20742-2431, USA
Session No. 7	Balmer	Dipl.-Ing., Hans Institut für Computer-Anwendungen Technische Universität Stuttgart Pfaffenwaldring 27, D-7000 Stuttgart 80

Session No. 5	Banerjee	Prof., P.K. Department of Civil Engineering State University of New York at Buffalo 212 Ketter Hall, Buffalo New York 14260, USA
Session No. 17	Banichuk	Prof., N.V. USSR Academy of Sciences Prospect Vernadskogo 101 II 7526 Moscow, USSR 43 44 341
	Becker	Dipl.-Phys., Ingo Institut für Technische Mechanik/Festigkeitslehre Universität Karlsruhe Kaiserstr. 12, D-7500 Karlsruhe
	Becker	Dipl.-Phys., Klaus FB 7, FG 7 Universität Duisburg Lotharstr. 1, D-4100 Duisburg
Session No. 3	Besseling	Prof., J.F. Laboratory for Engineering Mechanics Delft University of Technology P.O. Box 5033, 2600 CD Delft Mekelweg 2, 2628 Delft The Netherlands
Session No. 2	Brebbia	Dr., Carlos A. Director of Computational Mechanics Wessex Institute of Technology Ashurst Lodge, Ashurst GB-Southhampton SO4 2AA
Session No. 16	Bufler	Prof. Dr., Hans Institut für Mechanik (Bauwesen) Lehrstuhl II Universität Stuttgart Pfaffenwaldring 7, D-7000 Stuttgart

	Celigoj	Prof. Dr., Christian Institut für Werkstoffkunde, Festigkeitslehre und Materialprüfung Technische Universität Graz Kopernikusgasse 24, A-8010 Graz
Session No. 18	Chen	Dr., Zhen-Sheng Institut für Festigkeitslehre Technische Universtität Wien Karlsplatz 13, A-1040 Wien
Session No. 6	Cole	Mr., G. FEA Ltd., formerly Kingston Polytechnic Institute 15, Upper Gordon Road Camberley, Surrey GU152 HN U.K.
Session No. 6	Crisfield	Prof., Mike A. Department of Aeronautics Imperial College of Science and Technology London SW 7 2BY U.K.
Session No. 2	Cruse	Dr., Thomas A. Department of Engineering Mechanics Southwest Research Institute P.O. Drawer 28510 San Antonio, Texas 78238, USA
	Dallner	Dipl.-Ing., Rudolf Lehrstuhl für Technische Mechanik Universität Erlangen-Nürnberg Am Pestalozziring 20, D-8520 Erlangen
	Dinkler	Dr., Dieter Institut für Statik Technische Universität Braunschweig Beethovenstr. 51, D-3300 Braunschweig
	Eberhardsteiner	Dr., Josef Institut für Festigkeitslehre Technische Universtität Wien Karlsplatz 13, A-1040 Wien

Session No. 17	Fischer	Prof. Dr., Franz D. Institut für Mechanik Montanuniversität Leoben Franz Josef-Str. 18, A-8700 Leoben
	Gallagher	Prof., Richard H. President of Clarkson University Snell Hall Potsdam, New York 13676, USA
	Günther	Prof. Dr., Hubert Institut für Mechanik Akademie der Wissenschaften der DDR Reichenhainer Str. 88, DDR-9022 Karl-Marx-Stadt
Session No. 13	Hackbusch	Prof. Dr., Wolfgang Institut für Informatik und Praktische Mathematik Christian-Albrechts-Universität Olshausenstr. 40, D-2300 Kiel 1
	Heuer	Dr., Rudolf Institut für Allgemeine Mechanik Technische Universität Wien Wiedner Hauptstraße 8-10, A-1040 Wien
Session No. 12	Ingraffea	Prof., Anthony Department of Structural Engineering Hollister Hall, Cornell University Ithaca, New York 14853, USA
Session No. 18	Karaosmanoglu	Dipl.-Ing., Namik Institut für Technische Mechanik und Festigkeitslehre Universität Karlsruhe Kaiserstr. 12, D-7500 Karlsruhe
Session No. 14	Kleiber	Prof., Michał Institute of Fundamental Technological Research Polish Academy of Sciences Swiętokrzyska 21, PL-00-049 Warszawa

Session No.10	Krätzig	Prof. Dr., Wilfried B. Institut für Konstruktiven Ingenieurbau Lehrstuhl III - Statik und Dynamik Ruhr-Universität Bochum Universitätsstr. 150, D-4630 Bochum
Session No.10	Kröplin	Prof. Dr., Bernd Institut für Statik und Dynamik der Luft- und Raumfahrtkonstruktionen Universität Stuttgart Pfaffenwaldring 27, D-7000 Stuttgart
	Krzizek	Doz. Dr., Helmut Institut für Baubetrieb und Bauwirtschaft Technische Universität Wien Argentinierstr. 8, A-1040 Wien
	Kuhn	Prof. Dr., Günther Lehrstuhl für Technische Mechanik Universität Erlangen-Nürnberg Am Pestalozziring 20, D-8520 Erlangen
	Lehar	Dr., H. Institut für Festigkeitslehre und Flächentragwerke Universtität Innsbruck Technikerstraße 13, A-6020 Innsbruck
Session No. 18	Li	Mag.rer.nat., Zhong-Kui Institut für Festigkeitslehre Technische Universtität Wien Karlsplatz 13, A-1040 Wien
Session No. 9	Maier	Prof., Giulio Politecnico di Milano Dipartimento di Ingegneria Strutturale Piazza Leonardo da Vinci 32, I-20133 Milano
Session No. 18	Mang	Prof. Dr., Herbert A. Institut für Festigkeitslehre Technische Universtität Wien Karlsplatz 13, A-1040 Wien

	Meschke	Dr., Günther Institut für Festigkeitslehre Technische Universtität Wien Karlsplatz 13, A-1040 Wien
	Möhrmann	Dr., Walter Abt. E7CU c/o Daimler-Benz AG Postfach 60 02 02, D-7000 Stuttgart 60
Session No. 13	Mukherjee	Prof., Subrata Department of Theoretical and Applied Mechanics Kimball Hall, Cornell University Ithaca, NY 14850, USA
	Mukherjee	Dr., Yu Xi Department of Theoretical and Applied Mechanics Kimball Hall, Cornell University Ithaca, NY 14850, USA
Session No. 10	Nakagiri	Prof., Shigeru Institute of Industrial Science University of Tokyo 7-22-1, Roppongi, Minatoku Tokyo 106, Japan
Session No. 8	Noor	Prof., Ahmed K. Joint Institute for Advancement of Flight Sciences The George Washington University Mail Stop 269, Virginia 23665-5225, USA
Session No. 9	Novati	Ing., G. Politecnico di Milano Dipartimento di Ingegneria Strutturale Piazza Leonardo da Vinci 32, I-20133 Milano
Session No. 4	Nygård	Dr., Magne A.S Veritas Research P.O.Box 300, N-1322 Hoevik

Session No. 8	Oñate	Prof., Eugenio E.T.S. Ingenieros de Caminos, Canales y Puertos Universidad Politécnica de Cataluña Avda. Gregorio Maranón s/n, 08034 Barcelona, Spain
Session No.15	Orkisz	Prof., Janusz Institute of Structural Mechanics Technological University of Cracow Warsaw st. 24, PL-31 155 Kraków
Session No. 14	Owen	Prof., D. Roger J. Department of Civil Engineering University College of Swansea Applied Science Building, Singleton Park GB-Swansea SA2 8PP
	Parreira	Dr., Pedro Departamento Engenharia Civil Instituto Superior Técnico Av. Rovisco Pais 1096 Lisboa Codex, Portugal
Session No. 16	Pian	Prof., Theodeore H.H. Department of Aeronautics and Astronautics (Rm 33-311) Massachusetts Institute of Technology Cambridge, MA 02139, USA
Session No. 9	Polizzotto	Prof., Castrenze Dipartimento Ingegneria Strutturale e Geotecnica, DISEG Università di Palermo Viale delle Scienze, I-90128 Palermo
	Prochazka	Dr., Werner Institut für Festigkeitslehre Technische Universität Wien Karlsplatz 13, A-1040 Wien
	Quante	Dipl.-Ing., R. Institut für Konstruktiven Ingenieurbau Lehrstuhl III - Statik Postfach 10 21 48, D-4630 Bochum

Session No. 17	Rammerstorfer	Prof. Dr., Franz G. Institut für Leichtbau und Flugzeugbau Technische Universität Wien Gußhausstr. 25-29, A-1040 Wien
	Rank	Dr., Ernst ZTI DES 252 Siemens AG Otto-Hahn-Ring 6, D-8000 München 83
	Reimann	Dr., Ants Institut für Festigkeitslehre Technische Universtität Wien Karlsplatz 13, A-1040 Wien
	Reissner	Prof., Eric 422 Lucas Drive, Solana Beach CA 92075, USA
	Rieder	Prof. Dr.rer.nat., Georg Sandweg 37, D-5100 Aachen
Session No. 12	Rizzo	Prof., Frank J. Department of Engineering Science and Mechanics Iowa State University Ames, Iowa 50011, USA
	Rußwurm	Dipl.-Ing., Siegfried Lehrstuhl für Technische Mechanik Universität Erlangen-Nürnberg Am Pestalozziring 20, D-8520 Erlangen
Session No. 6	Samuelsson	Prof., Alf Chalmers University of Technology Department of Structural Mechanics Sven Hultins Gata 8, S-41296 Göteborg
	Sawaki	Dr., Yutaka Lehrstuhl für Technische Mechanik Universität Erlangen-Nürnberg Am Pestalozziring 20, D-8520 Erlangen

Session No. 18	Schnack	Prof. Dr., Eckart Institut für Technische Mechanik Fakultät für Maschinenbau Universität Karlsruhe Kaiserstr. 12, D-7500 Karlsruhe
Session No. 8	Simo	Prof., Juan C. Division of Applied Mechanics Department of Mechanical Engineering Stanford University Stanford, CA 94305, USA
Session No. 9	Sirtori	Prof., S. Politecnico di Milano Dipartimento di Ingegneria Strutturale Piazza Leonardo da Vinci 32, I-20133 Milano
	Springer	Dipl.-Ing., Hartmut Lehrstuhl für Statik Technische Universität München Arcisstr. 21, D-8000 München 2
Session No. 1	Stein	Prof. Dr., Erwin Universität Hannover Institut für Baumechanik und Numerische Mechanik Callinstr. 32, D-3000 Hannover
Session No. 11	Sumi	Prof., Yoichi Department of Naval Architecture and Ocean Engineering Yokohama National University Tokiwadai, Hodogaya-ku Yokohama 240, Japan
Session No. 18	Torzicky	Dipl.-Ing., Paul Institut für Festigkeitslehre Technische Universtität Wien Karlsplatz 13, A-1040 Wien

	Tralli	Prof., Antonio Dipartimento di Construzioni Facoltá di Architettura Università di Firenze Piazza Brunelleschi, 6, I-50121 Firenze
Session No. 3	Ugodchicov	Prof., A.G. Gorki State University Gagarina 23, Gorki 603022, UdSSR
	Wagner	Dr., W. Institut für Baumechanik und Numerische Mechanik Universität Hannover Appelstraße 9A, D-3000 Hannover
Session No. 14	Waszczyszyn	Prof., Zenon Politechnika Krakowska Institut Mechaniki Budowli ul. Warszawska 24, PL-31-155 Kraków
Session No. 18	Wendland	Prof. Dr., Wolfgang L. Lehrstuhl 6, Angewandte Mathematik Universität Stuttgart Pfaffenwaldring 57, D-7000 Stuttgart
	Werner	Prof. Dr., Heinrich Institut für Bauingenieurwesen I Technische Universität München Arcisstr. 21, D-8000 München 2
	Wriggers	Dr., Peter Institut für Baumechanik und Numerische Mechanik Universität Hannover Appelstraße 9A, D-3000 Hannover 1
Session No. 7	Wunderlich	Prof. Dr., Walter Institut für Baustatik Technische Universität München Arcisstr. 21, D-8000 München

Session No. 11	Yamamoto	Prof., Yoshiyuki Department of Mechanical Engineering Tokyo Denki University Ishzaka Hatoyama Saitama 350-03, Japan
	Zalten	Dipl.-Ing., W. Institut für Konstruktiven Ingenieurbau Lehrstuhl III - Statik Ruhr-Universität Bochum Postfach 10 21 48, D-4630 Bochum 1
Session No. 15	Ziegler	Prof. Dr., Franz Institut für Allgemeine Mechanik Technische Universität Wien Wiedner Hauptstr. 8-10, A-1040 Wien

List of Sessions

Monday, June, 5

Session No. 1 — **FEM: Mathematical Fundamentals**
Chairman: W.L. Wendland

Babuška, I. — *The h-p Version of the Finite Element Method*
(paper not submitted for publication)

Stein, E.
Lambertz, K.-H.
Plank, L. — *Error Estimators and Mesh Adaptation for Thin-walled Structures at Finite Rotations*

Session No. 2 — **BEM: Domain Integrals**
Chairman: S.N. Atluri

Brebbia, C.A.
Nowak, A.J. — *Treatment of Domain Integrals by Using The Dual and Multiple Reciprocity Methods*

Cruse, T.A.
Raveendra, S.T.
Thacker, B.H. — *BIE Solutions for Applied Thermoplasticity by Collocation*

Session No. 3 — **FEM: Mechanical Fundamentals**
Chairman: H. Günther

Besseling, J.F. — *State Characterization in Sampling Points and Representation of Velocity Fields by Nodal Values as a Basis for Discretization*

Ugodchicov, A.G.
Pautov, A.N. — *The Method of Consecutively Embedded Finite-Element Models*

Session No. 4	**FEM: Element Development** Chairman: T.H.H. Pian
Altenbach, J.	*Locally Blending Interpolation Technique in Finite Element Mesh Refinements and Improved Stress Analysis*
Nygård, M.K. Bergan, P.G.	*Nonconforming Finite Elements Based on the Free Formulation*
Andelfinger, U. Ramm, E.	*Assumed Strain Elements with Optimal Interpolation*

Tuesday, June, 6

Session No. 5	**BEM: Time-Dependent Problems** Chairman: F. Ziegler
Banerjee, P.K. Dargush, G.F.	*Linear and Nonlinear Transient Problems of Thermomechanics by BEM*
Antes, H.	*Dynamic Interaction Analysis in Wave Propagation Problems by a Time-Dependent Boundary Element Method*

Session No. 6	**FEM: Beams, Plates** Chairman: J.F. Besseling
Crisfield, M.A. Cole, G.	*Co-Rotational Beam Elements for Two- and Three-Dimensional Non-Linear Analysis*
Kjellman, M. Samuelsson, A.	*A High-Precision Adaptive Procedure for Solving Kirchhoff Plates*

Session No. 7	**FEM: Shells** Chairman: E. Reissner
Argyris, J. Balmer, H. Doltsinis, I.St.	*Natural Finite Differences in Shell Modelling for Impact Analysis*
Wunderlich, W. Springer, H. Goebel, W.	*Discretization and Solution Techniques for Liquid Filled Shells of Revolution under Dynamic Loading*
Session No. 8	**FEM: Nonlinear Analysis** Chairman: A. Samuelsson
Noor, A.K. Peters, J.M.	*Novel Computational Strategies for Solution of Large-Scale Structural Problems*
Oñate, E. Taylor, R.L. Zienkiewicz, O.C.	*Consistent Formulation of Shear Constrained Reissner-Mindlin Plate Elements*
Simo, J.C. Fox, D.D. Rifai, M.S.	*A Nonlinear Shell Theory with Through-The-Thickness Stretch*

Wednesday, June, 7

Session No. 9	**BEM: Nonlinear Analysis** Chairman: S. Mukherjee
Atluri, S.N.	*On Boundary-Domain-Element Methods in Nonlinear Structural Mechanics* (paper not submitted for publication)
Maier, G. Novati, G. Sirtori, S.	*On Symmetrization in Boundary Element Elastic and Elastic-Plastic Analysis*
Polizzotto, C. Zito, M.	*A Variational Formulation of the BEM for Elastic-Plastic Analysis*

Session No. 10 **FEM: Instability**
Chairman: W. Wunderlich

Krätzig, W.B.
Discretized Strategies for the Investigation of Instability Problems in Structural Mechanics

Kröplin, B.
Dinkler, D.
On Static and Dynamic Instability Analysis of Thin-Walled Structures

Nakagiri, S.
Probabilistic Finite Element Analysis of Unsymmetrical Buckling of Thin Shallow Spherical Shells

Thursday, June, 8

Session No.11 **Fracture Mechanics**
Chairman: A. Ingraffea

Sumi, Y.
Yamamoto, Y.
A Combined Analytical and Finite Element Method for Non-Collinear Crack Growth in Brittle Solids

Lin, S.-C.
Abel, J.F.
Numerical Simulation of Crack Propagation in Shells

Session No. 12 **BEM/FEM: Fracture Mechanics**
Chairman: T.A. Cruse

Krishnasamy, G.
Rudolphi, T.J.
Schmerr, L.W.
Rizzo, F.J.
Discretization Considerations with Hypersingular Integral Formulas for Crack Problems

Wawrzynek, P.
Martha, L.
Ingraffea, A.
FRANSYS: A Software System for the Simulation of Crack Propagation in Three Dimensions

Session No. 13	**BEM: Shape Optimization, Clustering Techniques** Chairman: I. Babuška
Zhang, Q. Mukherjee, S.	*Design Sensitivity Coefficients for Linear Elasticity Problems by the Derivative Boundary Element Method*
Hackbusch, W.	*The Panel Clustering Method for BEM*

Session No. 14	**FEM: Computational Strategies/Stability/ Instability** Chairman: W.B. Krätzig
Waszczyszyn, Z. Picczara, J.	*Exact Finite Elements for the Linear Buckling Analysis of Unidimenisonal Structural Problems*
Kleiber, M.	*Coupled Numeric-Symbolic Computations in Structural Mechanics*
Owen, D.R.J. Alves Filho, J.S.R. Mitchell, G.P. Kasper, K.	*The Implementation of Finite Element Computations in Transputer Based Systems*

Friday, June, 9

Session No. 15	**BEM: Plates/Nonlinear Mechanics** Chairman: G. Maier
Heuer, R. Irschik, H. Ziegler, F.	*A BEM-Formulation of Nonlinear Plate Vibrations*
Krok, J. Orkisz, J.	*A Unified Approach to the FE and Generalized Variational FD Methods in Nonlinear Mechanics, Concepts and Numerical Approach*

Session No. 16	**FEM: Shells and Membranes** Chairman: G. Kuhn
Pian, T.H.H. Li, M.	*Stress Analysis of Laminated Composites by Hybrid Finite Elements*
Bufler, H. Schneider, H.	*Liquid Loaded Membranes under Large Deformations*
Session No. 17	**FEM: Vibrations, Micromechanics** Chairman: A.G. Ugodchicov
Banichuk, N.V. Bratus. A.S.	*Stabilizing and Destabilizing Effects of Small Damping for Structures with Finite Number of Degrees of Freedom*
Rammerstorfer, F.G. Fischer, F.D. Böhm, H.J.	*Treatment of Micro-Mechanical Phenomena by Finite Elements*
Session No. 18	**Coupling of FEM and BEM** Chairman: E. Stein
Wendland, W.L.	*On the Coupling of Finite Elements and Boundary Elements*
Schnack, E. Becker, I. Karaosmanoglu, N.	*Three-dimensional Coupling of FEM and BEM in Elasticity*
Chen, Z.-S. Hofstetter, G. Li, Z.-K. Mang, H.A. Torzicky, P.	*Coupling of FE- and BE-Discretizations for 3D-Stress Analysis of Tunnels in Layered Anisotropic Rock*

Contents

Error Estimators and Mesh Adaptation for Thin-Walled Structures at Finite Rotations

E. Stein [1] K.-H. Lambertz L. Plank

Abstract

In this paper we develop an approach to the nonlinear analysis of folded plate structures. It includes a new formulation for the geometrically nonlinear analysis, an adaptive mesh refinement strategy and the use of multigrid methods.

1 Introduction

Despite considerable progress in recent years, the nonlinear analysis of folded plate structures is still a computationally expensive task. Starting point of our approach to this problem is a new formulation of the nonlinear equations for folded plate structures with small strains and finite rotations. As this formulation allows only small relative rotations in the element — besides finite rigid body rotations —, we control the out-of-plane distortion and additional a-priori and a-posteriori criteria, and refine the finite element mesh adaptively. The availability of several refinement levels is further exploited by the use of a multigrid solver.

2 The Nonlinear Equations for Folded Plate Structures

2.1 Kinematics

We assume the initial configuration to be composed of several plane surfaces. The plate thickness h is small compared with the other dimensions. We define the orthonormal basis in the plane (see Stein, Gruttmann, Lambertz [12], Gruttmann, Stein, Wriggers [7]). $\mathbf{t}_i$ (i=1,2,3) with the associated coordinates s_i in the initial configuration. The director $\mathbf{t}_3$ is normal to the undeformed middle surface. Further, the orthonormal basis $\mathbf{a}_i$ is introduced in the current configuration. Since transverse shear strains are allowed, the vector $\mathbf{a}_3$ is no longer normal to the deformed middle surface. As $\mathbf{t}_i$ and $\mathbf{a}_i$ are orthonormal vectors we can give the following transformation

$$\mathbf{a}_i = \mathbf{R}\mathbf{t}_i = \mathbf{R}^D\mathbf{t}_i^C, \tag{1}$$

where $\mathbf{t}_i^C$ denotes a corotational orthonormal basis. The orthogonal tensor $\mathbf{R}$ is described in matrix notation with three rotational degrees of freedom. To derive the material

[1] Prof. Dr.-Ing. E. Stein, Dr.-Ing. K.-H. Lambertz, Dipl. Math. L. Plank, Institut für Baumechanik und Numerische Mechanik, University of Hannover, Appelstraße 9a, D-3000 Hannover 1

G. Kuhn · H. Mang (Eds.)
Discretization Methods in Structural Mechanis
IUTAM/IACM Symposium Vienna, Austria, 1989

deformation gradient it is necessary to calculate the derivatives of the basis $\mathbf{a}_i$. Therefore we introduce a skew-symmetric tensor $\boldsymbol{\Omega}_\alpha$ by

$$\frac{\partial \mathbf{a}_i}{\partial S_\alpha} = \frac{\partial \mathbf{R}}{\partial S_\alpha}\mathbf{R}^T\mathbf{a}_i = \boldsymbol{\Omega}_\alpha \mathbf{a}_i = \boldsymbol{\omega}_\alpha \times \mathbf{a}_i \tag{2}$$

and the associated axial vector $\boldsymbol{\omega}_\alpha$. Using Reissner-Mindlin kinematics the position vector $\mathbf{x}$ of a point in the shell space in the current configuration is given by

$$\mathbf{x} = \mathbf{x}_0 + \xi \mathbf{a}_3 \quad ; \quad -\frac{h}{2} \leq \xi \leq +\frac{h}{2} \tag{3}$$

with the classical assumption of inextensibility of the director vector $\mathbf{a}_3$. Here,

$$\mathbf{x}_0 = \mathbf{X}_0 + \mathbf{u} = x_i\,\mathbf{t}_i = (X_i + u_i)\,\mathbf{t}_i \tag{4}$$

is the position vector of a point on the middle surface and $\xi = s_3$ is the coordinate in the thickness direction of the shell; the assumption of inextensibility is justified by the small strain assumption. To describe the deformation of the structure it is necessary to derive the material deformation gradient $\mathbf{F}$ which is defined by $d\mathbf{x} = \mathbf{F}\,d\mathbf{X}$. Making use of eq. 2 the variation of the deformation gradient can be computed with regard to the kinematics as

$$\delta\mathbf{F} \;=\; \left[\underbrace{\delta\mathbf{x}_{0,\alpha}}_{*} + \underbrace{\xi(\delta\boldsymbol{\omega}_\alpha \times \mathbf{a}_3 + \boldsymbol{\omega}_\alpha \times (\boldsymbol{\vartheta} \times \mathbf{a}_3))}_{**}\right] \otimes t_\alpha + \underbrace{\boldsymbol{\vartheta} \times \mathbf{a}_3}_{**} \otimes \mathbf{t}_3, \tag{5}$$

where * marks the terms which result from the variation of the middle surface and ** the terms resulting from the variation of the orthogonal matrices.

2.2 Principle of Virtual Work

In this section we define stress resultants and stress couple resultants from the three-dimensional theory. The stress resultants

$$\mathbf{n}_\alpha := \int_{-\frac{h}{2}}^{\frac{h}{2}} \mathbf{T}_\alpha d\xi \quad , \quad \mathbf{m}_\alpha := \int_{-\frac{h}{2}}^{\frac{h}{2}} (\mathbf{x} - \mathbf{x}_0) \times \mathbf{T}_\alpha d\xi \tag{6}$$

are computed by integration of the stress vector $\mathbf{T}_\alpha$ with respect to the coordinate ξ. Here $\mathbf{T}_\alpha$ is defined by $\mathbf{P}(s_1, s_2, \zeta) = \mathbf{T}_i(s_1, s_2, \xi) \otimes \mathbf{e}_i$, where $\mathbf{P}$ denotes the first Piola-Kirchhoff stress tensor. Following classical shell theories, we assume $\bar{M}_{\alpha 3} = \mathbf{m}_\alpha \cdot \mathbf{a}_3 = 0$. The back transformation of $\mathbf{n}_\alpha$ and $\mathbf{m}_\alpha$ with the rotation tensor $\mathbf{R}$ yields

$$\mathbf{N}_\alpha = \mathbf{R}^T\mathbf{n}_\alpha \quad \mathbf{M}_\alpha = \mathbf{R}^T\mathbf{m}_\alpha . \tag{7}$$

Since $\mathbf{R}$ is defined by the polar decomposition of the material deformation gradient, $\mathbf{N}_\alpha$ and $\mathbf{M}_\alpha$ are integrals of the non-symmetric Biot stress tensor defined by $\mathbf{R}^T\mathbf{P}$. With the strain measures

$$\boldsymbol{\Gamma}_\alpha(S_\alpha) \;=\; \mathbf{R}^T\mathbf{x}_{0,\alpha} - \mathbf{t}_\alpha \tag{8}$$

$$\mathbf{K}_\alpha(S_\alpha) \;=\; \mathbf{R}^T\boldsymbol{\omega}_\alpha \tag{9}$$

(see Gruttmann, Stein, Wriggers [7] or Lambertz [8]), the principle of virtual work can be written as

$$\delta W = \mathbf{P} \cdot \delta \mathbf{F} = \int_A (\mathbf{N}_\alpha \cdot \delta \mathbf{\Gamma}_\alpha + \mathbf{M}_\alpha \cdot \delta \mathbf{K}_\alpha) dA. \tag{10}$$

Introducing a corotational configuration (by the decomposition of the rotation tensor into the rigid body rotation $\mathbf{R}^C$ and the deformational part $\mathbf{R}^D$ (see Lambertz [8])) and the resultants of the corresponding rotated stress tensor as

$$\mathbf{N}_\alpha^C = (\mathbf{R}^D)^T \mathbf{n}_\alpha \quad \mathbf{M}_\alpha^C = (\mathbf{R}^D)^T \mathbf{m}_\alpha, \tag{11}$$

the weak form of equilibrium can be written as

$$\begin{aligned} G(\phi, \eta) \quad = \quad & \int_A \left\{ \mathbf{N}_\alpha^C \cdot \overbrace{(\mathbf{R}^D)^T (\eta_{0,\alpha}^C - \vartheta \times \mathbf{x}_{0,\alpha}^C)}^{\delta \mathbf{\Gamma}_\alpha^C} + \mathbf{M}_\alpha^C \cdot \overbrace{(\mathbf{R}^D)^T \vartheta_\alpha^C}^{\delta \mathbf{K}_\alpha^C} \right\} dA \\ & - \int_A (\overline{\mathbf{q}}^C \cdot \eta_0 + \overline{\mathbf{m}}^C \cdot \vartheta) dA \stackrel{!}{=} 0. \end{aligned} \tag{12}$$

The expressions $\delta \mathbf{\Gamma}_\alpha^C$ and $\delta \mathbf{K}_\alpha^C$ denote the variation of the strain measures in the corotational configuration. If the rotation tensor is linearized the strain measures are given by

$$\mathbf{\Gamma}_\alpha^C = \mathbf{x}_{0,\alpha}^C - \mathbf{t}_\alpha^C - \psi \times \mathbf{x}_{0,\alpha}^C \tag{13}$$

and

$$\mathbf{K}_\alpha^C = \psi_{,\alpha}^C. \tag{14}$$

This linearization depends on the assumption of small relative rotations $\mathbf{R}^D$ which must be controlled by an error estimation and, if necessary, adaptive mesh refinement (see chapter 4). If the relative rotation is neglected, i.e. $\mathbf{R}^D \to \mathbf{1}$, we obtain linear expressions and therefore linear element formulations.

2.3 Finite Element Discretization

A detailed description of the finite element formulation is given in [8]. In this section, we describe only the treatment of the rotation and curvature of the elements. Three rotational degrees of freedom are introduced in each element node. These unknowns are the components of the axial vector of a skew-symmetric matrix resulting from the linearization of the rotation tensor. We work with the four Euler-parameters (a unit-quaternion) to obtain a singularity-free description of finite rotations. The Euler-parameters, computed from the skew-symmetric matrix by an exponential map, are not independent. So, three independent Rodrigues-parameters are calculated from the Euler-parameters and computed by an isoparametric interpolation in the integration points of the element. From the Rodrigues-parameters, we obtain the components of the curvature vector and the rotation matrix (see Lambertz [8]).

The stiffness matrices and the residuals are integrated numerically. To avoid shear locking we use either selective reduced integration (see Malkus and Hughes [9]), which, however, can produce zero energy modes, or the assumed-strain approach (Bathe and Dvorkin [4]).

2.4 Plasticity

Neglecting thermal effects, we assume an isothermal elasto-plastic material law with von Mises yield condition and associated flow rule. We introduce a layer model, where the material law is integrated in several layer points by a projection method, and the stress resultants are computed from the projected stresses (see Stein, Lambertz, Plank [13]). This approach is more accurate than an Ilyushin-type material law for stress resultants. It is possible to differentiate the projected stresses with respect to the strains and obtain (consistent) tangent matrices (see Stein, Gruttmann [6]), which allows the application of Newton's method to the nonlinear equilibrium equation.

3 Solution Methods

In this chapter, we consider the solution of systems of equations as they arise from finite element discretizations of a folded plate structures. We focus our attention to the solution of very large systems and assume, that the matrix of the linearized system is symmetric and positive definite.

3.1 Direct and Iterative Methods for Linear Problems

The most popular method for this problem is the Gaussian elimination (in the variant given by Cholesky). The matrix is stored in band or profile form. The method is easily implemented, stable and entirely problem-independent. Furthermore, it can be vectorized, and optimized versions are available for supercomputers. The performance depends heavily on the node numbering. With optimal node numbering, the decomposition time normally increases like $o(n^2)$ and memory requirements like $o(n^{3/2})$, which means that the solution part eventually dominates the overall program requirements and limits the maximum adressable problem size. This motivates the investigation of iterative methods.

Iterative methods are superior to direct methods with respect to storage requirements, because no fill-in occurs and a sparse matrix storage scheme can be employed. This storage scheme slows down memory access and impedes automatic vectorization.

We first consider Jacobi- and Gauß-Seidel relaxation and it's variants. A detailed analysis of these methods is rather complicated, see [15]. For a typical model problem, the following estimates hold:

Jacobi–iteration	$o(n^2)$ operations
SOR with $\omega = 1$	$o(n^2)$ operations
SOR with $\omega = \omega_{opt}$	$o(n^{\frac{3}{2}})$ operations

A general theory for the computation of ω_{opt} is not available, one has to resort to numerical experiments. Numerical tests with finite element problems show that SOR with an approximate value for ω_{opt} is not competitive with the decomposition method for up to 20000 unknowns (on workstations and on vector computers). (We have not tested larger equations.)

Another promising iterative method is the method of conjugate gradients (cg). Without rounding errors, it terminates in at most n iteration steps, but is more expensive than Cholesky decomposition. As an iterative method, the convergence behaviour depends on the condition number κ of the matrix $\mathbf{A}$; for second order problems, $o(n^{\frac{3}{2}})$ operations are required for a certain accuracy. The cg-method can be accelerated by preconditioning with **V**: substitute **A** by $\mathbf{V}^{-1}\mathbf{A}\mathbf{V}^{-T}$ and choose $\mathbf{V}\mathbf{V}^T$ so, that $\kappa((\mathbf{V}\mathbf{V}^T)^{-1}\mathbf{A})$ smaller than $\kappa(\mathbf{A})$. Popular preconditioning methods are (see [1])

a) SSOR with optimal ω b) incomplete factorization of **A**	$o(n^{\frac{5}{4}})$ operations (in certain cases)

Again, the determination of the optimal value for ω in SSOR is difficult.

3.2 Multigrid Method for Linear Problems

From the foregoing discussion, the computing effort for the classical iterative methods grows superlinearly with the number of unknowns. An advantage of these methods is their problem-independence: only the sparsity of the matrix **A** is used.

The multigrid method promises optimal (linear) time complexity, but must be adapted to the finite element method. Starting point is the following observation about relaxation methods: the slow convergence of the SOR and Jacobi-iteration is caused by the low frequency (long wave) components of the residual, whereas the highly oscillatory terms converge fast (smoothing property). This motivates the multigrid idea:

— use a relaxation method for the oscillatory terms

— approximate the low frequency terms on a coarser mesh.

An optimal estimate for the contraction number ρ^k has the form

$$\rho^k = \frac{\|\mathbf{u}^k - \mathbf{u}\|}{\|\mathbf{u}^{k-1} - \mathbf{u}\|} \lesssim a < 1 \tag{15}$$

independent of meshsize h; it can be reached by combining the smoothing property of the relaxation method and the approximation property of finite element spaces. The approximation property requires special attention if non-standard element formulations are used.

Membrane Equation For a conforming FE-method (using bilinear shape functions) the standard multigrid theory is applicable. The efficiency of this method is documented by the following numerical results for a two-grid iteration; ν: no. of smoothing cycles.

smoothing method	ν	ρ_k
Gauß-Seidel	1	0.557
	3	0.159
cg, preconditioned by SGS	2	0.069

	CPU time	
	$n = 3184$	$n = 12512$
Cholesky (vectorized)	2.60 s	24.04 s
mg (optimal results)	0.90 s	3.46 s

Kirchhoff Plate Equation Although the analysis of these elements is more difficult, good convergence results have been obtained for some element formulations (e.g. discrete Kirchhoff theory elements, Morley's element (see [5]),...).

Mindlin-Reissner Plate Equation For displacement elements with low order (especially bilinear) shape functions, special techniques like reduced integration or the assumed-strain approach by Bathe and Dvorkin [4] are required to avoid shear locking. This poses special problems for the multigrid algorithm. Numerical tests with the assumed-strain element exhibit fast convergence for thick plates, a contraction factor of $\approx$ 0,88 for $\frac{h}{d}$ (=

elementsize/plate thickness) $= \frac{100}{16}$ and divergence for $\frac{h}{d} > 10$. This problem has not yet been solved, a possible solution – the application of a mixed method – is currently being investigated.

3.3 Multigrid Methods for Nonlinear Problems

Multigrid ideas can be applied to nonlinear problems in several ways: a linear multigrid method can be used to solve the linear problem in the Newton iteration or the multigrid method with nonlinear smoothing can be directly applied to the nonlinear problem. Furthermore, efficient start vectors for the iteration can be obtained by prolongation of a coarse grid solution (nested iteration).

The combination of a continuation method on the coarsest grid with nested iteration and a multigrid solver is especially promising.

4 Error Estimates

4.1 Linear Equations

For linear PDE's of second order, reliable error estimators have been developed by Babuška, Rheinboldt et al. (see [3] [2] [10]), so we restrict ourselves to a short overview. The energy norm of the error of a finite element approximation can be estimated by a sum of local (elementwise) error indicators:

$$c_1 \sum_{i=1}^{m} \| \wp_i(e^{(h)}) \|_E^2 \leq \| e^{(h)} \|_E^2 \leq c_2 \sum_{i=1}^{m} \| \wp_i(e^{(h)}) \|_E^2 \; . \tag{16}$$

The indicator $\| \wp_i(e^{(h)}) \|_E^2$ can be expressed as:

$$\begin{aligned} \| \wp_i(e^{(h)}) \|_E^2 \;\; \approx \;\; & h_i^2 \int_{\Omega_i} (Lu^{(h)} - f)^T (Lu^{(h)} - f) dx \\ & + h_i \int_{\partial\Omega_i} J(\sigma^{(h)})^T J(\sigma^{(h)}) ds, \end{aligned} \tag{17}$$

where

$$\begin{aligned} Lu^{(h)} - f &= \text{interior residual (can be neglected for regular loads)} \\ J(\sigma^{(h)}) &= \text{"jump" of stresses across adjacent elements.} \end{aligned}$$

Thus, it suffices to evaluate the resultants of the (Cauchy-)stresses

$$(n_1, n_2, n_{12}, m_1, m_2, m_{12}, q_1, q_2) \tag{18}$$

along the element sides and integrate the squares of the differences. The contributions from the membrane equation (n_1, n_2, n_{12}) and the plate equation $(m_1, m_2, m_{12}, q_1, q_2)$ must be evaluated separately, resulting in two indicators η_M and η_B. An element is refined, if one indicator exceeds the predefined limit. A mesh is optimal within a prescribed accuracy, if the error indicators are uniformly distributed.

4.2 Geometrically Nonlinear Equations

The relative rotation in the element is given by

$$\mathbf{R}^D = \mathbf{R}(\mathbf{R}^C)^T, \tag{19}$$

where $\mathbf{R}^C$ — defined by the displacements of the element nodes — describes approximately the rigid body rotation of the element. In the kinematical equations, the rotation tensor

$$\mathbf{R}^D = \mathbf{I} + \mathbf{\Psi} + \frac{1}{2}\mathbf{\Psi}\mathbf{\Psi} + \ldots \tag{20}$$

is approximated by

$$\tilde{\mathbf{R}}^D = \mathbf{I} + \mathbf{\Psi} \tag{21}$$

in each element. Thus, the magnitude of the relative rotation $||\tilde{\mathbf{R}}^D - \mathbf{I}||$ must be controlled. From

$$\tilde{\mathbf{R}}^D = \begin{pmatrix} 1 & \psi_3 & -\psi_2 \\ -\psi_3 & 1 & \psi_1 \\ \psi_2 & -\psi_1 & 1 \end{pmatrix} \tag{22}$$

we have (choosing the Euclidean norm)

$$||\tilde{\mathbf{R}}^D - \mathbf{I}|| = \sqrt{2}\sqrt{\psi_1^2 + \psi_2^2 + \psi_3^2} \tag{23}$$

and choose $\eta_{NL} = \max_{nodes} ||\tilde{\mathbf{R}}^D - \mathbf{I}||$ as indicator. Note the difference between the indicators η_M, η_B and η_{NL}: whereas in the former case an estimation of the energy norm of the error is given, η_{NL} only controls the linearization in equation (21). Again, elements with $\eta_{NL} > \overline{\eta}_{NL}$ are subdivided. ($\overline{\eta}_{NL}$ — measuring a rotation — should not exceed 0.1.) A reliable a-posteriori error estimation for non-linear shell equations is not yet available.

4.3 Material Nonlinearities

The same remark is true for the numerical treatment of inelastic material behaviour. We restrict ourselves to the practically important case of elasto-plastic material with von Mises yield condition and associated flow rule. As an initial-boundary-value problem, it requires adaptation in the space and in the time domain.

We propose an heuristic criterion for the mesh adaptation: choose the indicator $\eta_{PL} = \max J_2$ and refine all elements, where η_{PL} is sufficiently near the yield stress, (possibly excluding elements in an entirely plastified region). This strategy seems appropriate for the stability analysis of steel girders where the growth of plastic zones ultimately causes the failure of the structure. Different material laws and deformations with large strains (e.g. forming processes) probably require a different mesh adaptation algorithm (e.g. with remeshing). In addition to the mesh refinement, the length of the time steps must be controlled. This can be done using standard techniques developed for ODE's (see Stein and Mahnken [14]).

5 Software Aspects

A finite element program that combines mesh adaptation and a multigrid solver differs considerably from a more traditional system. In the following we comment briefly on the key points of our concept that has been realized in the finite element system **INA-SP** (**IN**elastic **A**nalysis – **S**hells and **P**lates), which has been developed at the IBNM during a research project, supported by Deutsche Forschungsgemeinschaft (DFG).

5.1 Representation of Geometry

In contrast to many current FE-systems, the problem geometry cannot be represented by discretization items (i.e. nodes and elements). For **INA-SP**, a rather simple problem definition that includes all important aspects as geometry, material properties, boundary conditions and loads has been developed. This allows a future combination with a CAD-system and an automatic mesh generation from the CAD data.

5.2 Adaptation Strategy

For the start of the refinement process, a conforming finite element mesh must be available. This mesh can be arbitrarily coarse and is currently entered by hand. The refinement step is prepared by the application of a-priori and a-posteriori refinement criteria. As a-priori criteria, we control bounds for element size and element shape (e.g. distortion of quadrilaterals) and the approximation of the problem geometry, if curved lines or faces are present. Elements that satisfy the a-priori conditions can be uniformly subdivided according to the a-posteriori criteria developed in section 4 (error indicators for linear membrane and bending problems, indicators for nonlinear equations).

There are several, partly contradictory constraints for the refinement procedure. From error estimation, it would be advantageous to allow strong local refinement (especially at a singularity of the solution). On the other hand, the multigrid solver requires that no more than four new elements should be generated per element on the coarse mesh and that the finite element spaces should be nested. Furthermore, we require that the element shapes should not deteriorate during refinement, that all meshes are conforming (i.e. without irregular nodes) and that quadrilateral elements can be handled (refinement of triangular elements is much easier). The procedure in **INA-SP** is:

- A uniformly refined element is subdivided into 4 new elements (For the a-priori criteria, only some element sides may be refined).
- If an element is affected by the refinement of a neighbour, temporary elements are introduced, which cannot be refined further; all necessary refinement is done on the original regular element.

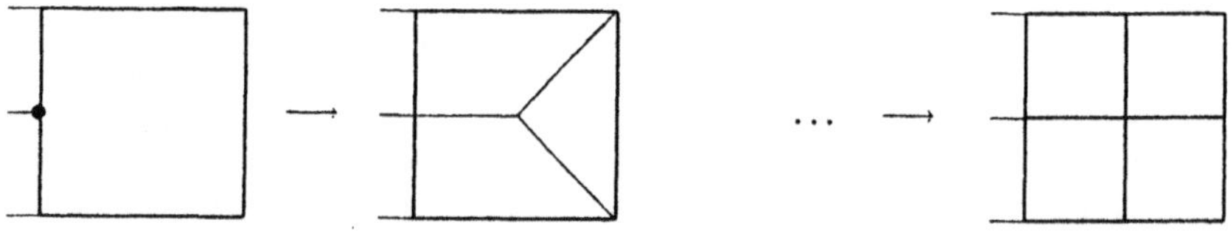

irregular node — adaptation with conforming, temporary elements — possibly further refinement

This refinement strategy satisfies most of the above mentioned constraints. Its shortcomings are the appearence of triangles from the refinement of quadrilaterals and the fact that several refinement cycles are required to obtain a strong local refinement. The latter is largely a problem of the storage strategy: the discretization data are stored seperately on each refinement level. For local refinement, a hierarchical storage method as in [11] would be more economical.

5.3 Combination with Multigrid

Apart from the sequence of finite element meshes, the multigrid algorithm needs information about the generation process in order to perform the prolongation and restriction operations. As it is difficult to provide this information by post-processing of the meshes, the refinement process must generate it. This information can also be used to restrict the smoothing iteration (the most time-consuming part of the mg-cycle) to the refined area of the mesh.

6 Numerical Examples

A large number of tests of the nonlinear formulation has been carried out; we refer to [12,8]. The first example is a plate girder under concentrated load which has been adaptively refined in 4 steps.

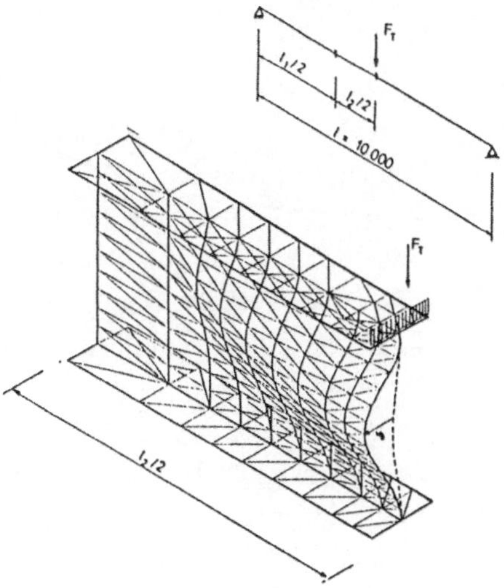

Loads and deformed geometry

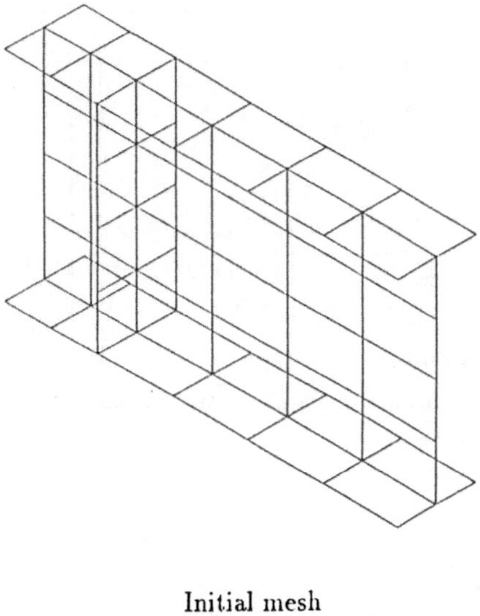

Initial mesh

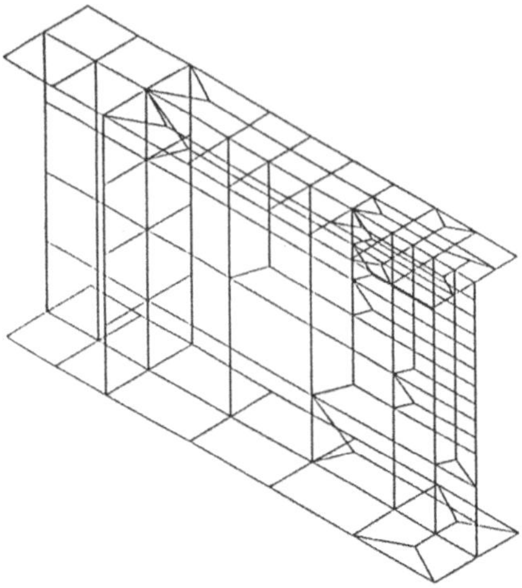

Refined mesh 3

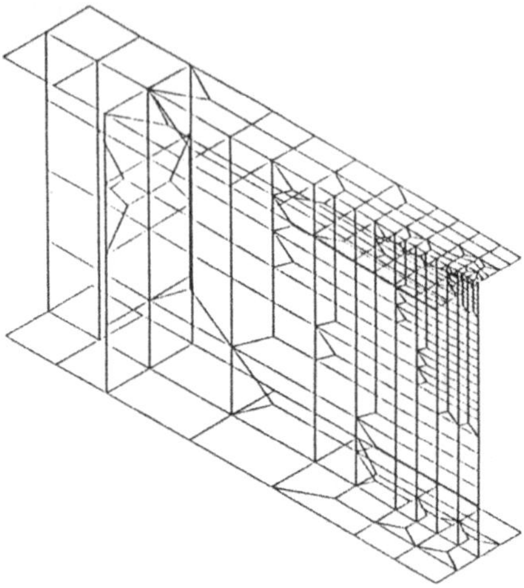

Refined mesh 5

Detailed investigations of this example will be published.

The second example is a clamped L-shaped frame with a point load at one end (for further details see [12]). Its strong nonlinear behaviour is shown in the load-displacement diagram below (we have plotted the out-of-plane displacement at the point where the load is applied).

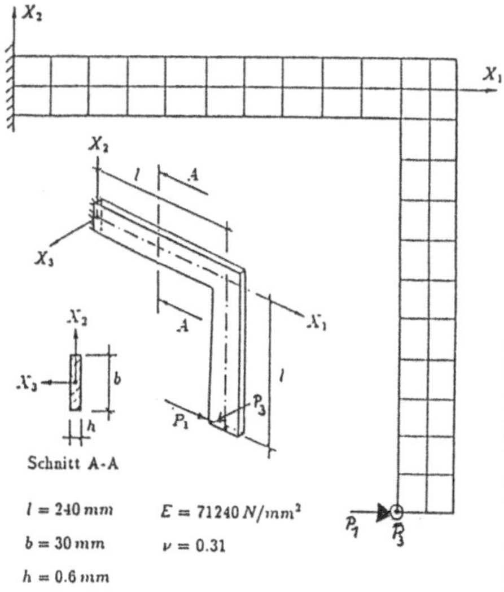

Clamped frame under point load

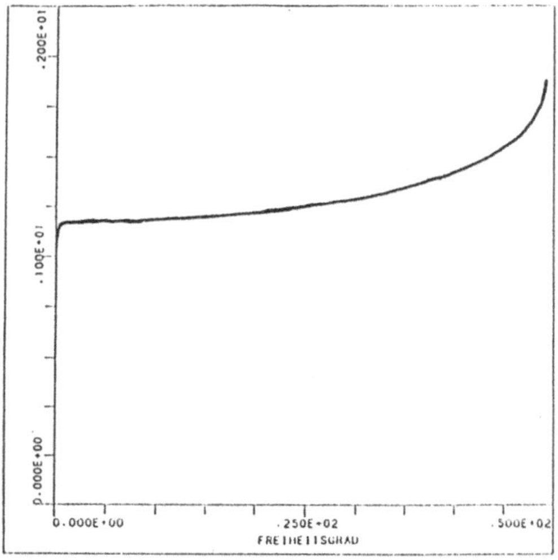

Load deflection curve of the frame

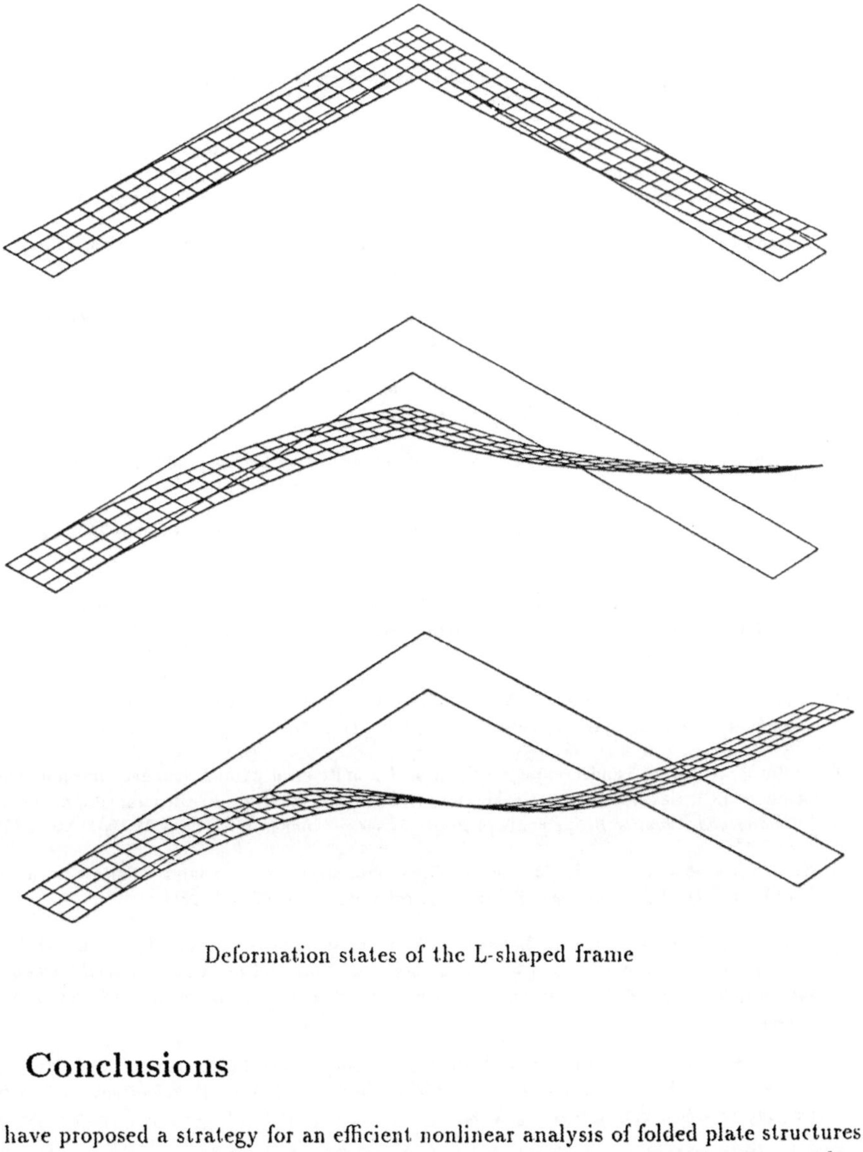

Deformation states of the L-shaped frame

7 Conclusions

We have proposed a strategy for an efficient nonlinear analysis of folded plate structures. The nonlinear formulation, combined with the control of the element distortion and an adaptive refinement, allows large load increments in stable regions of the system. While our error control for the nonlinear equations constitutes a necessary test, it is not a sufficient convergence criterion: a mathematical error analysis of the approximate solution of this type of nonlinear equations is highly desirable.

The multigrid method has been succesfully applied to linear and nonlinear membrane and Kirchhoff plate problems. Open questions arise in the treatment of Reissner-Mindlin plate elements. Despite these problems, we believe that the combination of mesh adaptation and multigrid methods is the most promising way to faster and more reliable computations.

References

[1] O. Axelsson and V. A. Barker. *Finite Element Solution of Boundary Value Problems: Theory and Computation*. Academic Press, 1984.

[2] I. Babuška. Feedback, adaptivity, and a-posteriori estimates in finite elements: Aims, theory, and experience. In I. Babuška, O. C. Zienkiewicz, J. Gago, and E. R. de A. Oliveira, editors, *Accuracy Estimates and Adaptive Refinements in Finite Element Computations*, John Wiley & Sons, 1986.

[3] I. Babuška and W. C. Rheinboldt. Error estimates for adaptive finite element computations. *SIAM Journal on Numerical Analysis*, 15:736–755, 1978.

[4] K. J. Bathe and E. Dvorkin. A continuum mechanics based four-node shell element for general nonlinear analysis. *Engineering Computations*, 1:77–88, 1985.

[5] D. Braess and P. Peisker. A conjugate gradient method and a multigrid algorithm for Morley's finite element approximation of the biharmonic equation. *Numerische Mathematik*, 50:567–586, 1987.

[6] F. Gruttmann and E. Stein. Tangentiale Steifigkeitsmatrizen bei Anwendung von Projektionsverfahren in der Elastoplastizitätstheorie. *Ingenieur-Archiv*, 58:15–24, 1988.

[7] F. Gruttmann, E. Stein, and P. Wriggers. Theory and numerics of thin elastic shells with finite rotations. *Ingenieur-Archiv*, 59:54–67, 1989.

[8] K. H. Lambertz. *Traglastberechnungen von Faltwerken mit elastoplastischen Deformationen*. Technical Report F 89/1, Forschungs- und Seminarberichte aus dem Bereich der Mechanik der Universität Hannover, 1989.

[9] D. S. Malkus and T. J. R. Hughes. Mixed finite element methods - reduced and selective integration techniques: a unification of concepts. *Comp. Meth. Appl. Mech. Eng.*, 15:63–81, 1978.

[10] E. Rank. Adaptivity and accuracy estimation for finite element and boundary integral element methods. In I. Babuška, O. C. Zienkiewicz, J. Gago, and E. R. de A. Oliveira, editors, *Accuracy Estimates and Adaptive Refinements in Finite Element Computations*, John Wiley & Sons, 1986.

[11] W. C. Rheinboldt and C. K. Mesztenyi. On a data structure for adaptive finite element mesh refinements. *ACM Transactions on Mathematical Software*, 6:166–187, 1980.

[12] E. Stein, F. Gruttmann, and K. Lambertz. Elastic-plastic analysis of thin shells and folded plate structures with finite rotations. In W. B. Krätzig and E. Oñate, editors, *Computational Mechanics of Nonlinear Behaviour of Shell Structures in Springer-Series on Computational Mechanics*, Springer-Verlag, 1989.

[13] E. Stein, K. Lambertz, and L. Plank. Verfahren zur Traglastberechnung dünnwandiger Strukturen mit elastoplastischen Deformationen. In E. Stein, editor, *Nichtlineare Berechnungen im Konstruktiven Ingenieurbau*, Springer-Verlag, 1989.

[14] E. Stein and R. Mahnken. Adaptive time-step control in creep analysis. *International Journal for Numerical Methods in Engineering*, 28:1619–1633, 1989.

[15] D. M. Young. *Iterative Solution of Large Linear Systems*. Academic Press, 1971.

Treatment of Domain Integrals by Using the Dual and Multiple Reciprocity Methods

C.A. BREBBIA and A.J. NOWAK

Wessex Institute of Technology
Ashurst Lodge, Ashurst
Southampton, SO4 2AA, UK

1. Introduction

The Boundary Element Method [1,2,3] is now well established as a very powerful engineering tool. There is a variety of practical boundary value problems which can efficiently and very accurately be solved by boundary elements.

In this paper the major problem of how to deal with domain integrals in boundary elements is discussed. These integrals can arise from body forces, time dependent effects or some non-linearities.

Two different types of approach can be employed. The first started at Computational Mechanics Institute in 1982 and is known now as the Dual Reciprocity Method (DRM) [4,5]. Proposed by Nardini and Brebbia for hyperbolic problems, this technique was later extended by Wrobel and Brebbia [6,7] to deal with parabolic problems, linear ones as well as non-linear.

The second approach developed by Nowak and Brebbia [8,9,10] is called the Multiple Reciprocity Method (MRM). Although it can be thought of as an extension of the idea of DRM, it is essentially different. Here instead of approximating the source term by the set of interpolating functions, a sequence of functions related to the fundamental solution is introduced. As a result this method gives in the limit the exact boundary only formulation of the original problem.

G. Kuhn · H. Mang (Eds.)
Discretization Methods in Structural Mechanis
IUTAM/IACM Symposium Vienna, Austria, 1989

2. Boundary Integral Equations

In order to illustrate the ways in which domain integrals can be computed in boundary elements the case of Poisson's equation will be considered as an example, i.e.

$$\nabla^2 u = b \qquad \text{in } \Omega \tag{1}$$

u is a potential function, Ω defines the domain under study and b represents body forces, which at present will be assumed to be known. Later on this b will be considered as time dependent function of the potential - $b(\dot{u})$ or $b(\ddot{u})$ - or a nonlinear term, such as $b(u)$.

A boundary integral equation can be set up by considering reciprocity relationships between the u state as given by equation (1) plus boundary conditions and a state corresponding to the fundamental solution u^*, such that

$$\nabla^2 u^* + \Delta^i = 0 \tag{2}$$

where Δ^i indicates a Dirac delta function applied at a source point 'i'.

The boundary integral equation corresponding to (1) and (2) for a particular point 'i' can now be written as

$$c^i u^i + \int_\Gamma q^* u \, d\Gamma + \int_\Omega b \, u^* \, d\Omega = \int_\Gamma u^* q \, d\Gamma \tag{3}$$

Γ indicates the boundary of the Ω domain, $q = \partial u/\partial n$ and $q^* = \partial u^*/\partial n$. The c^i constant depends on the type of boundary under consideration at point 'i'.

After considering n different positions of the source point 'i' one can write a system of equations which in matrix form results in the classical boundary element influence matrices **H** and **G** [1][2], i.e.

$$\mathbf{H}\,\mathbf{U} + \mathbf{D} = \mathbf{G}\,\mathbf{Q} \tag{4}$$

The vectors **U** and **Q** represent the potentials and derivatives at the different boundary nodes and the above matrix needs to be rearranged in accordance with their known values. The vector **D** results from the integration of the domain terms in (3).

This paper addresses itself to different ways in which the vector **D** can be computed, with special reference to two techniques by which it can be reduced to the boundary.

The classical manner in which the integral is computed consists of subdividing the domain into a series of cells over which numerical integrations are carried out. Unfortunately the whole domain requires integration for each position of the source point 'i' and hence the process is computationally inefficient in addition to requiring subdivision of the internal domain which can be cumbersome in three dimensional problems. In this case the domain integral in (3) for each 'i' is represented as,

$$\int_{\Omega} b\ u^*\ d\Omega = \sum_{j=1}^{M} \left\{ \int_{\Omega_m} b\ u^*\ d\Omega \right\} \tag{5}$$

Ω_m denotes the area or volume of each cell.

In certain cases vector **D** can be easily computed when the function b is such that one can satisfy it using a particular solution $\hat{u}$, such as

$$\nabla^2 \hat{u} = b \tag{6}$$

Hence one can write the domain integral as follows,

$$\int_{\Omega} b\ u^*\ d\Omega = \int_{\Omega} (\nabla^2 \hat{u})\, u^*\ d\Omega \tag{7}$$

Then the integral on the RHS of (4) can be taken to the boundary by a simple integration by parts, i.e.

$$\int_{\Omega} (\nabla^2 \hat{u})\, u^*\ d\Omega = \int_{\Gamma} \hat{q}\ u^*\ d\Gamma - \int_{\Gamma} \hat{u}\ q^*\ d\Gamma + \int_{\Omega} (\nabla^2 u^*)\, \hat{u}\ d\Omega \tag{8}$$

which taking into consideration the special character of the fundamental solution u* gives

$$\int_{\Omega} (\nabla^2 \hat{u}) u^* \, d\Omega = \int_{\Gamma} \hat{q} \, u^* \, d\Gamma - \int_{\Gamma} \hat{u} \, q^* \, d\Gamma - c^i \, \hat{u}^i \tag{9}$$

Note that the resulting integral form (9) is similar to the one for the original function u but in terms of the - known - particular solution. Taking all 'i' into consideration one can write (9) in matrix form, such that the **D** vector becomes,

$$\mathbf{D} = - \mathbf{H} \, \hat{\mathbf{U}} + \mathbf{G} \, \hat{\mathbf{Q}} \tag{10}$$

The vectors $\hat{\mathbf{U}}$ and $\hat{\mathbf{Q}}$ represent the potential and fluxes of the particular solution at the points under consideration. **H** and **G** are the same matrices and as computed earlier.

Equation (4) can now be written as:

$$\mathbf{H} \, \mathbf{U} - \mathbf{G} \, \mathbf{Q} = - \mathbf{D} \tag{11}$$

or

$$\mathbf{H} \, \mathbf{U} - \mathbf{G} \, \mathbf{Q} = \mathbf{H} \, \hat{\mathbf{U}} - \mathbf{G} \, \hat{\mathbf{Q}} \tag{12}$$

The above approach can unfortunately be used in only a few cases as it is difficult to find particular solutions for complex b functions. The idea will however be applied in a different way to generate the dual reciprocity approach which is one of the two discussed in this paper.

The two approaches are called Dual Reciprocity Method (DRM) and Multiple Reciprocity Method (MRM), and will be discussed in detail in what follows. Each can be seen as the adjoint of the other in so far as both start with the same term, i.e.

$$\int_{\Omega} b \, u^* \, d\Omega \tag{13}$$

but operate in opposite manners. While the DRM manipulates the b function, the MRM operates in the u* fundamental solution.

The main characteristics of both approaches are given in the following sections.

3. Dual Reciprocity Method (DRM)

The DRM starts by approximating the b function in equation (13) by the following

$$b \cong \sum_{j=1}^{M} f^j \alpha^j \tag{14}$$

where α^j are - still undetermined - coefficients and the f^j are known functions which can be applied at a series of M points called 'poles'. These poles usually consist of all the N boundary points plus a series of internal points L, such that M = N+L.

The form of the f^j function is well known now and comparatively simple, one can then propose finding the particular solution which satisfies, i.e.

$$\nabla^2 \hat{u}^j = f^j \tag{15}$$

Notice that $\hat{u}^j$ can be interpreted as a function localized at each 'j' pole and hence $\hat{u}^j$ is a 'localized particular solution'. The best form of f^j was found by Nardini and Brebbia [4] to be the simple function r (distance from the pole to all field points) plus some additional constants where necessary for rigid body conditions or to find more stable solutions.

Once the formulae for $\hat{u}^j$ are known, values of $\hat{q}^j = \partial \hat{u}^j / \partial n$ can be easily calculated. The integral (13) can be written as

$$\int_\Omega b\ u^*\ d\Omega = \sum_{j=1}^{M} \left\{ \int_\Omega [(\nabla^2 \hat{u}^j) u^*\ d\Omega] \alpha^j \right\} \tag{16}$$

Integrating by parts the $\nabla^2 \hat{u}^j$ terms one finds,

$$\int_\Omega b\ u^*\ d\Omega = - \sum_{j=1}^{M} \left\{ [c^j\ \hat{u}^j + \int_\Gamma q^*\ \hat{u}^j\ d\Gamma - \int_\Gamma u^*\ \hat{q}^j\ d\Gamma] \alpha^j \right\} \tag{17}$$

Equation (17) can be written in matrix form for all the poles under consideration and give the following system for **D**, i.e.

$$\mathbf{D} = -[\mathbf{H}\ \hat{\mathbf{U}} - \mathbf{G}\ \hat{\mathbf{Q}}]\boldsymbol{\alpha} \tag{18}$$

If the number of poles is identical and coincides with the boundary points matrices **H** and **G** are the same as the standard boundary element matrices already defined. $\hat{\mathbf{U}}$ and $\hat{\mathbf{Q}}$ are then squared matrices whose values consist of the localized particular solutions applying at each point under consideration. When a number of 'L' internal poles is also taken into account, additional terms in **H** and **G** are required for these points as well as more terms in $\hat{\mathbf{U}}$ and $\hat{\mathbf{Q}}$. This is a simple operation and does not complicate the problem.

The final step in the DRM is to define the $\boldsymbol{\alpha}$ in terms of the known values of the b function, which can easily be done by referring to formula (14), i.e.

$$b \cong \sum_{j=1}^{M} f^j \alpha^j \tag{19}$$

which in matrix form can be expanded to

$$\mathbf{B} = \mathbf{F}\,\boldsymbol{\alpha} \tag{20}$$

This system can be inverted to give,

$$\boldsymbol{\alpha} = \mathbf{F}^{-1}\,\mathbf{B} = \mathbf{E}\,\mathbf{B} \tag{21}$$

Hence the final system of equations can be expressed as

$$\begin{aligned} \mathbf{H}\,\mathbf{U} - \mathbf{G}\,\mathbf{Q} &= [\,\mathbf{H}\,\hat{\mathbf{U}} - \mathbf{G}\,\hat{\mathbf{Q}}\,]\boldsymbol{\alpha} \\ &= \mathbf{S}\,\boldsymbol{\alpha} \\ &= \mathbf{S}\,\mathbf{E}\,\mathbf{B} = -\,\mathbf{D} \end{aligned}$$

or simply

$$\mathbf{H}\,\mathbf{U} - \mathbf{G}\,\mathbf{Q} = -\,\mathbf{D} \tag{22}$$

where $\mathbf{S} = \mathbf{H}\,\hat{\mathbf{U}} - \mathbf{G}\,\hat{\mathbf{Q}}$ (23)

and $-\,\mathbf{D} = \mathbf{S}\,\mathbf{E}\,\mathbf{B}$ (24)

The DRM has been used by many authors to solve time dependent as well as nonlinear problems. Its main disadvantage is the need to approximate the function b in a way that one cannot control the error, particularly when using internal poles which are usually distributed.

However the technique gives excellent results and is recommended for engineering applications.

4. Multiple Reciprocity Method

This is a new approach, also developed at the Computational Mechanics Institute, WIT, more recently. It consists of working with the fundamental solution u* rather than the b term in equation (13).

In order to understand the technique better it is more convenient to use a subscript for the right hand side term b and the fundamental solution u*, such that equation (1) is now written as

$$\nabla^2 u = b_o \tag{25}$$

and equation (2) as

$$\nabla^2 u_o^* + \Delta^i = 0 \tag{26}$$

The governing boundary integral equation (3) can now be rewritten with the 'o' subscripts as follows,

$$c^i u^i + \int_\Gamma q_o^* \, u \, d\Gamma + \int_\Omega b_o \, u_o^* \, d\Omega = \int_\Gamma u_o^* \, q \, d\Gamma \tag{27}$$

One now proposes a higher order function in order to transform to the boundary the above integral in the Ω domain, i.e.

$$\int_\Omega b_o \, u_o^* \, d\Omega \tag{28}$$

Define now a u_1^* function such that

$$\nabla^2 u_1^* = u_o^* \tag{29}$$

Thus equation (28) can now be written as

$$\int_\Omega b_o (\nabla^2 u_1^*) \, d\Omega \tag{30}$$

Integrating by parts the above one finds that

$$\int_\Omega b_o(\nabla^2 u_1^*)d\Omega = \int_\Gamma \left\{b_o \frac{\partial u_1^*}{\partial n} - u_1^* \frac{\partial b_o}{\partial n}\right\}d\Gamma + \int_\Omega u_1^*(\nabla^2 b_o)d\Omega \quad (31)$$

Next, one can operate on the last integral on the RHS of (31) by defining a function b_1 such that

$$b_1 = \nabla^2 b_o \quad (32)$$

The domain integral on the RHS of (31) can be written as

$$\int_\Omega u_1^*(\nabla^2 b_o)d\Omega = \int_\Omega u_1^* b_1 \, d\Omega \quad (33)$$

Note that equation (33) has the same form as (28) but with the new subscript '1' instead of 'o'. Hence one can operate again on (33) after defining,

$$u_1^* = \nabla^2 u_2^* \quad (34)$$

Thus,

$$\int_\Omega b_1 u_1^* \, d\Omega = \int_\Omega b_1(\nabla^2 u_2^*)d\Omega \quad (35)$$

Integration by parts of the above gives,

$$\int_\Omega b_1 u_1^* \, d\Omega = \int_\Gamma \left\{b_1 \frac{\partial u_2^*}{\partial n} - u_2^* \frac{\partial b_1}{\partial n}\right\}d\Gamma + \int_\Omega u_2^*(\nabla^2 b_1)d\Omega \quad (36)$$

where the domain integral in the RHS can be expressed as

$$\int_\Omega u_2^*(\nabla^2 b_1)d\Omega = \int_\Omega u_2^* b_2 \, d\Omega \quad (37)$$

It is now evident that one can continue carrying out this process as many times as desired. The necessary recurrence relationships are

$$b_{j+1} = \nabla^2 b_j \quad (38)$$

$$u_j^* = \nabla^2 u_{j+1}^* \quad (39)$$

$$j = 0,1,2 \ldots \infty .$$

In the limit this process leads to the following expression for the domain integrals,

$$\int_\Omega b_o\, u_o^*\, d\Omega = \sum_{j=o}^{\infty} \int_\Gamma \left\{ b_j \frac{\partial u_{j+1}^*}{\partial n} - u_{j+1}^* \frac{\partial b_j}{\partial n} \right\} d\Gamma \tag{40}$$

One can apply the above relationship to compute the vector **D** such that

$$- \mathbf{D} = \sum_{j=o}^{\infty} \{ \mathbf{H}_{j+1}\, \mathbf{P}_j - \mathbf{G}_{j+1}\, \mathbf{R}_j \} \tag{41}$$

The vectors $\mathbf{P}_j$ and $\mathbf{R}_j$ represent the values of b_j and $\partial b_j/\partial n$ at the nodes, whereas $\mathbf{H}_{j+1}$ and $\mathbf{G}_{j+1}$ are new matrices which can be easily computed.

The final system can then be written as

$$\mathbf{H}_o\, \mathbf{U} - \mathbf{G}\, \mathbf{Q}_o = \sum_{j=1}^{\infty} \{ \mathbf{H}_{j+1}\, \mathbf{P}_j - \mathbf{G}_{j+1}\, \mathbf{R}_j \} \tag{42}$$

$\mathbf{H}_o$ and $\mathbf{G}_o$ are the normal boundary element matrices which were called **H** and **G** before, i.e. without a subscript. The right hand side expansion permits to express the domain integrals on the boundary only.

Higher order solutions

Some of the most interesting properties of the DRM are a consequence of the special character of the higher order functions used to represent the fundamental solutions. For two dimensional Laplace type problems these functions are

$$u_j^* = \frac{1}{2\pi}\, r^{2j}\, (A_j \ln r - B_j) \tag{43}$$

where

$$A_{j+1} = \frac{A_j}{4(j+1)^2} \quad ; \quad B_{j+1} = \frac{1}{4(j+1)^2} \left[\frac{A_j}{j+1} + B_j \right] \tag{44}$$

Notice the factorial parts in the A_{j+1} and B_{j+1} terms which ensure convergence properties in most cases once the solution has been properly scaled.

Some nice properties of the MRM are as follows.

i) Convergence The convergence of the solution is rapid in most cases as factorials are introduced into the denominators of A_j and B_j.

ii) Accuracy The contributions of different terms can be evaluated and hence the accuracy of the solution can be estimated.

iii) Computational Efficiency The integrals needed in the H_j and G_j $(j \geq 1)$ matrices are easy to compute as the u_j^* functions in these cases have no singularities. In practice the terms can be found using simple Gauss integration rules for instance.

Applications

The different cases will be considered here to illustrate how the MRM can be applied. They are

i) The case of Helmholtz Equation
ii) A time dependent parabolic equation

i) Helmholtz Equation

In this case the governing equation is

$$\nabla^2 u + k^2 u = 0 \tag{45}$$

where k^2 is a known positive value. Hence

$$b_o = -k^2 u \tag{46}$$

and then,

$$b_1 = \nabla^2(b_o) = -\nabla^2(k^2 u) = -k^2 \nabla^2 u = -k^2(-k^2 u) = +k^4 u$$

$$b_2 = \nabla^2(b_1) = \nabla^2(k^4 u) = k^4 \nabla^2 u = k^4(-k^2 u) = -k^6 u \tag{47}$$

etc.

This results in the following expression for (42)

$$(H_o - k^2 H_1 + k^4 H_2 - \ldots)U = (G_o - k^2 G_1 + k^4 G_2 - \ldots)Q \tag{48}$$

In a more compact form this can be written simply as

$$\mathbf{H}\ \mathbf{U} = \mathbf{G}\ \mathbf{Q} \tag{49}$$

where **H** and **G** are now different from the original ones - which are now called H_o and G_o.

Representative results are plotted in Figure 1 where the solid lines correspond to an analytical solution whereas the symbols are associated with numerical results obtained by the MRM. Namely, □ stands for first order approach (j=1), Δ represents second order approach (j=1,2) and o is related to the third order approach (j=1,2,3).

It is interesting to point out that for values of $k \leq 1$ all the approaches produce excellent results as the convergence of the higher order terms in equation (48) is very fast. The velocity of convergence of the coefficients of H_j and G_j is increased by the simultaneous convergence of the powers of k. For $k > 1$ the approach converges less rapidly as the effect of the factorial terms H_j and G_j is reduced by the effect of the increasing powers of k due to recurrence relationships. The example helps to demonstrate how the accuracy and convergence of a solution can be effectively controlled by the MRM. For a further explanation of how to implement MRM to solve Helmholtz problems the user is referred to [9].

ii) Parabolic Case

Next one can consider the case of the diffusion type equation, i.e.

$$\nabla^2 u = \frac{1}{\kappa} \dot{u} \tag{50}$$

In this case

$$b_o = \frac{1}{\kappa} \dot{u}$$

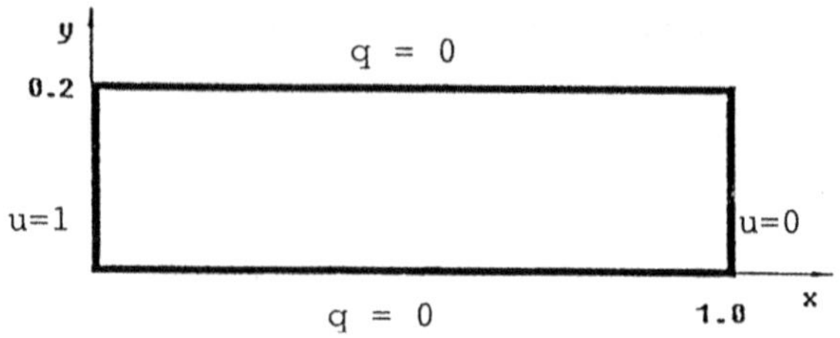

a) Geometry of the region under consideration

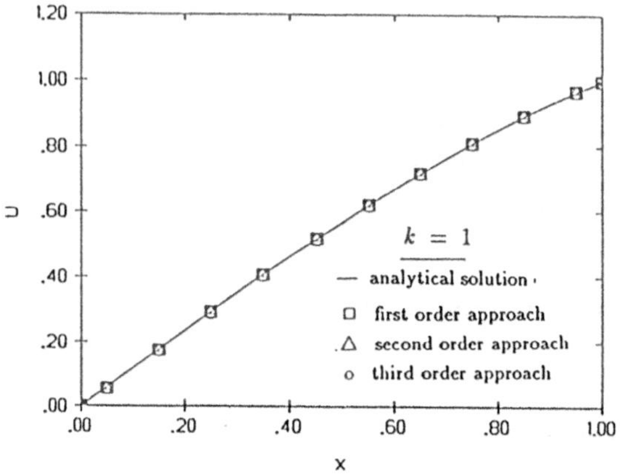

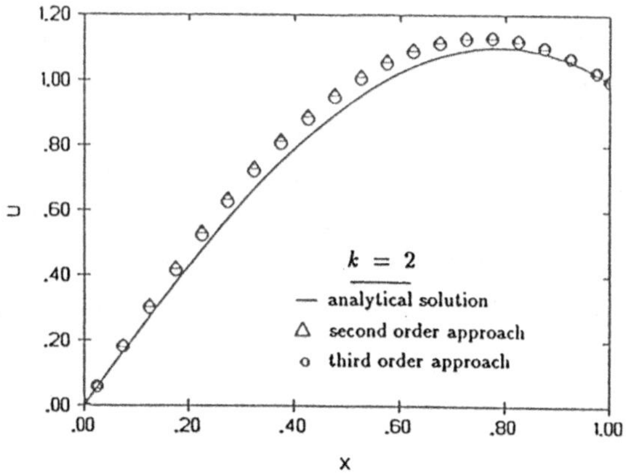

b) Results for Helmholtz equation (k=1 and k=2)

Fig.1 Helmholtz Equation Solution

and the recurrence form is

$$
\begin{aligned}
b_1 &= \nabla^2(b_o) = \nabla^2\left(\frac{1}{\kappa}\,\dot{u}\right) = \frac{1}{\kappa}\frac{\partial}{\partial t}(\nabla^2 u) = \frac{1}{\kappa}\,\ddot{u} \\
b_2 &= \nabla^2(b_1) = \nabla^2\left(\frac{1}{\kappa}\,\ddot{u}\right) = \frac{1}{\kappa}\frac{\partial^2}{\partial t^2}(\nabla^2 u) = \frac{1}{\kappa}\,\dddot{u}
\end{aligned}
\tag{52}
$$

etc.

This gives the following expression

$$
H_o\,U + \frac{1}{\kappa}H_1\,\dot{U} + \frac{1}{\kappa}H_2\,\ddot{U} + \ldots = G_o\,Q + \frac{1}{\kappa}G_1\,\dot{Q} + + \frac{1}{\kappa}G_2\,\ddot{Q} + \ldots \tag{53}
$$

The time derivatives can be approximated using some finite difference formulae and the system can then be solved.

Example

In order to demonstrate the accuracy of the proposed method, simple 2-D problems have been studied. Heat is conducted within the rectangle shown in Fig. 2. Heat conductivity k=1 W/mK and thermal diffusivity a = 1 m²/s.

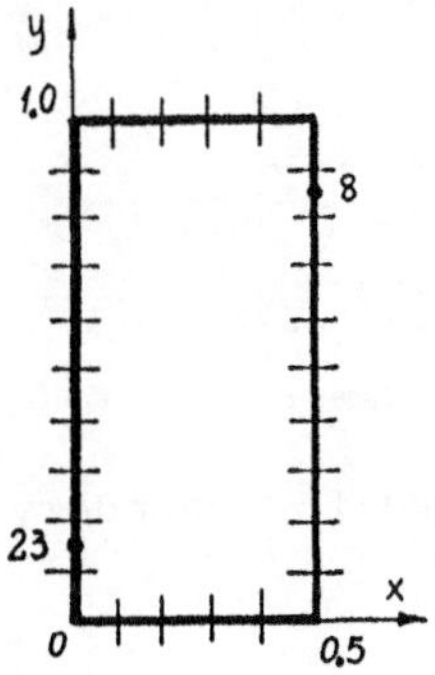

Fig.2 Geometry of the region

Faces x=0 and y=0 are insulated whereas faces x=0.5 and y=1 are subjected to the boundary condition of Robin's type with the heat transfer coefficient h=0.5 W/m² K and fluid temperature $u_f=0$.

Since the region was initially at temperature 1 the above boundary conditions cause thermal shock on the faces x=0.5 and y=1. Hence calculations were carried out applying smoothing algorithm in the first time step, e.g. Zienkiewicz [11]. The absolute error defined as the difference between numerical results and analytical solution for two selected boundary nodes is plotted in Figs 3 and 4. Results marked as curve 1

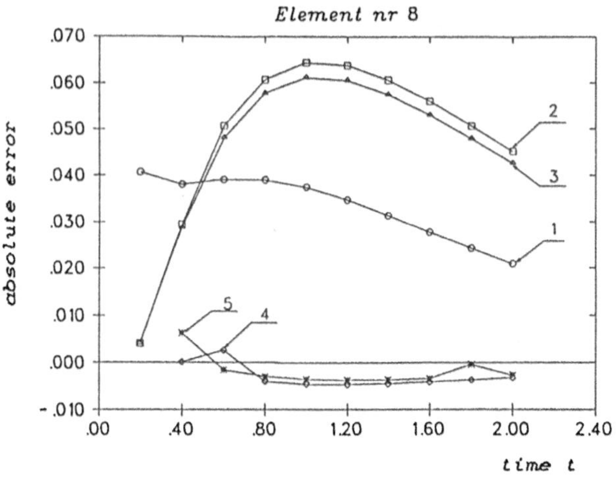

Fig. 3 Absolute error of the MRM and the DRM for boundary element nr 8

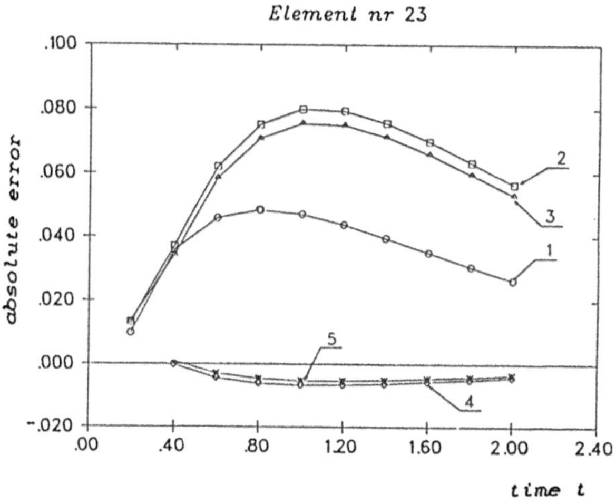

Fig. 4 Absolute error of the MRM and the DRM for boundary element nr 23

are obtained using the DRM formulation with one internal pole. The curve 2 is associated with first order approach of the MRM with no internal poles whereas curve 3 refers to the same approach but with one central internal pole. The second order approach of the MRM produces results marked as curve 4 - no internal poles, and curve 5 - one internal pole.

Calculations were carried out using constant elements with parameters $\Theta = 1$ and constant time step $\Delta t = 0.2$. For the second order approach parameter $p = 1.15$ was selected. Very good accuracy of the method, especially of the second order approach, is worth pointing out. Notice also that formulations with one central internal pole are more accurate than those without any internal poles.

5. Conclusions

In this paper two methods of transforming domain integrals into boundary were presented. Each approach can be thought of as the 'adjoint' of the other. Both techniques use the same fundamental solution u^* and start with the same domain integral representing the effect of a right hand side function b. However, the DRM expresses the b function in terms of interpolation functions, whereas the MRM works with higher order expressions for u^* and successive Laplacians of the function b.

It has been proved by many numerical examples that the DRM gives excellent results in many practical engineering problems. The MRM being a more recent idea is still the subject of further research. Nevertheless, the numerical examples studied so far indicate that the technique gives accurate results. In addition the technique has the important advantage of ensuring convergence of the solution by introducing successive terms in the series.

References

1. Brebbia, C.A. The Boundary Element Method for Engineers. London: Pentech Press; Boston: Computational Mechanics Publications 1978.
2. Brebbia, C.A.; Telles, J; Wrobel, L. Boundary Element Techniques - Theory and Applications in Engineering. Berlin and New York: Springer-Verlag, 1984.
3. Brebbia, C.A.; Dominguez, J. Boundary Elements - An Introductory Course. Southampton: Computational Mechanics Publications; New York: McGraw Hill, 1989.

4. Nardini, D.; Brebbia, C.A. A New Approach for Free Vibration Analysis using Boundary Elements. In Boundary Element Methods in Engineering. Springer-Verlag, 1982.
5. Nardini, D.; Brebbia, C.A. Boundary Integral Formulation of Mass Matrices for Dynamics Analysis. Chapter 7 in Topics in Boundary Element Research, Vol.2. Berlin and New York: Springer-Verlag, 1985.
6. Wrobel, L.C.; Brebbia, C.A. The Dual-Reciprocity Boundary Element Formulation for Non-Linear Diffusion Problems. Computer Methods in Applied Mechanics and Engineering, Vol. 65, 1987, 147-164.
7. Wrobel, L.C.; Brebbia, C.A. Boundary Elements for Non-Linear Heat Conduction Problems. Communications in Applied Numerical Methods, Vol.4 1988, 617-622.
8. Nowak, A.J.; Brebbia, C.A. The Multiple Reciprocity Method - A New Approach for Transforming BEM Domain Integrals to the Boundary. Eng. Analysis, Vol.6, No.3, 1989.
9. Nowak, A.J.; Brebbia, C.A. Solving Helmholtz Equation by Boundary Elements using the Multiple Reciprocity Method. In Computers and Experiments in Fluid Flow, (Eds G M Carlomagno and C A Brebbia). Southampton: Computational Mechanics Publications; New York: Springer-Verlag, 1989.
10. Nowak, A.J. The Multiple Reciprocity Method of Solving Transient Heat Conduction Problems. In Advances in Boudnary Elements (Ed. C A Brebbia) Berlin: Springer-Verlag, 1989.
11. Zienkiewicz, O.C. The Finite Element Method. New York: McGraw-Hill, 1977.

BIE Solutions for Applied Thermoplasticity by Collocation

T. A. CRUSE, S. T. RAVEENDRA, and B. H. THACKER

Department of Engineering Mechanics
Southwest Research Institute
San Antonio, Texas 78284 USA

SUMMARY

The design application of BIE methods to the stress analysis of advanced turbomachinery components involves four domain integral problems: transient thermal strains, temperature dependent material properties, steep thermal gradients due to thermal shock, and plastic strains caused by thermal shock. Application of the BIE method to such problems has been significantly hindered by the need to provide domain elements to model each of these effects. The goal of the current study is to include each of these effects through the use of collocation methods, using particular solutions corresponding to properly selected global interpolation functions.

The application of a collocation system for particular integral formulations to thermoplasticity problems in boundary element methods is presented. The particular solution is used to evaluate dual-boundary integral equations (BIE) for this inhomogeneous problem formulation. The dual-BIE replaces the usual domain integral formulations. Issues regarding numerical performance and the nature of the solution algorithm are discussed in the framework of progressively more complex application problems.

INTRODUCTION

Analytical methods have been developed for reducing certain types of body forces to equivalent boundary integrals. The first publication concerned three dimensional body forces due to steady state thermal loading and due to centrifugal acceleration [1]. This approach was then extended to the axisymmetric problem [2]. Subsequently, Danson [3] summarized this approach.

The limitation in the analytical approach is that many body force-like problems cannot be treated in the exact manner employed for the above studies. Examples of these important problems include non-steady thermal loading, vibration, temperature dependent properties, plasticity, etc. For such problems, the body forces have been modeled using domain elements, thereby voiding much of the BIE advantage of reducing problem dimensionality.

Relatively recently, an important development was published by Nardini and Brebbia [4]; their approach offers an alternative to the direct use of domain elements for these classes of problems. The method, which they have called the dual reciprocity method, uses global interpolation functions to replace standard domain interpolations, *a la* finite elements. The dual reciprocity formalism results in boundary integral equations that are approximately equivalent to the domain integrals of the standard BIE formulation. They have extended the method to transient dynamic problems [5].

Following the work of Nardini and Brebbia, Banerjee and co-workers [6, 7, 8, 9] have used a different formalism which results in an equivalent BIE formulation. The formalism draws very directly on the notion of particular integrals to the governing equations, and was first applied in the BIE context by Lachat [10]. The formalism of particular integrals results in a clearer understanding of the dual integral equation formulation, for a full range of body force problems.

G. Kuhn · H. Mang (Eds.)
Discretization Methods in Structural Mechanis
IUTAM/IACM Symposium Vienna, Austria, 1989

The dual BIE for a given domain problem is obtained (in principle) by the solution of the governing differential equations of equilibrium for the domain inhomogeneity terms. The solution, given as particular values of tractions and displacements, is imposed at the boundary of the region and corresponds exactly to the set of domain integrals in the original BIE formulation. Unfortunately, the exact particular solution cannot be obtained for any but the most trivial problems.

The use of a collocation algorithm with global shape functions provides approximate particular solutions to the appropriate equilibrium equation for the given domain problem. The approximate particular solution is the one used to define a dual BIE which then is used to approximate the domain integral of the original BIE formulation. The result is an approximate BIE in terms of boundary data only, for the original BIE with domain variables. For many problems, the approximation of the domain term results in surprising accuracy, using very simple global interpolation functions. The general approach has much to offer; further research into the selection of global interpolation functions and their corresponding BIE is still needed.

FORMULATION ISSUES

Dual Boundary Integral Equations

The procedure for eliminating the domain integrals is essentially the same for thermal and initial stress terms. The first step begins with the Navier equilibrium equation for the given domain problem. Letting the Navier equilibrium operator be denoted as $\underline{N}(\underline{x})$, we can write a general form of the governing equation for a given domain variable as

$$\underline{N} \cdot \underline{u}^{p} = \underline{F}(\underline{x}) \tag{1}$$

where $\underline{u}^{p}$ is the particular displacement solution to eq. (1) and $\underline{F}(\underline{x})$ is the general inhomogeneous term in the governing differential equation, such as thermal strains or plastic initial stresses. By means of Hooke's law, there is then also a corresponding particular stress $\underline{\sigma}^{p}$ and traction $\underline{t}^{p}$ solution to eq. (1).

The particular solutions to eq. (1) correspond to the domain terms in the BIE. Thus, solutions to the appropriate forms of eq. (1) - if they can be found - result in

$$\underline{u}^{p} \cdot \underline{C} + \int_{S} \underline{u}^{p} \cdot \underline{T} dS = \int_{S} \underline{t}^{p} \cdot \underline{U} dS + \int_{<R>} \underline{\epsilon}^{\theta} \cdot \underline{\Sigma} dV + \int_{<R>} \underline{\sigma}^{I} \cdot \underline{E} dV \tag{2}$$

That this is so follows directly from the well-known fact that the BIE is fully equivalent to the differential equilibrium equation (1) for the governing problem. In (2), the boundary conditions correspond to the values of the particular solution for the domain terms, not the physical boundary conditions.

Subtraction of eq. (2) from eq. (1) results symbolically in the dual BIE for the domain problems of thermoplastic behavior

$$\begin{aligned} &\underline{u} \cdot \underline{C} + \int_{S} \underline{u} \cdot \underline{T} dS - \int_{S} \underline{t} \cdot \underline{U} dS = \\ &\underline{u}^{p} \cdot \underline{C} + \int_{S} \underline{u}^{p} \cdot \underline{T} dS - \int_{S} \underline{t}^{p} \cdot \underline{U} dS \end{aligned} \tag{3}$$

Eq. (3) then constitutes a symbolically equivalent solution to the original domain problem, assuming that the correct particular solution can be found. The form of eq. (3) is in terms of dual BIE's for the original domain problem. Following the formulation of the dual BIE's, application of the usual discretization algorithms for the BIE method results in the following matrix form

$$[\underline{\Delta T}]\{\underline{u} - \underline{u}^{p}\} = [\underline{\Delta U}]\{\underline{t} - \underline{t}^{p}\} \tag{4}$$

Thus, the same BIE matrices used for the non-domain problem may be used to obtain boundary solutions for the domain problem, without resorting to domain integrations. The attractiveness of this approach is enormous, given the significant algorithmic and computational burden of domain integrations. The problem, of course, is to obtain suitable particular solutions. In general, these must be approximated.

The approach to obtaining approximate particular solutions relies on collocation methods. The domain variable is to be approximated with relatively simple spatial functions for which the particular solution to the governing differential equation is easily obtained. The spatial functions are superimposed and collocated with the specified domain variable(s) so as to approximately satisfy the governing Navier equation (1). The details of the collocation methodology will be discussed in the next section, dealing with the thermal loading problem.

Selected Particular Solutions by Collocation

Temperature-Independent Elastic Properties

We first seek to find a particular solution, $\underline{u}^{\theta}$, to the Navier equations of elastic equilibrium in the presence of thermal loading for which the governing Navier equation is given by

$$\frac{\nu}{1-2\nu}\nabla(\nabla\cdot\underline{u}^{\theta})+\nabla^{2}\underline{u}^{\theta}=\alpha\left(\frac{1+\nu}{1-2\nu}\right)\nabla\theta \tag{5}$$

Let eq. (5) be written symbolically as

$$\underline{N}\cdot\underline{u}^{\theta}=\beta\nabla\theta \tag{6}$$

where $\beta=\alpha\cdot(1+\nu)/(1-2\nu)$. The particular solution to (6) implies additional boundary displacements, $\underline{u}^{\theta}(Q)$, and boundary tractions, $\underline{t}^{\theta}(Q)$ which must be considered along with the physical boundary conditions.

The selected simple collocation solutions should be capable of accurate modeling of simple domain distributions, and must be convergent to the exact solution for an indefinite number of collocation points. The collocation system for the temperature field is given symbolically as

$$\tilde{\theta}(q)=\sum_{j=1}^{N}K(q,Q_j)\Phi(Q_j) \tag{7}$$

where $\tilde{\theta}(q)$ is the temperature interpolation at $q(\underline{x})$, $\Phi(Q)$ is a set of unknown coefficients, and $K(q,Q)$ is a global interpolation function. The global interpolation functions K(q,Q) are selected such that eq. (6) can be integrated. Thus, eq. (7) constitutes a collocation solution of eq. (1). The functions to be used are not uniquely defined by collocation modeling, but, in addition to being integrable, should provide accurate solutions of eq. (1) for simple thermal fields, and should be sufficiently smooth to satisfy eq. (1) exactly in the limit as the number of collocation points N increases indefinitely. The global interpolation function used by previous investigators (e.g., [5,7]) is given by

$$K(q,Q)=R_o[1-r(q,Q_j)/R_o] \tag{8}$$

where $r(q,Q_j)$ is the distance between the interpolated point q and the collocation point Q_j, and R_o is a scale constant such as the maximum size of the region. Generally, the point q is taken as arbitrary, while Q_j normally corresponds to the BEM surface mesh nodal locations. Collocation of the solution to additional temperature points (e.g., the interior) may be made.

The quality of the interpolation function for generalized thermal loads is discussed in a later Section. However, it is important to note that eq. (7) is generally well-behaved in terms of conditioning, due to the diagonal weighting given by the form of eq. (8); thus, we expect and find that the inversion of eq. (7), given by $K^{-1}(q,Q_j)$, is numerically stable.

The corresponding displacement field for (5) is given from, as

$$\underline{u}^{\theta}=\left(\frac{1+\nu}{1-\nu}\right)\alpha\sum_{j=1}^{N}R_o\left(4-3\frac{r(q,Q)}{R_o}\right)\frac{\underline{x}-\underline{\xi}}{12}\Phi(Q_j) \tag{9}$$

$$\underline{u}^{\theta}=\underline{G}\sum_{j=1}^{N}\underline{K}^{-1}(q,Q_j)\theta(Q_j)$$

where $\underline{x}-\underline{\xi}$ is the vector from $Q(\underline{\xi})$ to $q(\underline{x})$. Eq. (9) is then the solution to eq. (5), for the assumed global interpolation of temperatures. Stresses and surface tractions are derived from the displacement terms in eq. (9).

Temperature-Dependent Elastic Properties

For most applications in the design of thermally loaded structures, such as for gas turbine engine structures, the material properties are temperature dependent. Temperature dependence introduces inhomogeneities in terms of the boundary integral equation formulation that also appear as domain integrals. The treatment of the material inhomogeneity is through the definition of an initial stress term in the BIE formulation.

The derivation of the initial stress is obtained by separating the elastic properties into a temperature-independent and a temperature dependent set of values. The reason for separating the constant terms from the temperature dependency is to allow the direct use of Betti's reciprocal work theorem in formulating the boundary-integral equation. Beginning with the domain formulation, we obtain

$$\int_{<R>}(\underline{\sigma}-\underline{\sigma}'+\underline{\sigma}^{\theta})\cdot\underline{E}\,dV=\int_{<R>}\underline{\epsilon}^{T}\cdot\underline{\Sigma}\,dV \tag{10}$$

where the modified stress terms due to the thermal loading $\underline{\sigma}^{\theta}$ and inhomogeneity $\underline{\sigma}'$ are given as

$$\begin{aligned}\underline{\sigma}' &= \underline{C}'(\Theta)\cdot(\underline{\epsilon}^{T}-\underline{\epsilon}^{\theta})\\ \underline{\sigma}^{\theta} &= \underline{C}_{o}\cdot\underline{\epsilon}^{\theta}\end{aligned} \tag{11}$$

Application of the equilibrium requirement results in the following Navier equation of equilibrium for the displacement field in terms of the inhomogeneity field

$$\underline{N}\cdot\underline{u}' = \nabla\cdot(\underline{\sigma}'-\underline{\sigma}^{\theta}) \tag{12}$$

The inhomogeneous Navier equation appears in the same form as that for the thermal loading, with the material property and temperature terms replacing the temperature alone. In the case of eq. (6), a collocation solution is established by integrating eq. (6) prior to collocation. In general, such a solution strategy is not available for the case of temperature dependent properties.

The collocation approach for this case is to replace the product term in eq. (12) by the global interpolation

$$\underline{\tilde{\sigma}}'-\underline{\tilde{\sigma}}^{\theta} = \sum_{J=1}^{N}K(q,Q_J)\cdot\underline{\Phi}'(Q_J) \tag{13}$$

The global interpolation in (13) is then operated on to approximate the second term in (12). The solution algorithm to obtain the particular solution to eq. (12) is iterative. The first step is to assume $\underline{\sigma}'$ is zero. The elastic strains and the initial stresses are then calculated for the thermal strains (zero total strain) to set the first iteration step. The particular solution is collocated to the iterative values of $\underline{\sigma}'-\underline{\sigma}^{\theta}$ in eq. (12). Based on the particular displacement solution, new total strains are computed and the process is updated. Convergence is tested by comparison of the change in effective stress to the past value. For the current applications, convergence is quite rapid.

The resulting interpolation for $\underline{\tilde{u}}'$ is of lower order than the previous interpolations, in that the global function $K(q,Q_J)$ is differentiated prior to finding the particular solution to Navier's equation for this problem (12). The accuracy of the domain integral approximation is degraded slightly for our validation problems, as discussed below.

Temperature-Independent Thermoplasticity

The boundary element formulation for thermoplasticity derives most of its elements from temperature dependent thermoelastic analysis. In addition to the linear constitutive relationship for elastic materials, the plastic theory requires the relations between stress and strain during plastic flow. The incremental form of stress strain relationship may be written as

$$\dot{\underline{\sigma}} = \underline{C} \cdot (\dot{\underline{\epsilon}}^{T} - \dot{\underline{\epsilon}}^{n})$$
$$\dot{\underline{\epsilon}}^{n} = \dot{\underline{\epsilon}}^{\theta} + \dot{\underline{\epsilon}}^{p} \qquad (14)$$

where the superior dot indicates increment of variables. The increment of plastic strains in the above equation can be obtained from appropriate material models. Let the initial stress due to thermal and plastic fields be defined as

$$\dot{\underline{\sigma}}^{n} = \underline{C} \cdot \dot{\underline{\epsilon}}^{n} \qquad (15)$$

The equilibrium equation of the stress field gives the Navier equation for the increment of displacement field as

$$\underline{N} \cdot \dot{\underline{u}}^{n} = \nabla \cdot \dot{\underline{\sigma}}^{n} \qquad (16)$$

As explained in the previous section, the approach is to fit the global interpolation to the nonlinear stress field as in the thermally inhomogeneous case. The particular solution of the increment of displacement field is then obtained by the same iterative techniques employed in elastoplastic BEM [11]

SELECTED NUMERICAL RESULTS

Collocation of Temperature Fields

The simplest global interpolation function successfully used by previous publications was adopted in eq. (8), using the distance function r(q,Q) for variable points (q) relative to boundary collocation points (Q). Due to the collocation of the interpolation functions at selected points in eq. (7), the temperature field is satisfied at the collocation points, within computer accuracy. What should also be obvious is that the temperature field at non-collocation points will not be equal to the input temperature field (see eq. (7)).

To illustrate this fact, Fig. 1 shows the interpolation function temperatures at points on one end of a unit cube with constant temperatures. The result, in terms of relative error, is obtained by taking the point $q(\underline{x})$ in eq. (7) for points in the face of the cube. Clearly the magnitude of the error at non-collocation points can be quite significant.

The addition of collocation points other than the solution nodes will improve the collocation errors at points other than those used for collocation. The addition of collocation points may be made at the surface or at the interior. However, the strategy for selecting these additional points must take into account the BEM algorithm used to obtain the solution to eq. (4). More will be said on this topic in what follows.

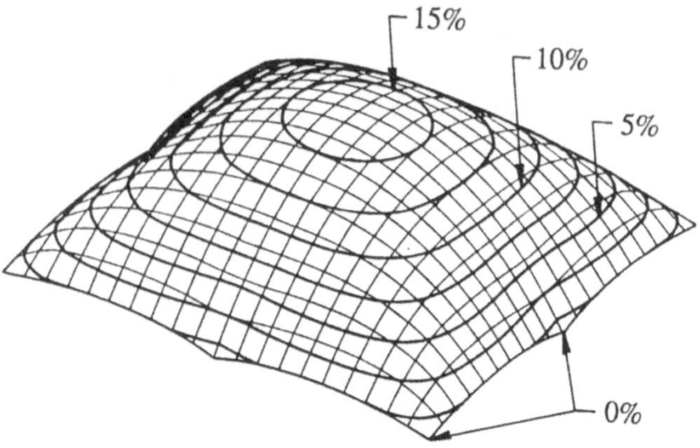

Fig. 1: Temperature Collocation Error for Unit Cube, Constant Temperature

Boundary Integral Equation Solutions for Thermal Loads

The simplest validation problem for the collocation algorithm with the dual BIE formulation is that for the linear temperature field in a cube. With quadratic isoparametric elements, the BEM mesh consists of six elements, eight corner nodes, and eight mid-side nodes. The errors in temperature collocation were discussed above. The dual BIE formulation with collocation was compared with that for the exact solution to eq. (5) for the linear temperature case.

Table I illustrates the level of accuracy that was obtained using collocation with the global interpolation function from eq. (8), the numerically exact linear solution, and the exact analytical solution. It is seen that the boundary solution points for the global interpolation method are nearly as accurate as those for the numerically exact solution, generally within two percent of the exact solution. The reason for the accuracy of the implementation of the collocation system to the solution of eq. (4) has been investigated in some detail.

Table I
Temperature Validation Results

Solution Condition	Solution Point	Global Interpolation	Global Linear	Exact Analytical
x-Displacement	Corner	1.0138	1.0000	1.0
x-Displacement	Mid-face	1.0114	1.0000	1.0
x-Displacement	Interior	0.1158	0.1200	0.12
$\sigma_{yy}/\alpha E$	Corner; x=1	1.0051	1.0000	1.0
$\sigma_{yy}/\alpha E$	Mid-Side; x=1	1.0097	1.0000	1.0
$\sigma_{yy}/\alpha E$	Interior	0.1000	0.1000	0.1

The next validation problem is the steep thermal gradient (thermal shock) formulation. In this problem, the temperatures are governed by the local temperature gradient normal to the surface. Fig. 2 shows the mesh that is used for this problem. Steep gradients are to be imposed normal to the face shown with nine elements; the transverse faces are constrained against growth in the normal direction to simulate one dimensional response. The temperature field is imposed by the following form

$$\Theta = \Theta_o e^{(f_o/\Theta_o)\xi} \tag{17}$$

where Θ_o is the surface temperature. The corresponding displacement particular solution is given by

$$u_i = \beta \frac{\Theta_o^2}{f_o} l_i e^{(f_o/\Theta_o)\xi} \tag{18}$$

In the validation problem, the surface temperature is taken as 200°F, and the temperature gradient is given as -5,000°F/inch. Since the cube is of unit dimension, the thermal field is confined to a very small surface layer.

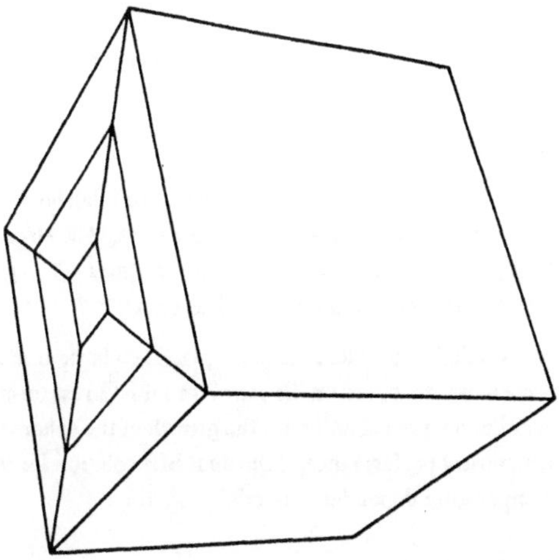

Fig. 2: BEM Mesh for Cube with Imposed Steep Thermal Gradient on One Face

Use of the particular solution for the steep thermal gradient case gave essentially exact stress results, as expected. The only discrepancy occurs at the nodes along the edges of the heated face, where the discontinuity of the boundary and stress averaging to obtain nodal stresses conspire to give algorithmically incorrect results. However, at all interior (to the edge) nodes on the heated face, the results are exact.

The same model was run using the standard global interpolation method without any mesh refinement near the heated face. The stress results at the interior nodes were only in error by up to five percent. The reasonable stress results stem from the fact that surface temperature rather than gradient is still the major driver in surface stress.

Thermoelasticity with Temperature Dependent Material Properties

A verification problem for temperature dependent thermoelasticity concerns the analysis of a hollow cylinder with radius ratio of 2 subjected to linear variation of temperature along the radius. A symmetric section of the cylinder was modeled using 18 elements (56 nodes). It was assumed that the Young's modulus varied linearly with the temperature given by $E = E_o\{1 - 10^{-4}\Theta(r)\}$. Comparison of normalized hoop stresses, using the current procedure, is made to the finite element method in Fig. 3 and shows good agreement between the results.

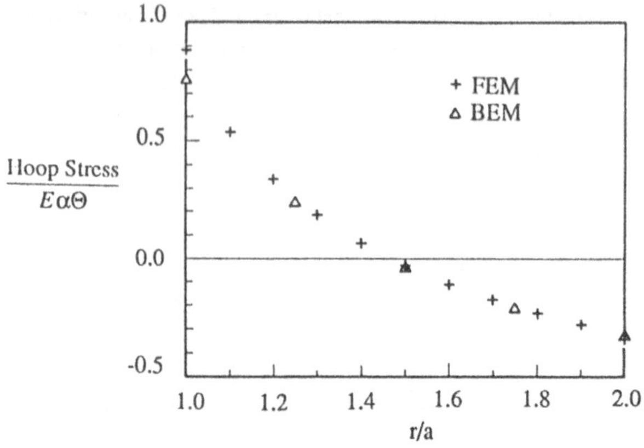

Fig. 3: Comparison of Transverse Stresses for Thick Cylinder with Temperature Dependent Material Properties

Modeling Thermally Induced Plasticity

The most basic validation problem was for a uniform slab heated between rigid walls. All thermal strains are transformed into plastic strains above the elastic limit. The stress-strain behavior of this problem can be obtained analytically by integrating the stress-strain relations. The agreement of the analytical and computed results was excellent, with a maximum error of less than 0.15%.

The second validation problem concerns plastic collapse of a hollow sphere due to internal pressure. The collocation schemes were based on the use of the initial stress formulation and collocation at surface points only. The problem has an exact solution for the growth of the sphere versus the applied pressure [12]. Figure 4 illustrates the numerical performance of the dual BIE solution for the same collocation particular solutions as used for the temperature dependent material properties.

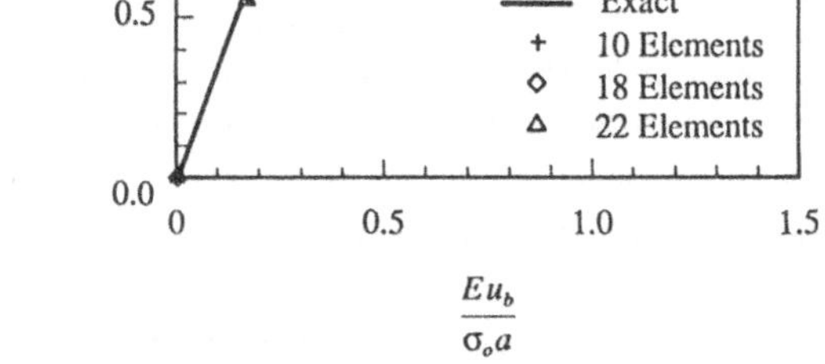

Fig. 4: Plastic Collapse of Hollow Sphere

The collapse load is not accurately predicted for the crude, 10 element map. The reason for this is that the collocation system is dependent on the number of boundary points for the BEM mesh. Addition of a few more elements clearly improves the collocation accuracy for the plastic initial stresses.

COMMENTS ON PARTICULAR SOLUTION ACCURACY

As noted in the numerical examples above, the collocation-dual BIE solutions demonstrate excellent accuracy, in spite of the fact that the collocation system itself involves significant error between collocation points. In an attempt to understand this, we studied the numerical implementations of the thermoelastic problems.

The key to the excellent performance of the solution strategy is contained in the implementation of the dual BIE, eq. (3), as a set of boundary element equations, eq. (4). To understand this we first calculated the actual thermoelastic particular solution to eq. (5) for the case of uniform heating. The values of $\underline{u}^p$ and $\underline{t}^p$ for the exact solution were computed, as were the values using the collocation field from eq. (7).

Remarkably, the two sets of particular solution variables were identical, at the computed collocation points. For points away from the collocation points, the two particular solutions do not agree. The lack of agreement is of the same type as that discussed in the interpolation of temperature fields. The exact particular solution for the uniform temperature rise is a set of linear displacements and constant tractions, while the collocation solution has a variation due to the radius term in eq. (8).

However, the BEM implementation to obtain eq. (4) from eq. (3) uses quadratic isoparametric shape functions for interpolating the boundary variables. Since, for the case of the uniform temperature increase, the collocation and exact particular solutions are the same at the collocation points, and because the collocation points we use are the same as the BEM mesh points, interpolation of both particular solutions by the BEM algorithm will produce the same boundary solution. In the case of the uniform temperature increase, the BEM solution is exact within computer accuracy for the numerical integrals.

As the domain field (e.g. temperature) becomes more non-uniform over the surface of the body, the exact particular solution (if it can be found) and the collocation solution are increasingly different. For the initial stress formulation in eq. (13), the collocation field is once differentiated before collocation. As a result, the accuracy of the method is reduced, although not significantly. Thus, the solution algorithm appears to be quite capable of modeling the full transient thermoplastic response of the turbine blade problem. The solution of that problem has been initiated.

Addition of boundary elements in the current algorithm is a very effective means for implementing improved collocation accuracy for higher-order fields. By adding collocation points, the accuracy of the particular solution increases in a manner consistent with the resolution capability of the BEM mesh. Numerical experiments on the trade-offs for adding interior collocation points (temperature-dependent properties example) vs. boundary collocation points (plastic collapse example) are not yet complete.

One additional point should be made in regard to the selection of particular solutions for collocation modeling. In the case of the uniform temperature field, it was found that the collocation scheme produced exact particular integral results at the collocation points. This is like a patch-test example used for FEM modeling in that the lowest order excitation is modeled exactly. Further work is needed to elucidate the role that such a lower order type of "patch-test" performance enhances the numerical accuracy of the collocation functions for higher order problems and for dynamic response problems, particularly those involving transient loading.

REFERENCES

1. Cruse, T. A., Boundary-Integral Equation Method for Three Dimensional Elastic Fracture Mechanics Analysis, *AFOSR-TR-75-0813* (1975).
2. Cruse, T. A., Snow, D. W., and Wilson, R. B., Numerical Solutions in Axisymmetric Elasticity, *Computers & Structures*, 7, 445-451 (1977).
3. Danson, D. J., A Boundary Element Formulation of Problems in Linear Isotropic Elasticity with Body Forces, *Recent Advances in Boundary Element Methods*, 3, 105-122 (1982).
4. Nardini, D. and Brebbia, C. A., A New Approach to Free Vibration Analysis Using Boundary Elements, *Boundary Element Methods in Engineering*, C. A. Brebbia (Ed.) 1982.
5. Nardini, D., and Brebbia, C. A., Boundary Integral Formulation of Mass Matrices for Dynamic Analysis, *Topics in Boundary Element Research*, 2, Springer-Verlag (1985).
6. Ahmad, S. and Banerjee, P. K., Free Vibration Analysis by BEM Using Particular Integrals, *J. Engrg. Mechs. ASCE*, 112, 682-695 (1986).
7. Pape, D. A. and Banerjee, P. K., Treatment of Body Forces in 2D Elastostatic BEM Using Particular Integrals, *Transactions of the ASME*, 54, 866-870 (1987).
8. Henry, D. P. and Banerjee, P. K., 'A New Boundary Element Formulation for Two- and Three-dimensional Thermoelasticity Using Particular Integrals, *International Journal for Numerical Methods in Engineering*, 26, 2061-2078 (1988).
9. Henry, D. P. and Banerjee, P. K., A New BEM Formulation for Two - and Three-dimensional Elastoplasticity Using Particular Integrals, *International Journal for Numerical Methods in Engineering*, 26, 2079-2096 (1988).
10. Lachat, J. C., *A Further Development of the Boundary Integral Technique for Elastostatics*, Ph.D. Thesis, University of Southampton, 1975.
11. Banerjee, P. K. and Raveendra, S. T., Advanced Boundary Element Analysis of Two- and Three-dimensional Problems of Elastoplasticity, *International Journal for Numerical Methods in Engineering*, 23, 985-1002 (1986)
12. Mendelson, A., *Plasticity : Theory and Application*, MacMillan Company (1968).

State Characterization in Sampling Points and Representation of Velocity Fields by Nodal Values as a Basis for Discretization

J.F. BESSELING

Laboratory for Engineering Mechanics
Delft University of Technology, Delft

1. Introduction

Continuum theory masks the fact that the concept of state pertains to a finite, be it very small neighbourhood of a material point. This neighbourhood is subjected to an average deformation, derived from the average motion of mass, that is described by the velocity field. In thermodynamic theory it is now generally realized that the stresses, dual to the local rates of deformation and that determine the rate of work of deformation, do not provide a sufficient description of the state of the material, in so far as this state is affected by deformation. Hidden, or internal state variables are needed in constitutive equations, that give a description of memory effects. It turns out that it is fruitful to enlarge the number of state variables without an increase of the number of independent kinematic variables. Apparently the description of the dissipative processes, responsible for the memory effects, tolerate a certain latitude in the description of the kinematics. In other words the effect of the thermodynamic processes in the different constituents overshadows the effect of disturbances of mechanical equilibrium among constituents, the number of equilibrium equations being equal to the number of kinematic variables.

If finite element theory is regarded as a modelling technique in its own right, we may also here make a distinction between the description of the kinematics and the description of the state. In a proper finite element representation, in which all rigid body motions of the individual finite elements are independently being described by the interpolations between the element nodal velocities, the loads that one element exerts on the system are selfequilibrating, provided the element does not carry external loads. Errors in selfequilibrating loads on the system, exerted by one element, damp out quickly, as is known in the mechanics of deformable bodies since the work of DE St. Venant. We may conclude that for the derivation of a discrete model it is essential to define for each finite element the necessary

G. Kuhn · H. Mang (Eds.)
Discretization Methods in Structural Mechanis
IUTAM/IACM Symposium Vienna, Austria, 1989

and sufficient conditions for rigid body motions in terms of the nodal velocities; or equivalently to define the deformation modes by properly chosen generalized strainmeasures.

The stress multipliers dual to the generalized strainrates must be related to the "observable" kinematics by constitutive equations, just as in the case of continuum theory, in order to arrive at a system of equations by which the physical deformation problem is fully described. Like in continuum theory, in finite element theory, and there even more so, the kinematics does not provide a sufficient basis for the description of the state. Either as a method of mathematical approximation of continuum theory considering the state in integration points, or as a modelling technique relating the state to appropriate deformations of material fractions, we introduce internal state variables. The stress multipliers dual to the generalized strainrates of the finite element are found as a weighted sum of internal stress variables in what we shall call sampling points [1].

After an analysis of the "observable" kinematics and statics of a finite element model of a deformable body we shall discuss the concepts of state and constitutive equations for a finite element. Some observations on the application of the general approach to the derivation of the properties of a shellelement will conclude the paper.

2. Discretization of continuity and equilibrium equations

In the case of material finite elements at interelement boundaries the adjoining elements must have common displacements in order to ensure preservation of continuity of the structure. These displacements are represented by interpolations between nodal displacements at the inteface between any two elements or between an element and the surroundings of the structure. Continuity is then ensured for all values of the nodal displacements and the displacementfield is represented by a finite dimensional vectorspace U with elements in $\mathbb{R}^n$.

Equilibrium may be formulated in terms of the principle of virtual power:

the power of the external forces must be equal to zero for all virtual rigid body motions.

In the continuum theory in the material description the deformation is appropriately described by the deformation tensorcomponents

$$E_{\alpha\beta} = \frac{1}{2}\left(\frac{\partial u_\alpha}{\partial a_\beta} + \frac{\partial u_\beta}{\partial a_\alpha} + \frac{\partial u_k}{\partial a_\alpha}\frac{\partial u_k}{\partial a_\beta}\right), \tag{1}$$

where a_α are the cartesian (material) coordinates in the undeformed reference configuration and where u_α are the components of the displacement vector.

Equilibrium according to the principle of virtual power is now expressed by

$$\int_{\partial B} t_i \dot{u}_i \, dA + \int_{B_o} \rho_o f_i \dot{u}_i \, dV_o = \int_{B_o} \sigma_{\alpha\beta} \dot{E}_{\alpha\beta} \, dV_o \quad \forall \dot{u}_i, \tag{2}$$

$\dot{E}_{\alpha\beta}$ being derived from (1). Note that the condition of rigidity is taken into account by the components of the symmetric tensor $\sigma_{\alpha\beta}$ as lagrangian multipliers. This tensor is easily identified as the pseudo stresstensor of Kirchhoff by application of (2) to a small tetreader.

In the finite element model a displacementfield for one element may be obtained by interpolation between the displacements at its surface, which are themselves interpolations between nodal values. The displacementfield must be such that all possible rigid body motions of the finite element are represented. Even for elements with curved boundaries this requirement is met by the isoparametric concept [2]. For the transformation to natural coordinates ξ_α of the finite element and for the interpolation between nodal displacements the same interpolation functions are used. For material finite elements the cartesian coordinates of the material points are then at any moment determined by

$$x_i^e = \psi_N(\xi_\alpha)(a_{Ni}^e + u_{Ni}^e), \quad \sum_{N=1}^{n^e} \psi_N(\xi_\alpha) = 1. \tag{3}$$

The nodal displacements of one element are elements of a subspace U^e of the vector space U of the structure as a whole. Rigid body motions of a finite element are isomorph with a subspace of U^e, that will be called the vectorspace of degrees of freedom of the element, V^e. It is appropriate to call the complementary subspace, E^e, the deformationspace of the finite element with elements in $\mathbb{R}^{me}$.

$$U^e = V^e \oplus E^e. \tag{4}$$

The conditions for a rigid body motion are expressed locally by $\dot{E}_{\alpha\beta}=0$. In other words, for the finite element model we may replace the contribution of the subsidiary conditions in the statement (2) by a finite set of conditions in terms of parameters obtained by local application of (1). With the aid of the interpolations (3) for each finite element, and thus for the structure as a whole, strainparameters are obtained, which are nonlinear functions of the nodal displacements.

$$\varepsilon_i = D_i(u_k). \tag{5}$$

By means of the velocity interpolations for the external forces the power equivalent duals of the nodal velocities can be determined and the equilibrium condition then reads

$$f_k \dot{u}_k = \sigma_i \dot{\varepsilon}_i = \sigma_i D_{i,k} \dot{u}_k \quad \forall\ \dot{u}_k \tag{6}$$

It follows that equilibrium requires the external forces to be the image of a vector of generalized stress,

$$f_k = \sigma_i D_{i,k}, \tag{7}$$

while continuity implies that the rate of strain is the image of the vector of nodal velocities.

$$\dot{\varepsilon}_i = D_{i,k} \dot{u}_k. \tag{8}$$

It will be clear that the number of independent strainparameters for every finite element of the structural model shall at least be equal to the dimension of its deformation space, i.e. equal to the number of nodal displacement components minus the number of degrees of freedom as a rigid body.

We shall now look at the linear mappings in (7) and (8) in more detail. By elementary row operations we determine the row canonical form, which, possibly after rearranging columns, can be written in the following form.

$$\dot{\varepsilon} = D\dot{u} \Rightarrow \begin{matrix}1\\ \vdots\\ p\\ \vdots\\ m\end{matrix}\overset{1\ldots p\ldots m}{\begin{bmatrix} D^{cc^{-1}} & 0 \\ -B^{T} & I \end{bmatrix}} \dot{\varepsilon} = \overset{1\ldots p\ldots n}{\begin{bmatrix} I & B^{*} \\ 0 & 0 \end{bmatrix}} \dot{u} , \qquad (9)$$

$$f^{T} = \sigma^{T} D \Rightarrow f^{T} \overset{1\ldots p\ldots n}{\begin{bmatrix} D^{cc^{-1}} & -B^{*} \\ 0 & I \end{bmatrix}}\begin{matrix}1\\ \vdots\\ p\\ \vdots\\ n\end{matrix} = \sigma^{T} \overset{1..p..n}{\begin{bmatrix} I & 0 \\ B^{T} & 0 \end{bmatrix}}\begin{matrix}1\\ \vdots\\ p\\ \vdots\\ m\end{matrix}$$

We note that in the case of continuity the velocities can be expressed in terms of strainrates and of rigid body velocities $\dot{u}^{r}$.

$$\dot{u} = \begin{matrix}1\\ \vdots\\ p\\ \vdots\\ n\end{matrix}\overset{1..p..m}{\begin{bmatrix} D^{cc^{-1}} & 0 \\ 0 & 0 \end{bmatrix}} \dot{\varepsilon} + \overset{1\,..\,n-p}{\begin{bmatrix} -B^{*} \\ I \end{bmatrix}} \dot{u}^{r} \qquad (10)$$

while in case m>p the strainrates must must obey the continuity equations

$$\begin{matrix}1\\ \vdots\\ m-p\end{matrix}\overset{1\,..\,p\,..\,m}{\begin{bmatrix} -B^{T} & I \end{bmatrix}} \dot{\varepsilon} = 0. \qquad (11)$$

In the case of equilibrium the generalized stresses are expressed in terms of external forces and statically indeterminate stressparameters σ^{r}

$$\sigma^{T} = f^{T} \overset{1\,\ldots\,p\,\ldots\,m}{\begin{bmatrix} D^{cc^{-1}} & 0 \\ 0 & 0 \end{bmatrix}}\begin{matrix}1\\ \vdots\\ p\\ \vdots\\ n\end{matrix} + \sigma^{rT} \overset{1\,..\,p\,..\,m}{\begin{bmatrix} -B^{T} & I \end{bmatrix}}\begin{matrix}1\\ \vdots\\ m-p\end{matrix} \qquad (12)$$

while in case n>p the external forces must obey the equilibrium equations

$$f^{T} \overset{1\,\ldots\,n-p}{\begin{bmatrix} -B^{*} \\ I \end{bmatrix}}\begin{matrix}1\\ \vdots\\ p\\ \vdots\\ n\end{matrix} = 0. \qquad (13)$$

The duality in the appearance of the matrices B and B* should be observed. It implies that the determination of the row canonical form has to be carried out only once, for one of the linear mappings (7) or (8).

The generalized stress parameters are to be determined by constitutive equations from the deformation history (or more generally from the state history) of the finite elements, these stress parameters being the duals of the strainrates, defined for each element independently.

It may be observed that almost traditionally the algebraic analysis of the finite element equations of continuity and equilibrium is only used in the force method of solution, while in the displacement method the finite element equations are established by introducing the constitutive equations in the integrals over the finite element bodies prior to numerical integration. In the latter case the generalized stresses do not appear explicitely in the equations, which are obtained directly from (2). The strain parameters of (5) could of course be looked upon as the straincomponents in the integration points.

3. State characterization of finite elements and constitutive equations

In order that the errors, due to the incomplete description of the deformations and of the state of a finite element, do damp out in accordance with the principle of De St. Venant, the element must be in equilibrium under the nodal forces, exerted by its surroundings. Hence each finite element may be considered as an independent mechanical system, for which the analysis of continuity and equilibrium of the previous paragraph is pertinent. If n^e denotes the number of independent nodal displacements and m^e the number of independent strain parameters, then $n^e-p=3$ for two-, and $n^e-p=6$ for three-dimensional finite elements, 3 and 6 being the number of degrees of freedom as a rigid body. If $m^e-p>0$, then dimension of the deformation space of the finite element exceeds the minimum number of strain parameters, required to enforce the condition of rigidity in the application of the principle of virtual power. As is intuitively clear, the matrix B depends only on the state of deformation and is independent of the rigid body displacements. On the other hand the matrices $D^{cc^{-1}}$ and B* do depend on the nodal displacements, the rigid body displacements included.

Since the redundant deformations and stress parameters can only be determined through the constitutive equations for the material of the finite element, the equilibrium conditions that have to be imposed on the nodal forces are exhausted by a number of independent strain parameters, equal to the dimension of the displacement space U^e minus the number of degrees of freedom as a rigid body. Deformations in excess of this minimum number may serve as a basis for "internal state variables" of the finite element for a more complete description of its state. Then the question of an optimum choice for the state variables of the finite element arises. This question was also addressed in [3] in the context of nonlinear elastic behaviour. Here in a more general approach we start out from the observation that the state of the finite element is mechanically affected solely through the rate of work of deformation, expressed fully by

$$f_k \dot{u}_k^e = \sigma_i^e \dot{\varepsilon}_i^e \ , \qquad \dot{\varepsilon}_i^e = D_{i,k}^e \dot{u}_k^e \ . \tag{14}$$

The concept of internal or hidden variables may be introduced by considering the absorbed work of deformation to be the sum of a number of material contributions, which number is taken to be finite for computational reasons. Thus in the continuum theory the so-called fraction model is being used to represent the anisotropic hardening in the Bauschinger effect. Then the "observable" deformation parameters are the components of the straintensor of the continuum in a material point, while the fraction stresses, σ_i^p, are the internal state variables. The introduction of layerstresses for inelastic bending of beams, plates and shells embodies the same concept. In all cases where an imposed overall deformation does not adequately describe the distribution of the absorbtion of the work of deformation, we may put with appropriate weighting factors α^p

$$\sigma_i^e \dot{\varepsilon}_i^e = \sum_p \alpha^p \sigma_i^p \dot{\varepsilon}_i^p \ , \quad \varepsilon_i^p = \varepsilon_i^p(\varepsilon_i^e). \tag{15}$$

It is clear that in a derivation of the finite element equations directly from (2) by numerical integration the redundant strain parameters of the individual elements are used as a basis for the internal state variables. This is however by no means the optimum choice, as is shown by the error due to the spurious shear energy in the case of the simple four node quadrangle as well as by the spurious shear and membrane energies, which spoil the performance of displacement-based shell elements. Before we return to this point, however, we briefly consider the constitutive equations.

For purely elastic deformation the work of deformation is absorbed as an elastic potential. How this elastic potential is distributed over the material is irrelevant, since in the reversible elastic processes the knowledge of the dependence of the potential on the observable strain parameters suffices, both in the continuum theory and in the finite element model. Let the elastic potential for a rubberlike material be given by

$$\rho_o e = \tfrac{1}{2}C\varepsilon_v^2 + \tfrac{1}{2}G\,[g^{1/3}\,C^{-1}_{\alpha\alpha} - 3 + \mu\{g^{-1/3}\,C_{\alpha\alpha} - 3\}^m]$$

$$C_{\alpha\beta} = 2E_{\alpha\beta} + \delta_{\alpha\beta}\;, \quad g = \det(C_{\alpha\beta}), \quad \varepsilon_v = g^{\frac{1}{2}} - 1. \tag{16}$$

The elastic potential for a finite element will depend only on the observable strainparameters of this element. The number of these independent strainparameters is determined by the dimension of the deformation space of the element, however it is common practice to apply numerical integration over the volume of the finite element to determine the contribution of the elastic potential to the equations. The pertinent integrals are

$$\int_{B_o} \sigma_{\alpha\beta}\dot{E}_{\alpha\beta}\,dV_o = \int_{B_o} \rho_o \frac{\partial e}{\partial E_{\alpha\beta}}\,\dot{E}_{\alpha\beta}\,dV_o\,,$$

$$\int_{B_o} \Delta\sigma_{\alpha\beta}\dot{E}_{\alpha\beta}\,dV_o = \int_{B_o} \rho_o \frac{\partial^2 e}{\partial E_{\alpha\beta}\partial E_{\gamma\delta}}\,\Delta E_{\gamma\delta}\dot{E}_{\alpha\beta}\,dV_o\,, \tag{17}$$

for the state and for the incremental equations respectively.

A simple and physically consistent procedure is obtained by elastic potential interpolation [2], if necessary preceded by strain interpolation. The elastic potential in a sufficient number of sampling points (comparable with integration points) is expressed by (16). The sum of these expressions, multiplied with the appropriate weighting factors as in numerical integration,

$$E = \sum_p \alpha^p \rho_o e^p\,, \tag{18}$$

forms the total elastic potential of the finite element. The components of the straintensors in the sampling points are determined by observable strainparameters of the element. If the latter are smaller in number than the first, then a strain interpolation is necessary, leading to the linear relationship

$$E^p_{\alpha\beta} = \chi^p_{i\alpha\beta}\,\varepsilon^e_i\,. \tag{19}$$

The integrals (17) are now replaced by

$$\sum_p \alpha^p \sigma_i^p \dot{\varepsilon}_i^p = \sum_p \alpha^p \Big(\rho_o \frac{\partial e}{\partial E_{\alpha\beta}} \chi_{i\alpha\beta}\Big)^p \varepsilon_i^e ,$$

$$\sum_p \alpha^p \Delta\sigma_i \dot{\varepsilon}_i^p = \sum_p \alpha^p \Big(\rho_o \frac{\partial^2 e}{\partial E_{\alpha\beta} \partial E_{\gamma\delta}} \chi_{i\alpha\beta}\chi_{i\gamma\delta}\Big) \Delta\varepsilon_i^e \dot{\varepsilon}_i^e . \tag{20}$$

The only essential difference between the more traditional numerical integration techniques and the present approach is the appearance of an explicite expression for the "observable" stress values, that determine the action of the finite element on its surroundings,

$$\sigma_i^e = \sum_p \alpha^p \Big(\rho_o \frac{\partial e}{\partial E_{\alpha\beta}} \chi_{i\alpha\beta}\Big)^p = \frac{\partial E}{\partial \varepsilon_i^e} , \tag{21}$$

and that appear in the geometrical stiffness matrix of the incremental equilibrium equations

$$(D_{i,k} S_{ij} D_{j,\ell} + \sigma_i D_{i,k\ell}) \Delta u_\ell = \Delta f_k . \tag{22}$$

The mechanical stability of the body is affected by the state solely through the values of these "observable" stress quantities, that in the case of elastic behaviour are given by the derivatives of the elastic potential of the finite element as a whole with respect to the "observable" strain quantities.

In the case of inelastic behaviour there is no such simple state independent constitutive law for the "observable" stresses. The development of the fraction stresses σ_i^p in (15) is governed by dissipative processes which are supposed to evolve independently, like the processes in different material points of the continuum or in different fractions in the same material point of the continuum [4]. The constitutive equations for fraction stresses σ_i^p as internal state variables can be given in the familiar form of creep and plasticity theory for metals.

Introducing a yieldfunction

$$\phi(\sigma_i^p) = 0 \tag{23}$$

and a creep dissipation function

$$\sigma_i^p \dot{\varepsilon}_i^{p''} = f(T, \sigma_i^p), \tag{24}$$

the normality rule leads to the rate equations

$$\dot{\sigma}_i^p = (S_{ij}^p - (1-h)Y_{ij}^p)(\dot{\varepsilon}_j^p - c_j^p), \tag{25}$$

with

$$c_j^p = f \frac{\partial f}{\partial \sigma_j^p} \Big/ \left(\sigma_m^p \frac{\partial f}{\partial \sigma_m^p}\right) \tag{26}$$

and where

if $(\phi(\sigma_i^p) = 0)$ and $\left(\frac{\partial \phi}{\partial \sigma_i^p} S_{ij}(\dot{\varepsilon}_j^p - c_j^p) >= 0\right)$ then

$$Y_{ij}^p = \frac{S_{im}^p \frac{\partial \phi^p}{\partial \sigma_m^p} \frac{\partial \phi^p}{\partial \sigma_n^p} S_{nj}^p}{\frac{\partial \phi^p}{\partial \sigma_k^p} S_{k\ell}^p \frac{\partial \phi^p}{\partial \sigma_\ell^p}} \tag{27}$$

else

$$Y_{ij}^p = 0.$$

Since each fraction, each set of internal state variables possesses a yieldfunction and a dissipation function of its own, while further the internal strainvariables ε_i^p will in general be different functions of the element deformations ε_i^e, these element deformations will result in an inhomogeneous energy dissipation, leading to locked-in stresses σ_i^p, responsible for memory effects like Bauschinger effect and elastic after-effects.

4. Shell element

A proper choice of the sampling points and avoidance of spurious deformations in the interpolations (19) is of utmost importance. This may be illustrated first by the four node quadrangle. The displacement interpolation here produces spurious shear deformations in the cornerpoints, which are easily avoided if the strain interpolation in (19) is based upon a linearly with y varying strain E_{xx}, a linearly with x varying strain E_{yy}, and a constant shearstrain E_{xy}.

A more important example is the initially curved shellelement with six nodal points. Together with 6 normalmoment multipliers, representing weak information on the rotations, the 18 nodal displacements give the element 24 deformations [5,6]. By representing the membrane deformations by interpolations between sampling points, and by deriving the independent bending deformations from linearly varying bending moments, the spurious membrane deformations are avoided and inextensional bending is represented perfectly, if at all possible.

Each shellelement is described on the basis of a flat triangular element with its own coordinate system. Six nodal points permit a quadratic interpolation between nodal values for all displacement components. With $\underline{e}_{\bar{x}}$ and $\underline{e}_{\bar{y}}$ as orthogonal base vectors in the base plane the initial geometry of the triangle with respect to the global cartesian system with base vectors $\underline{e}_x, \underline{e}_y, \underline{e}_z$, is given by the quantities

$$\begin{bmatrix} a_1 & a_2 & a_3 \\ -b_1 & -b_2 & -b_3 \end{bmatrix} = \begin{bmatrix} e_{\bar{x}x} & e_{\bar{x}y} & e_{\bar{x}z} \\ e_{\bar{y}x} & e_{\bar{y}y} & e_{\bar{y}z} \end{bmatrix} \begin{vmatrix} (x_3-x_2) & (x_1-x_3) & (x_2-x_1) \\ (y_3-y_2) & (y_1-y_3) & (y_2-y_1) \\ (z_3-z_2) & (z_1-z_3) & (z_2-z_1) \end{vmatrix} . \tag{28}$$

The initial curvature of the shellelement is represented by displacements $\bar{w}^o$ perpendicular to the base plane in the midpoints of the three sides of the element. Displacements with respect to the undeformed shellsurface are transformed to the local coordinate system by

$$\begin{vmatrix} \bar{u} \\ \bar{v} \\ \bar{w}-\bar{w}^o \end{vmatrix} = \begin{bmatrix} e_{\bar{x}x} & e_{\bar{x}y} & e_{\bar{x}z} \\ e_{\bar{y}x} & e_{\bar{y}y} & e_{\bar{y}z} \\ e_{\bar{z}x} & e_{\bar{z}y} & e_{\bar{z}z} \end{bmatrix} \begin{vmatrix} u \\ v \\ w \end{vmatrix} . \tag{29}$$

Using the Von Karman approximation the membrane deformations of the flat triangular element in the three cornerpoints can be written in the form

$$\varepsilon_i^m = D_{ik}^u \bar{u}_k + D_{ik}^v \bar{v}_k + \tfrac{1}{2} D_{ik\ell}\, \bar{w}_k\, \bar{w}_\ell \tag{30}$$

where $D_{ik}^{u,v}$ and $D_{ik\ell}$ can be expressed in terms of the quantities a_i and b_i of (28).

Since the undeformed shellgeometry is represented by $\bar{w}^o$ the membrane deformations of the shellelement in the three cornerpoints are given by ($\bar{w} = (\bar{w}-\bar{w}^o) + \bar{w}^o$)

$$\varepsilon_i^m - \varepsilon_i^{mo} = D_{ik}^u \bar{u}_k + D_{ik}\bar{v}_k + D_{ik\ell}\bar{w}_k^o(\bar{w}_\ell - \bar{w}_\ell^o) + \tfrac{1}{2}D_{ik\ell}(\bar{w}_k - \bar{w}_k^o)(\bar{w}_\ell - \bar{w}_\ell^o). \tag{31}$$

The 3 × 3 = 9 independent membrane deformations serve as a basis for the membrane deformation distribution. This distribution is obtained by linear interpolation between the corner values, thus making sure that no spurious membrane action may be caused by bending deformations.

These bending deformations are defined indirectly through the dual moment parameters in the expression for the rate of work of bending deformation. For bending the initial curvature of the shell is neglected. With bending moments $M_{\alpha\beta}$ and virtual angular velocities $\dot{\phi}$ about the sides of the triangle the rate of work on the basis of the Kirchhoff hypotheses of platebending is written in the form

$$- \int_{A_o^e} \frac{\partial^2 M_{\alpha\beta}}{\partial a_\alpha \partial a_\beta} \dot{w}\, dA_o + \sum_{i=1}^{3} \int_{s_i} \left[\left\{ \frac{\partial M_{\alpha\beta}}{\partial a_\alpha} n_\beta + \frac{\partial}{\partial s} (M_{\alpha\beta} n_\alpha t_\beta) \right\} \dot{w} + M_{\alpha\beta} n_\alpha n_\beta \dot{\phi} \right] ds$$
$$+ \sum_{i=1}^{3} [(M_{\alpha\beta} n_\alpha t_\beta)^{s_i^-} - (M_{\alpha\beta} n_\alpha t_\beta)^{s_i^+}]\, \dot{w}. \tag{32}$$

In the finite element model of the shell the moments $M_{\alpha\beta}$ are represented by a linear distribution with 3 × 3 = 9 parameters, while 3 × 2 = 6 angular velocity parameters $\dot{\phi}$ are taken to represent a weak information on the rotations in the form of multipliers, ensuring continuity of the normal moments at the sides of the elements, instead of the continuity of the normal slopes, which is much harder to obtain.

For the element displacement and rotation vector

$$u^{eT} = |\bar{u}_1 \ldots \bar{u}_6\ \bar{v}_1 \ldots \bar{v}_6 \bar{w}_1 \ldots \bar{w}_6\ \phi_4^-\ \phi_4^+\ \phi_5^-\ \phi_5^+\ \phi_6^-\ \phi_6^+| \tag{33}$$

and for the 3 moment parameters in 3 sampling points, σ_i^b, we derive from the rate of work expression

$$\sum_{i=1}^{3} \int \left[\left\{ \frac{\partial M_{\alpha\beta}}{\partial a_\alpha} n_\beta + \frac{\partial}{\partial s} (M_{\alpha\beta} n_\alpha t_\beta) \right\} \dot{w} + M_{\alpha\beta} n_\alpha n_\beta \dot{\phi} \right] ds$$
$$+ \sum_{i=1}^{3} \left[(M_{\alpha\beta} n_\alpha t_\beta)^{s_i^-} - (M_{\alpha\beta} n_\alpha t_\beta)^{s_i^+} \right] \dot{w} = \sigma_i^b D_{ik} \dot{u}_k^e = \sigma_i^b \dot{\varepsilon}_i^b .$$

Thus 3 × 3 = 9 independent bending deformations have been defined, which except for a weighting factor (= A/3) may be taken to be the dual curvatures in the sampling points.

By linear interpolation we now can obtain the membrane and bending deformations in any number of sampling points, but a sufficient description of the state of the finite element will be given with four sampling points, taken in the centers of the four subtriangles of the finite element. Even the minimum number of 3 sampling points will suffice in the case of homogeneous elasticity, while for inelastic behaviour or layered material of the shell the variation of the stress over the thickness will require the introduction of fraction stresses, related to a representation of deformation by

$$\varepsilon_i = \varepsilon_i^m + z\varepsilon_i^b \ . \tag{35}$$

A simpler alternative is the use of an Ilyushin's type of yieldfunction, representing the coupling of the "observable" membrane end bending stresses in inelastic deformation of plates and shells. For limit analysis this much cruder model will suffice, but for alternating loads it is unreliable.

References

[1] E. van der Giessen, FE thermomechanics and material sampling points, Comp. Methods in Applied Mechanics and Engineering, Vol. 64, 447-465 (1987).

[2] O.C. Zienkiewicz, The finite element method, third edition, McGraw-Hill Book Co., London (1977)

[3] J.F. Besseling, Finite element properties, based upon elastic potential interpolation, Hybrid and Mixed Finite Element Methods, Atluri S.N., Gallagher R.H., Zienkiewicz O.C. Eds., John Wiley and Sons, Ltd 253-266, (1983).

[4] J.F. Besseling, Models of metal plasticity: theory and experiment. Plasticity Today: Modelling, methods and applications, Sawczuk, A., Bianchi G. Eds., Elsevier Appl. Sci Publ., London/New York, 97-113, (1985).

[5] L.J. Ernst, A finite element-approach to shell problems, Dissertation T.H. Delft, WTHD-114 Dept. Mech. Eng., (1979).

[6] J.F. Besseling, Another look at the application of the principle of virtual work with particular reference to finite plate and shell elements. Proceedings of the Europe-U.S. Workshop, Wunderlich W., Stein E., Bathe K.-J. Eds, Springer Verlag, Berlin/Heidelberg/ New York, 11-27 (1980).

The Method of Consecutively Embedded Finite-Element Models

A. G. UGODCHICOV, A. N. PAUTOV

Gorky State University
Gagarin ave. 23, Gorky 603022

Summary

A multistage iteration scheme with consecutively embedded FE-models, alternative to Fedorenko method, is formulated in terms of displacements. The iteration scheme effectiveness is determined mainly by the proximity of the energy equivalence constants of the operators used.

1. General Scheme of the Method

For the initial linear boundary problem, formulated in displacements, a set of FE-models, $M^p(p = 0,1,\ldots,N)$, is constructed. Each model has a corresponding nodal point set, S^p, and a corresponding linear algebraic equation set:

$$A^p u^p = f^p, \tag{1}$$

where A^p is the stiffness matrix, u^p is the vector of nodal displacements, f^p is the vector of nodal loads. The equation set (1) may be treated as an operator equation with a linear self-conjugate, positively defined operator, A^p, acting in a Hilbert space, H^p.

D e f i n i t i o n. A model M^{p-1} is called embedded into a model M^p if $S^{p-1} \subset S^p$, and the vector $u^p \in H^p$ may be given as $u^p = \left\{u_1^p, u_2^p\right\}^T$, so that u_1^p is defined on S^{p-1} and $u_1^p \in H^{p-1}$ $(p = 1,2,\ldots,N)$.

Letting now M^p be a set of consecutively embedded models, (1) may be rewritten as

G. Kuhn · H. Mang (Eds.)
Discretization Methods in Structural Mechanis
IUTAM/IACM Symposium Vienna, Austria, 1989

$$\begin{bmatrix} A_{11}^p & A_{12}^p \\ A_{21}^p & A_{22}^p \end{bmatrix} \begin{Bmatrix} u_1^p \\ u_2^p \end{Bmatrix} = \begin{Bmatrix} f_1^p \\ f_2^p \end{Bmatrix} \qquad (2)$$

where u_1^p, $f_2^p \in H^{p-1}$.

To solve the system (2) we shall use a double-layer iteration scheme

$$A^{p-1}\, w_i^p = r_i^p, \qquad i = 0,1,2,\ldots \qquad (3)$$

with the correction

$$w_i^p = \frac{u_{1,i+1}^p - u_{1,i}^p}{\tau^p} \qquad (4)$$

where τ^p is the iteration parameter,

$$r_i^p = f_1^p - A_{11}^p\, u_{1,i}^p - A_{12}^p\, u_{2,i}^p \qquad (5)$$

being a residual. To calculate the residual, we must solve the system

$$A_{22}^p\, u_{2,i}^p = f_2^p - A_{21}^p\, u_{1,i}^p \qquad (6)$$

to determine $u_{2,i}^p$.

At each iteration step one has to solve two linear algebraic equation sets: the set (6) with the subsequent calculation of the residual according to (5), and the set (3) with the subsequent calculation of a new approximation

$$u_{1,i+1}^p = u_{1,i}^p + \tau^p\, w_i^p$$

The set (6) is defined much better than (3), so to solve it one may use any of the standard iteration procedures with the following approximation:

$$u_{2,i}^p = u_{2,i-1}^p + J_{p-1}^p\, \tau^p\, w_{i-1}^p, \qquad i = 1,2,\ldots$$

$$u_{2,0}^p = J_{p-1}^p\, u_{1,0}^p\,,$$

where $u_{1,0}^p$ is an initial approximation for the vector u_1^p, J_{p-1}^p is the interpolation operator from the S^{p-1} mesh to the S^p

mesh. If the set (6) is not A high-order one, it is expedient to solve it by a direct method.

To solve the set (3), an analogous double-layer iteration scheme may be designed using the A^{p-1} operator, and so on. In the coarsest model, M^0, the set (1) is not generally a high-order one and may be solved by a direct method.

2. The Main Convergence Properties

Dropping out the vector u_2^p, the equation set (2) may be put down as

$$\bar{A}^p u_1^p = \bar{f}_1^p , \tag{7}$$

where

$$\bar{A}^p = A_{11}^p - A_{12}^p (A_{22}^p)^{-1} A_{21}^p;$$

$$\bar{f}_1^p = f_1^p - A_{12}^p (A_{22}^p)^{-1} f_2^p$$

To solve the set (7) approximately we shall use a double-layer iteration scheme

$$A^{p-1} w_i^p = \bar{r}_i^p \tag{8}$$

$$\bar{r}_1^p = \bar{f}_1^p - \bar{A}^p u_{1,i}^p \tag{9}$$

with the correction w_i^p according to (4).

It can be readily shown that if the set (6) is solved exactly, the residuals defined by (5) and (9) are identical, i.e., the iteration scheme (3) in this case is equivalent to the scheme (8). On the other hand, the scheme (8) is an implicit, double-layer iteration scheme with the self-conjugate, positively defined operators $\bar{A}^p$ and A^{p-1} in H^{p-1}, with the general convergence properties completely defined by the γ_1^p and γ_2^p energy equivalence constants of these operators [1] (where γ_1^p and γ_2^p are the minimum and maximum eigenvalues of the generalized eigenvalue problem, $\bar{A}^p v = \lambda A^{p-1} v$).

In case of a simple iteration, when the iteration parameter τ^p does not depend on the iteration number, its optimal value is found by the formula

$$\tau^p = \tau_0^p = 2/(\gamma_1^p + \gamma_2^p)$$

With this optimal value of τ^p and an arbitrally set $\varepsilon^p > 0$ for the minimum iteratiom number, n^p, for which the inequality

$$\| \bar{A}^p \; u_{1,n}^p - \bar{f}_1^p \| \leqslant \varepsilon^p \; \| \bar{A}^p u_{1,0}^p - \bar{f}_1^p \|$$

holds, we get the unimprovable estimate

$$n^p \geqslant n_0^p \; (\varepsilon^p) = \ln \; \varepsilon^p / \ln \rho_0^p \tag{10}$$

Here $\rho_0^p = (1 - \xi^p) \; / \; (1 + \xi^p)$, $\xi^p = \gamma_1^p / \gamma_2^p$, $\|\cdot\|$ is the Euclidean norm.

It is evident that the above estimates are correct when (8) is solved exactly at every iteration step; now, turning back to the scheme (3), only the sets (6) and (3) are solved exactly at every iteration step. Naturally, a question arises: to what extend does the estimation error for these sets affect the set (3) convergence?

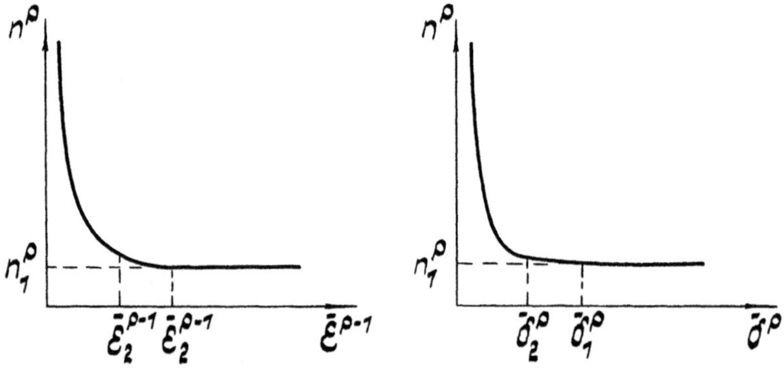

Fig.1. Characteristic dependence of the iteration number, n^p, on the parameter values of $\bar{\varepsilon}^{p-1} = 1/\varepsilon^{p-1}$ and $\bar{\delta}^p = 1/\delta^p$

Fig.1 shows the characteristic dependence of the iteration number, n^p, according to (3) for $\tau^p = \tau_0^p = 2/(\gamma_1^p + \gamma_2^p)$ and a preset relative error, ε^p, of the set (1) estimate, on the relative errors ε^{p-1} and δ^p of the estimates of (3) and (6), respectively. Here n_1^p is the minimum iteration number determined by the estimate of (10); $\bar{\varepsilon}^{p-1}$ and $\bar{\delta}^p_1$ are the threshold values, and $\bar{\varepsilon}^{p-1}_2$ and $\bar{\delta}^p_1$ are the optimal values in the sense of the minimal total arithmetic operation number of the parameters $\bar{\varepsilon}^{p-1} = 1/\varepsilon^{p-1}$, $\bar{\delta}^p = 1/\delta^p$.

The set of consecutively embedded FE-models $M^p (p = 0,1,2,...N)$ can be constructed by a variety of methods. The effectiveness of the N-stage iteration scheme (3) shall depend mainly on the value of the ratio $\xi^p = \gamma^p_1/\gamma^p_2$ of the energy equivalence constants γ^p_1 and γ^p_2 of the operators $\bar{A}^p$ and A^{p-1} ($p = 1,2,...N$), generated by these models. The closer ^{p} is to the unity, the higher is the convergence rate at each stage. The optimal value of the iteration parameter $^p_0$ is determined by the same quantities. When the matrices $\bar{A}^p$ and A^{p-1} are high-order ones, the solution of the generalized eigenvalue problem $\bar{A}^p v = \lambda^p A^{p-1} v$ to estimate $\gamma^p_1 = \lambda^p_{min}$ and $\gamma^p_2 = \lambda^p_{max}$ requires a substantial computational effort, though the numerical experiments show that reliable estimates of the ξ^p and τ^p_0 parameters may be obtained by solving simplified problems [2]; estimates for the ε^{p-1}_1, ε^{p-1}_2, δ^p_1, δ^p_2 parameters may be obtained by analysing the same problems.

In what follows some experimental numerical results for two construction methods of the M^p set are presented.

3. The Method of Embedded Meshes

Fig.2 depicts the meshes of a simplified problem with the generalized quadrilateral element consisting of the four mutually intersecting triangular elements with linear approximations of the displacements. The values of the constants γ^p_1 and γ^p_2 estimated by solving the problem $\bar{A}^p v = \lambda^p A^{p-1} v$ ($p = 1,2,3$) are listed in the Table along with the respective values of the parameters ξ^p and τ^p_0, as well as the iteration numbers n^p according to the estimate of (10) for $\varepsilon^p = 0.001$. Shown on the plot are the curves of the iteration number n^p estimated by (3) versus the values of τ^p for $\varepsilon^p = 0.001$, when solving the sets (3) and (6) by the direct method.

The results show a weak dependence of the ξ^p and τ^p_0 paramet-

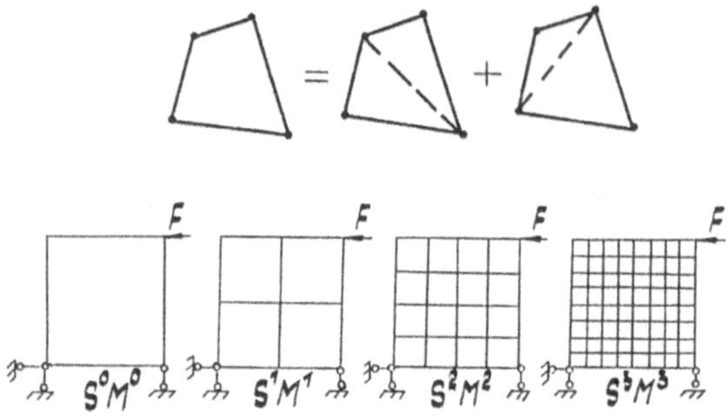

p	δ_1^p	δ_2^p	ε^p	τ_0^p	n^p
1	0.419	0.791	0.530	1.65	6
2	0.440	0.877	0.502	1.52	7
3	0.450	0.956	0.471	1.42	7

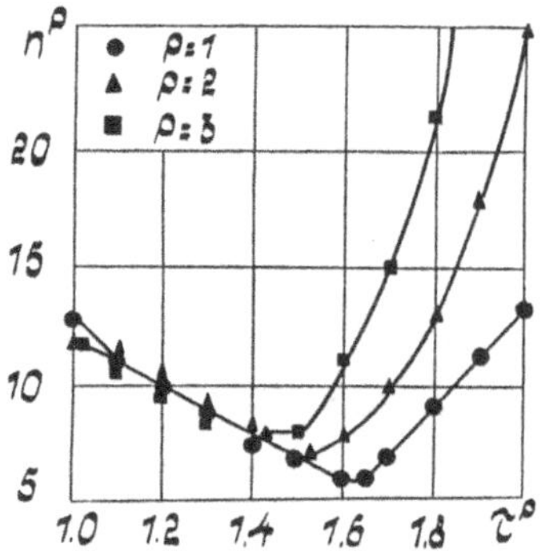

Fig.2. A simplified problem

ers on the mesh refinement degree, and insignificant variation of the iteration number, n^p, with a substantial deviation of the iteration parameter τ^p off its optimal value, τ_0^p.

Using the same simplified problem, it was found that the threshold and optimal values of the parameters $\mathcal{E}^{p-1}$ and δ^p practically do not depend on the mesh refinement degree and are of the order of $\mathcal{E}_2^{p-1} \approx \mathcal{E}_1^{p-1} \approx 0.25$, $\delta_2^p \approx \delta_1^p \approx 0.1$, the iteration number for solving the set (3) being $n^{p-1} = 2$, while for solving the set (6) by the conjugate gradient method, $m^p = 2 \div 3$.

A cantilever beam-strip bending problem (Fig.3), solved by two-stage iteration scheme, was used as a test. For $\mathcal{E}^2 = 0.001$, $\mathcal{E}^1 = 0.25$, $\delta^2 = \delta^1 = 0.1$ and $\tau_0^2 = \tau_0^1 = 1.4$, the iteration number was $n^2 = 8$, $\Sigma\ n^1 = 16$, $\Sigma\ m^2 = 18$, $\Sigma\ m^1 = 38$. The equation set (6) was solved by the conjugate gradient method. The results are in good agreement with the above estimates.

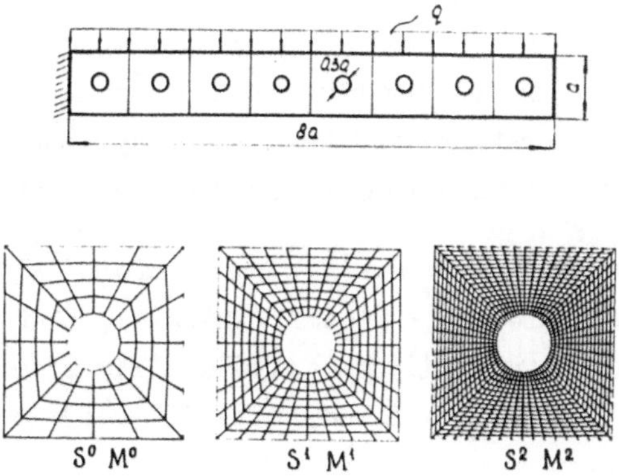

Fig.3. A test problem

4. Base Reduction

For complex structures the equation set (1) may have a high order even on the coarsest mesh. In such cases one can construct sequences of coarser FE-models by introducing the interpolation and compression operators. To this end, the structure may be decomposed into substructures, substructures into structural members, and the members into subregions.

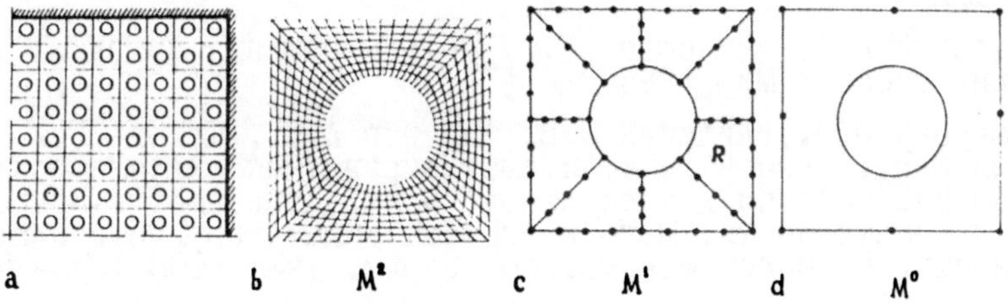

Fig.4. A test example

For the subregion mesh, a notion of the simple nodes (belonging only to a given subregion) and of the complex nodes (belonging to the interface between a given subregion and other ones) is introduced. Among the set of the complex nodes, a subset of the basical nodes is distinguished. By condensing the simple node displacements and by carrying out interpolation and compression of the complex node displacements the stiffness matrix, corresponding to the basical node displacement vector, is constructed. Analogous operations are carried out for the structural members and the substructures.

Fig.4 shows the discretization of a perforated square plate (Fig.4a) into structural members (Fig.4b), and the structural members into subregions (Fig.4c). The dots stand for the basical nodes in the structural members (Fig.4d) and in the subregions (Fig.4c).

The analyses of a single subregion R and of a single structural member of the plate, used as simplified problems, gave the following estimates: $\varepsilon^2 = 0.75$, $\tau_0^2 = 1.13$, $\varepsilon^1 = 0.56$, $\tau_0^1 = 1.45$, $\varepsilon_2^1 \approx \varepsilon_1^1 \approx 0.1$, $\delta_2^2 \approx \delta_1^2 \approx 0.3$, $\delta_2^1 \approx \delta_1^1 \approx 0.3$.
Used as a test was the problem of a perforated plate deformation (Fig.4a) heated up to a temperature T = Const, which was analyzed using the two-stage iteration scheme (3) and the models M^2, M^1 and M^0 (Fig.4). For $\varepsilon^2 = 0.001$, $\varepsilon^1 = 0.1$, $\delta^2 = \delta^1 = 0.3$, $\tau_0^2 = 1.13$, $\tau_0^1 = 1.45$, the iteration number was equal to $n^2 = 4$, $\Sigma n^1 = 7$, $\Sigma m^2 = 14$ and $\Sigma m^2 = 13$. The equation set (6) was solved by the conjugate gradient method.

References

I. Самарский А.А., Николаев Е.С. Методы решения сеточных уравнений. Москва: Мир 1978.

2. Крамарева И.В., Лиликин С.В., Паутов А.Н., Сухов М.Ф., Толкачев И.Н., Шуваев Д.Н. Автоматизированная система исследований многосеточных алгоритмов. Прикладные проблемы прочности и пластичности. Методы решения задач упругости и пластичности. Всесоюз. межвуз. сб. Горьк. ун-т (1985) 73-77.

Locally Blending Interpolation Technique in Finite Element Mesh Refinements and Improved Stress Analysis

J. ALTENBACH

Technische Universität "Otto von Guericke", Magdeburg

Summary

The paper concentrates on a blending function method and on the derivation of shape functions for a class of transition blended elements. All transition elements are compatible with the conventional standard displacement elements. With individual transition elements and so-called macro transition elements for a substructure-standard macro element concept intensive local mesh refinements can be undertaken without any element distorsion.

Introduction

Effective mesh refinements and coupling facilities for different individual and/or macro standard displacement elements and locally blended transition elements characterize significantly the discretization quality of finite element codes considering local and global structure response in the finite element structure analysis.

Various local refinement and coupling techniques, the h-, the p- or the h-p-versions of the finite element codes have been incorporated in commercial or research finite element codes, using /1/. The standard procedure used for coupling elements of different type or order in the COSAR general purpose finite element code /2/ is the penalty technique. However, special transition elements have been incorporated for plane stress, plane strain, thick plate and shell problems and for three-dimensional solid problems. Fig. 1 demonstrated mesh refinements with conventional and unconventional element coupling. The present paper concentrates on the unconventional element coupling with special transition elements. There are no constraints for the nodes in the coupling zone and so the accuracy

G. Kuhn · H. Mang (Eds.)
Discretization Methods in Structural Mechanis
IUTAM/IACM Symposium Vienna, Austria, 1989

of the finite element analysis for nodes in coupling regions and other nodes is of the same order.

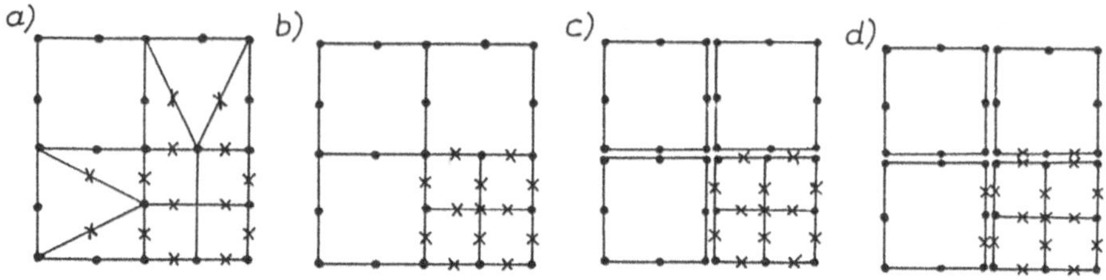

Fig. 1. Mesh refinement techniques a) conventional, b) unconventional, c) penalty function method (kinematic constraints d) two transition elements • original nodes x new nodes

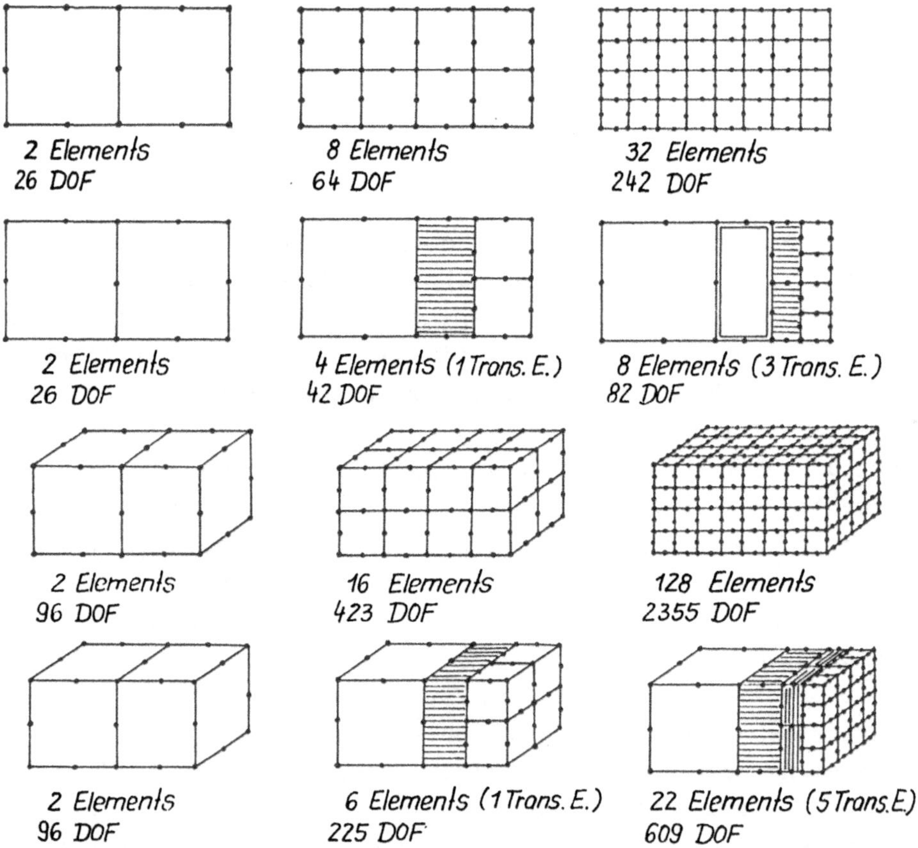

Fig. 2. Global and local mesh refinements

Fig. 2 demonstrates the great difference in the number of degrees of freedom using a general or a local mesh refinement technique, fig. 3 illustrates that only three transitions elements realize a satisfying quality of local mesh refinement in 3D-analysis.

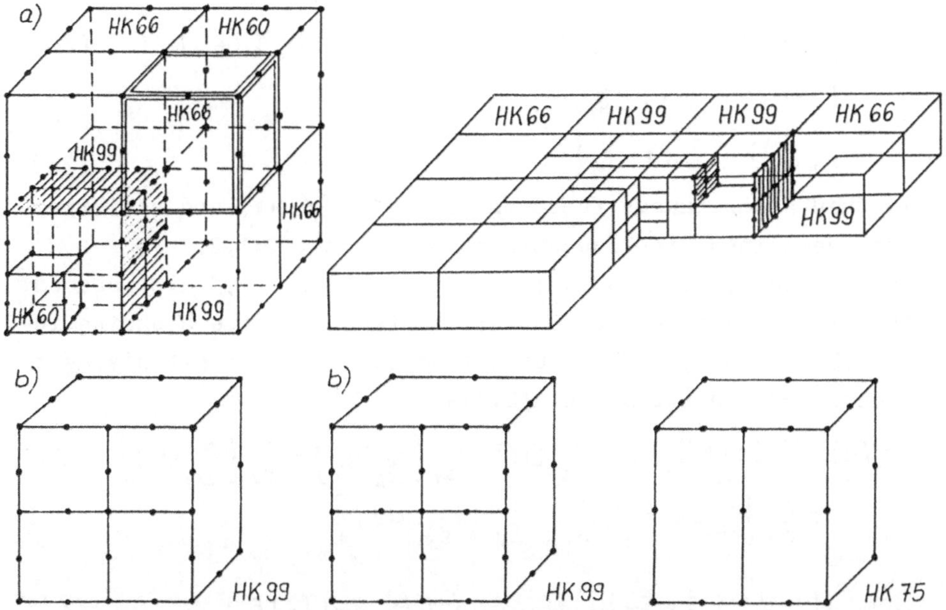

Fig. 3. Mesh refinements for 3D solid problems
a) Mesh refinements , b) 3D transition elements

Interpolation technique

The blending function method is a general technique of interpolation and has been used in construction of many finite elements /3.-8./. The following equations summarize the technique for two-dimensional, square, and threedimensional cube elements.

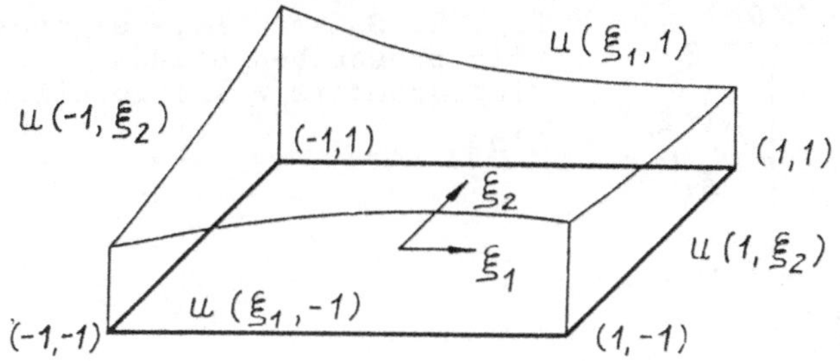

Fig. 4. Square elements, ξ_1, ξ_2 natural coordinates, $u(\xi_1,\pm1)$, $u(\pm1,\xi_2)$ prescribed boundary displacements

We consider first the square element of fig. 4.
On ∂G we have prescribed displacements $u(\xi_1, \pm 1)$, $u(\pm 1, \xi_2)$.
The blending interpolant $P[u]$

$$P[u] = P_{\xi_1}[u] + P_{\xi_2}[u] + P_{\xi_1}\left\{P_{\xi_2}[u]\right\} \tag{2.1}$$

$$P_{\xi_1}[u] = u(+1, \xi_2)\, I_1(\xi_1) + u(-1, \xi_2)\, I_2(\xi_1) \tag{2.2}$$

$$P_{\xi_2}[u] = u(\xi_1, +1)\, I_1(\xi_2) + u(\xi_1, -1)\, I_2(\xi_2) \tag{2.3}$$

$$I_1(\xi_i) = \frac{1}{2}(1 + \xi_i) \tag{2.4}$$

$$I_2(\xi_i) = \frac{1}{2}(1 - \xi_i) \tag{2.5}$$

is then a continuous function in $G: -1 \leqq \xi_i \leqq +1$; and satisfies the prescribed displacements on $\partial G: \xi_i = \pm 1$, $i = 1, 2$.
For the cube, $G: -1 \leqq \xi_i \leqq +1$; $\partial G: \xi_i = \pm 1$, $i = 1, 2, 3$ displacement functions are prescribed on all surfaces and the blending interpolant $P[u]$, a continuous function in G and satisfying the prescribed conditions on ∂G, is

$$P[u] = P_{\xi_1}[u] + P_{\xi_2}[u] + P_{\xi_3}[u] + P_{\xi_1}\left\{P_{\xi_2}[u]\right\} + P_{\xi_2}\left\{P_{\xi_3}[u]\right\} + P_{\xi_3}\left\{P_{\xi_1}[u]\right\} + P_{\xi_1}\left\{P_{\xi_2}\left(P_{\xi_3}[u]\right)\right\} \tag{2.6}$$

$$P_{\xi_1}[u] = u(+1, \xi_2, \xi_3)\, I_1(\xi_1) + u(-1, \xi_2, \xi_3)\, I_2(\xi_1) \tag{2.7}$$

$$P_{\xi_2}[u] = u(\xi_1, +1, \xi_3)\, I_1(\xi_2) + u(\xi_1, -1, \xi_3)\, I_2(\xi_2) \tag{2.8}$$

$$P_{\xi_3}[u] = u(\xi_1, \xi_2, +1)\, I_1(\xi_3) + u(\xi_1, \xi_2, -1)\, I_2(\xi_3) \tag{2.9}$$

In a simular way we can construct blending interpolants for functions $w(\xi_i)$, which are continuous in $G: -1 \leqq \xi_i \leqq +1$; and have prescribed displacements and rotations on $\partial G: \xi_i = \pm 1$.

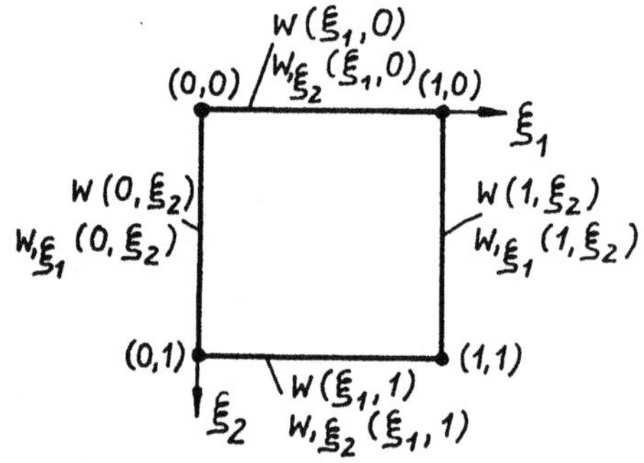

Fig. 5. Square plate element with prescribed boundary displacements w and rotations $w_{,n}$

We consider the square element, fig. 5, and get the following blending interpolant for $w(\xi_1,\xi_2)$ sufficient smooth in G and satisfies the conditions on ∂G:

$$P[w] = P_{\xi_1}[w] + P_{\xi_2}[w] - P_{\xi_1}\left\{P_{\xi_2}[w]\right\} \tag{2.10}$$

$$\begin{aligned} P_{\xi_1}[w] = {} & w(0,\xi_2)\,h_{00}(\xi_1) + w_{,\xi_1}(0,\xi_2)\,h_{01}(\xi_1) \\ & + w(1,\xi_2)\,h_{10}(\xi_1) + w_{,\xi_1}(1,\xi_2)\,h_{11}(\xi_1) \end{aligned} \tag{2.11}$$

$$\begin{aligned} P_{\xi_2}[w] = {} & w(\xi_1,0)\,h_{00}(\xi_2) + w_{,\xi_2}(\xi_1,0)\,h_{01}(\xi_2) \\ & + w(\xi_1,1)\,h_{10}(\xi_2) + w_{,\xi_2}(\xi_1,1)\,h_{11}(\xi_2) \end{aligned} \tag{2.12}$$

h_{00}, h_{01}, h_{10} and h_{11} are the Hermite polynoms in the interval $0 \leqq \xi_i \leqq 1$.

Transition blended elements

Following the concept of chapter 2, the shape functions for transition elements with piecewise second order boundary displacements have been derived /9, 10/. In connection with the isoparametric elements of second order of the COSAR element library these transition elements realize efficient mesh refinement conditions. Standard macro elements and transition macro elements in connection with substructuring are the basis for an automatic mesh generation /11, 12/.

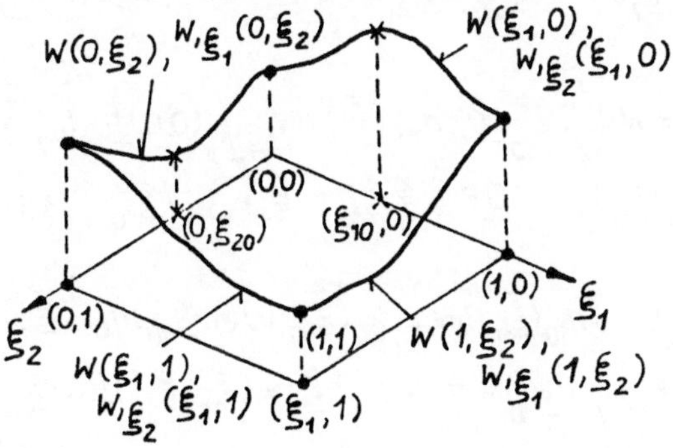

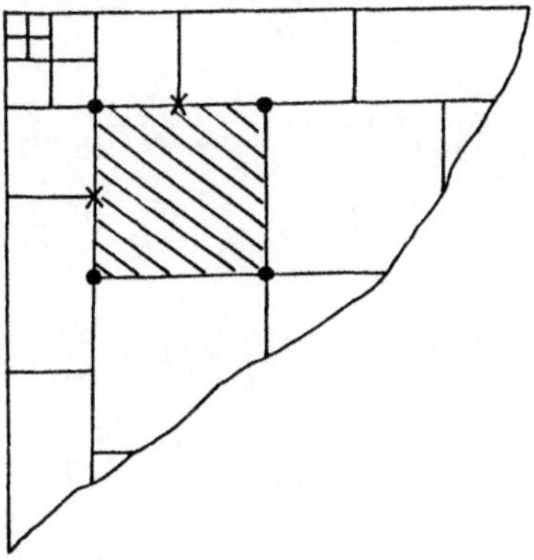

Fig. 6. Hermite plate element with two additional nodes
• nodes of the standard plate element, × additional nodes
(number and position of additional nodes may be as you like)

There are no difficulties in generalizing the derivation of C_o-transition elements to C_1-transition elements. We demonstrate this for a transition Hermite plate element with piecewise cubic boundary conditions. In fig. 6 a Hermite plate element with two additional nodes is given. For all nodes we have 4 degrees of freedom: w, $w_{,\xi_1}$, $w_{,\xi_2}$, $w_{,\xi_1\xi_2}$; the position of the additional nodes is $(\xi_{10}, 0)$; $(0, \xi_{20})$.

Now we interpolate piecewise the displacements and rotations by Hermite Interpolation Polynomials for the boundaries $(\xi_1,0)$; $(0,\xi_2)$ to garantee the demanded continuity in the additional nodes:

$$\begin{aligned} w(\xi_1,0) = {} & w(0,0)\,h_{00}(\tilde{\xi}_1) + w(\xi_{10},0)\,h_{10}(\tilde{\xi}_1) + w_{,\xi_1}(0,0)\,\xi_{10}\,\xi_{01}(\tilde{\xi}_1) \\ & + w_{,\xi_1}(\xi_{10},0)\,\xi_{10}\,h_{11}(\tilde{\xi}_1)\,; \qquad 0 \leqq \xi_1 \leqq \xi_{10}\,;\ \tilde{\xi}_1 = \xi_1/\xi_{10} \end{aligned} \tag{3.1}$$

$$\begin{aligned} w(\xi_1,0) = {} & w(\xi_{10},0)\,h_{00}(\tilde{\xi}_1) + w(1,0)\,h_{10}(\tilde{\xi}_1) + w_{,\xi_1}(\xi_{10},0)(1-\xi_{10})\,h_{01}(\tilde{\xi}_1) \\ & + w_{,\xi_1}(1,0)(1-\xi_{10})\,h_{11}(\tilde{\xi}_1)\,; \quad \xi_{10} \leqq \xi_1 \leqq 1\,;\ \tilde{\xi}_1 = (\xi_1-\xi_{10})/(1-\xi_{10}) \end{aligned} \tag{3.2}$$

$$\begin{aligned} w(0,\xi_2) = {} & w(0,0)\,h_{00}(\tilde{\xi}_2) + w(0,\xi_{20})\,h_{10}(\tilde{\xi}_2) + w_{,\xi_2}(0,0)\,\xi_{20}\,h_{01}(\tilde{\xi}_2) \\ & + w_{,\xi_2}(0,\xi_2)\,\xi_{20}\,h_{11}(\tilde{\xi}_2)\,; \qquad 0 \leqq \xi_2 \leqq \xi_{20}\,;\ \tilde{\xi}_2 = \xi_2/\xi_{20} \end{aligned} \tag{3.3}$$

$$\begin{aligned} w(0,\xi_2) = {} & w(0,\xi_{20})\,h_{00}(\tilde{\xi}_2) + w(0,1)\,h_{10}(\tilde{\xi}_2) + w_{,\xi_2}(0,\xi_{20})(1-\xi_{20})\,h_{01}(\tilde{\xi}_2) \\ & + w_{,\xi_2}(0,1)(1-\xi_{20})\,h_{11}(\tilde{\xi}_2)\,; \quad \xi_{20} \leqq \xi_2 \leqq 1\,;\ \tilde{\xi}_2 = (\xi_2-\xi_{20})/(1-\xi_{20}) \end{aligned} \tag{3.4}$$

$$\begin{aligned} w_{,\xi_2}(\xi_1,0) = {} & w_{,\xi_2}(0,0)\,h_{00}(\tilde{\xi}_1) + w_{,\xi_2}(\xi_{10},0)\,h_{10}(\tilde{\xi}_1) + w_{,\xi_1\xi_2}(0,0)\,\xi_{10}\,h_{01}(\tilde{\xi}_1) \\ & + w_{,\xi_1\xi_2}(\xi_{10},0)\,\xi_{10}\,h_{11}(\tilde{\xi}_1)\,; \qquad 0 \leqq \xi_1 \leqq \xi_{10}\,;\ \tilde{\xi}_1 = \xi_1/\xi_{10} \end{aligned} \tag{3.5}$$

$$\begin{aligned} w_{,\xi_2}(\xi_1,0) = {} & w_{,\xi_2}(\xi_{10},0)\,h_{00}(\tilde{\xi}_1) + w_{,\xi_2}(1,0)\,h_{10}(\tilde{\xi}_1) + w_{,\xi_1\xi_2}(\xi_{10},0)(1-\xi_{10})\,h_{01}(\tilde{\xi}_1) \\ & + w_{,\xi_1\xi_2}(1,0)(1-\xi_{10})\,h_{11}(\tilde{\xi}_1)\,; \quad \xi_{10} \leqq \xi_1 \leqq 1\,;\ \tilde{\xi}_1 = (\xi_1-\xi_{10})/(1-\xi_{10}) \end{aligned} \tag{3.6}$$

$$\begin{aligned} w_{,\xi_1}(0,\xi_2) = {} & w_{,\xi_1}(0,0)\,h_{00}(\tilde{\xi}_2) + w_{,\xi_1}(0,\xi_{20})\,h_{10}(\tilde{\xi}_2) + w_{,\xi_2\xi_1}(0,0)\,\xi_{20}\,h_{01}(\tilde{\xi}_2) \\ & + w_{,\xi_2\xi_1}(0,\xi_{20})\,\xi_{20}\,h_{11}(\tilde{\xi}_2)\,; \qquad 0 \leqq \xi_2 \leqq \xi_{20}\,;\ \tilde{\xi}_2 = \xi_2/\xi_{20} \end{aligned} \tag{3.7}$$

$$w_{,\xi_1}(0,\xi_2) = w_{,\xi_1}(0,\xi_{20})h_{00}(\tilde{\xi}_2) + w_{,\xi_1}(0,1)h_{10}(\tilde{\xi}_2) + w_{,\xi_2\xi_1}(0,\xi_{20})(1-\xi_{20})h_{01}(\tilde{\xi}_2)$$
$$+ w_{,\xi_2\xi_1}(0,1)(1-\xi_{20})h_{11}(\tilde{\xi}_2); \quad \xi_{20} \leqq \xi_2 \leqq 1 \; ; \; \tilde{\xi}_2 = (\xi_2 - \xi_{20})/(1-\xi_{20}) \tag{3.8}$$

For the boundaries $(\xi_1, 1)$ and $(1, \xi_2)$ we use the standard cubic interpolation without additional nodes. The blending interpolant

$$P[w(\xi_1,\xi_2)] = P_{\xi_1}[w(\xi_1,\xi_2)] + P_{\xi_2}[w(\xi_1,\xi_2)] - P_{\xi_1}\left\{P_{\xi_2}[w(\xi_1,\xi_2)]\right\} \tag{3.9}$$

is a continuous function in G and is satisfying all indicated boundary conditions for w and $w_{,n}$. Now we can derive the shape functions $G(\xi_1, \xi_2)$ for the transition plate element, g.e.

Node (0,0): $w(0,0) = 1 \qquad w(\xi_{1K}, \xi_{2K}) = 0 \qquad (\xi_{1K}, \xi_{2K}) \neq (0,0)$

$$w_{,\xi_i}(0,0) = 0 \qquad w_{,\xi_i}(\xi_{1K}, \xi_{2K}) = 0$$

$$w_{,\xi_1\xi_2}(0,0) = 0 \qquad w_{,\xi_1\xi_2}(\xi_{1K}, \xi_{2K}) = 0$$

The shape function $G_{(0,0)}(\xi_1, \xi_2)$ for the node (0,0) follows from (2.10)-(2.12) and (3.1)-(3.9).

$$P[w(\xi_1,\xi_2)] \equiv G_{(0,0)}(\xi_1,\xi_2) = h_{00}(\tilde{\xi}_1)h_{00}(\xi_2) + h_{00}(\xi_1)h_{00}(\tilde{\xi}_2) - h_{00}(\xi_1)h_{00}(\xi_2)$$
$$\tilde{\xi}_1 = \xi_1/\xi_{10} \; ; \quad \tilde{\xi}_2 = \xi_2/\xi_{20} \tag{3.10}$$

In an analoguos manner follow the shape functions $G_{\xi_1(0,0)}$; $G_{\xi_2(0,0)}$; $G_{\xi_1\xi_2(0,0)}$ for $w_{,\xi_1}(0,0)$; $w_{,\xi_2}(0,0)$; $w_{,\xi_1\xi_2}(0,0)$

$$G_{\xi_1(0,0)}(\xi_1,\xi_2) = \xi_{10}h_{01}(\tilde{\xi}_1)h_{00}(\xi_2) + h_{01}(\xi_1)h_{00}(\xi_2) - h_{01}(\xi_1)h_{00}(\xi_2) \tag{3.11}$$

$$G_{\xi_2(0,0)} = h_{00}(\tilde{\xi}_1)h_{01}(\xi_2) + \xi_{20}h_{00}(\xi_1)h_{01}(\tilde{\xi}_2) - h_{00}(\xi_1)h_{01}(\xi_2) \tag{3.12}$$

$$G_{\xi_1\xi_2(0,0)} = \xi_{10}h_{01}(\tilde{\xi}_1)h_{01}(\xi_2) + \xi_{20}h_{01}(\xi_1)h_{01}(\tilde{\xi}_2) - h_{01}(\xi_1)h_{01}(\xi_2) \tag{3.13}$$

The shape functions for the additional nodes between the edge nodes are more simple. There is no influence of the bicubic term $P_{\xi_1}\left\{P_{\xi_2}[w]\right\}$ and therefore the shape functions follow from

$$P_{\xi_1}[w] \quad (0 \leqq \xi_1 \leqq 1 \, ; \; \xi_2 = 0;1) \text{ or from } P_{\xi_2}[w] \quad (\xi_1 = 0;1 \, ; \; 0 \leqq \xi_2 \leqq 1)$$

For the transition Hermite plate element with 4 edge nodes and

2 additional nodes is then

$$
\begin{aligned}
W(\xi_1,\xi_2) = \sum_{n=0}^{1}\sum_{m=0}^{1} \Big[& W(n,m)\,G_{(n,m)}(\xi_1,\xi_2) + W_{,\xi_1}(n,m)\,G_{\xi_1(n,m)}(\xi_1,\xi_2) \\
& + W_{,\xi_2}(n,m)\,G_{\xi_2(n,m)}(\xi_1,\xi_2) + W_{,\xi_1\xi_2}(n,m)\,G_{\xi_1\xi_2(n,m)}(\xi_1,\xi_2)\Big] \\
& + W(\xi_{10},0)\,G_{(\xi_{10},0)}(\xi_1,\xi_2) + W_{,\xi_1}(\xi_{10},0)\,G_{\xi_1(\xi_{10},0)}(\xi_1,\xi_2) \\
& + W_{,\xi_2}(\xi_{10},0)\,G_{\xi_2(\xi_{10},0)}(\xi_1,\xi_2) + W_{,\xi_1\xi_2}(\xi_{10},0)\,G_{\xi_1\xi_2(\xi_{10},0)}(\xi_1,\xi_2) \\
& + W(0,\xi_{20})\,G_{(0,\xi_{20})}(\xi_1,\xi_2) + W_{,\xi_1}(0,\xi_{20})\,G_{\xi_1(0,\xi_{20})}(\xi_1,\xi_2) \\
& + W_{,\xi_2}(0,\xi_{20})\,G_{\xi_2(0,\xi_{20})}(\xi_1,\xi_2) + W_{,\xi_1\xi_2}(0,\xi_{20})\,G_{\xi_1\xi_2(0,\xi_{20})}(\xi_1,\xi_2)
\end{aligned}
\tag{3.14}
$$

Very interesting from engineering analysis are transition strip elements and possibilities of a combined mesh generation with transition strip and transition macros (fig. 7). Theoretical derivations and applications in shipbuilding you can find in /13/.

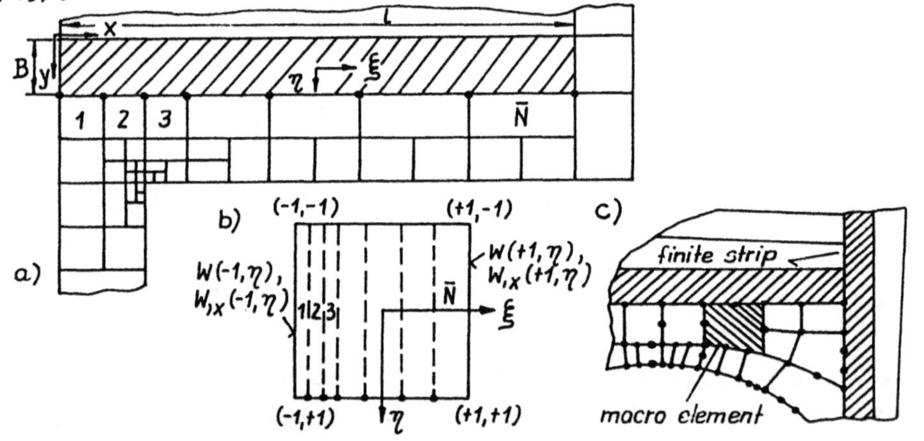

Fig. 7. Transition plate strips and plate macros
a) Transition strip and local mesh refinement, b) Transition strip in natural coordinates, c) Combined mesh generation with transition strips and macro elements

Conclusion

With transition elements based on locally blended interpolation techniques very intensive local mesh refinements are possible without any element distorsions. This was tested in the last years for plane stress and plane strain elements, for axis-symmetric elements, for thick plate and shells elements and for 3D membran and solid elements. In combination with the idea of so called ficticious surface membran elements significant improvements of the accuracy of surface stresses are possible /14/.

References

1. Babuska, J.; Suri, M.: The p- and h-p versions of the finite element method. An overview. Institut for Physical Science and Technology, Univ. of Maryland, BN-1101, 1989.
2. Altenbach, J.; Sacharov, A.: Finite Elements in Solid mechanics. Hanser Verlag, München 1982 (in German).
3. Gordon, W.J.: Blending functions method of bivariate and multivariate interpolation and approximation. Siam J. Num. Anal. 8, 158-177, 1971.
4. Gordon, W.J.; Hall, C.A.: Transfinite element methods blending function interpolation over arbitrary curved element domains. Numer. Math. 21, 109-129, 1973.
5. Cavendish, J.C.: Local mesh refinements using rectangular blended finite elements. J. Comp. Phys. 19, 211-228, 1975.
6. Cavendish, J.C.; Gordon, W.J.: Substructured macro elements based on a locally blended interpolation. Int. J. Numer. Meth. Eng. 11, 1405-1421, 1977.
7. Marshall, J.A.; Mitchell, A.R.: Blending interpolants in the finite element method. Int. J. Numer. Meth. Eng. 12, 1978.
8. Cavendish, J.C.; Hall, C.A.: A new class of transition blended finite elements for the analysis of solid structures. Int. J. Numer. Meth. Eng. 20, 241-252, 1984.
9. Scholz, E.; Altenbach, J.: Transition elements for local mesh refinements in 2D and 3D FE analysis (in German). Techn. Mech. 6, 72-78, 1985.
10. Altenbach, J.; Scholz, E.: Derivation of shape functions for standard and transition elements based on blending interpolations technique (in German). Techn. Mechanik 8, (in print).
11. Altenbach, J.; Horeschi, H.; Scholz, E.: Data generation in general purpose FE programme systems for standard and transition macro elements (in German). Techn. Mechanik 8, 1987.
12. Altenbach, J.; Scholz, E.: Local mesh refinements and macro Transition elements in stress analysis, in Math. of Finite Elements and Applications VI, Academic Press Ltd, London 1988.

13. Röhr, Udo: Development and Application of a combined Finite Elements Method for Structure Analysis in Shipbuilding. Dissertation B, Wilhelm-Pieck-Universität, Rostock 1988 (in German).

14. Altenbach, J.; Scholz, E.: Improved stressanalysis for surfaces- and transition elements in FE-structure modelling Technische Mechanik 8 (1987), 4, 41-46 (in German).

Nonconforming Finite Elements Based on the Free Formulation

M.K. Nygård and P.G. Bergan

A.S Veritas Research, 1322 Høvik, Norway

Abstract

The paper describes how the "individual element test" for finite elements may be expressed for linear as well as nonlinear mechanical applications. Further, the derivation of the "free formulation" of finite elements is obtained for linear and nonlinear problems. After a brief definition of a new class of nonlinear shell elements the paper goes on to demonstrate an application to a nonlinear cylindrical shell.

1. Introduction

The finite element technology is by now more than thirty years old; and it is in many ways a mature technology. However, the finite element research is confining with undiminished force even though most advances made today are minor, incremental improvements rather than great steps forward.

The theoretical basis for the finite element method is well established. Elements may be derived using virtual work, potential energy, complementary energy, the Reissner principle or "modified" or extended variational forms. An alternative to all this is the free formulation of the finite element method proposed by Bergan and Nygård [1]. This formulation is based on the individual element test proposed by Bergan and Hanssen [2]. A significant advantage by the free formulation is that allows for using nonconforming displacements functions as well as new types of element freedoms. A new class of finite elements has been derived on this basis. Elements that have been developed so far include membrane elements with drilling freedoms, nonconforming bending elements for thin and thick plates, shell elements with six freedoms per node, and nonconforming solid elements. Many of these elements have been recognized as "best performers" as compared to other elements of similar type.

The free formulation may also be used for nonlinear problems. This paper outlines the theoretical foundation for this extension and describes some triangular and quadrilateral nonlinear shell elements that are derived on this basis.

G. Kuhn · H. Mang (Eds.)
Discretization Methods in Structural Mechanis
IUTAM/IACM Symposium Vienna, Austria, 1989

2. The Individual Element Test

A displacement field $\mathbf{u}$ within an element may be expressed by a complete set of rigid-body and constant straining field expansion (rc-modes) and a set of higher order modes (h-modes):

$$\mathbf{u} = \mathbf{N}_{rc}\mathbf{q}_{rc} + \mathbf{N}_h\mathbf{q}_h \tag{1}$$

$\mathbf{N}_{rc}$ represents a complete polynomial to the degree which corresponds to the total number of rc-modes for the given type of problem, while $\mathbf{q}_{rc}$ are the associated coefficients. $\mathbf{N}_h$ expresses a set of higher order functions (h-modes) with $\mathbf{q}_h$ as the associated coefficients. The total number of rc- and h-modes equals the number of degrees of freedom for the element.

The kinematic relationship between these generalized modes and the nodal displacement vector $\mathbf{v}$ is established by inserting the appropriate nodal coordinates in (1), and, in case of freedoms involving derivatives, $\mathbf{N}_{rc}$ and $\mathbf{N}_h$ must be differentiated correspondingly.

$$\mathbf{v} = \mathbf{v}_{rc} + \mathbf{v}_h = \mathbf{G}_{rc}\mathbf{q}_{rc} + \mathbf{G}_h\mathbf{q}_h = \mathbf{G}\mathbf{q} \tag{2}$$

It is necessary that all nodal values of the generalized patterns are linearly independent and distinguishable through the nodal freedoms, hence, (2) must be invertible.

$$\mathbf{q} = \begin{bmatrix} \mathbf{q}_{rc} \\ \mathbf{q}_h \end{bmatrix} = \mathbf{G}^{-1}\mathbf{v} = \begin{bmatrix} \mathbf{H}_{rc} \\ \mathbf{H}_h \end{bmatrix}\mathbf{v} \tag{3}$$

The basic concept of the individual element test is that the element, when interacting with its neighbours, should be capable of reproducing identically an arbitrary rigid-body/constant strain field. This requirement is a consequence of the limit condition of convergence when elements become infinitesimally small [2]. Such a state for a single element may be expressed by the displacements of the first term of (1).

Considering two adjacent elements, equilibrium at the nodes requires the interelement forces transferred at nodes shall cancel out in a pairwise manner. The nodal forces, $\mathbf{T}_{rc}$, that will act on the element, may be obtained by a consistent transfer (lumping) of tractions acting on the element boundary to the nodes. In mathematical terms this may be written

$$\mathbf{T}_{rc} = \mathbf{L}\sigma_c = \mathbf{LCB}_{rc}\mathbf{q}_{rc} = \mathbf{P}_{rc}\mathbf{q}_{rc} \tag{4}$$

These forces depend on the strain-generalized displacement matrix $\mathbf{B}_{rc}$, the constitutive matrix $\mathbf{C}$, and the traction lumping procedure and the geometry of the element which affects the L-matrix.

Having a stiffness matrix **k** for the element, the nodal forces $\mathbf{S}_{rc}$, produced by this stiffness during a rc-state are

$$\mathbf{S}_{rc} = \mathbf{k}\mathbf{v}_{rc} = \mathbf{k}\mathbf{G}_{rc}\mathbf{q}_{rc} \tag{5}$$

The test [2] now requires that the forces $\mathbf{S}_{rc}$, produced by the stiffness, should be equal to the consistently lumped forces $\mathbf{T}_{rc}$ of (4) for any constant stress field, i.e. any $\mathbf{q}_{rc}$. This implies

$$\mathbf{k}\mathbf{G}_{rc} = \mathbf{P}_{rc} \tag{6}$$

This equation represents the fundamental set of constraints applicable to **k**. This test has been denoted the "individual element test" (IE-test) to emphasise that it applies to a single element rather than a patch of elements. It should be noted, however, that if an element satisfies the IE-test, it also satisfies Irons's patch test [4].

3. Linear Free Formulation

Establishing the lumping matrix

The virtual work equation on a total form is given as

$$\int_V \delta\epsilon^T \sigma \mathbf{dV} = \int_A \delta\mathbf{u}^T \mathbf{tdA} + \int_V \delta\mathbf{u}^T \mathbf{fdV} \tag{7}$$

σ and ϵ are the Cauchy stresses and engineering strains, respectively, **t** the surface tractions, **f** the body forces, and $\delta\mathbf{u}$ are the virtual displacements. No separation is here made between a deformed and an undeformed configuration due to the assumption of small deformations and displacements. During a state of constant stress/strain, the local equilibrium equation requires the body force to be zero. Hence, the part of the virtual work expression expressing the external virtual work made by the surface tractions are

$$\delta W_{ext}^{rc} = \int_A \delta\mathbf{u}^T \mathbf{t}_c dA \tag{8}$$

The displacements on the surface A may be given as

$$\mathbf{u} = \mathbf{N}_u \mathbf{v}_u \tag{9}$$

with $bldN_u$ and $\mathbf{v}_u$ as the interpolation functions and the nodal freedoms, respectively, associated with this surface. The traction is related to the Cauchy stress tensor via (using tensor notation form)

$$t_i = \hat{n}_j \sigma_{ji} \tag{10}$$

$\hat{n}_j$ is the outward normal vector of the surface area dA. For simplicity and for reasons of laer use, this equation is rewritten to a matrix form

$$\mathbf{t}_c = \bar{\mathbf{n}}\sigma_c \tag{11}$$

σ_c is here written as a column matrix, while $\bar{\mathbf{n}}$ is a two-dimensional matrix. E.g. for a two-dimensional state of stress with n_x and n_y as the x- and y-components of the outward unit normal vector, respectively, Eq. (11) becomes

$$\begin{bmatrix} t_x \\ t_y \end{bmatrix} = \begin{bmatrix} n_x & 0 & n_y \\ 0 & n_y & n_x \end{bmatrix} \begin{bmatrix} \sigma_{xx} \\ \sigma_{yy} \\ \sigma_{xy} \end{bmatrix} \tag{12}$$

Introducing Eqs. (9,11) into the virtual work expression yields

$$\delta W_{ext}^{rc} = \delta \mathbf{v}_u^T \int_A \mathbf{N}_u^T \bar{\mathbf{n}} dA \, \sigma_c = \delta \mathbf{v}_u^T \mathbf{T}_{rc} \tag{13}$$

σ_c is taken outside the integral since it is constant in a rc-state. $\bar{\mathbf{n}}$ is constant along straight or planar element sides. $\mathbf{T}_{rc}$ is the same set of nodal forces as in Eq. (4). By comparing these with Eq. (13) it is seen that the lumping matrix is

$$\mathbf{L} = \int_A \mathbf{N}_u^T \bar{\mathbf{n}} dA \tag{14}$$

Equation (14) gives the expression for $\mathbf{L}$ as used in the Update Lagrangian description of motion when A_n, the deformed area over which $\mathbf{t}_c$ acts, replaces A.

Stiffness formulation

The key requirement for a stiffness matrix of an element is that it must satisfy the constraints of equation (6) as well as giving a matrix with correct rank. A formulation that satisfies this is

$$\mathbf{k} = \int_V \mathbf{B}^T \mathbf{C} \mathbf{B} dV \tag{15}$$

A special form of the strain-displacement matrix $\mathbf{B}$ [5],[6] is applied in order to satisfy the IE-test

$$\mathbf{B} = \frac{1}{V} \mathbf{L}^T + \tilde{\mathbf{B}}_h \mathbf{H}_h \tag{16}$$

The contribution from the higher order modes in $\mathbf{B}$ is

$$\tilde{\mathbf{B}}_h = \mathbf{B}_h - \frac{1}{V} \int_V \mathbf{B}_h dV \tag{17}$$

This implies that $\tilde{\mathbf{B}}_h$ is energy orthogonal to the rc-modes, i.e. [7]

$$\int_V \tilde{\mathbf{B}}_h dV \equiv 0 \tag{18}$$

V is the volume of the element. The strains may now be found in a straight-forward manner as

$$\epsilon = \mathbf{Bv} \tag{19}$$

The proof that the stiffness expression of equation (15) satisfies the IE-test is simply to perform the test as specified in equation (1), i.e.

$$\mathbf{kG}_{rc} = \frac{1}{V}\mathbf{LCL}^T\mathbf{G}_{rc} + \mathbf{H}_h^T \int_V \mathbf{B}_h^T \mathbf{CB}_h dV \mathbf{H}_h \mathbf{G}_{rc} = \mathbf{LC}(\frac{1}{V}\mathbf{L}^T\mathbf{G}_{rc}) + 0 \tag{20}$$

The last term cancels because of $\mathbf{H}_h \mathbf{G}_{rc} = 0$. By comparing this equation with the expression for $\mathbf{P}_{rc}$ in (5), it is seen that satisfaction of the IE-test requires

$$\mathbf{B}_{rc} = \frac{1}{V}\mathbf{L}^T\mathbf{G}_{rc} \tag{21}$$

Let $2m$ be the order of the differential operator of the governing differential equation. Equation (21) then holds when the rc-modes satisfy C^{m-1}-compatibility, and $\mathbf{L}$ is constructed in consistency with the principle of virtual work [5]. Note also that the patch test itself requires the rc-modes to fulfil the same compatibility condition [8], thus equation (21) represents no loss of generality. Further, the only restriction imposed on the higher order modes is that they must be chosen such that they give linearly independent values at the freedoms, i.e. the **G**-matrix is rendered regular.

4. Nonlinear Free Formulation

Establishing the lumping matrix

The lumping matrix used for an Updated Lagrangian (UL) formulation is identical to that of Eq. (14) when integrations are taken over the deformed area A_n. The Total Lagrangian (TL) or Co-rotational Lagrangian (CL) concepts for description of motion requires a somewhat extended formulation. Applying the TL-description, a the pseudo-traction $\tilde{\mathbf{t}}$ (see e.g. p.222 in [9]) is introduced in order to obtain the expression for the lumping matrix of Eq. (4). A matrix form of this relation is

$$\mathbf{t}dA_n = \mathbf{t}_o dA_o = \mathbf{F}\tilde{\mathbf{t}}dA_o \tag{22}$$

The index o refers to the initial configuration C_o. **F** is the matrix expressing the deformation gradient from the initial configuration C_o to the current deformed configuration C_n defined as

$$d\mathbf{x} = \mathbf{F}d\mathbf{X} \tag{23}$$

where $d\mathbf{x}$ and $d\mathbf{X}$ are the vectors of the deformed and initial line segments. The second Piola-Kirchhoff stress tensor is related to $\tilde{\mathbf{t}}$ according to

$$\tilde{t}_j = \hat{N}_i S_{ij} \tag{24}$$

$\hat{\mathbf{N}}$ is the outward normal vector of the surface dA_o. As for the Cauchy stresses in Eqs. (11) and (12), the traction may be written on a matrix form

$$\tilde{\mathbf{t}} = \overline{\mathbf{N}}\mathbf{S} \tag{25}$$

Substitution of Eq. (25) into Eq. (22), the the external virtual work for an rc-state may now be written similarly as for the linear formulation in Eq. (13).

$$\delta W_{ext}^{rc} = \int_{A_o} \delta\mathbf{u}^T \mathbf{t}_{oc} dA_o = \delta\mathbf{v}_u^T \int_{A_o} \mathbf{N}_u^T \mathbf{F}_{rc} \overline{\mathbf{N}} dA_o \mathbf{S}_c = \delta\mathbf{v}_u^T \mathbf{T}_{rc} \tag{26}$$

The lumping matrix for TL-description is thus obtained as

$$\mathbf{L}_o = \int_{A_o} \mathbf{N}_u^T \mathbf{F}_{rc} \overline{\mathbf{N}} dA_o \tag{27}$$

Note that $\mathbf{F}_{rc}$ should be calculated using the rc-modes only. As an example, this means that $\mathbf{F}_{rc}$ for a membrane element, will be constant throughout the element, while for e.g. a plate element, $\mathbf{F}_{rc}$ will be linear since the rc-modes contain quadratic terms.

In the CL-formulation the lumping matrix, now denoted $\mathbf{L}_{on}$, takes the same form as that in Eq. (27). The following modifications must be done:

- $\mathbf{F}$ is the deformation gradient of C_n referred to C_{on}, with the components also given in the co-rotated coordinate system.
- The integration is performed over A_{on}, which again is identical to A_o.
- The components of $\overline{\mathbf{N}}$ are those of C_{on}.

Hence, the $\mathbf{L}_{on}$ matrix may be written as

$$\mathbf{L}_{on} = \int_{A_{on}} \mathbf{N}_u^T \mathbf{F}_{rc} \overline{\mathbf{N}} dA_{on} \tag{28}$$

Stiffness and internal force formulation, TL and CL formulation

According to the incremental virtual work equations the tangential element stiffness may be separated into two contributions, the material and geometric stiffness matrix, respectively. The material stiffness takes the form

$$\mathbf{k}_{mat} = \int_V \mathbf{B}^T \mathbf{C}_T \mathbf{B} dV \tag{29}$$

$\mathbf{C}_T$ is the tangential constitutive tensor expressing obtained from the material behaviour, while $\mathbf{B}$ is the strain-displacement matrix of Eq. (3.3.9). The integration is performed over the initial volume, V_o.

The geometric stiffness matrix, $\mathbf{k}_{geom}$, depends on the current state of stress. This matrix originates from the nonlinear strain terms and takes a traditional form. The use of generalized displacements also makes simplification of $\mathbf{k}_{geom}$ convenient by only applying the most basic modes for this stiffness. The tangential or incremental element stiffness matrix is now found to be

$$\mathbf{k}_T = \mathbf{k}_{mat} + \mathbf{k}_{geom} \tag{30}$$

The internal nodal reaction forces for an element, $\mathbf{S}_{int}$, are found from the left hand side of the virtual work expression.

$$\mathbf{S}_{int} = \int_{V_o} \mathbf{B}^T \mathbf{S} dV_o \tag{31}$$

The reaction forces are thus obtained simply by integration of the internal state of stress.

The $\mathbf{B}$-matrix is established in accordance with the description of motion that is being used, which determines the formulation to be applied for the $\mathbf{L}$- and $\mathbf{B}_{qh}$-matrices.

5. Application to Shell Elements

Shell elements

The Kirchhoff thin shell theory which is the most common theory for thin shell or plate elements is adopted here for three node and four node shell elements. This theory neglects transverse shear strains.

The CL-formulation is chosen due to its possibility for applying simplified strain terms. This removes the average rotations from the nonlinear strain terms which are left with representation of the local, large deformation of the element only. Each shell element has a reference base vector system attached to the co-rotated configuration of the element. All the kinematic assumptions are made entirely with reference to this flat ghost reference element [5],[6].

The elements have six degrees of freedom at each node, three translational and three rotational freedoms, all oriented in accordance with the base vector system of the reference element. These freedoms may be separated into two groups, $\mathbf{v}_m$ is connected to the in-plane or membrane behaviour of the element, and $\mathbf{v}_b$ with the bending action of the element.

For a single node these freedoms are given as

$$\mathbf{v}_{mi} = \begin{bmatrix} u_o \\ v_o \\ \frac{1}{2}(v_{,x} - u_{,y}) \end{bmatrix}_i = \begin{bmatrix} u_o \\ v_o \\ \Theta_z \end{bmatrix}_i , \quad \mathbf{v}_{bi} = \begin{bmatrix} w \\ \Theta_x \\ \Theta_y \end{bmatrix}_i = \begin{bmatrix} w_o \\ w_{,y} \\ -w_{,x} \end{bmatrix}_i \tag{32}$$

The rotational freedom Θ_z about the z-axis (the "drilling freedom") is defined as the mechanical rotation derived from the in-plane displacement gradients. This quantity is invariant with respect to the reference coordinates. However, there is no unique relationship between the freedom Θ_z and the rotations of the adjacent element sides, see discussion by Bergan and Felippa [10]. This "drilling freedom" in the context of free formulation elements was first utilized for linear triangular membrane elements by Bergan and Felippa [10], and followed by extensions to quadrilateral membrane elements [5,11] and nonlinear shells [5],[6] by Bergan and Nygård. This freedom gives highly improved membrane action of the elements [5],[6],[10],[11], and overcomes the normally stiff behaviour of lower order shell elements.

Two different elements are included in this study; a three node and a four flat shell element. Shell elements have a total of 12 rigid body/constant strain modes, 6 from the inplane action and 6 from the out-of-plane action. This leaves an additional 3 membrane modes and 3 plate bending modes to be selected for the triangular element to make a complete set of 18 modes. This gives significantly more flexibility for the membrane part as compared to a CST element which has no higher order modes. The corresponding number of higher order modes for the 4 node element is thus 6 for both membrane and plate bending action. The complete set of modes are given in [5],[12].

The elements possess no spurious zero energy modes when fully integrated. However, only 2x2 Gaussian integration for the quadrilateral element gives the same zero energy mode for the membrane action as obtained when using the same integration for a 8 node serendipity membrane element.

The current spatial position $\mathbf{x}$ of a nodal point is uniquely defined by the initial position vector $\mathbf{X}$ and the nodal displacement vector $\mathbf{v}$

$$\mathbf{x} = \mathbf{X} + \mathbf{v} \tag{33}$$

It is assumed that a node carries along with it an attached Cartesian base vector system $\mathbf{i}_i^{(n)}$, which initially is parallel to the global base vector system $\mathbf{I}_i$, and rotates along with the node. The orientation in space of these nodal base vectors in configuration C_n are defined with reference to $\mathbf{I}_i$ through

$$\mathbf{i}_i^{(n)} = T_{ij}^{(n)} \mathbf{I}_j \tag{34}$$

The transformation matrix $T_{ij}^{(n)}$ represents an orthonormal second-order tensor. The spatial motion of a node is thus uniquely defined by the nodal displacement vector **v** and the rotation tensor $T_{ij}^{(n)}$. The sequential position of this nodal base vector system, configuration C_{n+1}, is determined from configuration C_n and the incremental rotations, yielding

$$\mathbf{i}_i^{(n+1)} = T_{ik}^{(n)} \Delta T_{kj} \mathbf{I}_j \tag{35}$$

The ΔT_{ij} is given by the following expression

$$\Delta T_{ij} = \delta_{ij}\cos(\Delta\Theta) + \frac{\Delta\Theta_i}{\Delta\Theta}\frac{\Delta\Theta_j}{\Delta\Theta}(1 - \cos(\Delta\Theta)) + e_{ijk}\frac{\sin(\Delta\Theta)}{\Delta\Theta}\Delta\Theta_k \tag{36}$$

This simple expression for updating the transformation matrix was first given in [5],[6]

6. Numerical Examples

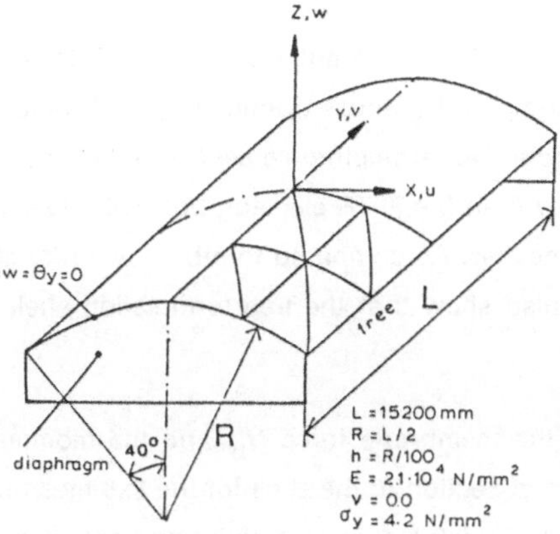

Figure 1. Geometry and material parameters for cylindrical shell roof.

The cylindrical "Scordelis-Lo's shell roof" supported on end diaphragms shown in Fig. 1. is a classical and very demanding test problem for shell elements. Both bending and membrane deformations contribute significantly to the total response, and in the nonlinear regime the geometrical as well as material nonlinearities are important.

The material is assumed elastic-perfectly-plastic with yield limit σ_y=4.2N/mm^2. Von Mises yield criterion is assumed, and ten integration points are used for integration through the thickness of the shell. The roof is subjected to a uniform gravity force with intensity increasing gradually up to 4.0 kN/m^2. One quarter of the cylindrical roof is modeled, and a nonlinear convergence study using nxn meshes is performed. The results in Fig. 2. shows the

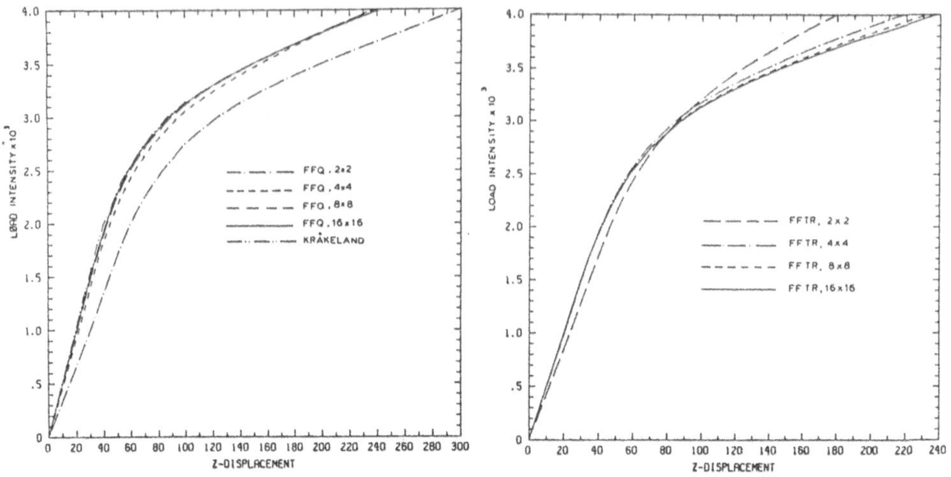

Figure 2. Vertical deflection at the midpoint of the free edge against the intensity of the distributed vertical load.

vertical deflection at the midpoint of the free edge against the intensity of the distributed vertical load. The results are compared to those obtained by Kråkeland [13], who employed a curved, Ahmad-type, eight node isoparametric element in a 4x4 mesh. The FFQ element shows a slightly better accuracy than the FFTR element, but both elements show very good results even for rather coarse meshes. As compared to other low order shell elements, more extensive investigations [5],[6] also show that the free formulation shell elements display a superior performance.

Fig. 3. shows the variation of the membrane force N_{yy} and the moment M_{xx}, respectively, along the symmetry line at the mid-section of the shell for the 8x8 mesh for the FFQ element. The quantities are calculated at the 2x2 Gaussian points, and are then extrapolated to the nodes. This may in some cases give nodal stresses which exceed the yield limit.

The force and the moment are plotted for five different load levels, as indicated on the figures. At load level $q=1.0\ kN/m^2$, the shell is still in the linear range, while the plastification gradually takes place for the succeeding load levels. For $q=4.0\ kN/m^2$, the shell yields in almost pure tension close to the free edge, leaving the moment capacity close to zero as shown in Fig. 3 for the same load level. This demonstrates clearly also that even if a mesh is fine enough for analyses in the linear range, it may still be too coarse when extensive yielding occurs.

A large number of analyses have been made with these shell elements. These tests range from simple academic verification problems to large complex buckling-postbuckling analyses of parts of ships and offshore structures. Some of these may be found in e.g. ref. [5],[6],[14].

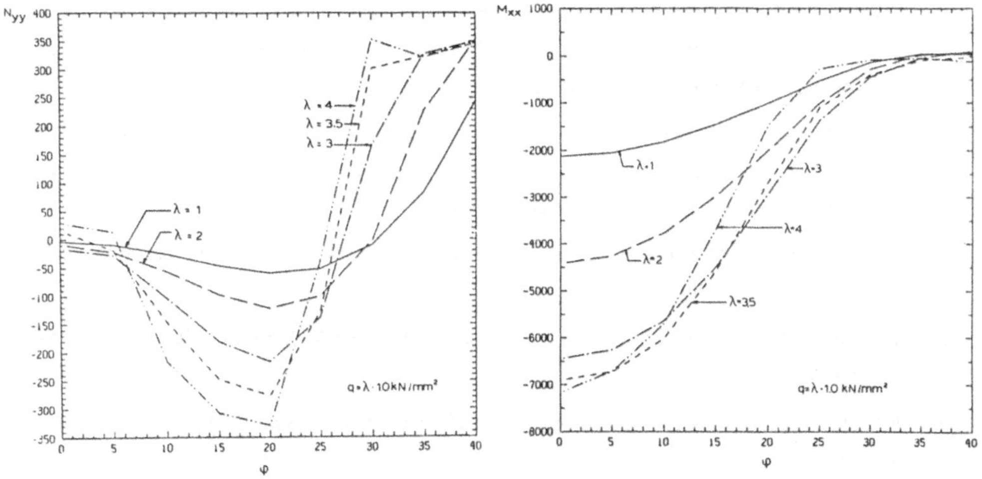

Figure 3. Plot of force N_{yy} and moment M_{xx} along symmetry line of the mid-section of cylindrical shell roof using 8x8 mesh for rectangular elements (FFQ).

7. Conclusions

The free formulation of finite elements has opened the door to deriving a series of new, improved elements in linear and nonlinear structural mechanics. The paper shows that a particularly useful application of this is for nonlinear analysis of thin shells. Planar triangular and quadrilateral elements with three translational and three rotational freedoms per node are used. The high efficiency of these elements is demonstrated herein through a nonlinear analysis of a thin, cylindrical shell.

8. References

1 Bergan, P.G. and Nygård, M.K.: Finite elements with increased freedom in choosing shape functions. *Int. J. for Numerical Methods in Engineering,* Vol. 20, 643-664, (1984).

2 Bergan, P.G. and Hanssen, L.: A new approach for deriving "good" finite elements. In J.R. Whiteman (ed.), *The Mathematics of Finite Elements and Applications.* Vol. II. 483-498, London: Academic Press (1976).

3 Hanssen, L.: Finite Elements Based on a new Procedure for Computing Stiffness Coefficients, *Report No. 76-1,* Division of Structural Mechanics, The Norwegian Institute of Technology, Trondheim, 1976.

4 Irons, B.M. and Razzaque, A.: Experience with the Patch Test for convergence of finite elements. In K. Aziz (Ed.): *Mathematical Foundations of the Finite Element Method with Applications to Partial Differential Equations. 557-587,* New York, London: Academic Press, 1972.

5 Nygård, M.K.: *The Free Formulation for Nonlinear Elements with Applications to Shells, Dr. ing. thesis, Report no. 86-2,* Division of Structural Mechanics, The Norwegian Institute of Technology, Trondheim, Norway, 1986.

6 Bergan, P.G. and Nygård, M.K.: Nonlinear shell analysis using free formulation finite elements. In P.G. Bergan, K.J. Bathe and W. Wunderlich (Eds.), *Finite Element Methods for Nonlinear Problems.* Springer Verlag 1985.

7 Bergan, P.G.: Finite elements based on energy orthogonal functions. *Int. J. for Numerical Methods in Engineering,* Vol. 15, 1541-1555, (1980).

8 Strang, G. and Fix, G.J.: *An Analysis of the Finite Element Method,* Prentice-Hall, Englewood Cliffs, N.J. (1973).

9 Malvern, L.E.: *Introduction to the Mechanics of a Continuous Medium.* Prentice-Hall, Inc. 1969.

10 Bergan, P.G. and Felippa, C.A.: A triangular membrane element with rotational degrees of freedom, *Computer Methods in Applied Mechanics and Engineering,* Vol. 50, 25-69, (1985).

11 Bergan, P.G. and Nygård, M.K.: A quadrilateral membrane element with rotational freedoms. In G. Yagawa and S. N. Atluri: *Computational Mechanics'86, Proc. Int. Conf. on Computational Mechanics,* Tokyo, Springer Verlag, 1986.

12 Bergan, P.G. and Nygård, M.K.: The individual element test: a basis for checking and deriving elements. Presented at "*2nd Int. Conference on Quality Assurance and Standards in Finite Element Methods*", 22-25 May, 1989, Stratford-upon-Avon, England

13 Kråkeland, B.: Large Displacement Analysis of Shells Considering Elasto-Plastic and Elasto-Viscoplastic Materials, *Report No. 77-6,* Division of Structural Mechanics, The Norwegian Institute of Technology, Trondheim, 1977.

14 Bergan, P.G., Nygård, M.K. and Bjærum, R.O.: Nonlinear Collapse Simulation of Anisotropic and Corrugated Plates ans Shells, In "*Spacecraft Structures and Mechanical Testing Conference*", ESA SP-289, ES.:C, October, 1988.

Assumed Strain Elements with Optimal Interpolation

U. Andelfinger and E. Ramm

Institut für Baustatik, Universität Stuttgart

Summary

The study investigates the extension of the classical degenerated solid elements for plate and shell analysis to the assumed strain concept. Different interpolation schemes for the strains are studied in order to balance the displacement variables on one side and the strains or stresses on the other side. The various results are compared and interpreted.

1. Introduction

The degenerated solid approach (C^0 - continuity) has been proved to be a successful and most general element formulation in thin and moderately thick plate and shell analysis. The original isoparametric displacement models however tend to the well known locking phenomena. Similarly the four node membrane or eight node brick element show an undesired stiff behaviour, especially if they are distorted or have a low aspect ratio.
Substantial improvements can be obtained by a modification assuming strain and / or stress fields [1], [2], [3], [4], [5]. Here, the so - called assumed strain technique is used, where in addition to the global displacements, the covariant tensorial strain fields are assumed through carefully chosen sampling points. For efficiency, the strains are decomposed into lower and higher order parts, where the lower order part is evaluated by an underintegration [6], [7]. The main objective with multifield elements is the choice of the optimal higher order strain (stress) distribution. In order to isolate the stiffening behaviour, first the displacement models of the four node membrane and plate elements and the eight node brick element are considered. It is the objective of the present study to investigate the individual displacement modes and the corresponding strain fields to filter out the necessary modifications of the strain assumption. Hereby, the existing assumed strain models can be partly improved, the four node shell element for example,

G. Kuhn · H. Mang (Eds.)
Discretization Methods in Structural Mechanis
IUTAM/IACM Symposium Vienna, Austria, 1989

by assuming the membrane and bending strains as well. The investigation has also been extended to higher order elements.

2. Four node membrane element

The displacement field **u** of the isoparametric element is given by

$$\mathbf{u} = \begin{bmatrix} u \\ v \end{bmatrix} = \mathbf{N} \cdot \mathbf{d} \tag{1}$$

where **N** are the bilinear shape functions and **d** is the vector of the nodal degrees of freedom (Fig. 1).

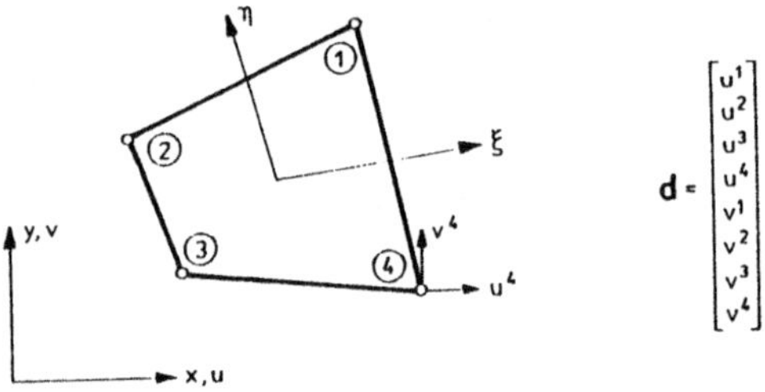

Figure 1: Four - node element with global and natural coordinate system

For closer inspection, we split **N** and **d** into

$$\mathbf{N} = \mathbf{P} \cdot \mathbf{A} = \begin{bmatrix} \mathbf{p} & \\ & \mathbf{p} \end{bmatrix} \cdot \begin{bmatrix} \mathbf{a} & \\ & \mathbf{a} \end{bmatrix} \tag{2}$$

$$\mathbf{a} = \frac{1}{4} \cdot \begin{bmatrix} 1 & 1 & 1 & 1 \\ 1 & -1 & -1 & 1 \\ 1 & 1 & -1 & -1 \\ 1 & -1 & 1 & -1 \end{bmatrix} \qquad \mathbf{p} = \begin{bmatrix} 1 & \xi & \eta & \xi\eta \end{bmatrix} \tag{3}$$

$$\mathbf{d} = \mathbf{Q} \cdot \boldsymbol{\alpha} = \begin{bmatrix} \mathbf{q} & \\ & \mathbf{q} \end{bmatrix} \cdot \boldsymbol{\alpha} \tag{4}$$

$$\mathbf{q} = \begin{bmatrix} 1 & x_1 & y_1 & x_1y_1 \\ 1 & x_2 & y_2 & x_2y_2 \\ 1 & x_3 & y_3 & x_3y_3 \\ 1 & x_4 & y_4 & x_4y_4 \end{bmatrix} \qquad \boldsymbol{\alpha}^T = [\alpha_1, \alpha_2, \ldots, \alpha_8] \tag{5}$$

which leads to

$$\mathbf{u} = \mathbf{P} \cdot \mathbf{A} \cdot \mathbf{Q} \cdot \boldsymbol{\alpha} \tag{6}$$

P defines the polynomials for the displacement assumption, **A** and **Q** contain the nodal coordinates of the reference and physical element respectively. α_i are the global displacement modes.
In view of the forthcoming derivation, the strains are defined in the natural coordinate system. In the linear case, the covariant components of the strain tensor are given by

$$\boldsymbol{\varepsilon} = \begin{bmatrix} u_{\xi|\xi} \\ u_{\eta|\eta} \\ u_{\xi|\eta} + u_{\eta|\xi} \end{bmatrix} \tag{7}$$

where | stands for the covariant derivative, for instance

$$u_{\xi|\eta} = \frac{\partial \mathbf{x}}{\partial \xi} \cdot \frac{\partial \mathbf{u}}{\partial \eta} = \frac{\partial \mathbf{x}}{\partial \xi} \cdot \mathbf{P}_{,\eta} \cdot \mathbf{A} \cdot \mathbf{Q} \cdot \boldsymbol{\alpha} \tag{8}$$

$\frac{\partial \mathbf{x}}{\partial \xi}$ or $\frac{\partial \mathbf{x}}{\partial \eta}$ are the covariant base vectors. From Eq.(8) it can be seen, that the order of the strain fields compatible to the displacements is given by the derivatives of the polynomials used in **P**. Eq.(7) together with Eq. (8) gives the strain

$$\boldsymbol{\varepsilon} = \mathbf{B} \cdot \boldsymbol{\alpha} \tag{9}$$

which by using the principle of the minimum of the total potential energy leads to the element stiffness matrix

$$\mathbf{k} = \int_v \mathbf{B}^T \, \mathbf{C} \, \mathbf{B} \; dv \tag{10}$$

C is the fourth - order constitutive tensor with contravariant components, defined with respect to the contravariant base vectors of the natural coordinate system. The assumed displacement model passes the patch test for the 2x2 - as well as for the 1 - point integration, provided that for the latter case the two hourglass modes are somehow stabilized. This means, that the three rigid - body modes and the three constant modes are exactly evaluated. Because each global mode α_i can be identified by the i - th column and row of the element stiffness matrix **k**, 60 of the 64 components of **k** are absolutely correct. The only components which can be improved by another formulation are $k_{4,4}$, $k_{4,8}$, $k_{8,4}$ and $k_{8,8}$.
In the present study, the so - called assumed strain technique is chosen, where the constant strains are evaluated by a one - point integration and higher order strain fields are based on additional integration points. The main task is the selection of the higher order strain fields. This will be

discussed using table 1. For simplicity but with no loss of generality, the axes ξ, η are assumed to coincide with the axes x and y.
In the first two rows, we have written the polynomial expansion for the displacements u and v, where each polynomial represents one displacement mode shown in Figure 2. The strains associated with each mode are calculated in rows 3 to 5 of table 1.

	(1)	(2)	(3)	(4)	(5)	(6)	(7)	(8)
u	1	ξ	η	$\xi\eta$				
v					1	ξ	η	$\xi\eta$
$\frac{\partial u}{\partial \xi}$		1		η				
$\frac{\partial v}{\partial \eta}$							1	ξ
$\frac{\partial u}{\partial \eta} + \frac{\partial v}{\partial \xi}$			1	ξ		1		η

Table 1: Strain assessment for a four - node membrane element

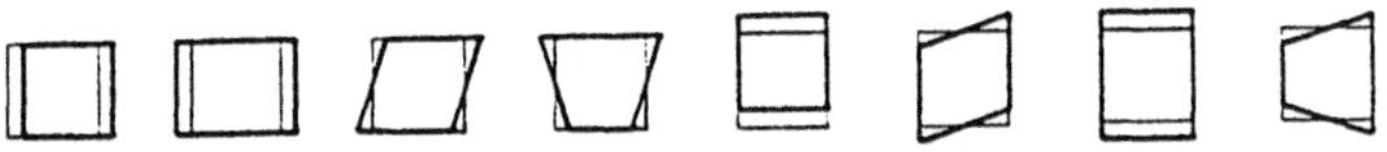

Figure 2: Displacement modes for a four - node membrane element

Besides the three rigid body modes ((3) minus (6) defines the rotation), we can identify the constant strain modes (2), (7) and (3) plus (6) and two linear modes (4) and (8). These two are the modes which make for a 2x2 integration the element too stiff. Columns (4) and (8) in table 1 show that a linearly varying normal strain is always combined with a linearly varying shear distribution. The ideal of an element being able to represent also pure modes cannot be satisfied. For a modification in the assumed strain concept, one has to decide either for two bending modes or for two linear shear modes.
Because the second option still exhibits severe stiffening in some cases, we take the first one. The stiffness matrix is obtained by the following procedure: First a one - point integration is performed leadind to the exact constant modes. In addition we use two sampling points A and B (Figure 3) to evaluate the linear part of $\varepsilon_{\xi\xi}$ and the points C and D for the linear part of $\varepsilon_{\eta\eta}$, which means that the shear terms in columns (4) and (8) of table 1 are omitted.

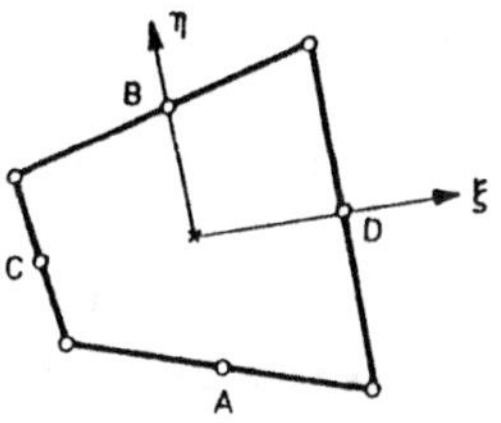

Figure 3: Location of additional sampling points

By comparing the strain distribution with the possible 8 displacement modes, one can see, that apart from the rigid body modes, each displacement mode represents only one pure strain distribution.
Although the polynomial expansion for the strains is not complete, we get an invariant element, due to the formulation in the natural coordinate system. Furthermore, we want to remark the similarities to elements like the hybrid stress element of Pian/Sumihara or the γ - elements of Belytschko et al. We have chosen the assumed strain technique, because for an optimal four node shell element, this is the only concept to handle the transverse shear part (if no additional shape functions shall be used).

3. Eight node solid element

Although the extension from a two - dimensional to a three - dimensional continuum element is straightforward, we want to discuss the optimal behaviour of the eight - node hexahedron, because it makes it easier to understand the problems of plate bending elements.

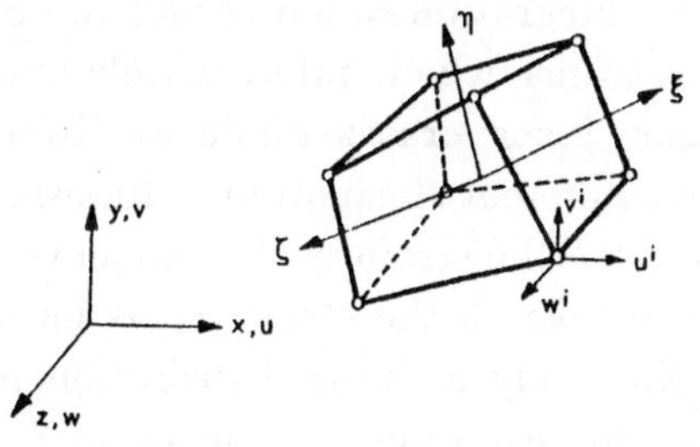

Figure 4: Eight - node solid element

The element has 24 degrees or freedom (Fig. 4) so the polynomials used for the trilinear interpolation are

$$\mathbf{p} = \begin{bmatrix} 1 & \xi & \eta & \zeta & \xi\eta & \xi\zeta & \eta\zeta & \xi\eta\zeta \end{bmatrix} \tag{11}$$

The displacement field is defined as in Eq.(6), where **P** is of the order 3x24, **A** and **Q** are matrices 24x24 and $\boldsymbol{\alpha}$ contains the 24 global displacement modes.

Table 2 shows the strain distribution of a pure displacement formulation. Again, for simplicity, we made the axes ξ, η, ζ parallel to the axes x, y and z respectively.

u	1	ξ	η	ζ	$\xi\eta$	$\xi\zeta$	$\eta\zeta$	$\xi\eta\zeta$																
v									1	ξ	η	ζ	$\xi\eta$	$\xi\zeta$	$\eta\zeta$	$\xi\eta\zeta$								
w																	1	ξ	η	ζ	$\xi\eta$	$\xi\zeta$	$\eta\zeta$	$\xi\eta\zeta$
$\frac{\partial u}{\partial \xi}$		1			η	ζ		$\eta\zeta$																
$\frac{\partial v}{\partial \eta}$											1		ξ		ζ	$\xi\zeta$								
$\frac{\partial w}{\partial \zeta}$																				1		ξ	η	$\xi\eta$
$\frac{\partial u}{\partial \eta} + \frac{\partial v}{\partial \xi}$			1		ξ		ζ	$\xi\zeta$		1			η	ζ		$\eta\zeta$								
$\frac{\partial u}{\partial \zeta} + \frac{\partial w}{\partial \xi}$				1		ξ	η	$\xi\eta$										1			η	ζ		$\eta\zeta$
$\frac{\partial v}{\partial \zeta} + \frac{\partial w}{\partial \eta}$												1		ξ	η	$\xi\eta$			1		ξ		ζ	$\xi\zeta$

Table 2: Strain assessment for the eight - node solid element

One - point integration gives the exact rigid body and constant strain modes (columns (1) to (4), (9) to (12) and (17) to (20)); however it cannot evaluate the 12 linear strain modes, thus leading to 12 zero - energy modes.

On the other side, 2x2x2 integration would render these modes as too stiff, mainly caused by shear strains, which unfortunately spoil the bending modes. Mode (8) for instance incorporates a bilinear distribution for the normal strains as well as two different shear strains. In order to avoid excessive stiffening, we have to reduce the strain fields somehow.

Again a good choice is to take all the strain terms for the normal strains, while for the shear strains only a linear distribution in one of the three directions is admitted (for example: $\varepsilon_{\xi\eta}$ linear in ζ). Therefore four sampling points for each of the normal strains and two sampling locations for the shear strains are used.

It is obvious that due to 18 different sampling points, this assumed strain concept cannot be efficient, so a stabilization procedure based on the hybrid stress method [8] is more suitable. However in this study the investigation of the solid element served primarely to prepare the following on plate bending elements. Before we continue let us state that despite the above mentioned optimal strain distribution, the three modes (7), (14) and (21) are still coupled.

4. Four node plate element

In order to get a balanced, non - stiffening four - node shell element based on the Reissner - Mindlin - theory the above given optimal strain distribution for the linear solid element (Table 2) could be applied together with the degeneration concept. However first the pure displacement model is investigated to localize the locking. A typical element with the nodal d.o.f is shown in Fig. 5.

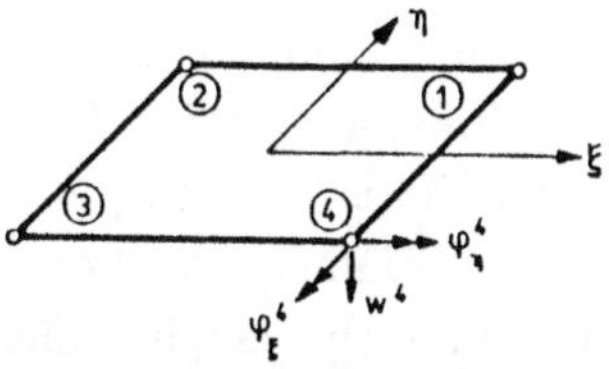

Figure 5: Notations for the four - node plate element

Again we use bilinear interpolation functions for each of the three variables. The corresponding strain distribution for a rectangular element is given in table 3.

The three rigid body modes are defined by mode (1), (2) plus (5) and (3) plus (9). Again they can be identified by zero - columns i.e. no strains will be developed for these modes. Checking the constant strain modes we see that the constant transverse shear modes ((2) minus (5), (3) minus (9)) are exactly represented. Mode (6) and (11) however fail to represent the pure bending, because it is always coming along with a linearly varying shear distribution which stiffens the element and finally leads to shear locking.

In order to pass the patch test, one has to reduce the bilinear tranverse shear strains to a constant - linear field. Therefore each of the shear strains is evaluated at only two sampling points, which for $\gamma_{\xi\zeta}$ are located at $\xi = 0$ and for $\gamma_{\eta\zeta}$ at $\eta = 0$. The modification results in a plate element that passes all constant patch tests [1], [2].

A further improvement can be achieved by limiting, in analogy to the shear strain of the optimaly integrated membrane element, the twist to a constant

w	1	ξ	η	$\xi\eta$								
φ_ξ					1	ξ	η	$\xi\eta$				
φ_η									1	ξ	η	$\xi\eta$
$\frac{\partial\varphi_\xi}{\partial\xi}$						1		η				
$\frac{\partial\varphi_\eta}{\partial\eta}$											1	ξ
$\frac{\partial\varphi_\xi}{\partial\eta}+\frac{\partial\varphi_\eta}{\partial\xi}$							1	ξ		1		η
$\frac{\partial w}{\partial\xi}-\varphi_\xi$		1		η	-1	$-\xi$	$-\eta$	$-\xi\eta$				
$\frac{\partial w}{\partial\eta}-\varphi_\eta$			1	ξ					-1	$-\xi$	$-\eta$	$-\xi\eta$
	(1)	(2)	(3)	(4)	(5)	(6)	(7)	(8)	(9)	(10)	(11)	(12)

Table 3: Strain assessment for the four - node plate element

distribution, leading (8) and (12) to pure linear bending modes.
As mentioned in Section 3, there is a remaining coupling within the displacement modes (4), (7) and (10). While it was a coupling among the three shear strains in the 3 - dim. element, it here is, due to the degeneration from three to two dimensions, a coupling between the transverse shear strains and the constant twist. This coupling is the reason, that the one - point integrated stabilized shell element cannot pass the patch test. The incorrect stabilization of the linearly varying shear always influences the constant twist.

5. Eigenmode investigation

5.1 Four node membrane element

Eigenvalues and eigenmodes of the stiffness matrix for a quadratic element with side lengths 1.0 are calculated (Table 4). Young's modulus E and Poissons's ratio ν are 1.0 and 0.0 respectively. It can be seen that for all interpolation schemes the three rigid body modes (zero - eigenvalues) and the three constant modes are obtained exactly; they would also be exact for any distorted geometry or any Poisson's ratio ν.
The two linear modes (mode (4) and (8) in Figure 2) however are too stiff when using a full integration and lead to the hourglass modes with no stiffness for a one - point integration. The results for the optimal interpolation scheme are shown in column three. Now, mode (4) and (8) (eigenvalues 0.333) are pure bending modes, which means that these modes

Integration	1 × 1	2 × 2	*assumed strain*
	0	0	0
	0	0	0
	0	0	0
$\varepsilon_{\xi\xi}$ = *linear*	0	0.5	0.333
$\varepsilon_{\eta\eta}$ = *linear*	0	0.5	0.333
$\varepsilon_{\xi\xi}$ = *const.*	1.0	1.0	1.0
$\varepsilon_{\eta\eta}$ = *const.*	1.0	1.0	1.0
$\varepsilon_{\xi\eta}$ = *const.*	1.0	1.0	1.0

Table 4: Eigenvalues of different four - node membrane elements ($a = b = 1.0, E = 1.0, \nu = 0.0$)

are not stiffened by any shear strain. The stiffening of the fully integrated displacement model is even more severe, the lower the aspect ratio of the element is.

5.2 Four node plate element

The eigenvalues of the linear plate element with 12 degrees of freedom are shown in Table 5. In the first column, reduced integration was used, which only renders the constant strain modes, according to the five possible strains. The necessary linear modes cannot be obtained, resulting instead in four additional zero - energy modes. A full integration leads to the correct number of zero - eigenvalues but now the constant and linear bending modes are spoiled by twist and transverse shear strains. The third column is identical with the element given in [1] and [2]. By reducing the transverse shear strains as described in Section 4 to constant - linear fields, we now get the pure bending modes (eigenvalues of 0.0833). The linear bending modes (0.041667) however are still coupled with the linear twist. This coupling can be avoided by reducing the twist to a constant field which results in pure linear bending modes (eigenvalue of 0.02777 in column 4). Again, the remaining coupling between linear transverse shear strains and constant twist can be recognized.

SH4RI (1 × 1)	SH4 (2 × 2)	SHELAA4	SHELBB4
0	0	0	0
0	0	0	0
0	0	0	0
0	0.0347222 $\gamma_{\xi\zeta}^{lin}$, $\gamma_{\eta\zeta}^{lin}$	0.0347222 $\gamma_{\xi\zeta}^{lin}$, $\gamma_{\eta\zeta}^{lin}$	0.027777 $\kappa_{\xi\xi}^{lin}$
0	0.0532407 $\kappa_{\xi\xi}^{lin}$, $\kappa_{\xi\eta}^{lin}$, $\gamma_{\xi\zeta}^{lin}$	0.0416667 $\kappa_{\xi\xi}^{lin}$, $\kappa_{\xi\eta}^{lin}$	0.027777 $\kappa_{\eta\eta}^{lin}$
0	0.0532407 $\kappa_{\eta\eta}^{lin}$, $\kappa_{\xi\eta}^{lin}$, $\gamma_{\eta\zeta}^{lin}$	0.0416667 $\kappa_{\eta\eta}^{lin}$, $\kappa_{\xi\eta}^{lin}$	0.0347222 $\gamma_{\xi\zeta}^{lin}$, $\gamma_{\eta\zeta}^{lin}$
0	0.0713352 $\kappa_{\xi\eta}^{c}$,($\gamma_{\xi\zeta}^{lin}$, $\gamma_{\eta\zeta}^{lin}$)	0.0713352 $\kappa_{\xi\eta}^{c}$,($\gamma_{\xi\zeta}^{lin}$, $\gamma_{\eta\zeta}^{lin}$)	0.0713352 $\kappa_{\xi\eta}^{c}$,($\gamma_{\xi\zeta}^{lin}$, $\gamma_{\eta\zeta}^{lin}$)
0.083333 $\kappa_{\xi\xi}^{c}$	0.118056 $\kappa_{\xi\xi}^{c}$, $\gamma_{\xi\zeta}^{lin}$	0.083333 $\kappa_{\xi\xi}^{c}$	0.083333 $\kappa_{\xi\xi}^{c}$
0.083333 $\kappa_{\eta\eta}^{c}$	0.118056 $\kappa_{\eta\eta}^{c}$, $\gamma_{\eta\zeta}^{lin}$	0.083333 $\kappa_{\eta\eta}^{c}$	0.083333 $\kappa_{\eta\eta}^{c}$
0.083333 $\kappa_{\xi\eta}^{c}$	0.324498 $\gamma_{\xi\zeta}^{lin}$, $\gamma_{\eta\zeta}^{lin}$, ($\kappa_{\xi\eta}^{c}$	0.324498 $\gamma_{\xi\zeta}^{lin}$, $\gamma_{\eta\zeta}^{lin}$, ($\kappa_{\xi\eta}^{c}$)	0.324498 $\gamma_{\xi\zeta}^{lin}$, $\gamma_{\eta\zeta}^{lin}$, ($\kappa_{\xi\eta}^{c}$)
0.520833 $\gamma_{\xi\zeta}^{c}$	0.520833 $\gamma_{\xi\zeta}^{c}$	0.520833 $\gamma_{\xi\zeta}^{c}$	0.520833 $\gamma_{\xi\zeta}^{c}$
0.520833 $\gamma_{\eta\zeta}^{c}$	0.520833 $\gamma_{\eta\zeta}^{c}$	0.520833 $\gamma_{\eta\zeta}^{c}$	0.520833 $\gamma_{\eta\zeta}^{c}$

Table 5: Eigenvalues and corresponding strains of a four - node plate element with different interpolation schemes
($a = b = 1.0, E = 1.0, \nu = 0.0$)

Conclusions

The stiffening behaviour of the differently interpolated four node membrane and plate element and the eight node brick element was investigated. By looking at the derivatives of the assumed displacement fields, zero - energy modes can be easily avoided and undesired 'non - pure' modes can be identified. The assumed strain technique is applied which by using an optimal strain distribution avoids the stiffening of most higher order modes. A very efficient formulation can be achieved by coupling an underintegration with the use of additional sampling points.
The same procedure was also applied to higher order elements. It can be extended to shell elements in order to avoid membrane locking (by reducing the membrane strains) or locking due to incompressible material (by reducing the strains, so that the constraint of no volumetric strain is satisfied). In all cases however it has to be insured that apart from the rigid - body modes, each mode develops at least one strain component to avoid zero - energy modes.

References

[1] Hughes, T.J.R.; Tezduyar, T.E.: Finite elements based upon Mindlin plate theory with particular reference to the four - node bilinear isoparametric element. J. Appl. Mech., 46, 587-594, 1981.

[2] Dvorkin, E.N.; Bathe, K.J.: A continuum mechanics based four - node shell element for general nonlinear analysis. Eng. Computations, Vol. 1, 77-88, 1984.

[3] Huang, H.C.: Implementation of assumed strain degenerated shell elements. Computers and Structures, Vol. 25, 147-155, 1987.

[4] Pian, T.H.H.; Sumihara, K.: Rational approach for assumed stress finite elements. Int. J. Num. Meth. Eng., 20, 1685-1695, 1984.

[5] Pian, T.H.H.; Tong, P.: Relations between incompatible displacement model and hybrid stress model. Int. J. Num. Meth. Eng., 22, 173-181, 1986.

[6] Belytschko, T.; Engelmann, B. E.: On flexurally superconvergent four - node quadrilaterals. Comp. and Structures, Vol.25, 909-918, 1987.

[7] Belytschko, T.; Wong, B. L.: Assumed strain stabilization procedure for the 9 - node Lagrange shell element. Int. J. Num. Meth. Eng., Vol.28, 385-414, 1989.

[8] Wang, X. J.; Belytschko, T.: An efficient flexurally superconvergent hexahedral element. Eng. Comput., Vol.4, 281-288, 1987.

Linear and Nonlinear Transient Problems of Thermomechanics by BEM

P.K. Banerjee
G.F. Dargush

Department of Civil Engineering
State University of New York at Buffalo
Buffalo, New York U.S.A.

Summary

The development of a general boundary element method (BEM) for transient problems of thermomechanics is discussed. The new formulation operates directly in the time domain. These formulations are implemented in a general purpose BEM system which includes higher-order conforming elements, self-adaptive integration, and multi-region capability. A number of detailed examples are presented to illustrate the accuracy and suitability of this boundary element approach for linear and nonlinear thermomechanical analyses.

Introduction

A complete derivation of thermoelastic theory can be found in the textbooks by Boley and Weiner [1] and Nowacki [2]. Consequently, only a few of the key assumptions will be mentioned before presenting the governing differential equations. In particular, the classical theory assumes infinitesimal deformations and linear isotropic materials. Thus, the linearized strain-displacement relationship

$$\varepsilon_{ij} = \frac{1}{2}(u_{i,j} + u_{j,i}) \tag{1}$$

is employed, along with the constitutive laws of Duhamel-Neumann,

$$\sigma_{ij} = \delta_{ij}\lambda\varepsilon_{kk} + 2\mu\varepsilon_{ij} - \delta_{ij}(3\lambda + 2\mu)\alpha(\theta - \theta_o) \tag{2}$$

and Fourier,

$$q_i = -k\theta_{,i} \tag{3}$$

In equations (2) and (3), θ is the present temperature, while θ_o represents the temperature of a zero stress reference state, and q_i is the heat flux vector. Additionally, λ and μ are the usual Lame's isothermal elastic constants, α is the coefficient of thermal expansion, and k represents the isotropic thermal conductivity.

Then, application of the laws of conservation of momentum and energy lead, in the absence of inertia, to the following set of equations:

$$(\lambda + \mu)u_{j,ij} + \mu u_{i,jj} - (3\lambda + 2\mu)\alpha\theta_{,i} + f_i = 0 \tag{4a}$$

G. Kuhn · H. Mang (Eds.)
Discretization Methods in Structural Mechanis
IUTAM/IACM Symposium Vienna, Austria, 1989

$$k\theta_{,jj} - \rho c_i\dot{\theta} - (3\lambda + 2\mu)\alpha\theta_o\dot{u}_{j,j} + \psi = 0 \tag{4b}$$

in which ρ is the mass density, c_e is the specific heat at constant deformation, while f_i and ψ are the body forces and sources, respectively. The theory portrayed by equations (4) is formally named Coupled Quasistatic Thermoelasticity (CQT). Notice, in particular, the appearance of displacement and temperature in both equations. Thus, in CQT not only do changes in temperature cause deformation, but also deformation produces temperature variations. In general, the set of equations (4) must be solved simultaneously. However, from a practical engineering standpoint, it has been determined by a number of researchers (e.g. Boley and Weiner [1], Day [3]) that the term involving displacement in equation (4b) is negligible, thus uncoupling the momentum and energy balance equations. Dropping that term, equation (4b) can first be solved independently for the temperature field. Subsequently, displacements are determined from equation (4a) with the known temperature distribution. This is, in fact, the accepted procedure for thermal stress analysis.

The Boundary Integral Formulations

A reciprocal theorem generally provides a convenient starting point for a direct boundary element formulation. For the coupled problem at hand, it appears that Ionescu-Cazimir [4] was the first to state explicitly an appropriate reciprocal theorem, although certainly the groundwork was laid earlier by Biot [5]. The theorem can be written, for a body of volume V and surface S with zero initial conditions, in the following time-domain form:

$$\int_s [\dot{t}_i^{(1)} * u_i^{(2)} + q^{(1)} * \theta^{(2)} - \dot{u}_i^{(1)} * t^{(2)} - \theta^{(1)} * q^{(2)}]dS$$
$$+ \int_v [\dot{f}_i^{(1)} * u_i^{(2)} + \psi^{(1)} * \theta^{(2)} - \dot{u}_i^{(1)} * f_i^{(2)} - \theta^{(1)} * \psi^{(2)}]dV = 0 \tag{5}$$

The superscripts $^{(1)}$ and $^{(2)}$ denote any two independent states existing in the body defined by $[\dot{u}_i^{(1)}, \dot{t}_i^{(1)}, \theta^{(1)}, q^{(1)}, \dot{f}_i^{(1)}, \psi^{(1)}]$ and $[u_i^{(2)}, t_i^{(2)}, \theta^{(2)}, q^{(2)}, f_i^{(2)}, \psi^{(2)}]$, respectively. The symbol $*$ indicates a Riemann convolution integral, where, for example,

$$q^{(1)} * \theta^{(2)} = \int_0^t q^{(1)}(X, t-\tau)\theta^{(2)}(X, t)d\tau = \int_o^t q^{(1)}(X, \tau)\theta^{(2)}(X, t-\tau)d\tau \tag{6}$$

This reciprocal theorem involves any two independent states existing in the body of interest. Let one of these states, say state (2), be the, as yet, unknown solution to a given boundary value problem. Then the remaining state may be chosen arbitrarily. However, by initially selecting for state (1), the infinite region response to a unit step force in the j-direction acting at ξ within V and beginning at time zero, equation (5) leads to (see Dargush [6], Dargush and Banerjee [7-10],)

$$u_j(\xi, t) = \int_s [\dot{G}_{ij}(X-\xi, t) * t_i(X, t) + \dot{G}_{\theta j}(X-\xi, t) * q(X, t)$$
$$- \dot{F}_{ij}(X-\xi, t) * u_i(X, t) - \dot{F}_{\theta j}(X-\xi, t-\tau) * \theta(X, t)]dS(X) \tag{7}$$

with no summation with respect to the subscript θ. Similarly by selecting the state (2) to be that generated by a heat source acting at time zero at ξ we get the temperature $\theta(\xi,t)$ as

$$\theta(\xi,t) = \int_s [\dot{G}_{i0}(X-\xi,t) * t_i(X,t) + \dot{G}_{00}(X-\xi,t) * q(X,t) - \dot{F}_{i0}(X-\xi,t) * u_i(X,t) - \dot{F}_{0\theta}(X-\xi,t-\tau) * \theta(X,t)]dS(X) \tag{8}$$

This, of course, is the desired statement for interior temperature in terms of boundary quantities.

Equations (7) and (8) can be written together in a convenient matrix notation as:

$$\begin{Bmatrix} u_j(\xi,t) \\ \theta(\xi,t) \end{Bmatrix} = \int_s \left(\begin{bmatrix} \dot{G}_{ij}\dot{G}_{i0} \\ \dot{G}_{0j}\dot{G}_{00} \end{bmatrix}^T * \begin{Bmatrix} t_i(X,t) \\ q(X,t) \end{Bmatrix} - \begin{bmatrix} \dot{F}_{ij}\dot{F}_{i0} \\ \dot{F}_{0j}\dot{F}_{00} \end{bmatrix}^T * \begin{Bmatrix} u_i(X,t) \\ \theta(X,t) \end{Bmatrix} \right) dS(X) \tag{9}$$

However, this can be compacted even further by generalizing the displacement and traction vectors to include temperature and flux, respectively, as an additional component. Thus, in three dimensions, for example, let

$$u_\alpha = \{u_1 \quad u_2 \quad u_3 \quad \theta\}^T \tag{10a}$$

$$t_\alpha = \{t_1 \quad t_2 \quad t_3 \quad q\}^T \tag{10b}$$

where the subscript α, and subsequently β, varies from 1 to 4. Obviously, for the two-dimensional counterpart

$$u_\alpha = \{u_1 \quad u_2 \quad \theta\}^T \tag{10c}$$

$$t_\alpha = \{t_1 \quad t_2 \quad q\}^T \tag{10d}$$

Then equation (9) becomes simply

$$u_\alpha(\xi,t) = \int_s [\dot{G}_{\beta\alpha} * t_\beta(X,t) - \dot{F}_{\beta\alpha} * u_\beta(X,t)]dS(X) \tag{11}$$

Equation (11) can be viewed as a generalized Somigliana's identity which can be expressed for a boundary point simply as

$$c_{\beta\alpha}(\xi)u_\beta(\xi,t) = \int_s [\dot{G}_{\beta\alpha} * t_\beta(X,t) - \dot{F}_{\beta\alpha} * u_\beta(X,t)]dS(X) \tag{12}$$

where $c_{\beta\alpha}(\xi)$ is a matrix of constants, dependent only upon the local geometry at ξ. Equation (12) can be extended by using an initial stress formulation for the inelastic thermal stress analysis and written in incremental form as

$$c_{\beta\alpha}(\xi)\dot{u}_\beta(\xi,t) = \int_s [\dot{G}_{\beta\alpha} * \dot{t}_\beta(X,t) - \dot{F}_{\beta\alpha} * \dot{u}_\beta(X,t)]dS(X) + \int_v [\dot{Q}_{ik\alpha} * \dot{\sigma}^o_{ik}(Z,t) + \dot{G}_{\theta\alpha} * \dot{\psi}^o(Z,t)]dV(Z), \tag{13}$$

where

$$\psi^o(Z,t) = \gamma \sigma_{ij} \dot{\epsilon}^p_{ij} \tag{14}$$

and γ is a dimensionless material parameter specifying the portion of the inelastic dissipation converted into heat. The kernel $\dot{Q}_{ik\alpha}$, appearing in (13) can be derived from $\dot{G}_{i\alpha}$ via

$$\dot{Q}_{ik\alpha} = \frac{1}{2}\left(\frac{\partial \dot{G}_{i\alpha}}{\partial x_k} + \frac{\partial \dot{G}_{k\alpha}}{\partial x_i}\right) \tag{15}$$

The interior and boundary stresses can be then derived in the usual manner. Explicit details of these as well as the kernel functions appearing in equations (12) and (13) can be found in Dargush [6], Dargush and Banerjee [7-10].

For the uncoupled thermoelastic as well as thermoplastic analyses, kernel functions G_{ij} and F_{ij} are time independent and functions $G_{i\theta}$ and $F_{i\theta}$ are zero; consequently only four of the convolution integrals remain. These integrals $\dot{G}_{\theta j} * q, \dot{G}_{\theta\theta} * q, \dot{F}_{\theta j} * \theta$, and $\dot{F}_{\theta\theta} * \theta$ are approximated by a piecewise constant approximation with respect to time and the integrals are then, evaluated analytically [11].

Numerical Implementation

The boundary integral equations (12) and (13) developed in the last section, represent exact statements. No approximations have been introduced other than those used to formulate the boundary value problem. However, in order to apply these to the solution of practical engineering problems, approximations are required in both time and space.

In this section, highlights of a general-purpose, state-of-the-art numerical implementation are presented. This entire implementation has been accomplished within the GP-BEST (General Purpose Boundary Element Solution Technique) analysis system, which, in addition to thermoelasticity and thermoplasticity contains a wide spectrum of capabilities in linear and non-linear static and dynamic problems of solid and fluid mechanics. As a result, many of the features and techniques used in this implementation were developed previously for elastostatics and elastodynamics (e.g., Banerjee et al [12-14], Banerjee and Ahmad [15] Ahmad and Banerjee [16], Dargush [6], Dargush and Banerjee [7-10], Henry and Banerjee [17], but are here adapted for the thermal analysis. Explicit details of this implementation in the context of the closely related poroelastic analysis can be found in Dargush [6] and Dargush and Banerjee [7,8]. In this subsection, a number of these advanced features of the general purpose implementation are briefly mentioned.

Perhaps the most significant of these items is the capability to analyze substructured problems. This not only extends the analysis to bodies composed of several different materials, but also often provides computational efficiencies. During the integration process, the individual regions remain separate entities. The geometric modeling regions (or GMR's) are brought together for the first time at the assembly stage, where compatibility relations are enforced on the common boundary between adjacent regions. This multi-GMR assembly process produces block-banded system matrices that can be solved in an efficient manner.

As another convenience, a high degree of flexibility is provided for the specification of boundary conditions. In general, time-dependent values can be defined in either global or local coordinates, with or without sliding. Not only can generalized displacements and tractions be specified, but also spring, convection, and contact resistance relations, respectively, are available. Furthermore, a comprehensive symmetry capability is included with provisions for both planar and cyclic symmetry.

Examples of Application

Circular Disc

As a first example,, transient thermal stresses in a circular disc are investigated. The disc of radius a initially rests at zero uniform temperature. The top and bottom surfaces are thermally insulated, and all boundaries are completely free of mechanical constraint. Then, suddenly, at time zero, the temperature of the entire outer edge (i.e., $r = a$) is elevated to unity and, subsequently, maintained at that level.

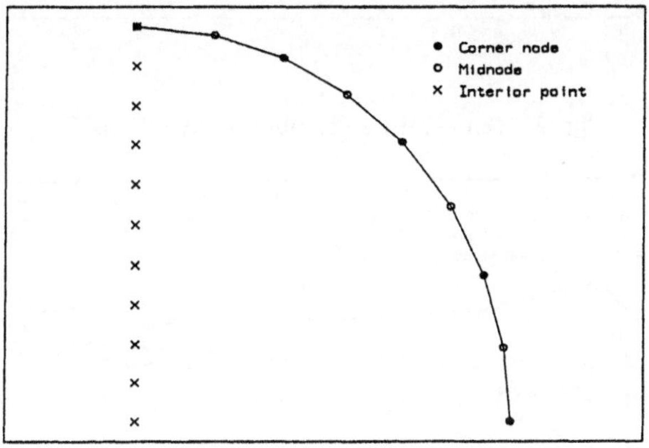

Fig. 1. Boundary element model of circular disc

A GP-BEST boundary element model of the disc with unit radius is shown in Fig. 1. Only four quadratic elements are employed, along with quarter symmetry. Ten interior points are also included strictly to monitor response. In addition, the following non-dimensionalized material properties are arbitrarily selected for the plane analysis:

$$E = 1.333 \qquad \rho c_\epsilon = 1.0$$

$$\nu = 0.333 \qquad k = 1.0$$

$$\alpha = 0.75$$

Results obtained under UQT (Uncoupled Quasistatic Thermal analysis) for a time step of 0.005 are compared, in Figs. 2 and 3, to the analytical solution presented in Timoshenko and Goodier. Notice that temperatures and stresses are accurately determined via the boundary element analysis. In particular from Fig. 3, even the tangential stress on the outer edge is faithfully reproduced.

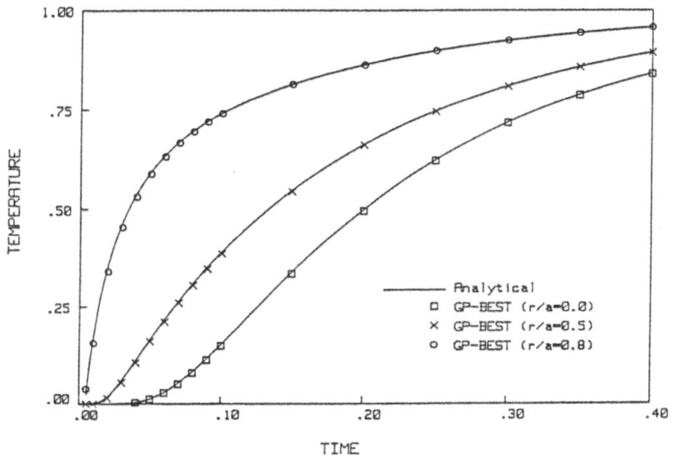

Fig. 2. Temperature distribution by GPBEST

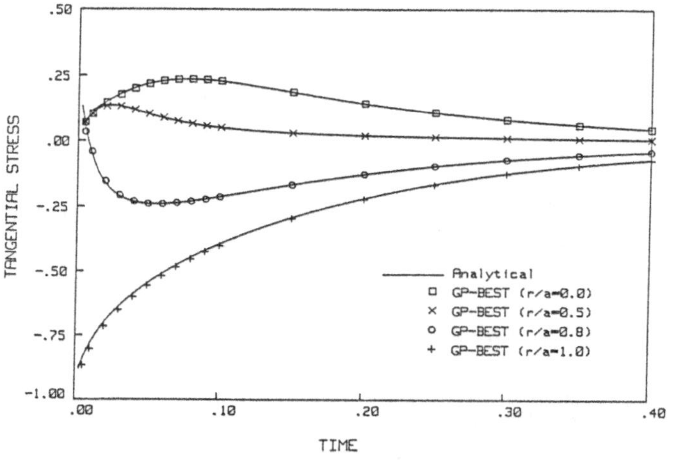

Fig. 3. Tangential stress distribution at various times

Residual Stress in a Steel Cylinder

The second example involves determination of the residual stresses resulting from the rapid cooling of a long, 2.5 in diameter steel cylinder. This cylinder is initially in a stress free state at 1250° F, and then suddenly exposed to an 80°F fluid. A film coefficient of $h = 12.25$ in-lb/sec in^2°F is specified on the cylindrical surface for the ensuing convective process.

The following standard thermoelastic material properties are employed for 1060 steel:

$$E = 30 \times 10^6 \text{psi} \qquad \rho c_\epsilon = 283 \text{in lb/in}^3{}^\circ\text{F}$$
$$\nu = 0.30 \qquad k = 5.8 \text{in lb/sec in}^\circ\text{F}$$
$$\alpha = 6 \times 10^{-6}/^\circ\text{F}$$

Additionally, a temperature-dependent Von Mises thermoplastic constitutive model is adopted. The yield stress at room temperature is 48 ksi reducing to zero at 1300°F, and zero strain hardening is assumed.

The boundary element surface model, displayed in Fig. 4, consists of 38 source points and 12 elements, while the complete description of the interior requires one extra node and three volume cells. After some numerical experimentation, a time step of 4 sec was selected.

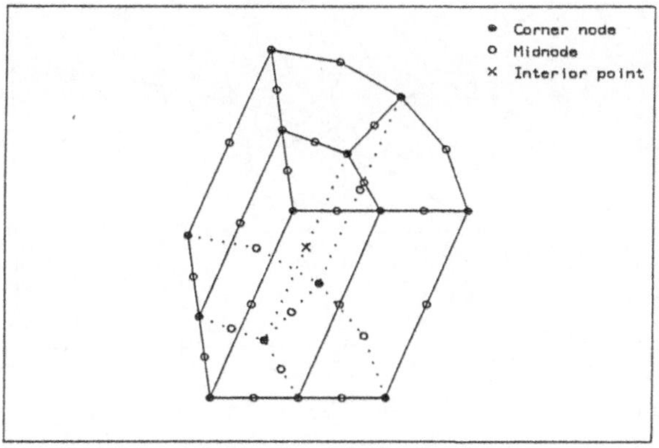

Fig. 4. BEM discretization for a segment of steel cylinder

Figure 5 shows the temperature response as a function of time for the surface ($r = 1.25in$) and center ($r = 0$) of the cylinder at mid-height. Meanwhile, axial stresses at those same locations are traced in Fig. 6. Notice that rapid cooling of the surface occurs during the first few seconds. This produces tensile stress in the outer portions and compression toward the center. However, throughout this initial period, the material remains quite hot, and consequently, has a relatively low yield stress, leading to significant plastic deformation.

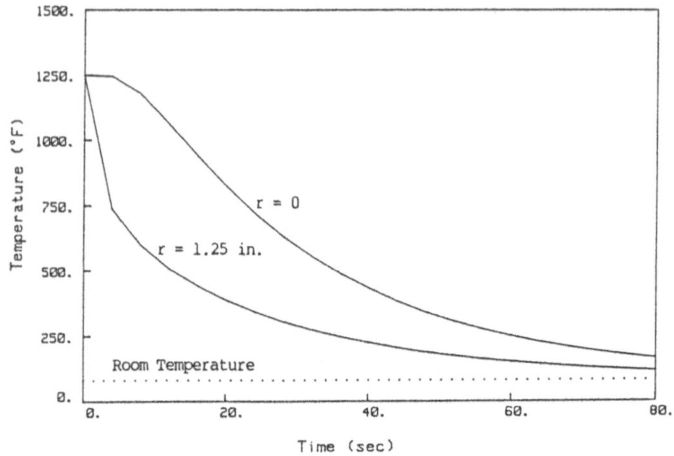

Fig. 5. Temperature distribution in the cylinder at different times

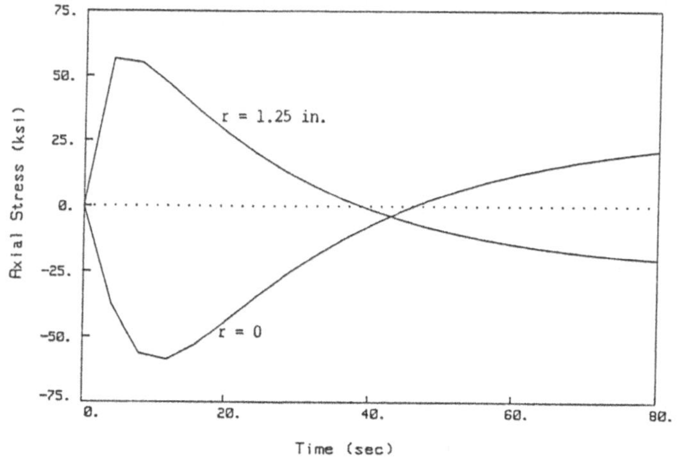

Fig. 6. Variation of axial stress with time in a steel cylinder

After the beginning ten seconds, the yield stress is elevated considerably and the all important thermal gradients diminish, so that no further plastic flow takes place. Instead, elastic unloading commences. At about the forty second mark, the axial stresses change signs. Thus, the outer fibers go into compression, while tension occurs in the central portion. This trend continues to steady-state. After the cylinder cooled for 80 seconds, the axial stress on the surface and at the center are –20.5 ksi and +21.1 ksi, respectively.

Conclusions

A time domain boundary element method has been developed for application to two- and three-dimensional problems of thermoelasticity and thermoplasticity. The general purpose, state-of-the-art boundary element implementation (GP-BEST) insures efficient integration, assembly and solution, while providing capability for the modeling of complicated multi-region geometry and the specification of arbitrary time-varying boundary conditions. Additionally, facilities are provided for symmetry and the calculation of interior and boundary values. The two simple numerical examples validate this new formulation and highlight the versatility of the implementation.

References

1. Boley, B.A. and Weiner, J.H.: *Theory of Thermal Stresses*, John Wiley, New York (1960).

2. Nowacki, W.: *Thermoelasticity*, Pergamon, Warsaw (1986).

3. Day, W.A.: Further justification of the uncoupled and quasistatic approximations in thermoelasticity, *Arch. Ration. Mech. Anal.*, **79**(1), 387-96 (1982).

4. Ionescu-Cazimir, V.: Problem of linear coupled thermoelasticity. Theorems on reciprocity for the dynamic problem of coupled thermoelasticity. II, *Bull. Acad. Pol. Sci., Sér. Sci. Tech.*, **12**(9), 473-88 (1964).

5. Biot, M.A.: New thermomechanical reciprocity relations with application to thermal stress analysis, *J. Aerospace Sci.* **26**(7), 401-8 (1959).

6. Dargush, G.F.: Boundary element methods for the analogous problems of thermomechanics and consolidation, Ph.D. thesis, State University of New York at Buffalo (1987).

7. Dargush, G.F. and Banerjee, P.K.: BEM in coupled problems of thermoelasticity and poroelasticity, *Developments in BEM*, Volume 5, Elsevier Applied Science, London, 126-136 (1989).

8. Dargush, G.F. and Banerjee, P.K.: A time domain boundary element method for poroelasticity, *Int. J. Numer. Methods Eng.*, in press (1989).

9. Dargush, G.F. and Banerjee, P.K.: Development of a boundary element method for time-dependent planar thermoelasticity, *Int. J. Solids Struct.*, in press (1989).

10. Dargush, G.F. and Banerjee, P.K.: Boundary element methods in three-dimensional thermoelasticity *Int. J. Solids Struct.*, in press (1989).

11. Banerjee, P.K. and Butterfield, R.: *Boundary Element Methods in Engineering Science*, McGraw-Hill, London (1981). Also US edition, New York (1983); Russian Edition, Moscow (1984).

12. Banerjee, P.K., Wilson, R.B. and Miller, N.M.: Development of a large BEM system for three-dimensional inelastic analysis, *Proc. ASME Conf. on Advanced Topics in Boundary Element Analysis*, Eds T.A. Cruse, A.B. Pifko and H. Armen, AMD Vol. 72, ASME, New York, 1-20 (1985).

13. Banerjee, P.K., Ahmad, S. and Manolis, G.D.: Transient elastodynamic analysis of three-dimensional problems by boundary element method, *Earthquake Eng. Structural Dyn.*, **14**, 933-49 (1986).

14. Banerjee, P.K., Wilson, R.B. and Miller, N.M.: Advanced elastic and inelastic three-dimensional analysis of gas turbine engine structures by BEM, *Int. J. Numer. Methods Eng.* **26**, 393-411 (1988).

15. Banerjee, P.K. and Ahmad, S.: Advanced dynamic analysis of 3-D problems by boundary element methods, *Proc. ASME Conf. on Advanced Topics in Boundary Element Analysis*, Eds. T.A. Cruse, A.B. Pifko and H. Armen, AMD Vol. 72, ASME, New York, 65-81 (1985).

16. Ahmad, S. and Banerjee, P.K.: Transient elastodynamic analysis of three-dimensional problems by BEM, *Int. J. Methods Eng.* **26**(8), 1560-80 (1988).

17. Henry. D.P. and Banerjee, P.K.: A variable stiffness type boundary element formulation for axisymmetric elastoplastic media *Int. J. Numer. Methods Eng.*, **26**, 1005-27 (1988).

Dynamic Interaction Analysis in Wave Propagation Problems by a Time-Dependent Boundary Element Method

H. ANTES

Institut für Angewandte Mechanik
Technische Universität Braunschweig
D-3300 Braunschweig, W.-Germany

Summary

The aim of this contribution is to inform about time-stepping boundary element procedures which can be used to analyze wave propagation in elastic media or in compressible fluids. Solutions of representative problems, such as dynamic interaction between the dam and the water in a reservoir, or unilateral contact between structures and the soil, are presented in order to demonstrate the wide variety of practical applications of the methods.

Introduction

The phenomenon of wave propagation is encountered in several engineering disciplines. In recent years, especially the dynamic interaction between elastic structures and adjacent media (soil or fluids) turned out to be very substantial in dynamic analyses and the design of structures. The catastrophic consequences of a dam failure, for example, make it important that dams can safely withstand strong earthquakes. The same is to urge for nuclear power plant structures, tanks in chemical industries etc. Thus, in earthquake analyses the knowledge of the details of wave propagation, e.g. of wave focussing due to special bedrock geometries [1], is essential.

Whenever problems involving infinite domains are under consideration, the most appropriate method seems to be the Boundary Element Method (BEM) as it automatically takes into account the effect of radiations damping. Moreover, it requires a discretization of the surface of the domain only, i.e. no elements in the interior are needed. In addition to those regular advantages of the BEM, a time domain BEM yields the non-steady, transient dynamic response of a structure directly (i.e., without the use of Fourier-Laplace transformations) and, therefore provides a firm basis for the extension of the technique to nonlinear problems.

G. Kuhn · H. Mang (Eds.)
Discretization Methods in Structural Mechanis
IUTAM/IACM Symposium Vienna, Austria, 1989

In practical problems related to geotechnical or earthquake engineering, material inhomogeneity or physical non-linearity may be essential. If these 'non-linear zones' are only finite parts of the system, the Finite Element Method (FEM) can be used to model therein the dynamic behavior, while the remaining domain is represented by boundary elements. Thus, a hybrid method, a combination of boundary and finite elements, can profit from the advantages of each of the two methods while evading their respective dis-advantages.

In order to analyze dynamic interaction in wave propagation problems, general time-stepping boundary element procedures have been developed [2,3,4]. This paper gives solutions of some representative problems [1,6,7,8,9,10] and reports on recent progress, e.g. in extending the method to unilateral contact problems [11,12], and in combining boundary and finite elements [13,14].

Time domain boundary integral equations

A time dependent integral representation of the dynamical initial-boundary value problem describing wave propagation in elastic media and compressible fluids, respectively, can be deduced by employing Graffi's reciprocal theorem and a subsequent integral transformation [2]. Finally, one obtains for elastodynamics (the overdot "•" indicates time differentiation, $t'=t-\tau$)

$$d_{ij}(\xi)u_j(\xi,t) = \int_0^t \Big\{ \int_\Omega \overset{*}{u}{}_j^{(i)}(\mathbf{x},\xi;t')b_j(\mathbf{x},\tau)d\Omega_{\mathbf{x}} + \int_\Gamma \Big[\overset{*}{u}{}_j^{(i)}(\mathbf{x},\xi;t')T_j(\mathbf{x},\tau) + $$
$$+ \overset{*}{P}{}_j^{(i)}(\mathbf{x},\xi;t')u_j(\mathbf{x},\tau) + \overset{*}{Q}{}_j^{(i)}(\mathbf{x},\xi;t')\dot{u}_j(\mathbf{x},\tau)\Big] d\Gamma_{\mathbf{x}}\Big\}\, d\tau$$
$$+ \rho\int_\Omega \Big[v_{jo}(\mathbf{x})\, \overset{*}{u}{}_j^{(i)}(\mathbf{x},\xi;t) - u_{jo}(\mathbf{x})\, \frac{\partial}{\partial t}\overset{*}{u}{}_j^{(i)}(\mathbf{x},\xi;t)\Big] d\Omega_{\mathbf{x}}, \tag{1}$$

and for compressible fluids in potential formulation

$$d(\xi)\Phi(\xi,t) = \int_0^t \Big\{ \int_\Omega \overset{*}{\Phi}(\mathbf{x},\xi;t')\ \gamma(\mathbf{x},\tau)d\Omega_{\mathbf{x}} + \int_\Gamma \Big[\overset{*}{\Phi}(\mathbf{x},\xi;t')\ \Psi(\mathbf{x},\tau) + $$
$$+ \frac{r_{,n}}{2\pi cR} H(t'-r/c)\Big(\frac{1}{ct'+r}\ \Phi(\mathbf{x},\tau) + \frac{1}{c}\ \dot{\Phi}(\mathbf{x},\tau)\Big)\Big]\ d\Gamma_{\mathbf{x}}\Big\}\ d\tau$$
$$+ \frac{1}{c^2}\int_\Omega \Big[\ \dot{\Phi}(\mathbf{x},0)\overset{*}{\Phi}(\mathbf{x},\xi;t) - \Phi(\mathbf{x},0)\frac{\partial}{\partial t}\overset{*}{\Phi}(\mathbf{x},\xi;t)\Big] d\Omega_{\mathbf{x}}. \tag{2}$$

On a smooth boundary ($\xi \in \Gamma$) we have $d_{ij} = 0.5\delta_{ij}$ and $d = 0.5$, while for interior points ($\xi \in \Omega$) $d_{ij} = \delta_{ij}$ and $d = 1.$ holds. $\overset{*(i)}{u_j}(x,\xi;t')$ and $\overset{*}{\Phi}(x,\xi;t')$ are known as "time dependent fundamental solutions of the full- space". $\overset{*(i)}{u_j}(x,\xi;t')$ gives the displacement response at a point x and time t due to a unit impulsive force at time τ in the direction x_i and located at point ξ [2,3], while $\overset{*}{\Phi}(x,\xi;t')$ is the velocity potential distribution of a pressure wave induced by a unit impulsive source at point ξ and time τ [3,5]:

$$\overset{*(i)}{u_j}(x,\xi;t') = \frac{1}{2\pi\rho}\left\{ \frac{H(t'-\frac{r}{c})}{c_1 r^2}\left[(2R_1 + \frac{r^2}{R_1})\, r_{,i}\, r_{,j} - R_1\delta_{ij}\right] - \frac{H(t'-\frac{r}{c})}{c_2 r^2}\left[(2R_2 + \frac{r^2}{R_2})\, r_{,i}\, r_{,j} - (R_2 + \frac{r^2}{R_2})\delta_{ij}\right]\right. \tag{3}$$

and

$$\Phi(x,\xi;t') = \frac{1}{2\pi c}\,\frac{1}{R}\, H(t' - \frac{r}{c}) \quad , \tag{4}$$

where, with $r = |x-\xi|$ and $t' = t-\tau$,

$$R_\alpha = (c_\alpha^2 t'^2 - r^2)^{1/2} \; ; \; \alpha = 1,2 \; ; \qquad R = (c^2 t'^2 - r^2)^{1/2} \; .$$

H is the Heaviside step-function, which guarantees the causality of the waves. The interior commas indicate space differentiation. The velocities of elastic dilatational and distortional waves are denoted by c_1 and c_2, whereas c stands for the pressure wave speed in the fluid. Also, ρ stands for the mass density and b_i for the body forces of the elastic medium, while γ and ρ_f give the source distribution and the density of the fluid. The lengthy kernels $\overset{*(i)}{P_j}$ and $\overset{*(i)}{Q_j}$ can be found in Ref. 2.

Moreover, it should be mentioned that the initial boundary value problem of wave propagation in compressible fluids is a special case of the more general wave propagation problem in elastic media. One obtains equation (2) from equation (1) by formally setting $c_1 = c_2 = c$ and considering scalar instead of vector-valued states. This holds for all analytical and numerical solution steps. Thus, for lack of space we may restrict ourselves in the following considerations to the more general case of elastic media.

The domain integrals in Equ. 1 can be transformed to boundary integrals in the case of constant body force distributions, and if constant initial velocity $v_{jo}(x)$ and zero initial displacement u_{jo} are prescribed (for additional information see Refs. 4, 5).

The time-stepping boundary element procedure

Equation (1) can be used to determine time dependent displacements in the interior of Ω as well as along the boundary Γ. However, one must first find values for the unknown boundary tractions T_j and displacements u_j. As a general analytical solution is impossible, a numerical evaluation is needed based on a discretization of both, geometry and time:

(i) the boundary is subdivided into a number k of line elements over which displacements and tractions are assumed to be constant

(ii) the whole observation period is discretized by equal time increments Δt and linear and constant shape functions are used to approximate the transient behavior of the displacements and tractions, respectively.

Collocation at each boundary node ξ_k and at all time steps $t_n = n\Delta t$ finally leads to a systems of algebraic equations (n > m)

$$[0.5\, \mathbf{I} + \overset{(1)}{\mathbf{T}}] \bullet \mathbf{u}^{(n)} + \sum_{m=1}^{n-1} \overset{(n-m+1)}{\mathbf{T}} \bullet \mathbf{u}^{(m)} = \sum_{m=1}^{n} \overset{(n-m+1)}{\mathbf{U}} \bullet \mathbf{t}^{(m)} + \mathbf{r}^{(n)}, \tag{5}$$

where $\mathbf{u}^{(m)}$ and $\mathbf{t}^{(m)}$ are vectors of all nodal displacements and tractions, respectively, at time step m. $\overset{(\mu)}{\mathbf{T}}$ and $\overset{(\mu)}{\mathbf{U}}$ represent the influence matrices whose elements are obtained by integrations over each boundary element (numerically) and over each time steps (analytically). The index $\mu = n\text{-}m$ indicates the difference between the observation time $t_n = n\Delta t$ and the impulse time $t_m = m\Delta t$. The vector $\mathbf{r}^{(n)}$ contains terms due to non-zero body forces b_j and/or non-zero initial conditions.

A system of algebraic equations as given in Equ. (5) has to be generated for each part of the domain with equal material properties, and a similar one for all fluid domains. For the dynamic interaction analysis these systems, the subdomains, have to be combined by imposing adequate interface conditions: for coupling of elastic domains [10], the compatibility of the displacements and the equilibrium of the tractions are required, while along the interface between an inviscid fluid and an deforming elastic structure tangential components of the traction are prescribed to be zero, and compatibility and equilibrium conditions are established in the normal direction [6].

The coupling of boundary and finite elements can be treated in the same way [13,14]. However, to obtain consistency between the FE and BE formulation, the boundary element interface tractions have to be transformed to resultant nodal forces. Moreover, also unilateral contact can be verified in this way by controling permanently, i.e. in each time step, the traction and the displacement compatibility along the interface [11, 12].

After the incorporation of these conditions the entire set of equations has to be reordered in accordance to the unknown and to the precribed or already determined boundary values. Finally, the resulting system of equations can be solved by a special time-stepping solution procedure. More details may be found in the publications on special interaction problems [e.g. 6, 10, 12, 13].

Numerical examples

In this section some representative examples are given in order to demonstrate the wide variety of possible applications of the time-stepping boundary element procedures.

Dam reservoir systems

(Soil/Structure - fluid interaction by time domain BEM)

In order to confirm the importance of the dynamic interaction between an elastic medium and an adjacent fluid, a dam reservoir system is considered.

First, a rigid dam of height H = 100 m storing an 'infinite' water reservoir of the same depth is examined (Figure 1), where the bottom is taken to be elastic soil. Adopting plane strain assumptions, the soil material is assumed to have a mass density ρ=2000 kg/m^3, Poisson's ratio υ=0.4 and a varying Young's modulus, while the velocity of the pressure waves in the water is 1438 m/s. Let this "dam" now be subjected to a horizontal impulsive ground acceleration $\ddot{u}_1 = q_0 H(t)[1-H(t-\Delta t)]/\rho$ with an intensity q_0=28.9 N/m^3 and a duration of one time step Δt=0.035 secs (corresponding to the boundary value ψ=28.9 m/s^3 along the rigid vertical dam face). Apparently, the consequences of the dynamic interaction on the total hydrodynamic forces F_{dyn} acting on the upstreams face of the dam are highly dependent on the soil material data (Figure 2). While in the case of a rigid bottom F_{dyn} will only be damped slowly by radiation to the 'infinite' fluid domain, the interaction with the flexible bottom medium reduces the amplitudes so fast, that F_{dyn} disappears almost immediately. The pressure waves produced by the impulsive excitation will be absorbed by the soft soil.

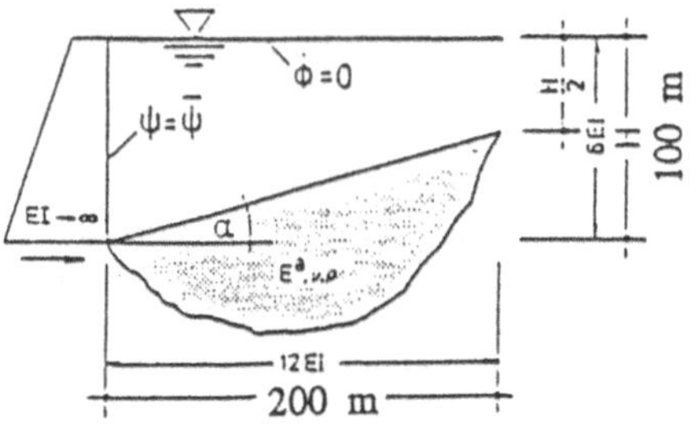

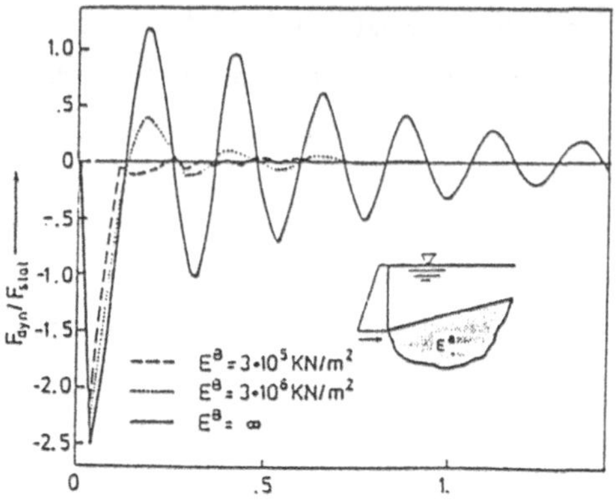

Figs.1 and 2. Geometry and discretization of an infinite reservoir and time history of F_{dyn}/F_{stat} - influence of elastic beds with a slope of $\alpha = 14°$

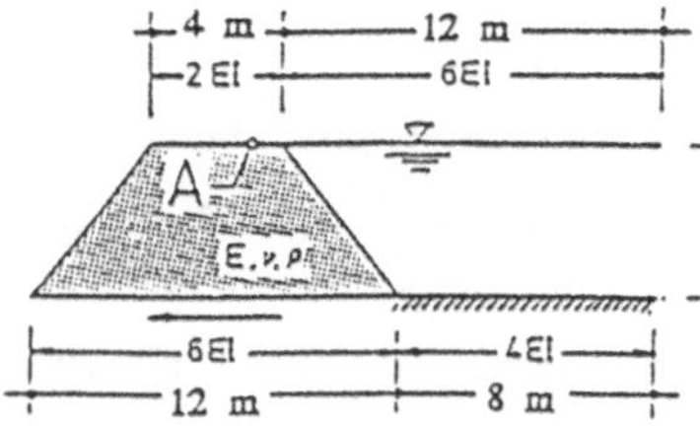

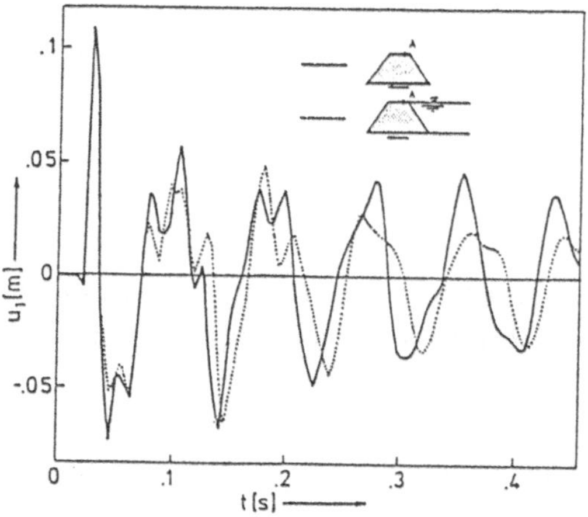

Figs. 3 and 4. Geometry and discretization of a dam - fluid system and time history of u_1 at point A due to a horizontal impulse at the dam base

Second, an elastic dam , having material parameters $E = 2 \cdot 10^5 KN/m^2$, $\upsilon = 0.25$ and $\rho = 2000\ kg/m^3$, is examined (Figure 3). It is subjected to a horizontal ground motion acting as a unit impulse of duration $\Delta t = 0.0046$ secs. Figure 4 gives the influence of structure-fluid coupling on the time history of the horizontal impulse response at point A of the dam crest.

Halfspace with a tunnel

(Soil-Structure Interaction by coupled time domain BEM-FEM)

A significant limitation of the BEM is that only linearly elastic homogeneous domains can be treated. A combination of the BEM with the FEM can eliminate this disadvantage. Thus, in order to investigate how the deformations of the surface and in the interior of a halfspace (soil) are affected by a tunnel, the tunnel and a small portion of the supporting soil is discretized by finite elements, while the remaining soil medium is represented by boundary elements (Figure 5). The halfspace (E^S=2.66•10^5 KN/m^2, υ=0.33, ρ=2000 kg/m^3) is subjected to a vertical distributed load whose time function is a rounded impulse. Directly beneath the load, an area of 10 m × 10 m, containing an rectangular hole (tunnel) of 6 m × 4 m is discretized by means of 17 square finite elements, 2 m × 2 m each. In addition, the liner of the tunnel is represented by 10 finite elements. Its Poisson's ratio and density are the same as for the halfspace, while its modulus of elasticity E^T is varied. The finite element region is coupled to 15 linear boundary elements of length 2 m, which model, together with those elements along the stress free surface, the remaining part of the soil.

Figure 6 depicts the transient vertical response at a point A located on the surface of the halfspace. Besides the displacement curves for two different liners (E^T= 6•10^6 KN/m^2 and E^T= 3•10^7 KN/m^2), the results corresponding to the halfspace without a tunnel, and to a tunnel without a liner (E^T= 0, ρ^T= 0) are plotted. Obviously, the displacements at point A are barely influenced by a rigid tunnel (E^T= 3•10^7 KN/m^2). With decreasing stiffness of the liner, however, the surface deformations increase significantly. Also, it can be seen that wave reflections at the tunnel surface affect the response at A.

The vertical displacements at the interior point C (below the tunnel) show different characteristics (Figure 7): the response is observed later with decreasing E^T, corresponding to the extra time it takes the waves to reach point C; the maximum displacement decreases with decreasing E^T.

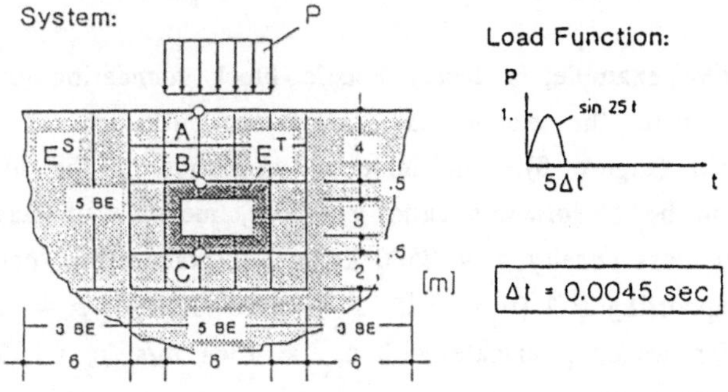

Fig.5. Geometry and discretization of the halfspace - tunnel system

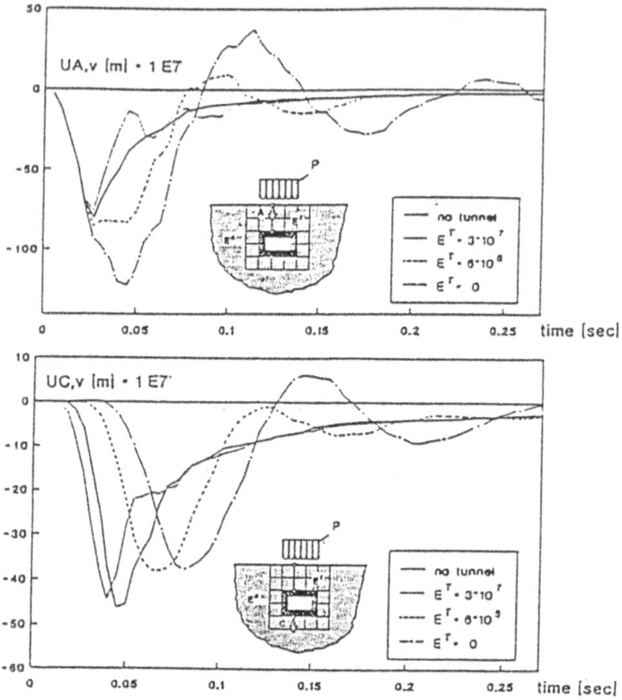

Figs.6 and 7. Vertical displacement at point A and C, respectively. Elastic halfspace with a tunnel - Influence of the liner stiffness E^T

Uplifting of heavy structures due to Rayleigh waves

(Unilateral boundary conditions by time domain BEM)

Several studies suggest that uplifting can be advantageous in reducing the dynamic responses of a structure caused by ground motions, e.g. incident waves. Since the response during the starting phase can differ considerably from that in the steady state, a procedure acting directly in the time domain - such as the proposed BEM - is particularly superior.

As a representative example, a heavy elastic block foundation in stick and unilateral contact with the elastic soil under the influence of a Rayleigh wave is considered (Figure 8). The material properties of the heavy elastic block are taken to be a Poisson's ratio $\upsilon_B = 0.2$, modulus of elasticity $E_B = 3 \cdot 10^7$ KN/m^2 and mass density $\rho_B = 2500$ kg/m^3. If plane strain conditions are assumed in the supporting soil ($\upsilon_S = 0.25$, $E_S = 3 \cdot 10^6$ KN/m^2, $\rho_S = 2000$ kg/m^3) P , S and Rayleigh waves propagate with $c_{1S} = 1341$ m/s, $c_{2S} = 774$ m/s and $c_R = 712$ m/s, respectively.

A Rayleigh wave $u_1^R = A_1 \cos \omega_R(t - x_1/c_R)$, $u_2^R = A_2 \sin \omega_R(t - x_1/c_R)$ with amplitudes $A_1 = 0.0028$ m, $A_2 = 0.0042$ m and frequency $\omega_R = 59.7$ rad/s, i.e. a wave of length $L_R = 11.9$ m is taken as actual excitation, and causes a horizontal and vertical acceleration of $\ddot{u}_1^R = 1.03$ g and $\ddot{u}_2^R = 1.51$ g, respectively. It starts acting after 0.12 secs., that is the time needed in order to reach the distribution of normal compressive stress resultants along the interface equilibrating the self-weight [11].

Figure 9 illustrates one important effect of the incident Rayleigh wave, propagating towards the positive x_1-direction, on the unilateral contact along the interface elements: The uplifting starts at the left hand side of the block and affects the adjacent elements. Then, depending on the frequency of the incident Rayleigh wave, this debonding and rebonding becomes periodic (in this case after 0.25 secs.).

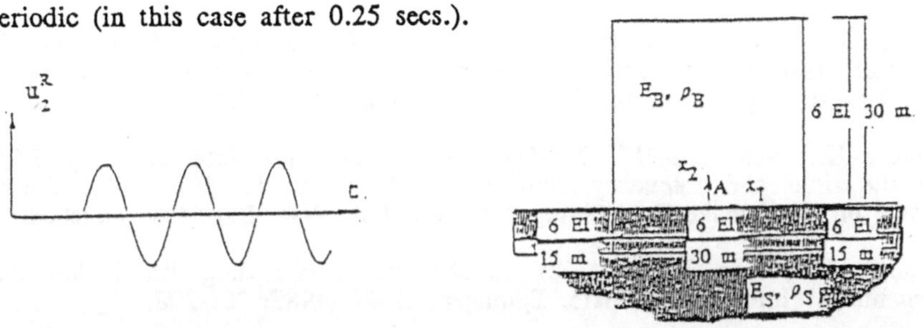

Fig.8. Geometry, discretization and "loading" of the block-halfspace system

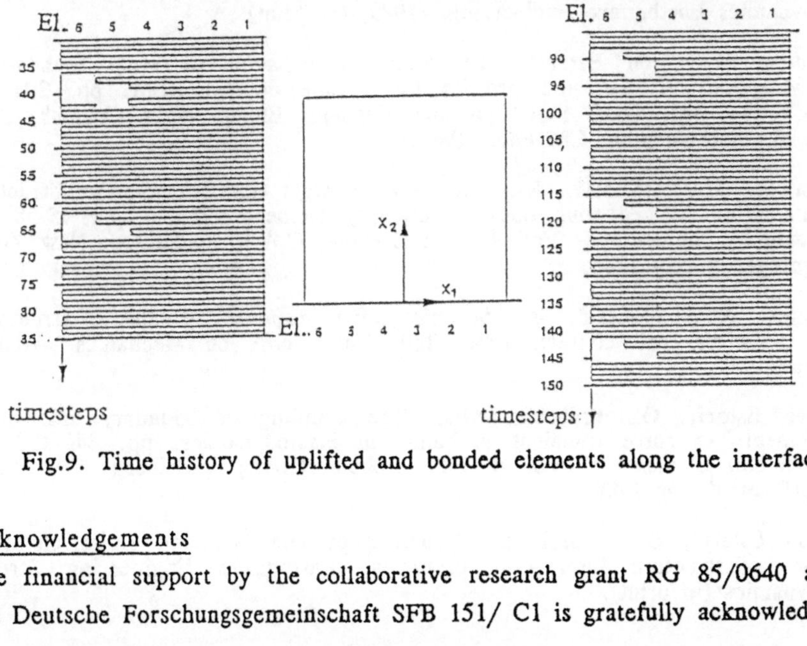

Fig.9. Time history of uplifted and bonded elements along the interface

Acknowledgements

The financial support by the collaborative research grant RG 85/0640 and by the Deutsche Forschungsgemeinschaft SFB 151/ C1 is gratefully acknowledged.

References

1. Antes, H., von Estorff, O.: Seismic response amplification due to topographic influences. Proc. 9th. World Conf. Earthqu. Engng., Tokyo / Kyoto 1988.

2. Antes, H.: A boundary element procedure for transient wave propagation in two-dimensional isotropic elastic media. Finite Elements in Analysis and Design 1 (1985) 313-322.

3. Antes, H.: Time domain boundary element solutions of hyperbolic equations for 2-D transient wave propagation. pp. 35-43 in: Notes on Numerical Fluid Mechanics 21, Braunschweig: Vieweg Verlag 1988.

4. Antes, H.: A boundary formulation for the elastodynamic self-weight domain integral. Computational Mechanics (submitted).

5. Antes, H: Eine vollständige Randintegralformulierung für den dynamischen Kontakt schwerer elastischer Strukturen. Z. Angew. Math. and Mechanik (in print).

6. Antes, H., von Estorff, O.: Analysis of absorption effects on the dynamic response of dam reservoir systems by boundary element methods. Earthquake Engineering and Structural Dynamics 15 (1987) 1023-1036.

7. Antes, H., von Estorff, O.: Dynamic soil-structure interaction by BEM in the time and frequency domain. pp. 5.5/30-40 in: Proc. 8th. Europ. Conf. on Earthquake Engineering 2 (1986), Lab. Nac. Eng. Civil, Lisbon.

8. Antes, H., von Estorff, O.: Erschütterungsausbreitung im Boden und dynamische Interaktionseffekte. Bauingenieur 62 (1987) 201-208.

9. Antes, H., von Estorff, O.: Dynamic response analysis of rigid foundations and of elastic structures by Boundary Element procedures. Soil Dynamics Earthquake Engineering. (1989) (in print).

10. Antes, H., von Estorff, O.: Transient behavior of strip foundations resting on different soil profiles by a time domain BEM. pp. 291-305 in: Proc. 3rd. Conf. Soil Dynamics Earthqu. Engng. (Ed.: Cakmak, A.S. et al.) Southampton: CM Publ. 1987.

11. Antes, H., Steinfeld, B.: Unilateral contact in dynamic soil-structure interaction by a time domain boundary element method. pp. 45-58 in: Boundary Elements X, Vol. 4 (Ed.: Brebbia, C.A.), Heidelberg, New York: Springer Verlag 1988.

12. Antes, H., Steinfeld, B.: Seismic wave response of elastic structures in unilateral soil contact. Proc. 2nd. Nat. Cong. on Mechanics, Athens, June 1989.

13. von Estorff, O., Prabucki, M.J.: The coupling of boundary and finite elements to solve transient problems in elastodynamics. pp. 447-450 in: Boundary Elements X, Vol. 4 (Ed.: Brebbia, C.A.), Heidelberg, New York: Springer Verlag 1988.

14. von Estorff, O., Kausel, E.: Coupling of boundary and finite elements for soil-structure interaction problems. Earthquake Engineering Struct. Dynamics (in print).

Co-Rotational Beam Elements for Two- and Three-Dimensional Non-Linear Analysis

M.A. CRISFIELD; Dept. of Aeronautics, Imperial College, London,
G. COLE; FEA Ltd., formerly Kingston Polytechnic, Surrey

Summary
The paper describes the development of both Kirchhoff and Timoshenko beam elements which are embedded in a continuously rotating frame. Emphasis is placed on the consistent derivation of both the internal force vector and the tangent stiffness matrix.

INTRODUCTION

Three-dimensional beam elements [1–8] and theories [9–11] have been proposed by a number of authors. In order to avoid over-stiff solutions, co-rotational formulations have been proposed [12–17]. In the present work, we use the term co-rotational in the sense of Belytschko [12,14] and use a single rotating frame coupled with conventional small-deflection, small-strain, beam-theory. Elements of this type have previously been derived by Belytschko [12,14], Rankin and Brogan [17] and Oran [16] with Belytschko using an explicit dynamic approach without a tangent stiffness matrix. In a three-dimensional context, it appears that previous work of this type has not used internal force-vectors as well as tangent stiffness matrices that are consistently derived from the "co-rotational strain measures".

Before addressing the three-dimensional problem, we will detail the two-dimensional formulation which allows the essential features of the technique to be more simply explained. The two-dimensional formulation has much in common with that of Oran [18].

TWO-DIMENSIONAL THEORY

Fig. 1 shows the beam element in a three-dimensional framework, but we will use the same notation for a two-dimensional formulation in which the element can only move within the x–y or 1–2 frame.

G. Kuhn · H. Mang (Eds.)
Discretization Methods in Structural Mechanis
IUTAM/IACM Symposium Vienna, Austria, 1989

Co-rotational "displacements" and local stress-resultants

The local axial displacement is u_ℓ which is simply the difference between the new length, ℓ_n and the old length, ℓ_o but is best computed from [12,13]:

$$u_\ell = \frac{2}{(\ell_n + \ell_o)}\left(x_{21} + \frac{1}{2}d_{21}\right)^T d_{21} \tag{1}$$

where the notation is illustrated in Fig. 1 and, for example, $x_{21} = x_2 - x_1$. The equivalent co-rotational rotations (Fig. 1) are given by:

$$\theta_{\ell 2} = -\sin^{-1} t_2^T e_1; \quad \theta_{\ell 5} = -\sin^{-1} q_2^T e_1 \tag{2}$$

where the unit vector t_2 is computed from the global rotation, α_3 at node 1 (Fig. 1) via:

$$t_2^T = \{-\sin(\alpha_3 + \alpha_{3o}), \cos(\alpha_3 + \alpha_{3o})\} \tag{3}$$

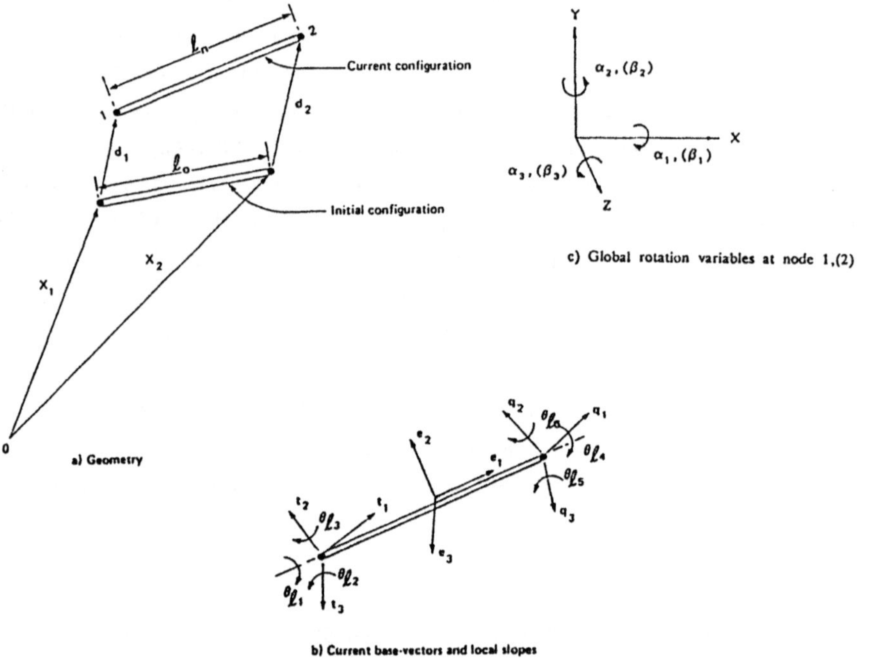

Fig.1 The Element

where α_{3o} is the initial value of α_3. A similar procedure, involving q_2 and β, is used for node 2.

In order to define the local slopes, $\theta_{\ell 2}$ and $\theta_{\ell 5}$, we also require the "base-plane" unit-vector, c_l which is given by:

$$c_l = (x_{21} + d_{21})/\ell_n \tag{4}$$

The local in-plane stress resultant is given by:

$$N = EA\epsilon_\ell = \frac{EA}{\ell_o}\left[u_\ell + \text{fn.}(\theta_{\ell 2}, \theta_{\ell 5})\right] \tag{5}$$

where the second "fn." term allows the introduction of "Green strain terms". These terms will be omitted from the present developments but have been considered in [19]. For the Kirchhoff element, the local bending moments are given by:

$$\begin{bmatrix} M_{\ell 2} \\ M_{\ell 5} \end{bmatrix} = \frac{EI}{\ell_o}\begin{bmatrix} 4 & 2 \\ 2 & 4 \end{bmatrix}\begin{bmatrix} \theta_{\ell 2} \\ \theta_{\ell 5} \end{bmatrix} \tag{6}$$

while for the Timoshenko beam-element, we have:

$$Q_\ell = EA\gamma = EA(\theta_{\ell 2} + \theta_{\ell 5})/2 \tag{7}$$

$$M_\ell = EI\aleph = \frac{EI}{\ell_o}(\theta_{\ell 5} - \theta_{\ell 2}) \tag{8}$$

Collectively, the local stress-resultants are called q_ℓ. For the thin-beam element q_ℓ contains (N, $M_{\ell 2}$, $M_{\ell 5}$) while for the Timoshenko-beam element it contains (N, Q_ℓ, M_ℓ).

Virtual work

In order to compute the global internal forces, q_i, we apply virtual work so that:

$$q_i^T \delta p = q_\ell^T \delta p_\ell \tag{9}$$

where: $$\delta p^T = (\delta u_1, \delta v_1, \delta\alpha_3, \delta u_2, \delta v_2, \delta\beta_3) \tag{10}$$

are the "global" virtual displacements and δp_ℓ are the equivalent local (co–rotational) "displacements" given by:

$$\delta p_\ell = (\delta u, \delta\theta_2, \delta\theta_5)_\ell \tag{11}$$

The constituent terms in (11) are obtained by differentiating (1) and (2) so that:

$$\delta u_\ell = c_1{}^T \delta d_{21} \tag{12}$$

$$\cos\theta_{\ell 2}\delta\theta_{\ell 2} = c_1{}^T t_1 \delta\alpha_1 - t_2{}^T \Lambda \delta d_{21} \tag{13}$$

with an equivalent expression for $\delta\theta_{\ell 5}$. From (4), the Λ matrix in (14) is given by:

$$\delta c_1 = \Lambda \delta d_{21} = \frac{1}{\ell_n}\left[I - c_1 c_1{}^T\right]\delta d_{21} \tag{14}$$

Collectively, equations (13) and (14) and the equivalent for $\delta\theta_{\ell 5}$ can be expressed as:

$$\delta p_\ell = \begin{bmatrix} \delta u_\ell \\ \delta\theta_{\ell 2} \\ \delta\theta_{\ell 5} \end{bmatrix} = \begin{bmatrix} g^T \\ f_2{}^T \\ f_5{}^T \end{bmatrix} = B^T \delta p \tag{15}$$

where: $$g^T = \{-c_1{}^T, 0, c_1{}^T, 0\} \tag{16}$$

$$f_2{}^T = \{-t_2{}^T \Lambda, -c_1{}^T t_1, t_2{}^T \Lambda, 0\} \tag{17}$$

From (9) and (15), the "global" internal forces are given by:

$$q_i = B q_\ell \tag{18}$$

The tangent stiffness matrix

Differentiation of (18) leads to:

$$\delta q_i = K_t \delta p = B \delta q_\ell + \delta B q_\ell = B D B^T \delta p + K_g \delta p \tag{19}$$

where D relates the "local forces", q_ℓ to the local "displacements", p_ℓ so that:

$$\delta q_\ell = D \delta p_\ell = \frac{1}{\ell_o}\begin{bmatrix} EA & 0 & 0 \\ 0 & 4EI & 2EI \\ 0 & 2EI & 4EI \end{bmatrix} \delta p_\ell \quad \text{or} = \frac{1}{\ell_o}\begin{bmatrix} EA & 0 & 0 \\ 0 & GA\ell_o/2 & GA\ell_o/2 \\ 0 & -EI & EI \end{bmatrix} \delta p_\ell \tag{20}$$

where the first D matrix relates to the Kirchhoff element and the second to the Timoshenko element.

The contributions to the geometric stiffness, K_g in (19) are best obtained by differentiating individual columns of B from (15–17) so that:

$$\delta B_1 q_\ell(1) = N \delta g = N \begin{bmatrix} A & 0 & -A & 0 \\ 0^T & 0 & 0^T & 0 \\ -A & 0 & A & 0 \\ 0^T & 0 & 0^T & 0 \end{bmatrix} \delta p \tag{21}$$

where δB_1 is the first column of B. From the second column, f_2 (see (15) and (17)) one obtains:

$$\delta B_2 q_\ell(2) = q_\ell(2) \delta f_2 = \frac{-q_\ell(2)}{\cos\theta_{\ell 2}} \begin{bmatrix} -V & At_1 & V & 0 \\ t_1^T A & -c_1^T t_2 & -t_1^T A & 0 \\ V & -At_1 & -V & 0 \\ 0^T & 0 & 0^T & 0 \end{bmatrix} \delta p \tag{22}$$

where: $$V = (At_2c_1^T + c_1t_2^T A + c_1^T t_2 A)/\ell_n \tag{23}$$

The third column of B is differentiated in a similar manner while the remaining part of K_g is given by:

$$K_g = K_g + q_\ell(2)\tan\theta_{\ell 2} f_2 f_2^T + q_\ell(3)\tan\theta_{\ell 5} f_5 f_5^T \tag{24}$$

THREE DIMENSIONAL ELEMENTS

The main difficulty with three–dimensional analyses relates to the non–vectorial nature of rotations [9] so that one cannot simply update α and β (Fig. 1) using, say, $\alpha = \alpha + \delta\alpha$. Instead, one must update the rotation triad

$T(\alpha) = [t_1,t_2,t_3]$ which is a function of the pseudo-vector, α and is given by the Rodriquez formula:

$$T = \left[I + \frac{\sin\alpha}{\alpha}S(\alpha) + \frac{(1-\cos\alpha)}{\alpha^2}S(\alpha)^2\right] \tag{25}$$

where the skew-symmetric matrix, $S(\alpha)$, is given by:

$$S(\alpha) = \begin{bmatrix} 0 & -\alpha_3 & \alpha_2 \\ \alpha_3 & 0 & -\alpha_1 \\ -\alpha_2 & \alpha_1 & 0 \end{bmatrix} \tag{26}$$

The rotation matrix, T, maps a vector x_o to x_1 via:

$$x_1 = T(\alpha)x_o \tag{27}$$

For compound rotations, if $x_1 = T_1(\alpha_1)x_o$ is followed by $x_2 = T_2(\alpha_2)x_1$,

$$x_2 = T_2(\alpha_2)T_1(\alpha_1)x_o = T_{12}(\alpha_{12})x_o \tag{28}$$

with $\alpha_{12} \neq \alpha_1 + \alpha_2$. More effective methods of updating the rotations are available and, in practice, the authors use Euler parameters which are unit quarternions [19]. The small rotation version of (25) is $[I + S(\alpha)]$. Hence, with the aid of (28),

$$\delta x = S(\delta\alpha)x = -S(x)\delta\alpha \tag{29}$$

This relationship will be repeatedly used in the subsequent developments.

Co-rotational "displacements"

The local axial displacement, u_ℓ, is computed using the same formula (1) as in two-dimensions. While, in place of (2) for example, we have:

$$2\sin\theta_{\ell 2} = -t_2^T e_1 + e_2^T t_1 \tag{30}$$

In contrast to (2), (30) contains the base vector e_2. In order to compute this vector (and e_3), we firstly compute a triad $R(r_1,r_2,r_3)$ which is obtained by rotating the left-hand nodal triad $T(t_1,t_2,t_3) = T(\alpha)$ (see Fig. 1) through half the angle between the right hand triad $Q(\beta)$ and this left hand triad. Hence:

$$R = \Delta R(\frac{\beta - \alpha}{2})T \tag{31}$$

where: $$\Delta R(\beta - \alpha) = QT^T \tag{32}$$

Computational details are given in [13].

Having obtained the R matrix, the unit vector, r_1, will generally not coincide with c_1. Hence the R matrix must be rotated through an angle γ given by:

$$\gamma = \cos^{-1}(r_1^T c_1)\frac{r_1 \times c_1}{|r_1 \times c_1|} \tag{33}$$

so that, for example:

$$c_2 = \Delta R(\gamma) r_2 \tag{34}$$

With a view to the subsequent differentiation for the virtual work, Crisfield [13] has proposed the following approximation to (33)–(34) whereby:

$$c_2 = r_2 - \frac{r_2^T c_1}{2}(c_1 + r_1) \tag{35}$$

Virtual work

The principle of virtual work is applied in a very similar manner to that previously discussed for the two–dimensional case. In particular, equations (9), (15) and (18) still apply although with a different B matrix. In addition, (12) applies as before for the variation of the axial co–rotational displacement, u_ℓ, while in place of (13), we have:

$$2\cos\theta_{\ell 2}\delta\theta_{\ell 2} = -\ \delta t_2^T c_1 - t_2^T \delta c_1 + \delta c_2^T t_1 + c_2^T \delta t_1 \tag{36}$$

The variation of c_1 has already been expressed in (14) while using (29),

$$\delta t_1 = S(\delta\alpha)t_1 = -S(t_1)\delta\alpha \tag{37}$$

Differentiation of (36) follows in a similar fashion but requires δr_2 which is approximated as:

$$\delta r_2 = S(\frac{\delta\alpha + \delta\beta}{2})r_2 = -S(r_2)\frac{\delta\alpha + \delta\beta}{2} \tag{38}$$

A more exact derivation [20] shows that the error is small.

Using equations of this type similar internal force vectors to (18) can be derived [13]. In addition the step to produce consistent tangent matrices is similar to that given in (19) although using expression of the form (37)–(39). Full details are given in [13]. In line with Simo et al [7,8], the tangent stiffness matrix is found to be non–symmetric away from equilibrium, but in the absence of applied moments and non–conservative loadings, is symmetric at equilibrium. However, numerical experiments have indicated that, in these circumstances, the device of artificially symmetrising the matrix leads to an almost inconsequential deterioration in the iterative performance.

Example: Forty–five degree bend

This example [2] involves a genuinely three–dimensional response and has been used by a number of authors [2,3,4,8]. Fig. 2 shows the initial geometry and the response under increasing load. The bend is of unit cross–section and E was taken as 10^7. It was modelled using eight elements (as in [8]). The computed response has been shown [13] to be in close agreement with the responses quoted by other authors [2,3,4,8]. Table 1 compares the incremental/iterative performances. Such a comparison cannot be definitive because of different convergence criteria but, nonetheless, gives an overall trend. For the present solutions, convergence was assumed once the residual norm was less than 0.1 % of the applied force norm. It will be noticed that the introduction of an artificially symmetrised tangent stiffness matrix (Table 1) has had almost no effect on the convergence characteristics.

CONCLUSIONS AND FUTURE DEVELOPMENTS

It has been shown that the proposed co–rotational procedure can be applied to both Kirchhoff and Timoshenko beam elements. The method adopts small–deflection, small–strain, beam theory but embeds it within a single continuously rotating frame. In contrast to much previous work in this area, both the internal force vector and tangent stiffness matrix have been consistently derived from the "co–rotational strain measures". This has been shown to lead to very efficient elements. Further work [19] describes the the addition of "Green–strain terms" and includes a more extensive range of numerical tests.

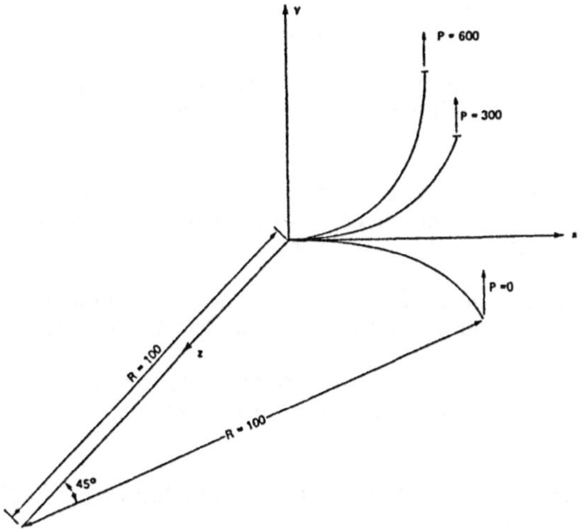

Fig.2 Initial and deformed geometries for 45 degree bend

Table 1: Incremental/iterative performance for 45^0 bend (No. of iterations/increment)

	Load level		
	300	450	600
Kirchhoff non–symmetric	8	5	3
Kirchhoff symmetric	9	5	3
Timoshenko symmetric	12	8	6
Simo & Vu–Quoc[13]	13	8	6
Bathe & Bolourchi[2]		60 equal increments	
Cardona & Geradin[4]		6 equal incs., average 7.8 its/inc.	

REFERENCES:

1. Argyris, J.H., Balmer, H., Doltsinis, J. St., Dunne, P.C., Haase, M., Klieber, M., Malejannakis, G.A., Mlejenek, J.P., Muller, M. & Scharpf, D.W., Finite element method – the natural approach, Comp. Meth. in Appl. Mech. & Engng., 17/18, 1979, 1–106

2. Bathe, K–J. and Bolourchi, S., Large displacement analysis of three–dimensional beam structures, Int. J. for Num. Meth. in Engng., 14, 1979, 961–986.

3. Cardona, A. and Geradin, M., A beam finite element non–linear theory with finite rotations, Int. J. for Num. Meth. in Engng., 26, 1988, 2403–2438.

4. Dvorkin, E.N., Onate, E. & Oliver, J., On a nonlinear formulation for curved

Timoshenko beam elements considering large displacement/ rotation increments, Int. J. for NUm. Meth. in Engng., 26, 1988, pp.1597–1613.

5. Besseling, J.F., Large–rotations in problems of structural mechanics, Finite Element Methods for Nonlinear Problems, Europe– US Symposium, Trondheim, Norway, ed. Bergan et al., Springer, 1986.

6. Shi, G. & Atluri, S.N., Elasto–plastic large deformation analysis of space–frames: a plastic–hinge and stress–based explicit derivation of tangent stiffnesses, Int. J. for Num. Meth. in Engng., 26, 1988, pp.589–615.

7. Simo, J.C., A finite strain beam formulation. The three–dimensional dynamic problem, Part 1, Comp. Meth. in Appl. Mech. & Engng., 49, 1985, pp.55–70.

8. Simo,J.C. & Vu–Quoc, L., A three–dimensional finite strain rod model: Part 2: Computational aspects, Comp. Meth. in Appl. Mech. & Engng., 58, 1986, pp.79–116

9. Argyris, J., An excursion into large rotations, Comp. Meth. in Appl. Mech. & Engng., 32, 1982, pp.85–155.

10. Rankin, C.C. and Nour–Omid, B., The use of projectors to improve finite element performance, Symp. on Advances and Trends in Computational Struct. Mechanics and Fluid Dynamics, Washington, Oct. 1988.

11. Reissner, E., On one–dimensional, large–displacement, finite–strain beam theory, Stud. Appl. Math., 52, 1973, pp.87–95.

12. Belytschko, T. & Schwer, L., Large displacement, transient analysis of space frames, Int. J. for Num. Meth. in Engng., 11, 1977, pp. 65–84.

13. Crisfield, M.A., A consistent co–rotational formulation for nonlinear three–dimensional beam elements, submitted to Computer Methods in Applied Mechanics & Engng., 1989.

14. Belytschko, T. & Hseih, J., Non–linear transient finite element analysis with convected co–ordinates, Int. J. for Num. Meth. in Engng., 7, 1973, pp.255–271.

15. Oran, C. & Kassimali, A., Large deformations of framed structures under static and dynamic loads, Computers & Structures, 6, 1976, pp. 539–547.

16. Oran, C., Tangent stiffness in space frames, ASCE, J. of Engng. Mech. Div., 99, ST6, 1973, pp. 987–1001.

17. Rankin, C.C, & Brogan, F.A., An element independent corotational procedure for the treatment of large rotations, in: Collapse analysis of Structures, ed. L/H. Sobel & K. Thomas, New York, ASME, 1984, pp.85–100.

18. Oran, C., Tangent stiffness in plane frames, ASCE, J. of Struct. Div., 99, ST6, 1973, pp.973–985.

19. Cole, G., Co–rotational beam elements consistently formulated for geometrically nonlinear problems involving large rotations, Ph. D. Thesis, to be submitted, Kingston Polytechnic, 1990.

20. Rankin, C.C., Private communication, 1988.

A High-Precision Adaptive Procedure for Solving Kirchhoff Plates

M. Kjellman, and A. Samuelsson

Department of Structural Mechanics
Chalmers University of Technology
412 96 Gothenburg, Sweden

Summary

A procedure for numerical, high precision finite element calculation of stresses in Kirchhoff plates is proposed. Constant curvature triangles in mixed form are used. They represent a minimum of complementary energy for concentrated loads at the vertices. Moment isolines have been used for steering of a remeshing design code based on uniform distribution of a priori error estimates. The successive meshes are smoothed by an r-method and finally locally enriched guided by local a posteriori error estimates.

Introduction

Structural engineers have traditionally calculated the elastic stresses at bending of thin plates such as floor slabs according to the Kirchhoff theory. This theory was originally (1809) developed by Sophie Germain and Lagrange as an extension of the Euler-Bernoulli beam theory to two dimensions. It got its final form including correct boundary formulations by Kirchhoff in 1850.

The Kirchhoff theory can be regarded as an approximation of 3D-linear elasticity that is valid for a thin plate when the thickness $t \to 0$, see e.g. Morgenstern [1]. The well-known approximation implies bending and twisting deformations but disregards shear deformations through the thickness. Closed solutions and effective series solutions have been developed for simple forms and boundary conditions as demonstrated in the famous book by Timoshenko [2].

Babuska and Scaloppa [3] have in recent paper reported calculation of elastic energy of rhombic simply supported plates with

G. Kuhn · H. Mang (Eds.)
Discretization Methods in Structural Mechanis
IUTAM/IACM Symposium Vienna, Austria, 1989

uniform load both according to 3D-linear elasticity and according to the Kirchhoff theory. They have in both cases used very accurate h-p finite element methods.

At the boundaries they assumed in the 3D-calculation the displacements normal to the plate surface to be constrained over the thickness and the displacements along the boundaries to be free (weak support) or constrained (hard support). For $t/L = 0.01$ and $\nu = 0$ they found for the weak support the difference in energy to be 0.12 of the Kirchhoff solution for the square plate and 0.23 for the 30° rhombic plate. For hard support the corresponding figure for the square plate was 0.02 so the Kirchhoff approximation can be used for right-angled thin plates with hard support.

Babuska and Scaloppi have tested three different potential energy finite elements for triangles. They found the element with the highest degree, the complete fifth degree polynomial, to give the best accuracy for the same computational cost. In this paper we report testing of two mixed (complementary energy) finite elements, see below.

A more accurate approximation to the plate problem is obtained by taking the shear deformations into account. This was proposed by Reissner [4] and Mindlin [5]. Babuska and Scaloppi and Zienkiewicz et al. [6] show that this improved approximation gives much better results than the Kirchhoff one for the soft support case and so even for skew plates. However, skew plates with $t/L \geq 0.1$ can not be solved accurately even with this approximation. 3D-analysis must be used.

Finite element approximation of the Reissner-Mindlin theory has been a not so easy task. This problem appears when the theory is applied to thin plates (say $t/L = 0.01$). Without a careful choice of the approximation (e.g. in normal displacement w and rotation θ_x, θ_y around the midplane xy) the solution for thin plates "locks" or becomes unstable. Arnold and Falk [7] and Zienkiewicz et al. [8] have investigated this problem and designed new Reissner-Mindlin elements that are valid also for small t/L.

Two mixed (complementary energy) elements for Kirchhoff plates

Two mixed triangular elements were designed in the late 60th. One, which was designed independently by L. Herrman [9] and Hellan [10], has linear normal displacement w and constant moment components M_x, M_y, M_{xy}. The 6 degrees of freedom are chosen as three corner displacements and three normal bending moments along the sides. The other which was designed by Visser [11], has quadratic w and linear M_x, M_y, M_{xy}. The 15 degrees of freedom are chosen as three corner and three midside displacements and three moment components at each corner in the direction of the sides.

The two elements were analysed by Johnson [12] for convergence and stability. Their low degree make them suitable for an adaptive, highly graded procedure. For load only at the nodes the two elements correspond to a minimum of complementary energy.

Error indicators

We will consider and use two types of error indicators. One is based on an a priori asymptotic error estimate and the other is based on an a posteriori estimate of the residual type.

The a priori estimates given by Johnson [12] for the two finite element approximations over a domain divided into triangles are

$$\| \tilde{M}(w) - \tilde{M}^h \| \leq C(\max_K h_K^m)\, |w|_{m+3,K}, \quad m = 0,1 \tag{1}$$

where $\tilde{M}$ is the vector of moment components calculated from the exact solution w, $\tilde{M}^h$ the vector of approximate moment components and h_K the diameter of the triangle K. Further $\|\cdot\|$ is the complementary energy and $|w|_{n,K}$ is the nth seminorm over the domain. In a recent paper by Eriksson and Johnson [13] an a priori estimate in maxnorm for Poisson's equation approximated over a domain by constant strain triangles is given as

$$\| \nabla(u-u^h) \|_{L\infty} \leq C \max_K [h_K \| D^2u \|_{L\infty(K)}] \tag{2}$$

Without proof we assume that a similar estimate is valid for our constant curvature triangles, that is

$$\| \tilde{M}(w) - \tilde{M}^h \|_{L\infty} \leq C \max_K [h_K \| D^3 w \|_{L\infty(K)}] \tag{3}$$

An optimal mesh has the same error in all elements. This means that

$$Ch_k \| DM_0 \|_{L\infty,K} = \underset{K}{\text{const}} \leq \delta \qquad (4)$$

where we have replaced D^3w with DM_0, where

$$M_0 = (M_x{}^2 + M_y{}^2 + M_{xy}{}^2)^{1/2} \qquad (5)$$

and $\| DM_0 \|_{L\infty}$ for a triangle K is taken as

$$\max \left| \frac{M_0{}^i - M_0{}^K}{H_i} \right|, \quad i = 1,2,3 \qquad (6)$$

for the three triangles incident to K with H_i being the components of the distances between the center of K and the centers of the other triangles. Further δ in (4) is chosen as a low number.

The criterium (4) is used for the remeshing procedure, see below. For the refinement an a posteriori estimate deviced by Rank and Zienkiewicz [14] is used.

We calculate the average of each moment component in triangles having a common node and regard this average as an approximation of the component at that node. By interpolating all such node values over the triangles an improved approximation is obtained. The complementary energy norm of the difference between the first and the improved approximation over a triangle is an estimator e_k, k = x, y, xy of the local error in that moment component. We then use the scalar

$$e = (e_x{}^2 + e_y{}^2 + e_{xy}{}^2)^{1/2}$$

as an indicator for judging where a refinement is needed.

Adaptive remeshing and refinement

For the adaptive remeshing a code developed by Jin and Wiberg [15] has been used. It is based on the isolines of M_0 from a first trial triangulation. From an estimated maximal error given δ the number of isolines is determined. The distances between the isolines are then measures of the sizes of the triangles that make (4) reasonably satisfied.

The code is designed according to an advancing front technique, see e.g. Zhu and Zienkiewicz [16] that gives triangles with reasonably equal side lengths. Unfortunately, the differences between the side lengths sometimes become too large. For that reason a smoothening technique according to the spring method, see Löhner et al. [17] is added to the code.

Finally, when the error is small and almost equally distributed except for some small regions an automatic refinement procedure is used based on the indicator (7).

Examples

The proposed procedure is tasted on a rectangular plate with two opposite sides clamped and the other sides simply supported. The plate is loaded by a concentrated load, see Fig. 1. The plate has been calculated with Fourier series by Åkesson [18]. The constant curvature triangles are used. In Fig. 1 the moments in point 1 and 2 are given with the series solution, with a start net, with the net after 3 remeshings and, finally, after three refinement procedures. The moments under the load (point 2) tend smoothly to the series solution while the moments (point 1) at the boundary show a small perturbation. It can also be seen that the ratio between M_x and M_y here is not exactly $\nu = 0.16$ as it should be, either in the series or the finite element solution. The explanation is probably that two different boundary conditions meet here. Koiler and Alblas [19] found a perturbation in the exact solution in a similar case.

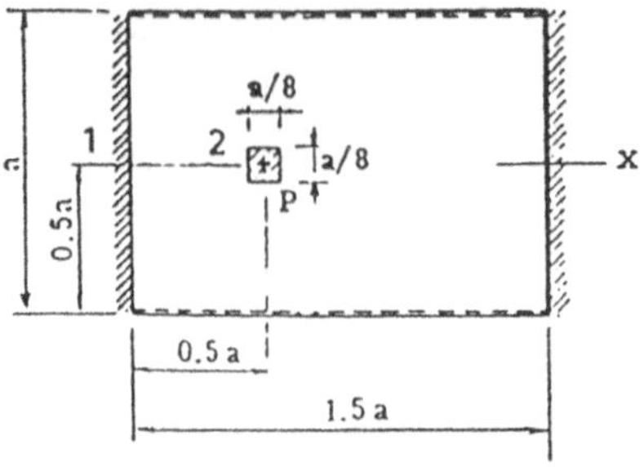

	M_x^*	M_y^*
1	-0.209	-0.034
2	0.212	0.215

Series solution (Åkesson)

a = 1, t = 1/8
ν = 0.16
P = 1

Fig. 1 Thin plate. Exact moments at 1 and 2

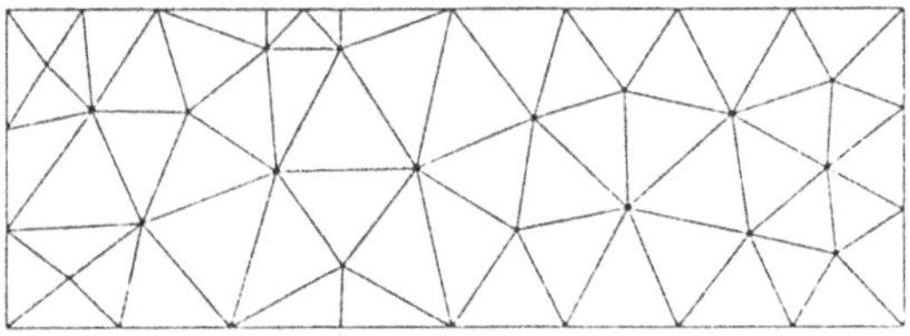

	M_x	M_y
1	-0.162	-0.004
2	0.166	0.190

(a) Start net

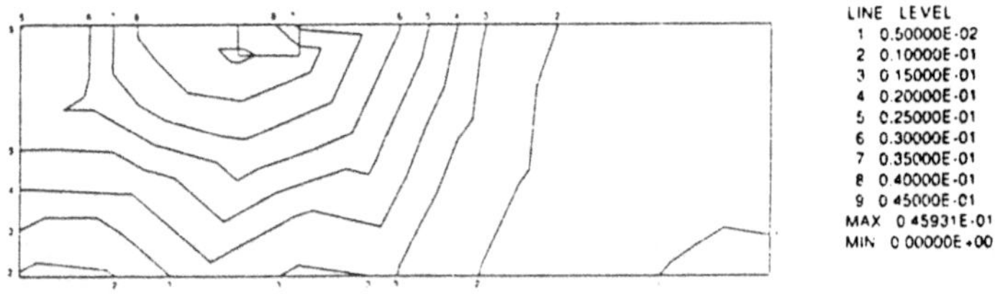

(b) Error indicators net (a)

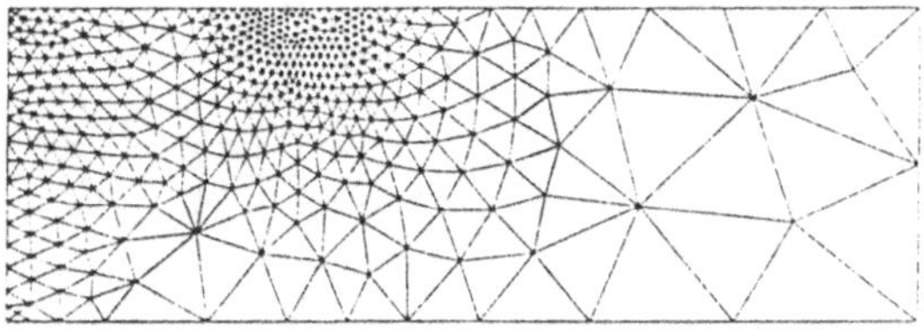

	M_x	M_y
1	-0.208	-0.031
2	0.211	0.216

(c) Net after 3 remeshings

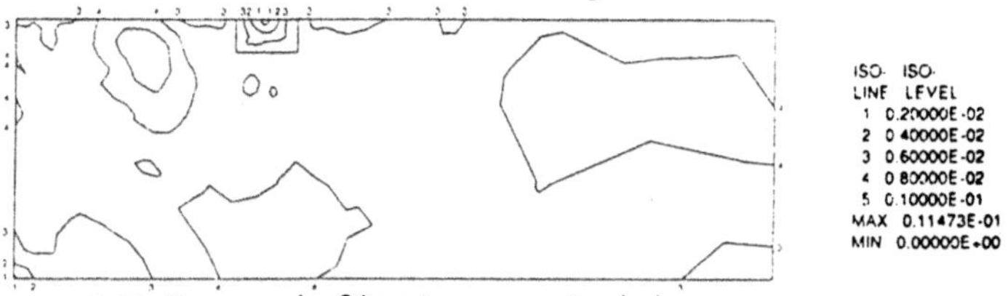

(d) Error indicators net (c)

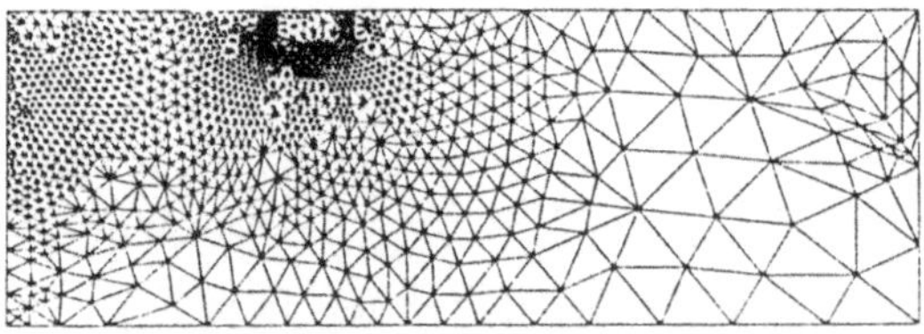

	M_x	M_y
1	-0.202	-0.030
2	0.212	0.215

(e) Net after 3 remeshings and 3 refinements

Fig. 2 Numerical solutions

Conclusions

The method proposed seems to work reasonably. In order to be able to calculate skew plates the procedure should be supplemented by a way to take shear deformations into account where these are important.

References

1. Morgenstern, D: Herleitung der Plattentheorie aus der dreidimensionalen Elasticitätsteori, Arc. Rat. Mech. Anal. 4 (1959) 145-152.

2. Timoshenko, S.: Theory of Plates and Shells, Mc Graw Hill, New York 1940.

3. Babuska, I. and Scaloppa, T.: Benchmark Computation and Performance Evaluation for a Rhombic Plate Bending Problem, Int. J. Numerical Meth. in Eng. 28 (1989) 155-179.

4. Reissner, E.: The effect of transverse shear deformation on the bending of elastic plates, J. Appl. Mech. 12 (1945) 69-76.

5. Mindlin, R.D.: Influence of rotatory intertia and shear in fluxural motions of isotropic elastic plates. J. Appl. Mech. 18 (1951) 31-38.

6. Zienkiewicz, O.C.; Taylor; R.L.; Papadopolous, P.: Oñate, E.: Plate bending elements with discrete constraints: new triangular elements. Int. J. Num. Meth. Eng. 1989.

7. Arnold, D.N.; Falk, R.S.: A uniformly accurate finite element method for Mindlin-Reissner plate. Num. Math. 1989.

8. Zienkiewicz, O.C.; Zhu, J.Z.: The Essential Ingredients of Engineering FEM Analysis. Benchmark 1989, p. 9.

9. Herrmann, L.R.: Finite-element bending analysis for plates. J. Eng. Mech. Div. ASCE, V. 93, EM5 (1967) 13-26.

10. Hellan, K.: Analysis of elastic plates in flexure by a simplified finite element method. Acta Polytechnica Scandinavica, CI46, Trondheim 1967.

11. Visser, W.: A Refined Mixed-Type Plate Bending Element. AIAA Journal, 7, (1969) 1801-1804.

12. Johnson, C.: On the convergence of some mixed element methods in plate bending problems, Num. Meth. 21, 1973.

13. Eriksson, K. and Johnsson, C.: An adaptive finite element method for linear elliptic problems. Math. of Comp. 50 (1988) 361-383.

14. Rank, E; and Zienkiewicz, O.C.: A simple error estimator in the finite element method. Comm. Appl. Num. Meth. 5, 1988.

15. Jin, H.; Wiberg, N-E.: Two-dimensional mesh generation, adaptive remeshing and refinement, under publication.

16. Zienkiewicz, O.C. and Zhu, J.Z.: A simple error estimator and adaptive procedure for practical engineering analysis, Int. J. Num. Eng. 24 (1987) 337-357

17. Löhner, R.; Morgan, K.; Zienkiewicz, O.C.: Adaptive grid refinement in finite element computations, in Accuracy Estimates, Wiley (1986) 281-297.

18. Åkesson, B.Å.: Rationalization of Lévy's Plate Solution. Trans. Chalmers Univ. of Technology 252, Göteborg 1961.

19. Koiter, W.T. and Alblas, J.B.: On the Bending of Cantilever, Rectangular Plates, Proc. k. Ned. Ak. Wet. (B) 1954.

Natural Finite Differences in Shell Modelling for Impact Analysis

J. Argyris, H. Balmer, I.St. Doltsinis

Institute for Computer Applications
University of Stuttgart

1. Introduction

This article presents an outline of a simple shell model which is applicable to the elasto-plastic large deflection analysis of thin shells under impact. The shell model evolves naturally out of an original lumped beam model [1], the main feature of which is that the bending deformation is described by (finite) folding angles between adjacent elements. These angles are expressed in terms of displacement degrees of freedom; rotational degrees of freedom do not appear explicitly. This is in contrast to a distributed deformability in bending in a finite element approximation. The present shell model may be considered as a consistent generalization of the original one-dimensional beam model. In Section 2, the general dynamic scheme of calculation which provides the background for the extension to the shell model is demonstrated on the beam model. In Section 3, a description of the shell model is given together with a discussion of the underlying natural finite difference theory. The accuracy and reliability of the shell model is tested in Section 4 by comparing numerical predictions with experiments conducted with a cylindrical panel exposed to the detonation of a sheet explosive. For brevity, the particular handling of inelastic effects by the shell model, as well as the numerical integration of the differential equations of motion, are not discussed in this article (for details see [2]).

2. Basic beam model

The (plane) beam structure is represented by a chain of hinges or joints connected by links which are rigid in bending and act merely as flanges (Fig. 1). The extensional deformation is assigned to the links while the bending deformation is concentrated at the joints and is defined by the change of the angle between adjacent links. These angular changes can be related directly to the second-order difference quotients of the displacements.

In dynamic applications, discrete masses are attached to the joints, and the connecting links are considered massless. The motion of the structure is defined in terms of the trajectories of the individual mass points, starting from a given configuration at time t_0. An explicit time integration of the equations of motion is used and is based on a simple difference scheme.

The dynamic calculation scheme proceeds as follows:

(1) Given, for example, the initial velocities of the mass points, calculate the displacement increments at time $t_1 = t_0 + \tau$, where τ is the constant time increment.

G. Kuhn · H. Mang (Eds.)
Discretization Methods in Structural Mechanis
IUTAM/IACM Symposium Vienna, Austria, 1989

(2) Knowing the displacement increments at time t_1, calculate the incremental changes of link length and angle which jointly define the state of strain increment.

(3) Use an appropriate constitutive relation to determine the stress increments and by means of addition update the already existent stresses.

(4) Calculate the stress resultants, i.e. the axial forces and bending moments.

(5) Deduce the shear forces in the links from the equilibrium of the massless links.

(6) Use the known axial forces, shear forces and, if they are present, the externally applied loads to determine the accelerations of the mass points. If these accelerations are written in difference form, they will in turn furnish the displacement increments at time t_2.

The current time cycle has in this way been terminated and the process is now repeated starting from step (2). Note that the displacements (coordinates) themselves do not enter directly into the calculation. They are obtained as by-products by summation of the increments.

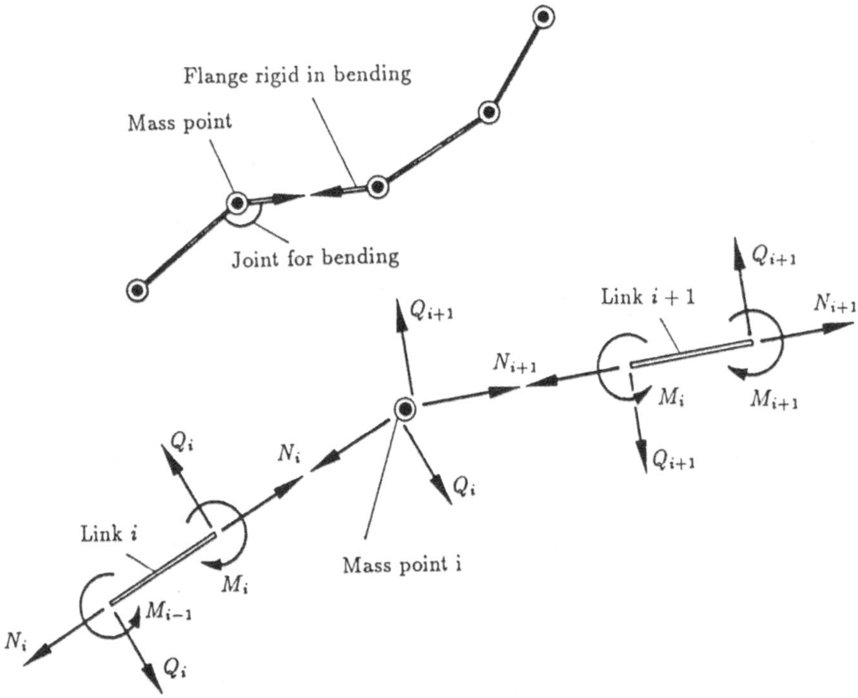

Fig. 1 - Beam model

3. A natural extension of the beam model to a shell

The shell is idealized by a triangular lattice structure composed of (massless) strings or flanges which are rigid in bending, whereas the bending deformation is concentrated at the nodal (mass) points (Fig. 2). We note that each nodal point may be considered as the

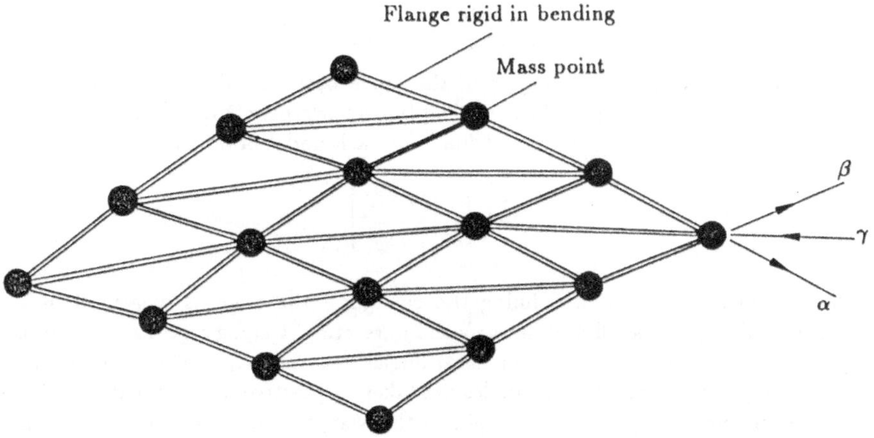

Fig. 2 - Lay-out of shell model

meet of *three* one-dimensional beam models as given in Section 2, pointing in different directions on the middle surface of the shell. The strings thereby define three families of natural (supernumerary) surface curves (α, β, γ-curves). The membrane deformation of the shell may be defined by the elongations of three concurrent flanges pointing in the α, β, γ-directions. The bending deformation on the other hand, which is concentrated at the nodal point, is expressed for each string by the change of the angle between adjoining flanges. Thus, we obtain three angular changes per nodal point. These can once again be related to second-order difference quotients, now in the three directions associated with the triangular pattern of the lattice. In fact, the present bending model may be construed as the natural (homogeneous) version of the ordinary cartesian two-dimensional difference scheme.

3.1 Kinematics

Owing to the homogeneous representation of the kinematics and statics in the natural method it suffices to consider merely one direction, e.g. the direction α. The respective relations for the remaining directions β and γ follow by cyclic permutation (indicated in the text by dots).

Membrane deformation: The elongation increment of a flange is obtained in the usual way by projection of the difference of the nodal displacement increments on the axis of the flange. The elongation increments $\delta l_\alpha, \ldots$ of the three concurrent flanges at a given nodal point form the column vector

$$\delta \boldsymbol{l} = \{\delta l_\alpha \quad \delta l_\beta \quad \delta l_\gamma\} \tag{1}$$

The total (actual) natural membrane strain increments associated with that point may then be expressed as

$$\delta \boldsymbol{\gamma}_N = \boldsymbol{l}_0^{-1} \delta \boldsymbol{l} \tag{2}$$

where the diagonal matrix

$$\boldsymbol{l}_0 = \lceil l_\alpha \quad l_\beta \quad l_\gamma \rfloor_0 \tag{3}$$

defines the lengths of the flanges at the initial (undeformed) configuration of the shell. This conforms with the small strain assumption within the present large deflection analysis. The assignment of the strain increment as given in (2) to one of the nodal points of the flange inevitably involves an approximation.

Bending deformation: The change of the angle between adjoining flanges is used to define the curvature over a finite arc length. Accordingly, the curvature increment in the direction of the α-string,... is defined for each nodal point by

$$\delta K_\alpha = \left[\frac{\delta(\Delta\varphi)}{\Delta s_0}\right]_\alpha, \ldots \tag{4}$$

The increment $\delta(\Delta\varphi)_\alpha, \ldots$ defining the change of the angle is measured in a plane perpendicular to the shell middle surface, intersecting it along an α-line,... and is given by the algebraic sum of the incremental angles $\delta\varphi_i$ and $\delta\varphi_{i+1}$ of two adjacent flanges (Fig. 3 top). These incremental angles are taken as positive if they are directed towards the interior (negative) side of the shell. In this way, the curvature increment is positive if the fiber on the positive side of the surface is stretched.

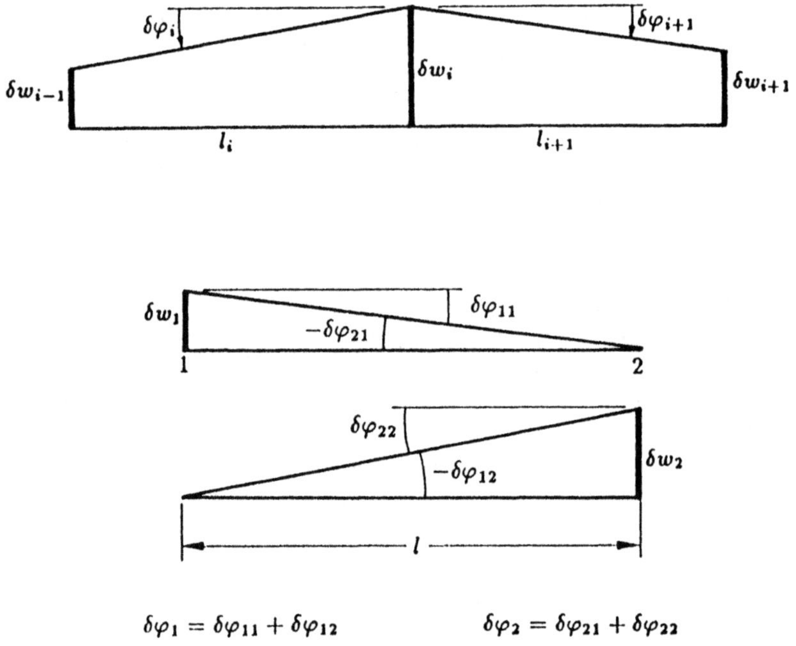

Fig. 3 - Shell model. Bending kinematics for α, β, γ-directions

Considering an individual flange (Fig. 3 bottom) the incremental angles $\delta\varphi_1, \delta\varphi_2$ appertaining to its ends 1, 2 are composed of the contributions arising from the normal (transverse) components $\delta w_1, \delta w_2$ of the nodal displacement increments. The normal to the shell surface is constructed simply by normalizing the vector sum of the unit normals on the two triangles adjacent to the flange. In matrix form, the incremental angles read

$$\delta\boldsymbol{\varphi}_\alpha = \{\delta\varphi_1 \quad \delta\varphi_2\}_\alpha = \boldsymbol{a}_\alpha\, \delta\mathbf{w}_\alpha, \ldots \tag{5}$$

where

$$\boldsymbol{a}_\alpha = \begin{bmatrix} 1/l & -1/l \\ -1/l & 1/l \end{bmatrix}_\alpha, \ldots \tag{5a}$$

and $l_\alpha, \ldots$ is the current length of the flange.

The reference arc length $\Delta s_{0\alpha}, \ldots$ of the curvature increment in (4) should obviously be taken as the average length of two adjoining flanges. The (3 x 1) vector of the total (actual) natural curvature increments at a nodal point is thus given by

$$\delta\boldsymbol{K}_N = \{\delta K_\alpha \quad \delta K_\beta \quad \delta K_\gamma\} \tag{6}$$

We note that its entries represent three difference quotients in accordance with (4) in the α, β, γ-directions of a triangular finite difference mesh.

3.2 Statics

The forces corresponding to the elongation increments in (1) are the (self-equilibrating) natural flange forces in the α, β, γ-directions (Fig. 4)

$$\mathbf{P}_N = \{\mathrm{P}_\alpha \quad \mathrm{P}_\beta \quad \mathrm{P}_\gamma\} \tag{7}$$

The static (equilibrium) relations for the flange are dual to the kinematic (compatibility) relations (5)

$$\boldsymbol{Q}_\alpha = \boldsymbol{a}_\alpha^t \boldsymbol{M}_\alpha, \ldots \tag{8}$$

where

$$\boldsymbol{M}_\alpha = \{M_1 \quad M_2\}_\alpha, \ldots \tag{9}$$

define the bending moments and

$$\boldsymbol{Q}_\alpha = \{Q_1 \quad Q_2\}_\alpha, \ldots \tag{10}$$

the associated transverse shear forces at the ends of the flange. Relation (8) may be checked with the aid of the equilibrium diagram of Fig. 4.

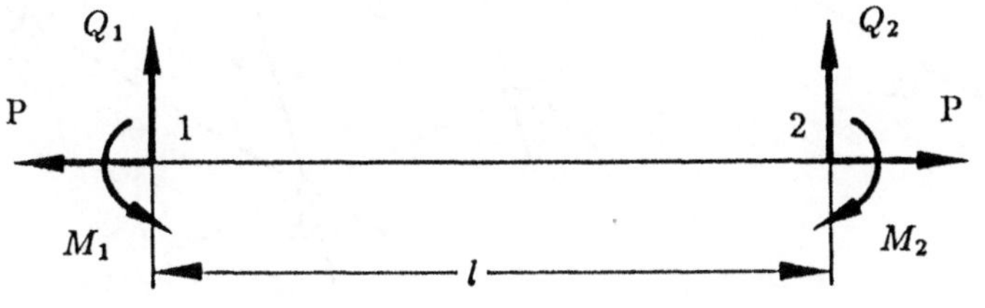

Fig. 4 - Shell Model. Statics in α, β, γ-directions

It is important to note that the simple (one-dimensional) beam equilibrium relations (8), if applied independently to three different directions, are consistent with the notion of *component* bending moments

$$M_N = \{M_\alpha \quad M_\beta \quad M_\gamma\} \tag{11}$$

— corresponding to the *total* curvature increment $\delta \boldsymbol{K}_N$ in (6) — which by superposition furnish the plane state of stress in the shell, together with the component membrane forces.

3.3 Natural method and finite difference scheme

In Section 3.1 we were led to a natural two-dimensional finite difference scheme for the curvature increments of the shell. We started from the original beam model, interpreted it as a second order difference operator in the displacement increments and subsequently applied it to the three directions of a triangular finite difference mesh covering the middle surface of the shell. These considerations suggest the use of the natural method in a related problem, namely that of deriving from a given differential operator the associated difference equation for a triangular mesh.

As an example we will examine the differential equation of a vibrating membrane prestressed by a constant force per unit length S normal to the boundary

$$S\left(\frac{\partial^2 w}{\partial x^2} + \frac{\partial^2 w}{\partial y^2}\right) = S\Delta w = m\ddot{w} \tag{12}$$

where w is the transverse displacement and m the mass per unit area. Our goal is to deduce from the Laplace operator Δw the difference equation associated with a triangular mesh in a purely formal manner. This requires first the conversion of Δw from cartesian to natural coordinates at the differential level. To this end, we take advantage of the fact that Δw is the first invariant of the curvature state of the deformed membrane.

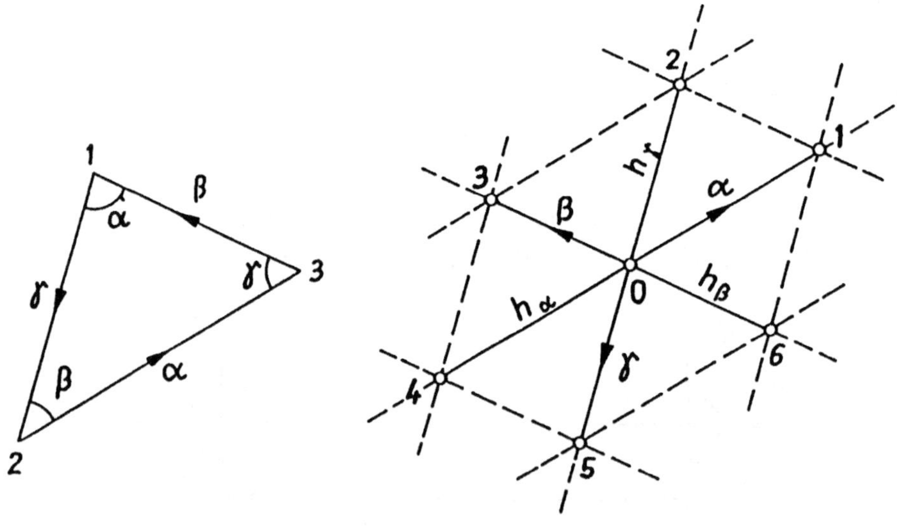

Fig. 5 - Arbitrary triangular mesh. Finite difference cell

In the natural method, the essence of which is its adaptivity to the particular form of the triangle, a distinction is, as we have already seen, made between *total* and *component* quantities. For example, we have come across the vector $\boldsymbol{K}_N$ of the total (actual) curvatures in three directions which are directly measurable with a strain gage. The corresponding vector $\boldsymbol{M}_N$ of the component bending moments represents three independent uniaxial stress states which yield jointly the actual plane state of stress.

For the following we will specify total and component quantities by subscripts t and c, respectively. Thus $\boldsymbol{K}_N$ becomes $\boldsymbol{K}_t$ and $\boldsymbol{M}_N$ becomes $\boldsymbol{M}_c$. We may equally define a vector $\boldsymbol{K}_c$ of component curvatures which represent, in analogy to the bending moments $\boldsymbol{M}_c$, three independent uniaxial strain states — the lateral curvature and torsion being identically equal to zero — which yield jointly the actual strain state. In the natural method it is shown [3] that the relation between total and component quantities, e.g. in the case of the above curvatures, is given by

$$\boldsymbol{K}_t = \boldsymbol{A}\boldsymbol{K}_c \quad , \quad \boldsymbol{K}_c = \boldsymbol{A}^{-1}\boldsymbol{K}_t \tag{13}$$

where the matrix

$$\boldsymbol{A} = \begin{bmatrix} 1 & \cos^2\gamma & \cos^2\beta \\ & 1 & \cos^2\alpha \\ \text{sym.} & & 1 \end{bmatrix} \tag{14}$$

contains the angles α, β, γ of the triangle (Fig. 5 left) and is independent of the coordinate system used.

The first invariant Δw of the curvature state in a cartesian coordinate system is given as in (12) by the sum of the second derivatives of w with respect to x and y. It is again shown [3] that the first invariant Δw in the natural system is given by the sum of the three component curvatures, namely

$$\Delta w = \mathbf{e}^t\boldsymbol{K}_c = \mathbf{e}^t\boldsymbol{A}^{-1}\boldsymbol{K}_t \tag{15}$$

where $\mathbf{e} = \{1\,1\,1\}$ and $\boldsymbol{K}_t$ contains the second derivatives of w in the directions α, β, γ

$$\boldsymbol{K}_t = \{\partial^2 w/\partial s_\alpha^2 \quad \partial^2 w/\partial s_\beta^2 \quad \partial^2 w/\partial s_\gamma^2\} \tag{16}$$

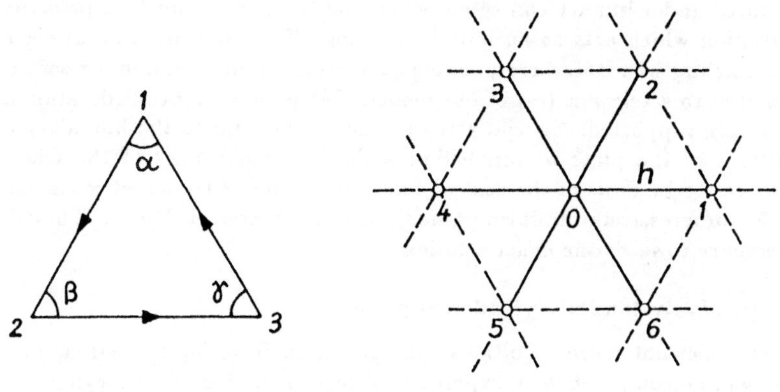

Fig. 6 - Equilateral triangular mesh. Finite difference cell

With reference to (16), relation (15) reproduces the natural form of the Laplace operator in (12). Formation of the associated difference equation merely requires the replacement of the derivatives in (16), the natural curvatures, by their finite counterparts

$$K_\alpha = \partial^2 w / \partial s_\alpha^2 \simeq (w_1 - 2w_0 + w_4)/h_\alpha^2, \ldots \tag{17}$$

These may be deduced directly from Fig. 5 which shows a cell of an arbitrary triangular mesh. The directions α, β, γ of the coordinate lines meeting at the center of the cell conform with the sides of an adjacent triangle.

As an example of an explicit evaluation of (15) we take an equilateral triangular mesh ($\alpha = \beta = \gamma = 60°$) as shown in Fig. 6. We obtain

$$\boldsymbol{A} = \begin{bmatrix} 1 & 1/4 & 1/4 \\ 1/4 & 1 & 1/4 \\ 1/4 & 1/4 & 1 \end{bmatrix}, \quad \boldsymbol{A}^{-1} = \frac{2}{9}\begin{bmatrix} 5 & -1 & -1 \\ -1 & 5 & -1 \\ -1 & -1 & 5 \end{bmatrix}, \quad \mathbf{e}^t \boldsymbol{A}^{-1} = \frac{2}{3}[1 \ 1 \ 1] \tag{18}$$

and for the finite difference approximation of the curvature (column) vector

$$\boldsymbol{K}_\iota \simeq \frac{1}{h^2}\begin{bmatrix} (w_1 - 2w_0 + w_4) \\ (w_3 - 2w_0 + w_6) \\ (w_5 - 2w_0 + w_2) \end{bmatrix} \tag{19}$$

where h is the mesh size. Thus, following (15) in conjunction with (18) and (19), the difference equation for the Laplace operator becomes

$$\Delta w_0 \simeq \frac{2}{3h^2}(w_1 + w_2 + w_3 + w_4 + w_5 + w_6 - 6w_0) \tag{20}$$

Expression (20) is seen to coincide with that quoted in the literature [4, 5]. For an arbitrary triangular mesh the weighting factors belonging to the terms w_0 through w_6 are supplied automatically by a similar (numerical) evaluation of (15). In the case of right-angled triangles the contribution of the diagonal to the weighting factors becomes zero and the well-known cartesian difference equation results.

3.4 Test on the consistency of the model

Structures under impact can often be simulated by exposure to a prescribed velocity distribution which acts as an initial condition. The calculation of an eigenmode of a plate, starting with the corresponding exact velocity distribution for zero deflection is well suited to a relevant test of our model. The plate under consideration is quadratic and simply supported. Special attention has to be paid to the boundary (symmetry) conditions of the plate in connection with the oblique lines of the triangular finite difference mesh. Plots of the central deflection of the first symmetric eigenmode versus time for progressively diminishing mesh sizes are shown in Fig. 7. They demonstrate convergence towards the exact solution.

4. Explosively loaded cylindrical panel

In this somewhat more ambitious example taken from [6] predictions of the present model will be compared with experimental results conducted on a cylindrical panel exposed to the detonation of a high-explosive (HE) layer. Figure 8 illustrates schematically

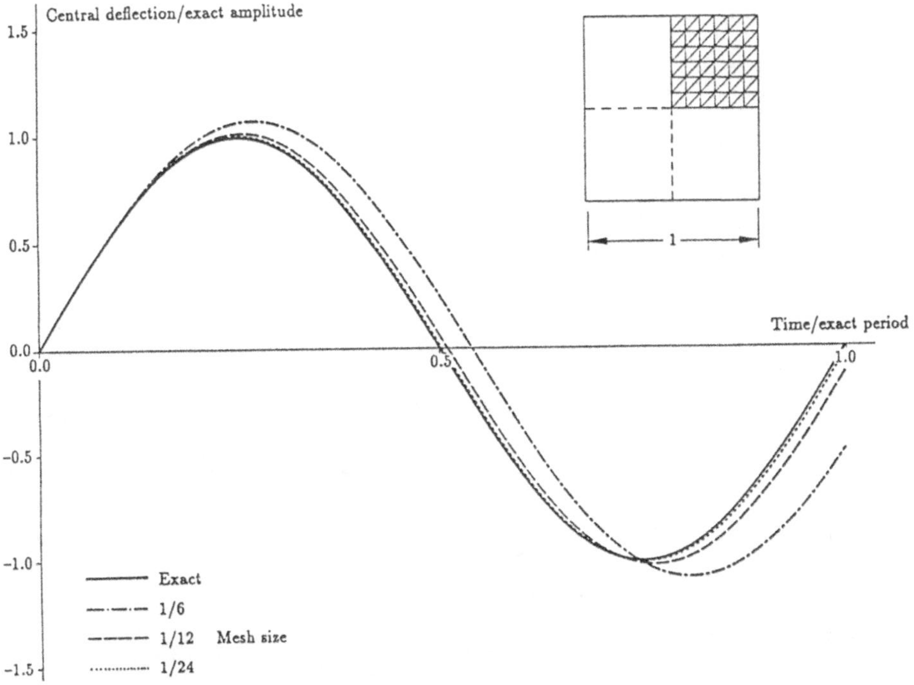

Fig. 7 - Shell model. Convergence test (first symmetric eigenmode)

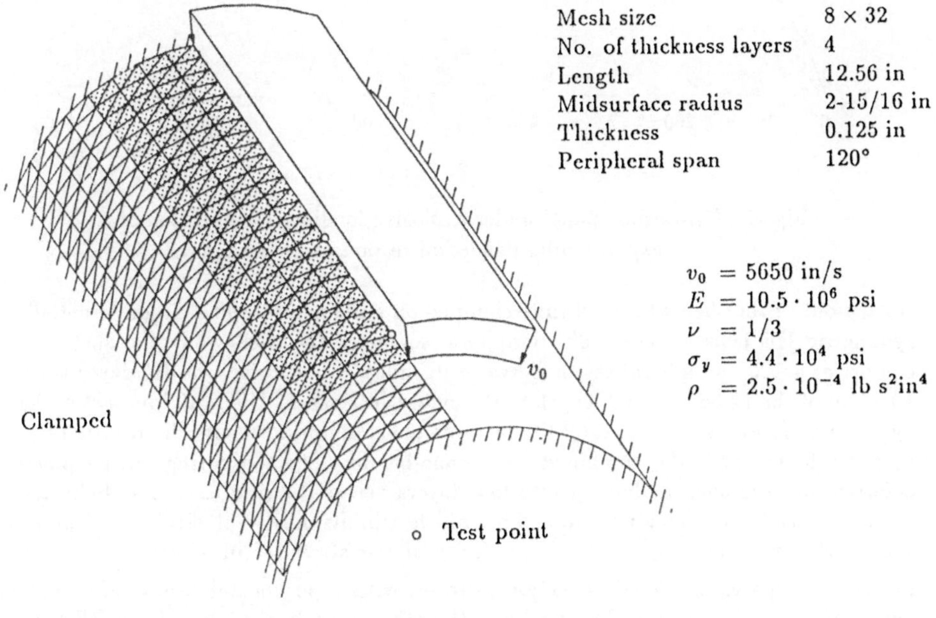

Fig. 8 - Cylindrical panel under explosive loading. Problem description

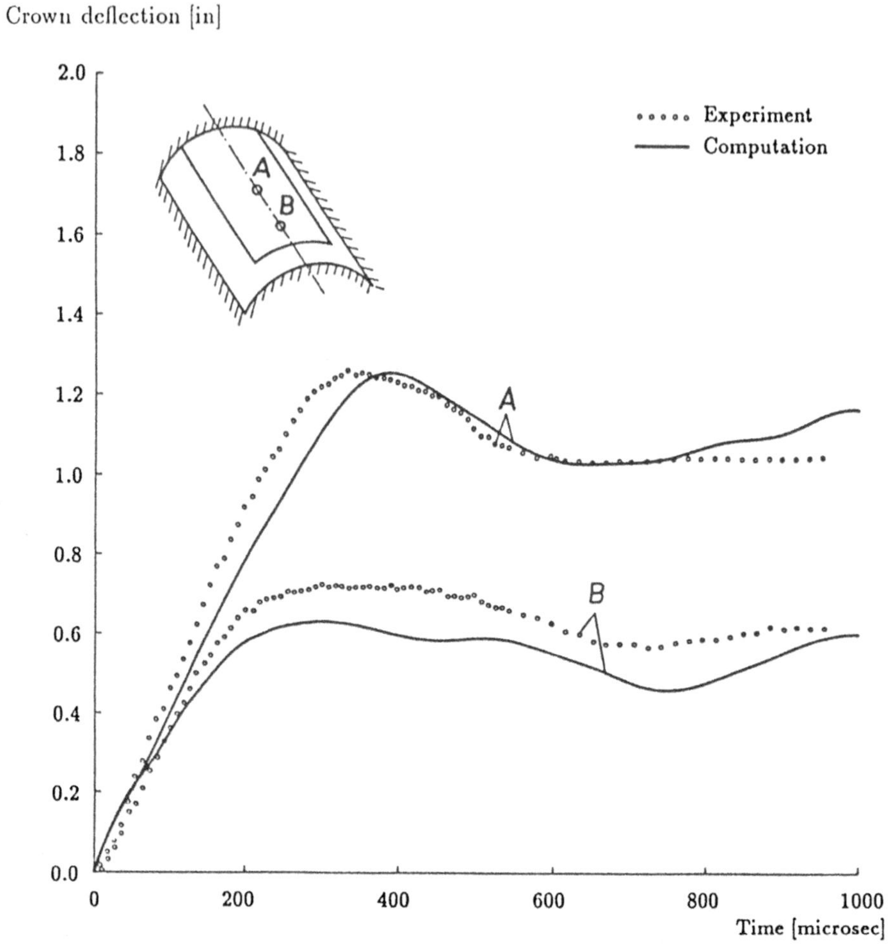

Fig. 9 - Cylindrical panel under explosive loading. Comparison of experimental deflection responses at the crown

the lay-out of the shell which is firmly clamped on all four sides. The circumferentially-symmetric HE layer covers a 60°-peripheral sector of the shell while the shell itself comprises a 120°-peripheral sector between its sides. Lengthwise, the HE layer covers 10.21 in. of the 12.56 in. axial length of the cylindrical panel. The initial impact caused by the detonation is simulated by the instantaneous uniform velocity distribution v_0. In order to simulate the combined membrane-bending effect occuring during plastic deformation, the shell is divided into four layers across its thickness. The behaviour of the material, 6061-T6 aluminium alloy, which exhibits only weak strain-hardening is adequately reproduced by an elastic, perfectly plastic stress-strain relation.

Figure 9 compares predicted deflection histories with experimental results of two test points, one at the center of the panel and the other close to the edge of the HE layer. The experimental-theoretical agreement at least for this first example is remarkably good. Figure 10 shows deformed shapes (actual geometry) of the panel after 100 and

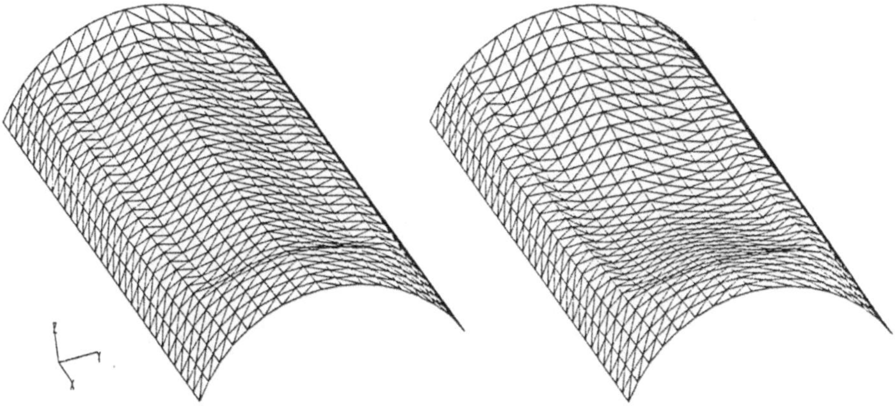

After 100 microseconds After 387 microseconds

Fig. 10 - Cylindrical panel under explosive loading. Deformed shapes

387 microseconds, the latter time corresponding to the peak of the deflection (Fig. 9). We observe the strong plastic hinge developed at the upper edge which is in contact with the HE layer. On the other hand, at the opposite edge which is free of HE, the bend is much smoother. Moreover, comparison of the crosswise bend of the two shapes at the crown (symmetry) line illustrates how it changes locally from convex to concave.

References

1. Witmer, E.A.; Balmer, H.A.; Leech, J.W.; Pian, T.H.H.: Large dynamic deformations of beams, rings, plates, and shells. AIAA Journal 1 (1963) 1848-1857.
2. Argyris, J.; Balmer, H.; Doltsinis, I.St.: On shell models for impact analysis. Proceedings ASME Winter Annual Meeting, San Francisco, Dec. 10-15, 1989 (to appear).
3. Argyris, J.H.; Scharpf, D.W.; Spooner, J.B.: Die elastoplastische Berechnung von allgemeinen Tragwerken und Kontinua. Ing. Arch. 37 (1969) 326-352.
4. Collatz, L.: Eigenwertaufgaben mit technischen Anwendungen. Leipzig: Akademische Verlagsanstalt 1949.
5. Marcus, H.: Die Theorie elastischer Gewebe und ihre Anwendung auf die Berechnung biegsamer Platten. Berlin: Springer-Verlag 1924.
6. Leech, J.W.: Finite-difference calculation method for large elastic-plastic dynamically-induced deformations of general thin shells. Ph.D. Thesis, Dept. of Aeronautics and Astronautics, Massachusetts Institute of Technology, 1966.

Discretization and Solution Techniques for Liquid Filled Shells of Revolution under Dynamic Loading

W. WUNDERLICH, H. SPRINGER, W. GOEBEL

Institute of Structural Analysis
Technical University Munich
Munich, West-Germany

Summary

For the analysis of liquid filled shells of revolution a problem–oriented discretization method is described, based on nonlinear ring elements for the shell structure and linear potential–ring elements for the fluid region. Non–axisymmetric loads and the variation of forces, displacements and fluid–pressures in the circumferential direction are approximated by Fourier series. To avoid the direct solution of the fully coupled, nonsymmetric matrix equations certain reduction techniques and iterative solution schemes are discussed. Supplementing the reduction of bandwith and/or the number of nodal unknowns via substructuring the so–called reduced basis technique is shown to be an efficient alternative. The methods are verified by the numerical simulation of vibration experiments. As an application to liquid storage tanks the nonlinear behavior of a shell due to the interaction of horizontal and vertical base excitation is investigated exhibiting the 'elephant–footing' failure phenomenon of tanks under strong earthquakes.

1. Introduction

The realistic assessment of the nonlinear vibration characteristics of liquid storage tanks, i.e. under strong earthquake excitation, requires the inclusion of the liquid in the computational model. In many practical cases a description of an inviscid, incompressible fluid according to linear potential flow theory is sufficient [1,2], and it has to be coupled with a theory for shells undergoing large displacements and plastic strains to describe the interaction between structure and fluid properly. Usual techniques for the numerical solution employ descriptions with hydrodynamic pressure and displacements as primary unknowns for the fluid region and the structure respectively. They lead to nonsymmetric and fully coupled matrix equations for dynamic equilibrium [3,4]. In this paper problem–oriented numerical methods for the efficient solution of this problem are described.

The benefits of the semi–analytic FEM–approach are well known for the linear analysis of structures with rotational geometry [5], in which non–axisymmetric distributions of displacements, internal and external forces are described by Fourier series, and this leads to a 'diagonalized' global stiffness matrix, the matrix blocks for the Fourier harmonics are decoupled, sparse and banded. The numerical advantage of reducing the solution space in the semi–analytic approach is accompanied by less modelling and discretization effort as well.

For static and dynamic analysis of shells of revolution the semi–analytic approach was also extended into the nonlinear regime [7–10]. On the basis of an incremental shell theory expressions for the tangential and secant stiffness were derived, which could be used to

G. Kuhn · H. Mang (Eds.)
Discretization Methods in Structural Mechanis
IUTAM/IACM Symposium Vienna, Austria, 1989

solve the nonlinear equilibrium equations via a standard or modified Newton/Raphson algorithm. As a consequence of the mutually coupled Fourier harmonics an iterative solution of the linearized incremental equations in each Newton step was proposed, using the linear stiffness matrix as a preconditioner [7]. Due to the 'diagonalization' by the Fourier series this preconditioning turned out to be very efficient.

The basic idea, employed here for the investigation of the coupled fluid–shell problem, is to retain the 'diagonalized' structure of the shell equations and to treat all disarranging fluid terms with the same preconditioned iteration algorithm, supplemented by the so-called reduced basis technique [11,12].

2. Basic Equations of Shell Dynamics

For the derivation of the underlying theory of shells as well as for the evaluation of the corresponding element matrices generalized variational principles with stresses and displacement as independent unknowns were used as starting point. The shell equations in incremental form – given in Table 1 – were developed on the basis of a mixed variational principle of Hellinger–Reissner type [6] . They are valid for large displacements and moderate rotations and are presented in the usual tensor notation using e.g. a vertical bar for the covariant derivative. The superscript (^{o}) denotes given state variables of the previous reference configuration at time t, while all other variables without superscript refer to unknown finite incremental quantities.

Kinematics

$$\alpha_{\alpha\beta} = \frac{1}{2}\left(v_\alpha|_\beta + v_\beta|_\alpha - 2B_{\alpha\beta}\, v_3 + \underline{\overset{o}{\phi}_\alpha\, \phi_\beta + \overset{o}{\phi}_\beta\, \phi_\alpha + \phi_\alpha\, \phi_\beta}\right)$$

$$\beta_{\alpha\beta} = \frac{1}{2}\left(\phi_\alpha|_\beta + \phi_\beta|_\alpha - B^\lambda_\alpha\, v_\lambda|_\beta - B^\lambda_\beta\, v_\lambda|_\alpha + 2B^\lambda_\alpha\, B_{\lambda\beta}\, v_3\right)$$

$$\phi_\alpha = -v_3|_\alpha - B^\lambda_\alpha\, v_\lambda \qquad \text{(normality hypothesis)}$$

Equilibrium

$$\left(n^{\alpha\beta} - B^\beta_\lambda\, m^{\alpha\lambda}\right)|_\alpha - B^\beta_\alpha\, q^\alpha + \bar{p}^\beta \underline{- \varrho\, t\, \ddot{v}^\beta + \varrho\, \frac{t^3}{12}\, B^\alpha_\alpha\, \ddot{\phi}^\beta} = 0$$

$$q^\alpha|_\alpha + B_{\alpha\beta}\left(n^{\alpha\beta} - B^\beta_\lambda\, m^{\alpha\lambda}\right) + \bar{p}^3 \underline{- \varrho\, t\, \ddot{v}^3} = 0$$

$$m^{\alpha\beta}|_\alpha - \left(\underline{n^{\beta\lambda}\, \overset{o}{\phi}_\lambda + \overset{o}{n}{}^{\beta\lambda}\, \phi_\lambda + n^{\beta\lambda}\, \phi_\lambda}\right) \underline{- \varrho\, \frac{t^3}{12}\left(\ddot{\phi}^\beta - B^\alpha_\alpha\, \ddot{v}^\beta\right)} = q^\beta$$

Constitutive Equations $\gamma_{\alpha\beta} = \gamma^E_{\alpha\beta} + \gamma^P_{\alpha\beta}$ (continuum)

$$n^{\alpha\beta} = t\, E^{\alpha\beta\varrho\lambda}\left(\alpha_{\varrho\lambda} - \underline{\frac{1}{t}\int \gamma^P_{\varrho\lambda}\, d\Theta^3}\right) \qquad m^{\alpha\beta} = \frac{t^3}{12}\, E^{\alpha\beta\varrho\lambda}\left(\beta_{\varrho\lambda} - \frac{12}{t^3}\underline{\int \gamma^P_{\varrho\lambda}\, \Theta^3\, d\Theta^3}\right)$$

Table 1: Incremental Shell Equations

As described in some detail in [7,8] our numerical discretization procedure is based on a canonical set of first-order ordinary differential equations of mixed type which contain all state variables along the circumferential boundary of the ring element as unknowns (see Fig. 1). An asymptotically exact integration of the differential equations yields, after some transformations, the various element matrices and load vectors of the ring element. They

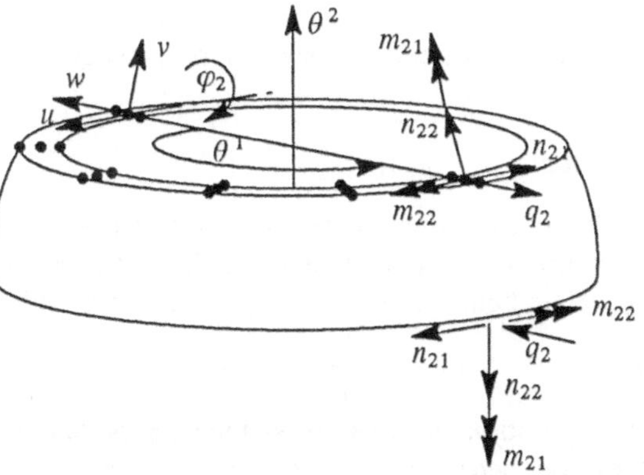

$$\mathbf{Z}(\theta^1,\theta^2) = \begin{bmatrix} u \\ v \\ w \\ \varphi_2 \\ \\ n_{21} \\ n_{22} \\ q_2 \\ m_{22} \end{bmatrix} = \sum_{n=0}^{N} \begin{bmatrix} u^{(n)} \sin n\theta^1 \\ v^{(n)} \cos n\theta^1 \\ w^{(n)} \cos n\theta^1 \\ \varphi_2^{(n)} \cos n\theta^1 \\ \\ n_{21}^{(n)} \sin n\theta^1 \\ n_{22}^{(n)} \cos n\theta^1 \\ q_2^{(n)} \cos n\theta^1 \\ m_{22}^{(n)} \cos n\theta^1 \end{bmatrix}$$

Fig. 1: Ring Element and Vector of Finite Incremental State Variables

are assembled in the usual manner of the stiffness approach to obtain (N+1) sets of nonlinear dynamic equilibrium equations for the finite displacement increments

$$\mathbf{M}\ddot{\mathbf{v}} + \mathbf{C}\dot{\mathbf{v}} + \mathbf{K}_L\mathbf{v} = \overline{\mathbf{P}}(t) + \mathbf{P}_S[\overset{o}{E}_t\,, e_t, \overset{o}{\mathbf{V}}\,, \mathbf{v}]\ . \tag{1}$$

All nonlinear terms are treated as pseudo-load vectors $\mathbf{P}_S[\overset{o}{E}_t\,, e_t, \overset{o}{\mathbf{V}}\,, \mathbf{v}]$, to leave the matrix blocks $\mathbf{M},\mathbf{C},\mathbf{K}_L$ for the different harmonics uncoupled. In addition, the dynamic forces are also treated as inhomogeneous terms. This makes it possible to derive a consistent element mass matrix from the particular solution of the numerical integration procedure. The use of such a consistent mass matrix, which relates directly to the asymptotically exact linear stiffness, seems to be important to satisfy the convergence requirements of the underlying variational principle.

The linear stiffness matrix $\mathbf{K}_L$ and the pseudo-loads $\mathbf{P}_S$ imply the identities

$$\mathbf{K}_S[\overset{o}{E}_t\,, e_t, \overset{o}{\mathbf{V}}\,, \mathbf{v}] = \mathbf{K}_L - \mathbf{P}_S[\overset{o}{E}_t\,, e_t, \overset{o}{\mathbf{V}}\,, \mathbf{v}] \tag{2a}$$

$$\mathbf{K}_T[\overset{o}{E}_t\,, \overset{o}{\mathbf{V}}\,] = \mathbf{K}_L - \mathbf{P}_T[\overset{o}{E}_t\,, \overset{o}{\mathbf{V}}\,, \mathbf{v}] \tag{2b}$$

where $\mathbf{K}_T$ is the familiar tangential stiffness matrix, which is the linearized form of the 'secant' stiffness matrix $\mathbf{K}_S$. Finally, the coupling of the shell with the fluid leads to the additional load vector

$$\mathbf{P}_F = \mathbf{L}_v^T\mathbf{p} \tag{3}$$

which results from the projection of the fluid pressure $\mathbf{p}$ normal onto the shell surface. Thus, a short-hand form of the discrete nonlinear equations of motion is given by

$$\mathbf{M}\ddot{\mathbf{v}} + \mathbf{C}\dot{\mathbf{v}} + \mathbf{K}_S[\overset{o}{E}_t\,, e_t, \overset{o}{\mathbf{V}}\,, \mathbf{v}]\ \mathbf{v} = \overline{\mathbf{P}}(t) + \mathbf{L}_v^T\mathbf{p}. \tag{4}$$

The inclusion of plastic deformations implicitly accounts for the hysteretic damping of the shell material. This in turn requires a constitutive description which correctly predicts the time history of the materials tangent moduli $\overset{o}{E}_t$ and its finite increment e_t appearing in the pseudo-load vector $\mathbf{P}_S$. This is achieved by a two-surface plasticity model, which accounts for non-proportional loading conditions in multi-axial stress states in an efficient manner [13]. The plastic strains are evaluated at discrete stations in the shell-continuum as can be seen in Fig. 1. Their contribution to the pseudo-loads will be calculated by a Fourier synthesis and subsequent numerical integration. Additional damping in the elastic range and due to unpredictionable energy–dissipation mechanisms may be approximated by the usual Rayleigh–damping.

3. Basic Equations of Fluid Dynamics

By combining the linearized Navier–Stokes equations for an inviscid and irrotational fluid with the continuity relation the well known governing equations for the hydrodynamic pressure within the fluid and the appropriate boundary conditions [3,4] are obtained:

$$\nabla^2 p - \frac{1}{k_F}\,\varrho_F\,\ddot{p} = 0 \quad in\ V_F\ ; \qquad \mathbf{n}\ddot{\mathbf{v}} = -\frac{1}{\varrho_F}\,\nabla p \quad on\ S_F \tag{5}$$

where ρ_F is the density of the fluid, p the hydrodynamic pressure, k_F the bulk modulus for compressible fluids, and $\ddot{\mathbf{v}}$ stands for the prescribed boundary accelerations, or alternatively for the unknown values on the fluid–structure interface or the free surface, and where $\mathbf{n}$ is the unit normal vector on the respective boundaries. The free surface wave boundary condition can be expressed following [3]. The basic equations (5) can now be discretized in a standard manner by a Galerkin procedure. Approximating the free surface boundary conditions with the hydrostatic relation $p = \rho g v_{fs}$ as an additional constraint equation in the weighted residual integral, using the wave amplitudes v_{fs} as Lagrange multipliers, leads to (see also [4])

$$\mathbf{H}\ddot{\mathbf{p}} + \mathbf{T}\mathbf{p} - \mathbf{L}_v\ddot{\mathbf{v}} - \mathbf{L}_{fs}\ddot{\mathbf{v}}_{fs} - \overline{\mathbf{A}} = 0 \tag{6}$$

$$\mathbf{K}_{fs}\mathbf{v}_{fs} + \mathbf{L}_{fs}^T\,\mathbf{p} = 0 \tag{7}$$

with $\mathbf{H}$ representing the influence of compressibility k_F. $\mathbf{T}$ is due to the inertia forces and $\mathbf{L}_v\ddot{\mathbf{v}}$ and $\overline{\mathbf{A}}$ represent boundary accelerations on the shell–structure interface and the solid boundaries, respectively. The vertical free surface displacements $\mathbf{v}_{fs}$ appear as additional unknowns coupled to the pressure $\mathbf{L}^T_{fs}\mathbf{p}$ by the 'wave stiffness' $\mathbf{K}_{fs}$.

4. Coupling of Fluid and Shell

The coupled free surface wave–shell–fluid equations can be written as the nonsymmetric matrix system

$$\begin{bmatrix} \mathbf{M} & 0 & 0 \\ -\mathbf{L}_v & 0 & -\mathbf{L}_{fs} \\ 0 & 0 & 0 \end{bmatrix} \begin{bmatrix} \ddot{\mathbf{v}} \\ \ddot{\mathbf{p}} \\ \ddot{\mathbf{v}}_{fs} \end{bmatrix} + \begin{bmatrix} \mathbf{K}_s & \mathbf{L}_v^T & 0 \\ 0 & \mathbf{T} & 0 \\ 0 & \mathbf{L}_{fs}^T & \mathbf{K}_{fs} \end{bmatrix} \begin{bmatrix} \mathbf{v} \\ \mathbf{p} \\ \mathbf{v}_{fs} \end{bmatrix} = \begin{bmatrix} \mathbf{P} \\ \overline{\mathbf{A}} \\ 0 \end{bmatrix} \tag{8}$$

where the first row refers to the shell, equ. (4), and the others to the incompressible fluid, equ. (6,7). All submatrices are linear and decoupled in the Fourier harmonics except the secant stiffness matrix $\mathbf{K}_S$ of the shell, for which in numerical calculations the identity (2a) is used. Viscous damping terms are neglected here but may be introduced easily.

The incompressibility of the fluid allows static condensation of the pressure $\mathbf{p}$ yielding equations of motion for a 'dry' shell which now obtain a symmetric structure (see also [3])

$$\left[\begin{array}{c|c} \mathbf{M} + \mathbf{L}_v^T\mathbf{T}^{-1}\mathbf{L}_v & \mathbf{L}_v^T\mathbf{T}^{-1}\mathbf{L}_{fs} \\ \hline \mathbf{L}_{fs}^T\mathbf{T}^{-1}\mathbf{L}_v & \mathbf{L}_{fs}^T\mathbf{T}^{-1}\mathbf{L}_{fs} \end{array}\right] \begin{bmatrix} \ddot{\mathbf{v}} \\ \ddot{\mathbf{v}}_{fs} \end{bmatrix} + \left[\begin{array}{c|c} \mathbf{K}_S & 0 \\ \hline 0 & \mathbf{K}_{fs} \end{array}\right] \begin{bmatrix} \mathbf{v} \\ \mathbf{v}_{fs} \end{bmatrix} = \begin{bmatrix} \tilde{\mathbf{P}} \\ \tilde{\mathbf{A}} \end{bmatrix}. \tag{9}$$

If the individual Fourier harmonics of the shell displacements $\mathbf{v}_{(n)}$ and the free surface displacements $\mathbf{v}_{fs(n)}$ are combined into the vector $\overset{*}{\mathbf{v}}{}^T_{(n)} = [\mathbf{v}^T_{(n)}\ \mathbf{v}^T_{fs(n)}]$ and all vectors again are collected into the vector $\mathbf{x}^T = [\overset{*}{\mathbf{v}}^T_{(1)}\ \overset{*}{\mathbf{v}}^T_{(2)} \dots \overset{*}{\mathbf{v}}^T_{(N)}]$ the nonlinear equations of motion of the complete fluid–shell system may be written as

$$\overset{*}{\mathbf{M}}\ddot{\mathbf{x}} + \overset{*}{\mathbf{C}}\dot{\mathbf{x}} + \overset{*}{\mathbf{K}}_L\mathbf{x} = \overset{*}{\mathbf{P}} + \overset{*}{\mathbf{P}}_S[\overset{o}{E}_t, e_t, \overset{o}{\mathbf{X}}, \mathbf{x}] \tag{10}$$

which is analogous to (1). The superscript (*) indicates, that the various matrices and vectors have a different dimension than the original ones from (1). The mass matrix $\overset{*}{\mathbf{M}}$ contains additional terms which account for the interaction and lead to a complete coupling of all components $\overset{*}{\mathbf{v}}_{(n)}$ in each fourier harmonic. In a formal sense, these additional terms may be viewed as 'added masses' although, strictly speaking, they are 'added pressures'.

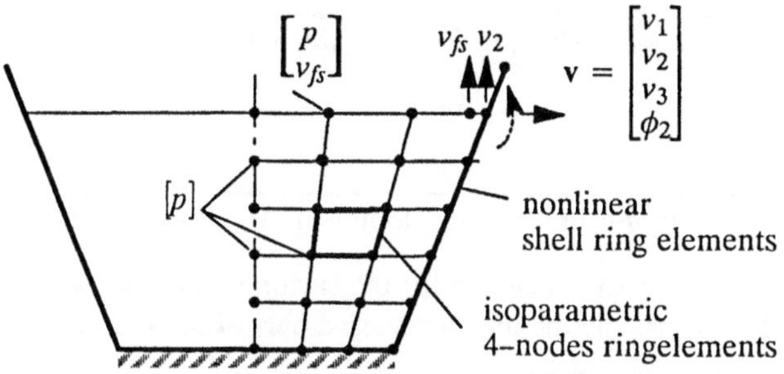

Fig. 2: Discretization of Fluid–Shell System

In summary, the approach is characterized by unknown nodal variables, which are the harmonic components of the hydrodynamic pressure in the fluid, of the structural displacements on the shell including the interface points, and of the displacements at the free surface due to gravity waves, as can be seen in Fig. 2.

5. Numerical Methods

When the Newmark temporal operator is used to discsretize (10) with respect to time the following system (11) of implicit nonlinear algebraic equations is obtained

$$\overset{*}{\mathbf{K}}_{Leff}\,\mathbf{x} = \overset{*}{\mathbf{P}}_{eff} + \overset{*}{\mathbf{P}}_S[\overset{o}{E}_t, e_t, \overset{o}{\mathbf{X}}, \mathbf{x}] - a_0\overset{*}{\mathbf{M}}_F\mathbf{x} \tag{11}$$

with

$$\overset{*}{\mathbf{K}}_{Leff} = a_0\overset{*}{\mathbf{M}}_L + a_1\overset{*}{\mathbf{C}} + \overset{*}{\mathbf{K}}_L \tag{11a}$$

$$\overset{*}{\mathbf{P}}_{eff} = \overset{*}{\mathbf{P}} + \overset{*}{\mathbf{M}}\,(a_2\dot{\overset{o}{\mathbf{X}}} + a_3\ddot{\overset{o}{\mathbf{X}}}) + \overset{*}{\mathbf{C}}\,(a_4\dot{\overset{o}{\mathbf{X}}} + a_5\ddot{\overset{o}{\mathbf{X}}}) \tag{11b}$$

$$\overset{*}{\mathbf{M}} = \overset{*}{\mathbf{M}}_L + \overset{*}{\mathbf{M}}_F \tag{11c}$$

where $\overset{*}{\mathbf{K}}_{Leff}$ and $\overset{*}{\mathbf{P}}_{eff}$ are the 'effective dynamic' stiffness matrix and load increment, respectively, and the mass matrix $\overset{*}{\mathbf{M}}$ is partitioned into a diagonal matrix $\overset{*}{\mathbf{M}}_L$ with the same bandwidth as $\overset{*}{\mathbf{K}}_L$ and a second one $\overset{*}{\mathbf{M}}_F$ with the off–diagonal terms. The constants $a_0 - a_5$ are functions of the time step Δt and the Newmark–parameters β and γ .

The solution of the nonlinear equations (11) is most efficiently achieved by the simple direct iteration scheme presented in Table 2 in contrast to static problems, where corresponding linearized algebraic equations are solved, e.g. by means of a conjugate gradient iteration in each Newton–step [7]. Only for the partitioned mass matrix (11c) it is recommended to perform a nested iteration, which reduces the number of equilibrium iteration

Initialize $\overset{*}{\mathbf{R}}_0 = \overset{*}{\mathbf{P}}_{eff}\ ;\ \mathbf{x}_0 = 0$

$$\overset{*}{\mathbf{K}}_{Leff}\,\delta\mathbf{x}_i = \overset{*}{\mathbf{R}}_{i-1} - a_0\overset{*}{\mathbf{M}}_F\delta\mathbf{x}_i$$

← inner loop, preconditioned conj. gradient iteration

update of state variables:

$$\mathbf{x}_i = \mathbf{x}_{i-1} + \delta\mathbf{x}_i \;\rightarrow\; \dot{\mathbf{x}}_i\ ;\ \ddot{\mathbf{x}}_i\ ;\ \overset{*}{\mathbf{P}}_S[\overset{o}{E}_t, e_{t_i}, \overset{o}{\mathbf{X}}, \mathbf{x}_i]$$

Residuum from (10):

$$\overset{*}{\mathbf{R}}_i = \overset{*}{\mathbf{P}} - \overset{*}{\mathbf{M}}\ddot{\mathbf{x}}_i - \overset{*}{\mathbf{C}}\dot{\mathbf{x}}_i - \overset{*}{\mathbf{K}}_L\mathbf{x}_i + \overset{*}{\mathbf{P}}_S[\overset{o}{E}_t, e_{t_i}, \overset{o}{\mathbf{X}}, \mathbf{x}_i]$$

$$\frac{\| \mathbf{x}_i - \mathbf{x}_{i-1} \|}{\| \mathbf{x}_i \|} = \varepsilon$$

outer loop if $\varepsilon > \varepsilon_{TOL}$

Table 2: Nested Iteration of Nonlinear Algebraic Equations

steps (outer loop). By this way the effort for the calculation of the 'secant' pseudo–loads $\overset{*}{\mathbf{P}}_S$ is reduced, which is usually the most expensive part of the iteration.

The solution algorithm can be supplemented by a reduced base method [11,12]. The displacements x are approximated by a combination of orthonormal base vectors

$$\mathbf{x} = \hat{\mathbf{U}}\beta. \tag{12}$$

The base vectors $\hat{\mathbf{U}}_i$ are generated at a time t from the recurrence relationship

$$\begin{aligned} \hat{\mathbf{U}}_1 &= \overset{*}{\mathbf{K}}{}_L^{-1}\,\mathbf{P}_{dyn} \\ \hat{\mathbf{U}}_i &= \mathbf{K}_L^{-1}(\overset{*}{\mathbf{M}}\hat{\mathbf{U}}_{i-1} + \overset{*}{\mathbf{P}}_T[\overset{o}{E}_t, \overset{o}{\mathbf{X}}, \hat{\mathbf{U}}_{i-1}]) \end{aligned} \tag{13}$$

where the implicit form of the tangential stiffness matrix (2b) was used to create approximations to the lowest 'tangential eigenmode', assuming, that they contain essential information on the solution characteristics in the neighborhood of the reference configuration. The basis generation should be started with the dynamic load vector

$$\overset{*}{\mathbf{P}}_{dyn} = \overset{*}{\mathbf{P}} - \overset{*}{\mathbf{M}}\ddot{\mathbf{x}} - \overset{*}{\mathbf{C}}\dot{\mathbf{x}} \tag{14}$$

which employs actual inertia and damping forces as well as external loads. Hence, the nonlinear equations (10) take the reduced form

$$\hat{\mathbf{M}}\ddot{\beta} + \hat{\mathbf{C}}\dot{\beta} + \hat{\mathbf{K}}_L\beta = \hat{\overset{*}{\mathbf{P}}} + \hat{\mathbf{P}}_S \tag{15}$$

where

$$\hat{\mathbf{M}} = \hat{\mathbf{U}}^T\overset{*}{\mathbf{M}}\hat{\mathbf{U}} \qquad \hat{\mathbf{C}} = \hat{\mathbf{U}}^T\overset{*}{\mathbf{C}}\hat{\mathbf{U}} \qquad \hat{\mathbf{K}}_L = \hat{\mathbf{U}}^T\overset{*}{\mathbf{K}}_L\hat{\mathbf{U}} \tag{15a}$$

$$\hat{\overset{*}{\mathbf{P}}} = \hat{\mathbf{U}}^T\overset{*}{\mathbf{P}} \quad ;\ \hat{\mathbf{P}}_S = \hat{\mathbf{U}}^T\overset{*}{\mathbf{P}}_S[\overset{o}{E}_t, e_t, \overset{o}{\mathbf{X}}, \mathbf{x}]\ . \tag{15b}$$

As the orthonormal displacement vectors $\hat{\mathbf{U}}_i$, also called 'Ritz–vectors', are not eigenmodes, the reduced stiffness matrix $\hat{\mathbf{K}}_L$ retains fully populated, but a complete diagonalization by an eigensolution in the subspace can easily be performed as the number of base vectors usually is quite limited.

For the iterative solution of the reduced nonlinear equations of motions essentially the same algorithm as described above may be employed. A nested iteration is not necessary anymore because the mass matrix will be diagonal automatically. Only the pseudo–loads have to be transformed into the subspace in each iteration step.

The accuracy of the dynamic response depends on the number of base vectors and the frequency of base change [9,10]. The number of base vectors is choosen in such a way, that that the truncation error of the loads $\overset{*}{\mathbf{P}}_{dyn}$ does not exceed a prescribed error tolerance. A similar truncation error is observed when the pseudo–loads are transformed into the subspace during the iteration. Even after convergence of the equilibrium iteration in the subspace, the residuum of the balanced forces (10) does not vanish completely and its increase during the time history indicates a deterioration of the basis, requiring a base change.

6. Numerical Examples

6.1 Vibration Analysis of Ring–Stiffened Cones Submerged in Water

The performance of the coupled fluid and shell ring elements in connection with the numerical procedures described in the previous chapters is first demonstrated by an eigen–solution of a submerged, ring–stiffened cone. The structure shown in Fig. 3 was already investigated theoretically and experimentally in [14].

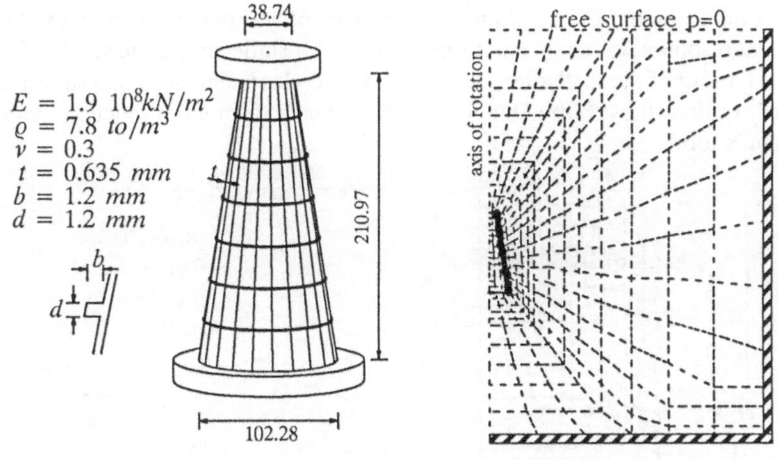

Fig. 3: Geometry of the Shell and Fluid–Discretization

The lobar eigenmodes predicted in [14] are in good agreement with our results (Fig. 4), and it is observed that the eigenfrequencies are nearer to the experimental results than those of the reported FEM–analysis.

6.2 Multi–axial Base Excitation of a Cylindrical Liquid Storage Tank

A water–filled liquid storage tank is investigated under combined vertical and horizontal base excitation. For the system in Fig. 5, which was also analysed in [15] an elastic–ideally plastic material behavior is assumed (structural steel). The amplitudes of the ground accelerations a_V and a_H are given by (λ_V g) and (λ_H g) – where g denotes the gravitational constant and λ is a scaling parameter indicating the 'load–intensity' –, and its time history is taken to be a periodic saw–tooth like function.

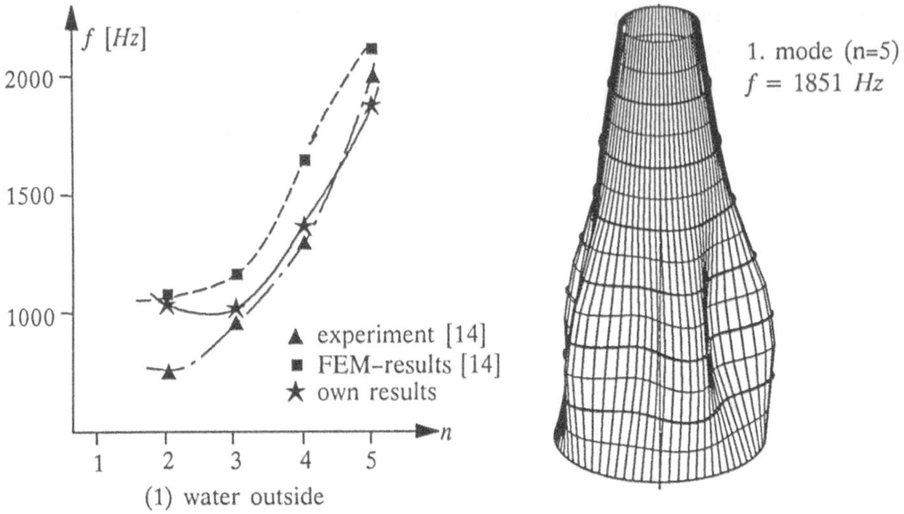

Fig. 4: Eigenfrequencies and 1st Eigenmode

Varying the ratio and the amplitudes of the vertical and horizontal base excitation the structural response was analysed. It turns out that for sufficiently large values of λ the deflection pattern of the shell after several response cycles is similar to that shown in Fig. 7. It clearly exhibits the localized bulge deformation near the bottom which is usually called 'elephant's foot'.

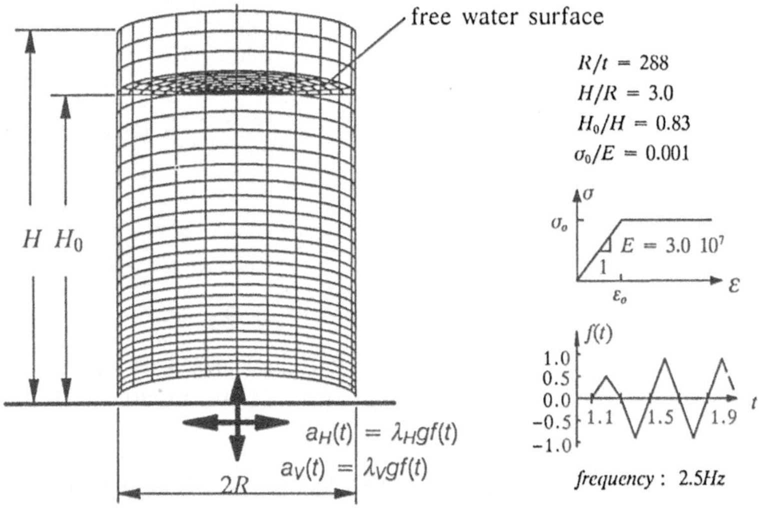

Fig. 5: Water-Filled Liquid Storage Tank under Multi-Axial Base Excitation

The development of the bulge deformation under increasing load intensities may be observed from the time histories of the radial displacement u_z in Fig. 6. The diagram shows the time histories for horizontal excitation only ($\lambda_V = 0$), but for other ratios λ_V/λ_H they are similar. Exceeding a critical value of λ, the mean position of the oscillations change,

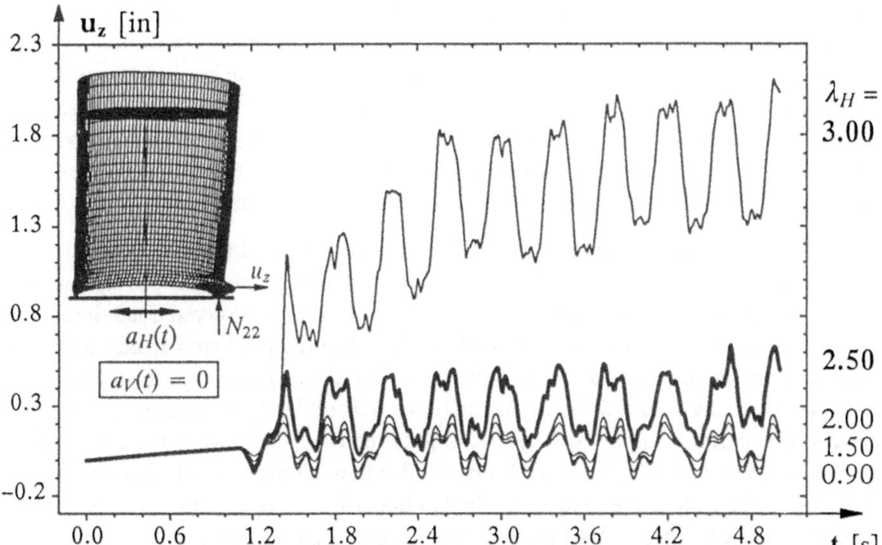

Fig. 6: Time-Histories of the Radial Displacement u_z

indicating the irreversible bulge deformation. To extrapolate the critical load intensity the maximum values of u_z and the corresponding maximum values of the (compressive) axial stress resultant N_{22} at the base have been plotted as functions of the loading parameter λ in Fig. 8. The diagram indicates a nearly linear behavior for u_z and N_{22} up to approximately $\lambda_{cr} = 2.4$, whereas for $\lambda > 2.4$ the value of N_{22} remains constant. This indicates that the appearance – and growth – of the plastic bulge coincides with a reduction of the associated axial stiffness to zero.

Of course, the rapid growth of u_z, as can be seen from Fig. 6, is superimposed on the – much slower – global bending displacements which also lead to a considerable increase

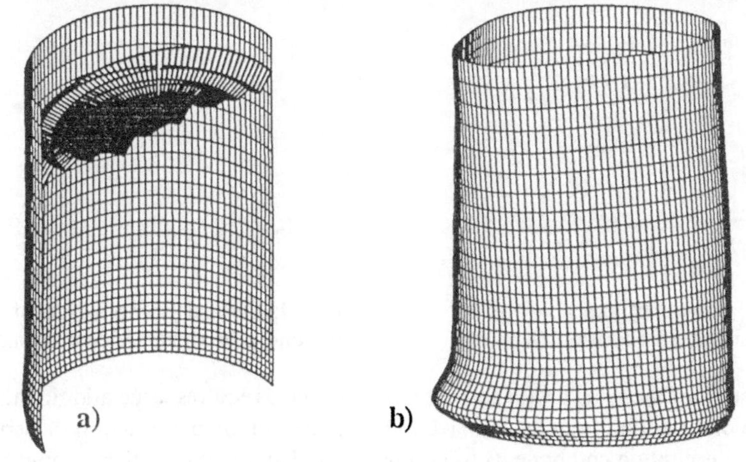

Sloshing Motion of the Water Surface

Deformed Shell - 'Elephant-Footing" (magnification factor: 10.0)

Fig. 7: Deformations of the Shell Wall and the Water Surface ($\lambda_H = \lambda_V = 0.9$)

in the overall axial, radial and tilting deformation. This is more graphically illustrated in Fig. 7. It shows not only the 'elephant–footing' near the base but also a considerable ovalization of the upper part of the shell. Since the free surface motion is much larger than the tank deformation the plot of the tank is magnified by a factor of 10 in Fig. 7b. The fundamental sloshing period – with approximately 6 seconds – is much longer than the excitation periods. Thus higher sloshing modes with relatively short–wavelength ripples are excited, whereas the circumferential distribution is dominated by the first Fourier harmonic.

The predicted dynamic response of the shell and the liquid, and in particular the development of the localized permanent deformations near the base are qualitatively similar to those observed on tanks subjected to severe earthquakes. The observed 'load–intensities' however, are usually less than 1g and much smaller than the predicted critical amplitude of $\lambda_{cr}=2.4g$ for horizontal excitation only. Inclusion of a vertical excitation component reduces the critical values drastically, as can be seen from Fig. 9. In contrast to the former case, the other extreme of pure vertical excitation predicts dynamic buckling at a load level of $\lambda_{cr}=0.6g$ which is much closer to the observed values. Combinations of a_V and a_H lead to a further reduction of the load level, the nonlinear interaction of the two components can be approximated nearly by a straight line for practical purposes.

Several calculations were performed twice, using the reduced–basis approach and as reference the direct time–integration procedure. Good results were obtained with the first method using in the subspace about 15% of the degrees of freedom originally present, which amounts to a considerable reduction of cpu–time of about 30–50%. In all cases it was possible to detect the initialization of dynamic buckling and the critical load intensities, even when the results in the strong nonlinear postbuckling regime were only approximate.

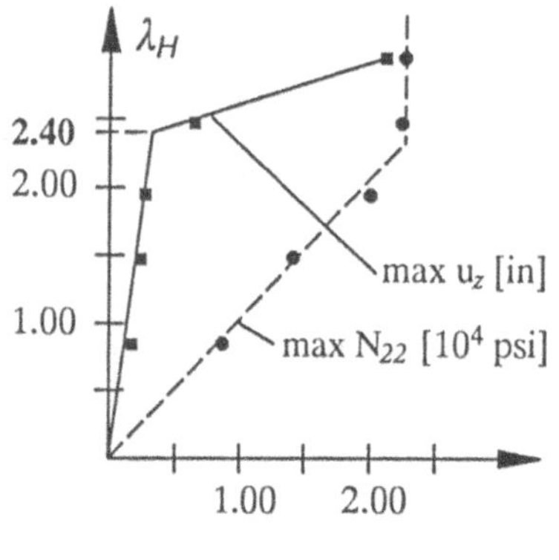

Fig. 8: Max u_z and max N_{22} for Horizontal Excitation ($\lambda_V=0$)

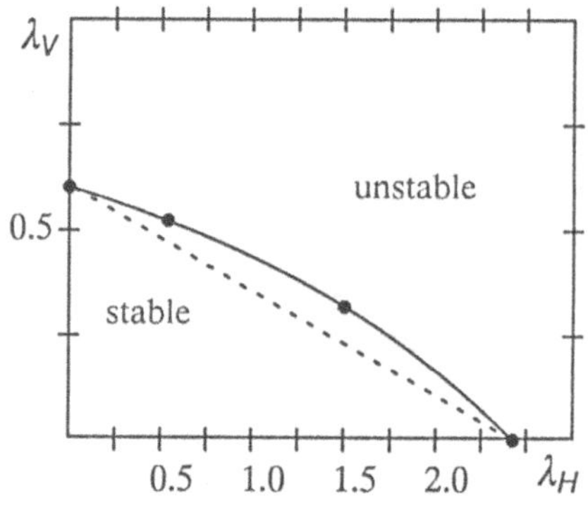

Fig. 9: Critical Combination of Vertical and Horizontal Accelerations

An evaluation of the results with respect to design rules requires large additional effort. The range of excitation frequencies have to be varied, out–of–phase action of horizontal and vertical excitation components to be investigated and realistic seismic input data as well as imperfections of the shell to be treated, too. This should be possible with the numerical methods proposed in this paper more effectively than with standard algorithms in general purpose FEM–computer programs.

7. ACKNOWLEDGEMENT

The authors greatly acknowledge the financial support of the Deutsche Forschungsgemeinschaft (German National Science Foundation) under grant SFB 151 / Project B4.

8. REFERENCES

[1] Housner, G.W.: Dynamic Pressures on Accelerated Fluid Containers. Bull. Seism. Soc. Am., Vol. 47 (1957), 15–35.

[2] Veletsos, A.S.: Seismic Effects in Flexible Liquid Storage Tanks. Proc. 5th World Conf. Earthqu. Eng., Vol. 1 (1974), 630–639.

[3] Zienkiewicz, O.C., Bettess, P.: Fluid–Structure Dynamic Interaction and Wave Forces. An Introduction to Numerical Treatment. Int. J. Num. Meth. Eng., Vol. 13 (1978), 1–16.

[4] Dubois, J.J, De Rouvray, A.L.: An Improved Fluid Superelement for the Coupled Solid–Fluid–Surface Wave Dynamic Interaction Problem. Earthqu. Eng. Struct. Dyn., Vol. 6 (1978), 235–245.

[5] Wunderlich, W.: On the analysis of shells of revolution by transfer matrices. Ing.–Archiv, Vol. 36 (1967), 262–279 (in German)

[6] Wunderlich, W.: On a consistent shell theory in mixed tensor formulations. In 'Theory of Shells', W. Koiter, G.K. Mikhailow (eds.), North–Holland Publishing Company (1980), 607–633.

[7] Wunderlich, W., Cramer, H., Obrecht, H.: Application of Ring Elements in the Nonlinear Analysis of Shells of Revolution Under Non–axisymmetric Loading. Comp. Meth. Appl. Mech. Eng., Vol. 51 (1985), 259–275.

[8] Wunderlich, W., Obrecht, H., Springer, H., Lu, Z.: A Semi–Analytical Approch to the Nonlinear Analysis of Shells of Revolution. In 'Analytical and Computational Models for Shells – ASME Winter Annual Meeting 1989', Noor, A.K., Belytschko, T, Simo, J.C. (eds.), ASME, New York 1989 (in press).

[9] Obrecht, H., Goebel, W., Wunderlich, W.: Nonlinear Dynamic Analysis of Shells of Revolution. In 'Proceedings of the Euromech–Colloquium 219 on Refined Dynamical Theories of Beams, Plates and Shells and Their Applications', Elishakoff, I., Irretier, H. (eds.), Springer Verlag, Berlin (1987), 402–419.

[10] Wunderlich, W., Obrecht, H., Goebel, W.: Nonlinear Dynamic Analysis of Shells of Revolution Using Reduced–Basis Techniques. In 'Proceedings of the International Conference on Numerical Methods in Engineering: Theory and Applications – NUMETA 87 ', Pande, G.N., Middleton, J. (eds.), M. Nijhoff Publ., Dordrecht (1987), T7/1–13.

[11] Wilson, E.L., Yuan, M.W., Dickens, J.M.: Dynamic Analysis by Direct Superposition of Ritz Vectors. Earthquake Eng. Struct. Dyn., Vol. 10 (1982), 813–821.

[12] Wilson, E.L.: A New Method of Dynamic Analysis for Linear and Nonlinear Systems. Fin. Elem. Anal. Design, Vol. 1 (1985), 21–23.

[13] Tseng, N.T., Lee, G.C.: Simple Plasticity Model of Two–Surface Type. Proc. ASCE, J. Eng. Mech., Vol. 109 (1983), 795–810.

[14] Ross, C.T.F., Johns., T., Johns, E.: Vibrations of Ring–Stiffened Cones submerged in Water. In 'Proceedings of the International Conference on Numerical Methods in Engineering: Theory and Applications – NUMETA 87 ', Pande, G.N., Middleton, J. (eds.), M. Nijhoff Publ., Dordrecht (1987), T18/1–8.

[15] Liu, W.K.: Finite Element Procedures for Fluid–Structure Interactions and Applications to Liquid Storage Tanks. Nucl. Eng. and Des., Vol. 65 (1981), 221–238.

Novel Computational Strategies for Solution of Large-Scale Structural Problems

Ahmed K. NOOR and Jeanne M. PETERS

George Washington University
NASA Langley Research Center
Hampton, Virginia

Summary

Novel computational strategies are presented for the analysis of large and complex structures. The strategies are based on generating the response of the complex structure using *large perturbations* from that of a simpler model, associated with a simpler structure (or a simpler mathematical/discrete model of the original structure). Numerical examples are presented to demonstrate the effectiveness of the strategies developed.

1. Introduction

Considerable effort has been devoted to reducing the cost and/or time of solving large structural problems. Most of these efforts can be thought of as applications of multilevel computational strategies, which include partitioning methods, hybrid modeling/analysis techniques, and reduction methods (see, for example, Park and Felippa [1], Kobayashi [2], Atluri and Nishioka [3], and Noor [4]). The present study is an attempt to unify and realize the full potential of a number of multilevel computational strategies for solution of large-scale structural problems. Specifically, the objective of the present paper is to report the progress made in developing computational strategies which combine the following three major characteristics:

1. give physical insight about the response;
2. help in assessing the adequacy of the computational model; and
3. are highly efficient.

2. Basic Idea and Key Elements of the Proposed Strategies

The strategies developed are based on generating the response of a large and complex structure using *large perturbations* from that of a simpler model associated with a simpler structure (or a simpler mathematical/discrete model of the original structure). The key elements of the proposed strategies are: 1) mixed (primitive variable) formulation with the fundamental unknowns consisting of generalized displacements and stress parameters; 2) operator splitting

G. Kuhn · H. Mang (Eds.)
Discretization Methods in Structural Mechanis
IUTAM/IACM Symposium Vienna, Austria, 1989

or restructuring of the discrete equations of the original (complex) structure to delineate the contributions of the simpler model and the correction terms; and 3) iterative process, or a reduction method, for the efficient generation of the response of the complex structure starting from that of the simpler structure. Although in some applications the strategies can be used in conjunction with the single-field displacement formulation, there are several applications in which the use of the mixed formulation is essential.

3. Relations Between the Original and Simpler Models

The crux of the proposed strategies is the proper selection of the simpler structure and/or the simpler model. For a given complex structure, a multitude of choices of the simpler structure can be made. The best choice is that which satisfies, to the greatest extent possible, the three characteristics listed in the preceding section. For each choice, the relations between the simpler model and the original structure can be established, and an efficient computational procedure can be developed for generating the response of the original structure using *large perturbations* from that of the simpler model. Two general approaches are described subsequently for selecting the simpler model and establishing the relations between the original and simpler models. The two approaches are: *partitioning or decomposition*, and *hierarchical modeling strategy*. For simplicity, the two approaches are described with reference to their application to linear stress analysis problems. However, they are equally applicable to nonlinear and dynamic problems.

The governing equations for the discrete model of the original structure can be written in the following compact form:

$$[K]\{Z\} = \{Q\} \tag{1}$$

where $\{Z\}$ is the vector of stress parameters and generalized displacements; $[K]$ is the global structure matrix which includes the flexibility and strain-displacement matrices; and $\{Q\}$ is the global right-hand side vector.

3.1 Decomposition or Partitioning Strategy

In the decomposition strategy the vector of fundamental unknowns $\{Z\}$ is partitioned into smaller subvectors. The governing discrete equations, Eqs. 1, are partitioned accordingly. The simpler model is associated with the uncoupled equations in the partitioned variables. For the case of two partitions, the process can be described by embedding Eqs. 1 in a single-parameter family of equations as follows:

$$\left(\begin{bmatrix} K_{11} & \cdot \\ \cdot & K_{22} \end{bmatrix} + \lambda \begin{bmatrix} \cdot & K_{12} \\ K_{21} & \cdot \end{bmatrix}\right) \begin{Bmatrix} Z_1 \\ Z_2 \end{Bmatrix} = \begin{Bmatrix} Q_1 \\ \cdot \end{Bmatrix} + \lambda \begin{Bmatrix} \cdot \\ Q_2 \end{Bmatrix} \tag{2}$$

where $\{Z_1\}$, $\{Z_2\}$ and $\{Q_1\}$, $\{Q_2\}$ are the partitions of the original vectors $\{Z\}$ and $\{Q\}$; λ is a tracing parameter which identifies all the correction terms needed in going from the simpler model to the discrete model of the original structure; and a dot (•) refers to a zero submatrix or a zero subvector. The simpler model corresponds to $\lambda=0$ (uncoupled equations in $\{Z_1\}$ and $\{Z_2\}$), and the discrete model of the original structure corresponds to $\lambda=1$ (fully coupled equations). The solution corresponding to $\lambda=1$ is generated from the corresponding solution at $\lambda=0$ using a reduction method, or an iterative process such as the Preconditioned Conjugate Gradient (PCG). Note that the basis vectors in the reduction method, and the correction vectors of the iterative process, provide a direct measure of the sensitivity of the response quantities to the complicating factors (the terms associated with the tracing parameter λ in Eqs. 2). Among the effective partitioning strategies developed in the present study are the following two:

a) *Uncoupling of Load-Carrying Mechanisms.* For example, if the original structure is associated with a two-dimensional shell structure, the simpler structure can be associated with the corresponding plate structure in which the membrane and bending load-carrying mechanisms are uncoupled.

b) *Symmetry Transformations.* This is accomplished by selecting $\{Z_1\}$ and $\{Z_2\}$ to represent the symmetric and antisymmetric components of the response vector (each is approximately half the size of the original vector, $\{Z\}$). The simpler model ($\lambda=0$) corresponds to a symmetrized structure in which the symmetric and antisymmetric components of the response vector are uncoupled. This approach can be thought of as a *physical domain decomposition.* If the reduction method is used in generating the solution at $\lambda=1$, each of the basis vectors, generated at $\lambda=0$, is either symmetric or antisymmetric. This is also true for the correction vectors in the iterative method. The details of the symmetry transformations are described in Noor and Peters [5].

3.2 Hierarchical Modeling

The simpler model is selected as one with considerably fewer degrees of freedom than those of the discrete model of the original structure. The governing equations of the simpler model can be written in an analogous form to those of the original structure, as follows:

$$[k]\{z\} = \{q\} \tag{3}$$

The response vectors of the original and simpler models are related through the interpolation operator $[\Gamma]$ as follows:

$$\{Z\} = [\Gamma]\{z\} \tag{4}$$

where $[\Gamma]$ is a rectangular transformation matrix (*interpolation operator*).

Based on Eqs. 4, the relations between $[k]$, $[K]$ and $\{q\}$, $\{Q\}$ are given by:

$$[k] = [\Gamma]^t[K][\Gamma] \tag{5}$$

$$\{q\} = [\Gamma]^t\{Q\} \tag{6}$$

where superscript t denotes transposition, and $[\Gamma]^t$ is referred to as the *restriction operator*.

Among the possible choices of the simpler model in this category are the following two:

a) *Discrete model, based on the same mathematical model and discretization procedure used for the original structure*, but with considerably fewer degrees of freedom. The resulting numerical process is similar to the classical multigrid technique.

b) *Discrete model, based on a mathematical model of lower dimensionality than that used for the original structure.* For example, if the original structure is modeled using a two-dimensional plate or shell theory, the simpler model can be associated with a one-dimensional thin-walled beam theory (with the effects of flexural-torsional coupling included). The interpolation operator, $[\Gamma]$, then reflects the basic assumptions made in the dimensionality reduction (i.e., in going from a two-dimensional shell/plate theory to a one-dimensional thin-walled beam theory; namely, the projection of each cross-section on a plane normal to the initial centroidal axes does not distort during deformation). This approach can, therefore, be thought of as a *physical multigrid method.*

4. Numerical Studies

To assess the effectiveness of the foregoing strategies, a number of linear and nonlinear stress analysis, free vibration, buckling and dynamic problems of complex structures have been solved by these strategies. For each problem the solutions obtained by the foregoing strategies were compared with those obtained by the direct analysis of the original (complex) structure. Some of these applications are reported in Noor and Whitworth [6] and [7], and Noor and Peters [8]. Two applications of the strategies are presented herein: 1) linear stress analysis of cantilevered composite plate with unsymmetrical lamination in the thickness direction; and 2) postbuckling of a flat unstiffened composite panel subjected to uniform end shortening, and having an unsymmetric postbuckling response. The two applications are discussed subsequently.

Both structures were modeled by using mixed finite element models with the stress resultants allowed to be discontinuous at interelement boundaries. A 12×12 grid was used for the first structure and a 26×8 grid was used for the second structure. Biquadratic shape functions were used for approximating each of the generalized displacements, and bilinear shape functions were used for approximating each of the stress resultants. The characteristics of the finite element model are given in Noor and Andersen [9].

4.1 Linear Stress Analysis of Cantilevered Composite Plate

As an application of the strategy based on uncoupling of the load-carrying mechanisms, consider the graphite-epoxy composite plate with trapezoidal planform shown in Fig. 1. The plate is subjected to uniform normal loading and has an unsymmetric lamination in the

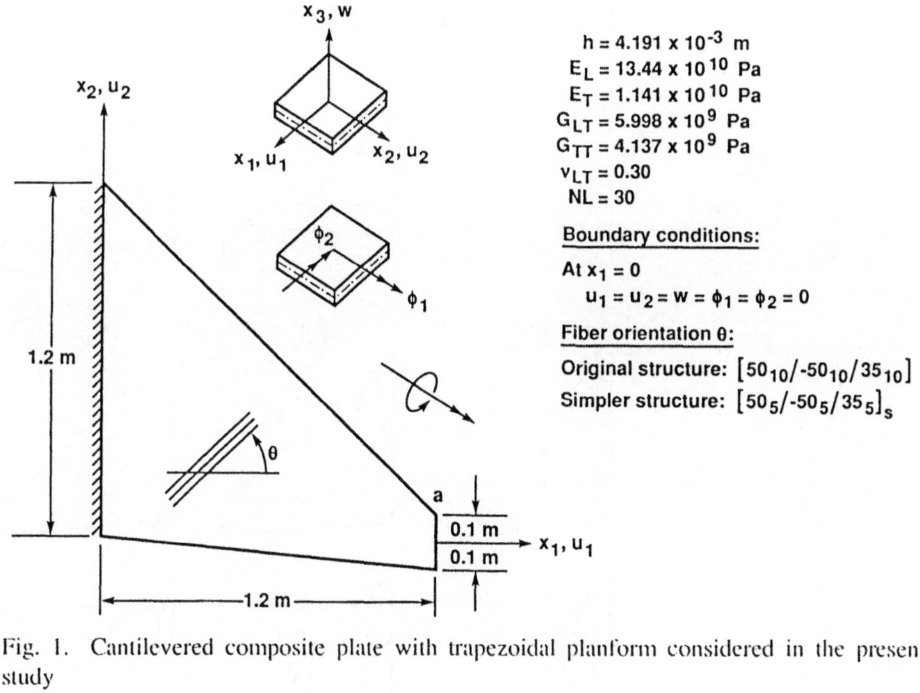

Fig. 1. Cantilevered composite plate with trapezoidal planform considered in the present study

thickness direction. The bending and membrane load-carrying mechanisms of the structure are, therefore, coupled. The simpler structure is selected to be a composite plate with the same planform and total thickness, but with symmetric lamination (see Fig. 1). The bending and membrane load-carrying mechanisms in the simpler structure are, therefore, uncoupled.

The same uncoupling is exhibited by the basis vectors, evaluated at $\lambda=0$. This fact is depicted in Fig. 2 in which normalized contour plots are shown for the generalized displacements of the first six basis vectors. At $\lambda=0$, the in-plane displacements vanish for the even-order

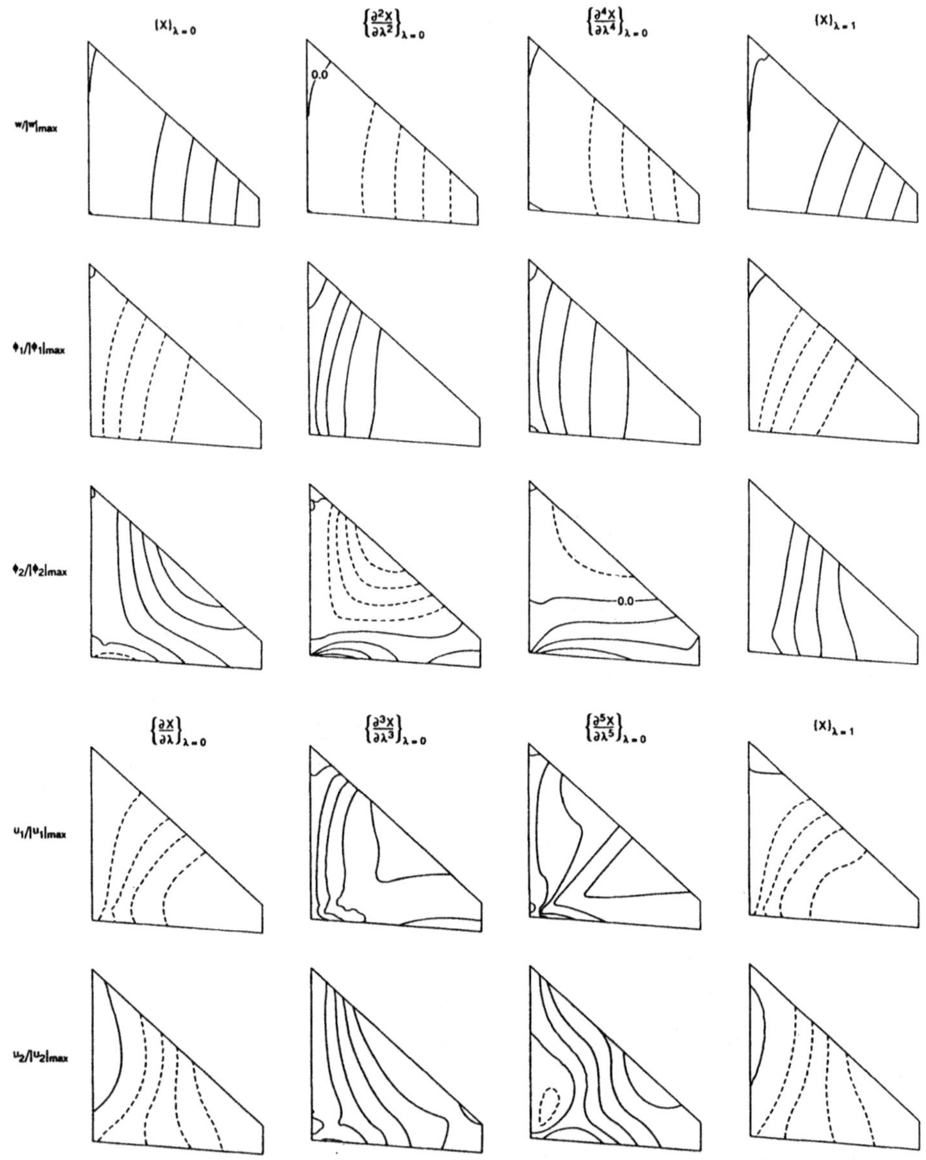

Fig. 2. Normalized contour plots for the nonzero basis vectors at $\lambda=0$, and the generalized displacements evaluated at $\lambda=1$. Cantilevered composite plate shown in Fig. 1. Spacing of contour lines is 0.2 and dashed lines represent negative contours

for the odd-order derivatives with respect to λ. The uncoupling between the membrane and bending degrees of freedom of the basis vectors can be used to reduce the computational effort in their evaluation.

Figure 2 also shows the normalized contour plots for the generalized displacements in the original structure. The strong coupling between the in-plane and bending degrees of freedom is evident from this figure. An indication of the accuracy and convergence of the solutions obtained by both the foregoing strategy and the Taylor series expansion, with the increase in the number of basis vectors is given in Fig. 3. The standard of comparison is taken to be the direct solution of the original structure. As can be seen from Fig. 3, the transverse displacement and total strain energy obtained by the foregoing strategy have a very rapid convergence. On the other hand, the Taylor series solutions diverge.

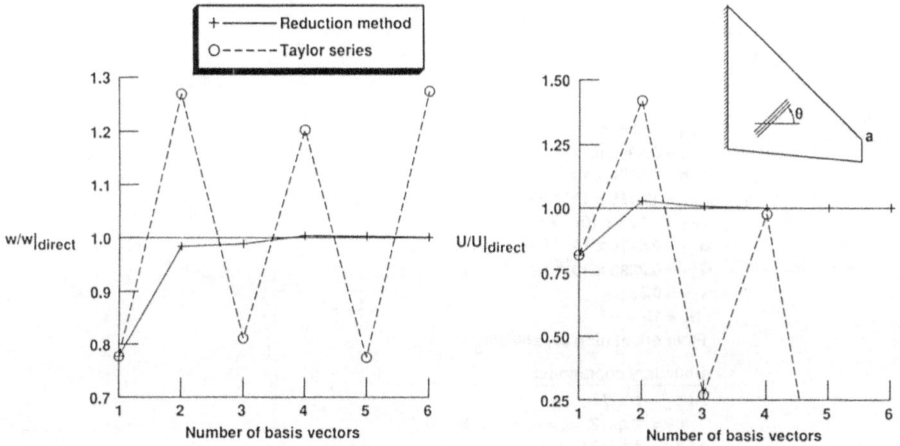

Fig. 3. Convergence of transverse displacement, w, at point a and total strain energy, U, obtained by reduction method and Taylor series expansion. Cantilevered composite plate shown in Fig. 1

4.2 Postbuckling of Composite Panel Subjected to Uniform End Shortening

As an application of the strategy based on symmetry transformations consider the composite panel shown in Fig. 4. The panel has fifteen layers and is subjected to uniform end shortening, q_e.

The response of the panel is depicted in Figs. 4 and 5. In Fig. 4 plots of the load versus end shortening and transverse displacement at point a, are shown. The response curves are in close agreement with the experimental results presented in Jeffrey [10]. Because of the bending-extensional coupling of the panel, the transverse (out-of-plane) displacements are not zero in the prebuckling range. In Fig. 5 normalized contour plots for the transverse displacement as well as its symmetric/antisymmetric components at $\frac{N_t L_2}{E_T h^3} = 26.59$ (where N_t is the total axial force at the edge), are shown. Each generalized displacement is divided by its maximum absolute value. As can be seen from Fig. 5, the postbuckling response is unsymmetric (see Noor and Peters [11]).

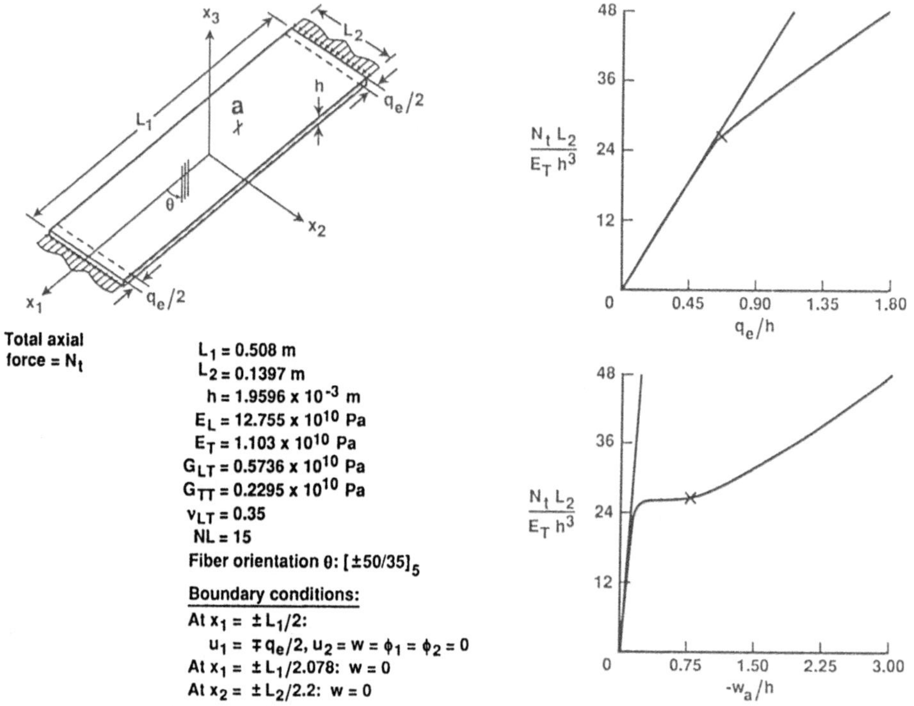

Fig. 4. Response of fifteen-layer graphite-epoxy panel subjected to uniform end shortening. N_t= total axial force acting on the edges $x_1 = \pm L_1/2$

The foregoing strategy was applied to the generation of the post-bifurcation equilibrium configurations. The panel was divided into four substructures, and symmetry transformations were used to reduce the analysis model to one quarter of the panel. Reflection symmetry (and antisymmetry) was assumed with respect to the centerlines of the panel (see Fig. 5).

Based on the displacement incrementation approach, an initial nonlinear solution was obtained in the vicinity of the bifurcation point (corresponding to a prescribed displacement u_1=0.317 h at x_1=-0.46 L_1 and x_2=0). The prescribed displacement was decomposed into four symmetric/antisymmetric components. Eleven Newton-Raphson iterations and an average of 25 preconditioned conjugate gradient (PCG) iterations, per Newton-Raphson iteration, were required to obtain convergence to within six significant digits for the response quantities. The number of PCG iterations represent the total for the two right-hand sides associated with the displacement incrementation algorithm (see Noor and Peters [4]). Eight path derivatives were generated using the foregoing strategy. An average of eleven PCG iterations were required to

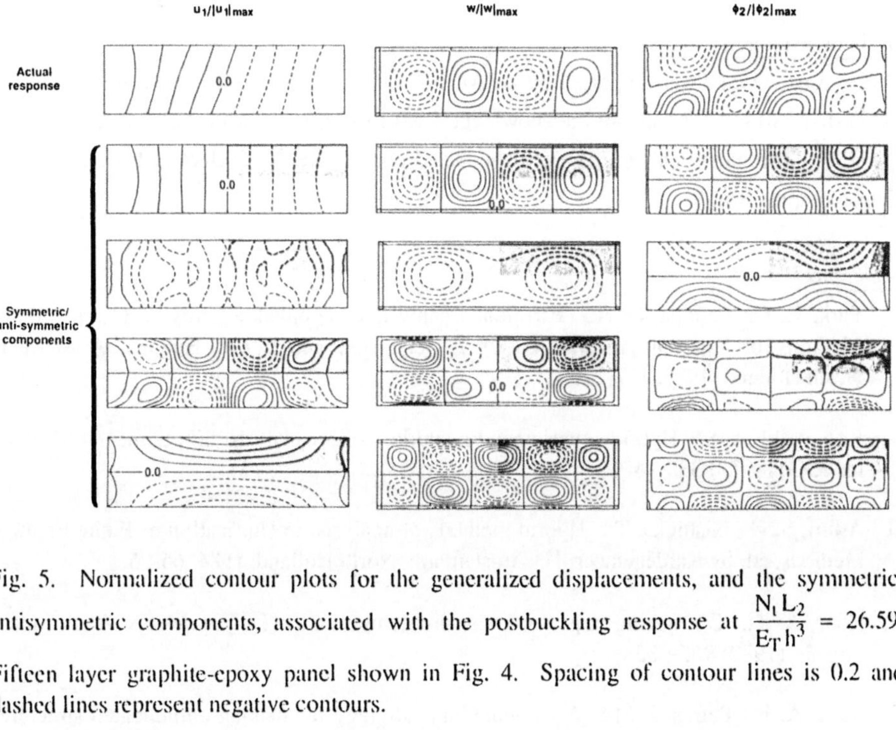

Fig. 5. Normalized contour plots for the generalized displacements, and the symmetric/antisymmetric components, associated with the postbuckling response at $\frac{N_t L_2}{E_T h^3} = 26.59$. Fifteen layer graphite-epoxy panel shown in Fig. 4. Spacing of contour lines is 0.2 and dashed lines represent negative contours.

obtain six significant digits of accuracy for each of the path derivatives. Reduced equations were formed and used in conjunction with the arc-length incrementation approach to advance the solution to a value of the normalized axial force, $\frac{N_t L_2}{E_T h^3} = 37.65$, at which the error of the reduced solution exceeded the prescribed tolerance. The reduced solution was used as a predictor and Newton-Raphson technique was applied to obtain a corrected (improved) solution. The corrected value for $\frac{N_t L_2}{E_T h^3}$ was 37.21. Five Newton-Raphson iterations and an average of 25 PCG iterations, per Newton-Raphson iteration, were required to obtain six significant digits of accuracy in the response quantities. Again, the number of PCG iterations represent the total for the two right-hand sides associated with the arc-length incrementation algorithm. An updated set of path derivatives was generated. An average of twelve PCG iterations, for each of the path derivatives, was required to obtain the same degree of accuracy. Note that *the PCG iterations are computationally inexpensive*; and *the most time-consuming tasks in the solution process are the decomposition of the left-hand side matrices in the Newton-Raphson iterations. The foregoing strategy reduces the size of the equations to about one quarter of the corresponding ones used in the direct analysis.*

Acknowledgment

The present research is supported by an Air Force Office of Scientific Research Grant No. AFOSR-87-0115. The authors acknowledge useful discussions with Anthony K. Amos of AFOSR. The numerical results were conducted on the CRAY X-MP/48 of the San Diego Supercomputer Center.

References

1. Park, K. C.; Felippa, C. A.: Partitioned analysis of coupled systems, in Computational Methods for Transient Analysis, ed. by T. Belytschko and T. J. R. Hughes. Amsterdam: North Holland, 1983, 157-219.

2. Kobayashi, A. S.: Hybrid experimental-numerical stress analysis. Experimental Mechanics 23(3) (1983) 338-347.

3. Atluri, S. N.; Nishioka, T.: Hybrid methods of analysis, in Unification of Finite Element Methods, ed. by Kardestuncer, H. Amsterdam: North Holland, 1984, 65-95.

4. Noor, A. K.: On making large nonlinear problems small. Comput. Meths. Appl. Mech. Eng. 34 (1982) 955-985.

5. Noor, A. K.; Peters, J. M.: A computational strategy for making complicated structural problems simple. Comput. Meths. Appl. Mech. Eng. 71(2) (1988) 167-182.

6. Noor, A. K.; Whitworth, S. L.: Computational strategy for analysis of quasi-symmetric structures. J. Eng. Mech. Div., ASCE 114(3) (1988) 456-477.

7. Noor, A. K.; Whitworth, S. L.: Vibration analysis of quasi-symmetric structures. Finite Elements in Analysis and Design 3 (1987) 257-276.

8. Noor, A. K.; Peters, J. M.: A new partitioning strategy for efficient nonlinear finite element dynamic analysis on multiprocessor computers. Comput. & Struct. 31(5) (1989) 795-810.

9. Noor, A. K.; Andersen, C. M.: Mixed models and reduced/selective integration displacement models for nonlinear shell analysis. Int. J. Num. Meths. Eng. 18 (1982) 1429-1454.

10. Jeffrey, G. L.: Postbuckling of laminated anisotropic panels. NASA TM-100509 (1987).

11. Noor, A. K.; Peters, J. M.: Buckling and postbuckling analyses of laminated anisotropic structures. Int. J. Num. Meths. Eng. (to appear).

Consistent Formulation of Shear Constrained Reissner-Mindlin Plate Elements

E. OÑATE*, R.L. TAYLOR**, O.C. ZIENKIEWICZ***

*Universidad Politécnica de Catalunya, 08039 Barcelona, Spain.
**University of California, Berkeley, U.S.A.
***University College of Swansea, U.K.

Summary

In this paper the necessary requeriments for the good behaviour of shear constrained Reissner-Mindlin plate elements for thick and thin plate situations are discussed and a simple explicit form of the substitute shear strain matrix is derived. This is applied to the development of some new triangular plate elements which show promising features. Some examples of the good behaviour of one of these elements are given.

Introduction

Recent work in the development of efficient plate finite elements has been mostly based in the so called Reissner-Mindlin thick plate theory [9],[11]. This by-passes the difficulties caused by the C^1 requirement of the classic Kirchhoff theory [15]. However, its direct aplications to thin plate situations induces locking and various artifices to eliminate this effect, like the introduction *of reduced or selective integration* procedures [1],[6], [7],[10],[12] or the use of *substitute shear strain fields* [2]–[5] have been proposed.

It is now clear that both these approaches can been re-interpreted in the more general framework of a mixed formulation in which shear forces and displacements are approximated independently, [13]. Moreover, the mixed form provides the necessary requirements that the elements should satisfy to be applicable for both thick and thin plate situations . This has allowed to define a general methodology for the development of successful shear constrained plate elements and indeed some of these have been recently reported by the autors [14],[17].

In this paper we re-examine the problem of shear constrained thick plate elements. We show that the condition of vanishing shear strain for the thin plate limit can only be naturally achieved if the coefficients defining the shear strain polynomial are a linear function of nodal rotations and deflections. This explains the success of reduced integration techniques and substitute transverse shear strain fields which lead to the satisfaction of such a condition. In the paper, the methodology for defining adequate substitute shear strain fields and the obtention of the corresponding substitute shear strain matrix are detailed. Some of the new triangular plate elements developed in

G. Kuhn · H. Mang (Eds.)
Discretization Methods in Structural Mechanis
IUTAM/IACM Symposium Vienna, Austria, 1989

[17] are discussed and examples of the good behaviour of the linear/quadratic discrete Reissner-Mindlin (DRM) triangle are given.

Basic Concepts

The basic expressions of the Reissner-Mindlin plate theory are the following:

Plate curvatures

$$\boldsymbol{\chi} = \begin{Bmatrix} \chi_x \\ \chi_y \\ \chi_{xy} \end{Bmatrix} = \begin{Bmatrix} \frac{\partial \theta_x}{\partial x} \\ \frac{\partial \theta_y}{\partial y} \\ \frac{\partial \theta_x}{\partial y} + \frac{\partial \theta_y}{\partial x} \end{Bmatrix} = \begin{bmatrix} \frac{\partial}{\partial x} & 0 \\ 0 & \frac{\partial}{\partial y} \\ \frac{\partial}{\partial y} & \frac{\partial}{\partial x} \end{bmatrix} \begin{Bmatrix} \theta_x \\ \theta_y \end{Bmatrix} = \mathbf{L} \; \boldsymbol{\theta} \tag{1}$$

Shear strains

$$\boldsymbol{\gamma} = \begin{Bmatrix} \gamma_x \\ \gamma_y \end{Bmatrix} = \begin{Bmatrix} \frac{\partial \omega}{\partial x} + \theta_x \\ \frac{\partial \omega}{\partial y} + \theta_y \end{Bmatrix} = \begin{bmatrix} \frac{\partial}{\partial x} & 1 & 0 \\ \frac{\partial}{\partial y} & 0 & 1 \end{bmatrix} \begin{Bmatrix} \omega \\ \theta_x \\ \theta_y \end{Bmatrix} = \mathbf{S} \; \mathbf{u} \tag{2}$$

In above ω, θ_x and θ_y are the deflection and rotations of a point in the plate mid-surface.

For sign convenion see Figure 1.

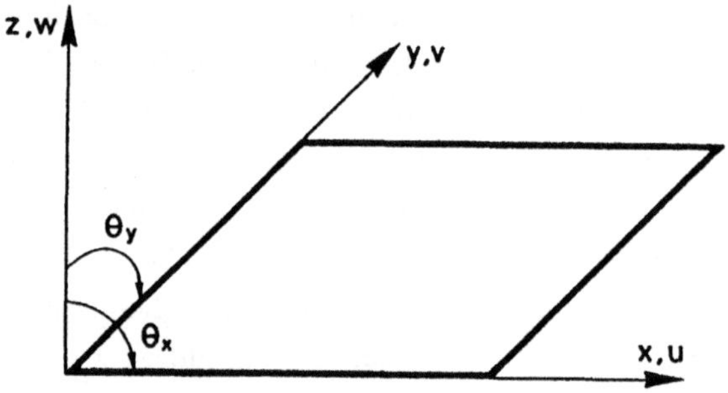

Figure 1. Definition of displacement variables in a plate.

Bending moments/curvatures relationship

$$\mathbf{m} = \begin{Bmatrix} M_x \\ M_y \\ M_{xy} \end{Bmatrix} = \frac{Et^3}{12(1 - \nu^2)} \begin{bmatrix} 1 & \nu & 0 \\ \nu & 1 & 0 \\ 0 & 0 & \frac{1-\nu}{2} \end{bmatrix} \begin{Bmatrix} \chi_x \\ \chi_y \\ \chi_{xy} \end{Bmatrix} = \mathbf{D}_b \; \boldsymbol{\chi} \tag{3}$$

Shear forces/shear strains relationship

$$\mathbf{s} = \begin{Bmatrix} Q_x \\ Q_y \end{Bmatrix} = G\alpha t \begin{bmatrix} 1 & 0 \\ 0 & 1 \end{bmatrix} \begin{Bmatrix} \gamma_x \\ \gamma_y \end{Bmatrix} = \mathbf{D}_s \, \boldsymbol{\gamma} \tag{4}$$

In above E, G, ν ,t are the Young's modulus, shear modulus, the Poisson's ratio and the plate thickness, respectively and α is the warping coefficient (usually taken as $\alpha = 5/6$).

Total plate energy

$$\Pi = \frac{1}{2}\int\int_A \boldsymbol{\chi}^T \mathbf{m} \; dA + \frac{1}{2}\int\int_A \boldsymbol{\gamma}^T \mathbf{s} \; dA - \int\int_A wq \; dA \tag{5}$$

The first two integrals of (5) represent the bending and shear energies, respectively, whereas the third represents the external energy due to a distributed force q acting over the plate area A.

Finite Element Discretization

We consider a finite element discretization of the plate using isoparametric elements. The deflection ω and the rotations θ_x, θ_y can be interpolated using a different approximation as

$$\omega = \mathbf{N}_\omega \bar{\omega} \quad , \quad \boldsymbol{\theta} = \mathbf{N}_\theta \bar{\boldsymbol{\theta}} \tag{6}$$

where $(\bar{\cdot})$ denotes nodal values ($\bar{\boldsymbol{\omega}}_i = [\omega_i]$, $\bar{\boldsymbol{\theta}}_i = [\theta_{x_i}, \theta_{y_i}]^T$).

Eqs(6) can be combined to give

$$\mathbf{u} = [\omega, \theta_x, \theta_y]^T = \mathbf{N}\bar{\mathbf{u}} \qquad \text{with} \qquad \bar{\mathbf{u}}_i = [\omega_i, \theta_{x_i}, \theta_{y_i}]^T \tag{7}$$

Substitution of eqs(6)–(7) in (1)–(2) yields

$$\boldsymbol{\chi} = \mathbf{L}\boldsymbol{\theta} = \mathbf{L}\mathbf{N}_\theta \bar{\boldsymbol{\theta}} = \mathbf{B}_b \bar{\mathbf{u}} \tag{8}$$

$$\boldsymbol{\gamma} = \mathbf{S}\mathbf{u} = \mathbf{S}\mathbf{N}\bar{\mathbf{u}} = \mathbf{B}_s \bar{\mathbf{u}} \tag{9}$$

where $\mathbf{B}_b$ and $\mathbf{B}_s$ are the standard bending and shear strain matrices, respectively [5]. Substitution of eqs(8)–(9) in the expression of the potential energy (5) yields, after minimization, the discretized equilibrium expressions and the standard form of the bending and shear stiffness matrices is obtained [5],[10],[12].

Thin plate limit

In the thin plate limit Kirchhoff's assumptions of vanishing transverse shear strain must be satisfied. This implies $\boldsymbol{\gamma} = \mathbf{0}$. From eq. (9) we can write for rectangular or straight sides triangular elements with n nodes

$$\gamma \;=\; \mathbf{B}_s\bar{\mathbf{u}} \;=\; \alpha_1(\omega_i,\theta_i) \;+\; \alpha_2(\omega_i,\theta_i)\xi \;+\; \alpha_3(\omega_i,\theta_i)\eta \;+\; \cdots \;+\; \alpha_n(\omega_i,\theta_i)\xi^p\eta^q \;=\; 0 \quad (10)$$

Satisfaction of (10) implies

$$\boldsymbol{\alpha}_j(\omega_i,\boldsymbol{\theta}_i) \;=\; 0 \qquad ; \qquad j=1,n \tag{11}$$

Eq(10) imposes a linear relationship between nodal deflections and rotations which usually can also be interpreted on physical grounds. Elements satisfying (11) can therefore reproduce naturally the limit thin plate conditions with absence of locking.

However, in many elements the $\boldsymbol{\alpha}_j$'s are a function of the nodal rotations only, and the condition $\boldsymbol{\alpha}_j(\boldsymbol{\theta}_i) = 0$ becomes too restrictive (and even unnatural) thus leading to locking.

A simple way to avoid locking is thus to sample the shear strains at points where the spurious $\boldsymbol{\alpha}_k(\theta_i)$ vanish.This is the basis of reduced integration procedures widely used in practice [10],[12]. However, reduced integration techniques have proved not to be generally sufficient for the development of robust isoparametric thick plate elements, [10],[12]. An alternative procedure is to impose "a priori" a shear strain field which satisfies the condition (10), thus allowing the natural satisfaction of the thin plate condition. The shear strains are now written as

$$\boldsymbol{\gamma} \;=\; \sum_{k=1}^{m} \mathbf{N}_{\gamma_k}\,\boldsymbol{\gamma}_k \;=\; \mathbf{N}_\gamma\;\bar{\boldsymbol{\gamma}} \tag{12}$$

Where $\boldsymbol{\gamma}_k$ are the values of the shear strains at some selected points within the element. Combining eqs(12) and (9)yields

$$\boldsymbol{\gamma} \;=\; \sum_{k=1}^{m} \mathbf{N}_{\gamma_k}\mathbf{B}_{s_k}\mathbf{u}_k \;=\; \hat{\mathbf{B}}_s\bar{\mathbf{u}} \tag{13}$$

It is easy to check that eq (12) can be written in the form (11) which guarantees the absence of locking.

The expression of the total potential energy can now be written using eqs. (8) and (12) as

$$\Pi = \frac{1}{2}\int\int_A [\mathbf{L}\boldsymbol{\theta}]^T \mathbf{D}_b \mathbf{L}\boldsymbol{\theta}\, dA + \frac{1}{2}\int\int_A [\mathbf{N}_\gamma\, \bar{\boldsymbol{\gamma}}]^T \mathbf{D}_s \mathbf{N}_\gamma\, \bar{\boldsymbol{\gamma}}\, dA - \int\int_A \omega q\, dA \qquad (14)$$

Note that in (14) C_o continuity is required for the rotations $\boldsymbol{\theta}$, whereas the deflection ω, and the substitute shear strains $\bar{\boldsymbol{\gamma}}$ can be discontinuous.

Choice of substitute shear strain fields

Adequate substitute shear strain fields can be successfully chosen by direct observation, having in mind the objectives of obtaining strain fields satisfying eq (11).

Recently, the authors have shown [13],[16] that the choice of the substitute strain field must satisfy certain basic requeriments. Thus, starting with the mixed form in which deflections, rotations and shear strains are independently interpolated the conditions to be satisfied are the following:

$$\begin{aligned} n_\theta + n_\omega &\geq n_\gamma \\ n_\gamma &\geq n_\omega \end{aligned} \qquad (15)$$

Where n_ω, n_θ and n_γ stand for the number of variables in each set of interpolating parameters $\bar{\boldsymbol{\omega}}$, $\bar{\boldsymbol{\theta}}$ and $\bar{\boldsymbol{\gamma}}$, respectively.The above inequalities have to be satisfed for any element patches as a condition which is necessary (although not always sufficient) for convergence [16].

In reference [17] the authors have examined a number of currently used plate elements and found that all those proving to be successfull in practice pass this stringent test. Also in [17] the authors have proposed new triangular plate elements which show very promising features. In next section a methology for the obtention of the substitute shear strain matrix for all these elements is presented.

A methodology for the obtention of the substitute shear strain matrix $\bar{\mathbf{B}}_s$

We consider the obtention on the substitute shear strain matrix of an isoparametric plate element of n nodes with an independent interpolation of deflections, rotations and shear strains defined by eqs. (6) and (12), and satisfying conditions (15).

Step 1. The starting point is the interpolation of the natural shear strains in the natural coordinate system ξ, η as follows

$$\boldsymbol{\gamma}' = \begin{Bmatrix} \gamma_\xi \\ \gamma_\eta \end{Bmatrix} = \begin{bmatrix} 1 & \xi & \eta & \xi\eta & \cdots & \xi^p\eta^q & | & 0 & 0 & 0 & \cdots & 0 \\ 0 & 0 & 0 & 0 & \cdots & 0 & | & 1 & \xi & \eta & \cdots & \xi^r\eta^s \end{bmatrix} \begin{Bmatrix} \alpha_1 \\ \alpha_2 \\ \vdots \\ \alpha_{n_\gamma} \end{Bmatrix} = \mathbf{A}\boldsymbol{\alpha} \qquad (16)$$

The cartesian shear strains are directly obtained as

$$\boldsymbol{\gamma} = \begin{Bmatrix} \gamma_x \\ \gamma_y \end{Bmatrix} = \mathbf{J}^{-1}\boldsymbol{\gamma}' \tag{17}$$

Where **J** is the standard jacobian matrix of the transformation $x, y \rightarrow \xi\ ,\ \eta$.

Step 2. The shear strains are prescribed in n_γ points along some particular natural directions $\bar{\xi}_i$ such that

$$\gamma_{\bar{\xi}_i} = \cos\beta_i \gamma_\xi + \sin\beta_i \gamma_\eta \tag{18}$$

β being the angle that the $\bar{\xi}$ forms with the natural coordinate ξ.

Substituting eq. (16) in (18) and sampling the resulting equation at the n_γ shear strain points we obtain the following system of equations

$$\mathbf{P}(\xi_i, \eta_i, \beta_i)\boldsymbol{\alpha} = \boldsymbol{\gamma}_{\bar{\xi}} \tag{19}$$

where $\boldsymbol{\gamma}_{\bar{\xi}} = [\gamma^1_{\bar{\xi}}, \gamma^2_{\bar{\xi}}, \cdots, \gamma^{n_\gamma}_{\bar{\xi}}]^T$ contains the prescribed shear strains at the n_γ points. From eq. (19) it can be found

$$\boldsymbol{\alpha} = \mathbf{P}^{-1}\boldsymbol{\gamma}_{\bar{\xi}} \tag{20}$$

Step 3. The shear strains along the $\bar{\xi}_i$ directions are related to the natural shear strains at the n_γ points by the simple transformation

$$\boldsymbol{\gamma}_{\bar{\xi}} = \mathbf{T}(\beta_i)\hat{\boldsymbol{\gamma}}' \tag{21}$$

with $\hat{\boldsymbol{\gamma}}' = [\gamma^1_\xi, \gamma^1_\eta, \gamma^2_\xi, \gamma^2_\eta \cdots, \gamma^{n_\gamma}_\xi, \gamma^{n_\gamma}_\eta]^T$.

Step 4. The natural and cartesian shear strains at each sampling point are related by

$$\hat{\boldsymbol{\gamma}}' = \begin{bmatrix} \mathbf{J}^1 & & 0 \\ & \ddots & \\ 0 & & \mathbf{J}^{n_\gamma} \end{bmatrix} \begin{Bmatrix} \hat{\boldsymbol{\gamma}}_1 \\ \vdots \\ \hat{\boldsymbol{\gamma}}_{n_\gamma} \end{Bmatrix} = \mathbf{C}\,\hat{\boldsymbol{\gamma}} \tag{22}$$

Where $\hat{\gamma}_i = [\gamma^i_x\ ,\ \gamma^i_y]$ and $\mathbf{J}^i$ is the jacobian matrix at the ith sampling point.

Step 5. The cartesian shear strains at each sampling point are related with the nodal displacement by

$$\hat{\boldsymbol{\gamma}} = \begin{Bmatrix} \hat{\gamma}_1 \\ \vdots \\ \hat{\gamma}_{n_\gamma} \end{Bmatrix} = \begin{Bmatrix} \mathbf{B}_s^1 \\ \vdots \\ \mathbf{B}_s^{n_\gamma} \end{Bmatrix} \bar{\mathbf{u}} = \bar{\mathbf{B}}\bar{\mathbf{u}} \tag{23}$$

where $\mathbf{B}_s^i$ is the standard shear strain matrix computed at sampling point i.

Step 6. Combining steps 1-5 we finally obtain

$$\boldsymbol{\gamma} = \mathbf{J}^{-1}\,\mathbf{A}\,\mathbf{P}^{-1}\,\mathbf{T}\,\mathbf{C}\,\bar{\mathbf{B}}\,\bar{\mathbf{u}} = \hat{\mathbf{B}}_s\,\bar{\mathbf{u}} \tag{24}$$

$$\text{with} \qquad \hat{\mathbf{B}}_s = \mathbf{J}^{-1}\,\mathbf{A}\,\mathbf{P}^{-1}\,\mathbf{T}\,\mathbf{C}\,\bar{\mathbf{B}} \tag{25}$$

being the substitute shear strain matrix.

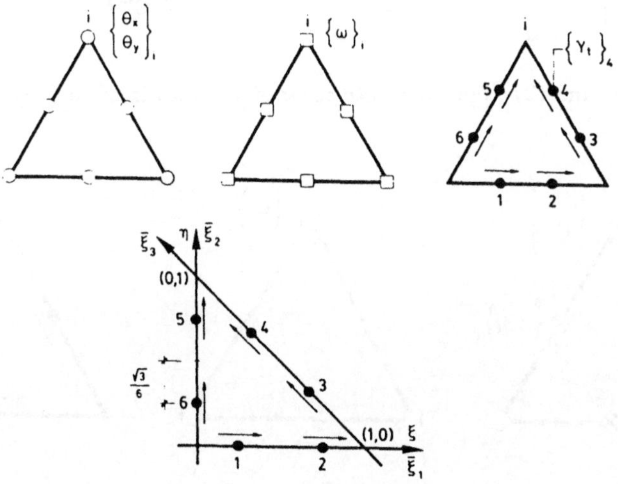

Figure 2. Six node quadratic triangular plate element with linear shear strain field.

Application to the six node quadratic triangular plate element with linear shear strain field

Recently [17] the authors have shown the good performance of the six node triangular element with quadratic variations for the deflection ω and the rotation fields $\boldsymbol{\theta}$ and a linear interpolation of the shear strain as

$$\begin{aligned} \gamma_\xi &= \alpha_1 + \alpha_2\xi + \alpha_3\eta \\ \gamma_\eta &= \alpha_4 + \alpha_5\xi + \alpha_6\eta \end{aligned} \qquad \text{i.e} \quad \mathbf{A} = \begin{bmatrix} 1 & \xi & \eta & 0 & 0 & 0 \\ 0 & 0 & 0 & 1 & \xi & \eta \end{bmatrix} \tag{26}$$

The location of the six shear sampling points is shown in Figure 2. It is easy to check that the element passes successfully the conditions (15) for all element patches [16],[17].

The local directions $\bar{\xi}_i$ are shown in Figure 2. It is easy to find for this case

$$\mathbf{P} = \begin{bmatrix} 1 & \xi_1 & \eta_1 & 0 & 0 & 0 \\ 1 & \xi_2 & \eta_2 & 0 & 0 & 0 \\ -a & -a\xi_3 & -a\eta_3 & a & a\xi_3 & a\eta_3 \\ -a & -a\xi_4 & -a\eta_4 & a & a\xi_4 & a\eta_4 \\ 0 & 0 & 0 & 1 & \xi_5 & \eta_5 \\ 0 & 0 & 0 & 1 & \xi_6 & \eta_6 \end{bmatrix} \quad ; \quad \mathbf{T} = \begin{bmatrix} 1 & 0 & & & & & & & & & & 0 \\ & & 1 & 0 & & & & & & & & \\ & & & & -a & a & & & & & & \\ & & & & & & -a & a & & & & \\ & & & & & & & & 0 & 1 & & \\ & & 0 & & & & & & & & 0 & 1 \end{bmatrix}$$

$$\mathbf{C} = \begin{bmatrix} \mathbf{J}^1 & & & & & 0 \\ & \mathbf{J}^2 & & & & \\ & & \mathbf{J}^3 & & & \\ & & & \mathbf{J}^4 & & \\ & & & & \mathbf{J}^5 & \\ & 0 & & & & \mathbf{J}^6 \end{bmatrix} \quad , \quad \bar{\mathbf{B}}_s = \begin{Bmatrix} \mathbf{B}^1 \\ \mathbf{B}^2 \\ \vdots \\ \mathbf{B}^6 \end{Bmatrix} \quad , \quad a = \frac{\sqrt{2}}{2} \tag{27}$$

From eqs. (26) and (27) the substitute shear strain matrix of eq (25) can be readily obtained.

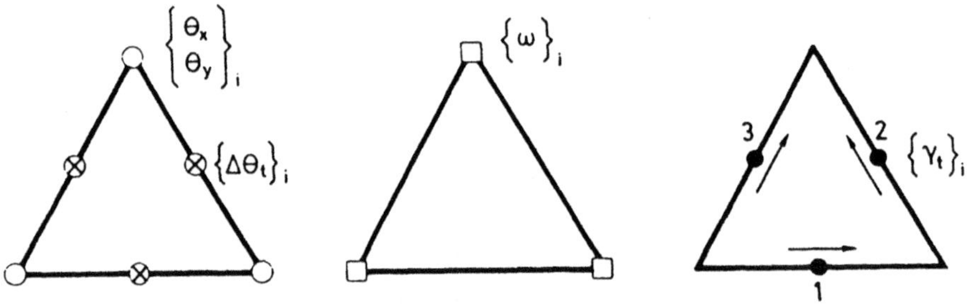

Figure 3. Linear/quadratic triangular element.

Linear/Quadratic triangle

An improved version of the quadratic triangle of the preceeding section, also presented by the authors in [17], is the following:

1) An incomplete quadratic variation of the rotations within the element is obtained according to the following interpolation

$$\boldsymbol{\theta} = \sum_{i=1}^{3} L_i \bar{\boldsymbol{\theta}}_i + 4\, L_2 L_3 \mathbf{e}_1 \Delta\bar{\theta}_1 + 4 L_1 L_3 \mathbf{e}_2 \Delta\bar{\theta}_2 + 4 L_1 L_2 \mathbf{e}_3 \Delta\bar{\theta}_3 \tag{28}$$

Where L_i are the standard linear shape functions of the three node triangle and $\Delta\bar{\theta}_k$ is a hierarchical tangential rotation parameter at the element mid-side (see Figure 3).

2) The deflection ω varies linearly over the element as

$$\omega = \sum_{i=1}^{3} L_i \bar{\omega}_i \tag{29}$$

3) The transverse shear strain field is again assumed to be linear in each element. However, the tangential shear strains are assumed constant along each side. This can be easily accomplished by setting $\gamma_\xi^1 = \gamma_\xi^2$, $\gamma_\xi^3 = \gamma_\xi^4$ and $\gamma_\xi^5 = \gamma_\xi^6$ (see Figure 3), thus leading to a three parameter interpolation.

The substitute strain matrix for this element (called DRM for discrete Reissner-Mindlin [17]) can be easily found by eq (25) noting that **A** and **P** matrices are identical to those of the 6 node quadratic element of preceeding section and now

$$\mathbf{T} = \begin{bmatrix} 1 & 0 & & & & 0 \\ 1 & 0 & & & & \\ & & -a & a & & \\ & & -a & a & & \\ & & & & 0 & 1 \\ & 0 & & & 0 & 1 \end{bmatrix} \quad ; \quad \mathbf{C} = \begin{bmatrix} \mathbf{J}^1 & & \\ & \mathbf{J}^2 & \\ & & \mathbf{J}^3 \end{bmatrix} \quad ; \quad \bar{\mathbf{B}}_s = \begin{Bmatrix} \mathbf{B}_s^1 \\ \mathbf{B}_s^2 \\ \mathbf{B}_s^3 \end{Bmatrix} \tag{30}$$

Matrix $\bar{\mathbf{B}}_s^i$ must be obtained using the displacement field of eqs (28)–(29).

It is worth noting that the DRM element has the distinct feature of satisfying the bending moment-shear forces equilibrium equations ($\mathbf{L}^T\mathbf{m} + \mathbf{s} = 0$), which explains its exellent behaviour in the examples analyzed as shown in next section.

Examples

A series of numerical tests have been conducted to assess the performance of the new triangular elements derived. Here some examples of the excellent behaviour of the DRM triangle are presented. More information can be found in [17].

Uniform loading on simply supported square plate

A quadrant of a square plate is modeled with different meshes of DRM triangular plate elements. Both soft ($\omega = 0$) and hard ($\omega = \theta_s = 0$) boundary conditions are considered. The structural properties of the plate are $E = 10.92$, $\nu = 0.3$, side length=10, $q = 1.0$. Numerical results obtained with different meshes for both thick ($a/t = 10$) and thin ($a/t = 100$) cases are shown in Figure 4 where analytical results are also shown for comparison [17]. Note the remarkable precision of the element for both cases even for course meshes.

	Simply supported square plate. Center displacement ω			
	$a/t = 100(\omega \times 10^{-4})$		$a/t = 10(\omega \times 10^{-1})$	
Mesh	Hard	Soft	Hard	Soft
2×2	4.0582	4.0903	4.8665	4.3992
4×4	4.0671	4.0737	4.2829	4.4600
8×8	4.0659	4.0719	4.2739	4.5393
16×16	4.0649	4.0756	4.2728	4.5906
Series	4.0623		4.2728	

Figure 4. Square plate under uniform loading. Results for the center displacementer for hard and soft boundary condition for different meshes (DRM plate element).

Point loading on thin square plate ($a/t = 100$)

Numerical results obtained for different meshes and boundary conditions are shown in Figure 5. The analytical solution exixting for the thin plate limit is also shown in the figure for comparison. Again, excellent results are obtained for all meshes

	Square Plate. Center Point loading ($a/t = 100$) Center displacement ($\omega \times 10^3$)		
Mesh	Simply supported		Clamped
	Soft	Hard	
1×1	1.2493	1.2492	0.5676
2×2	1.1703	1.1700	0.5867
4×4	1.1656	1.1649	0.5723
1×1	1.1645	1.1632	0.5660
1×1	1.1650	1.1627	0.5643
Series	1.160		0.560

Figure 5. Square plate under uniform loading. Results for the center displacement for various meshes and boundary conditions (DRM plate element).

Circular plate under uniform loading

Due to symmetry one quadrant of the circular plate has been discretized with five different meshes. Results for clamped and simply supported boundary conditions are computed for a thick ($R/t = 5/1$) and a relatively thin ($R/t = 5/0.1$) plate. Numerical results obtained are shown en Figure 6. We note that the error of the 6 element solution for all cases is less than 5%.

	Circular plate. Uniform loading. Center displacement					
	$t = 0.1$			$t = 1.0$		
Mesh	Clamp	Soft	Hard	Clamp	Soft	Hard
6 element	10315.6	37857.7	37857.7	11.9924	39.6043	39.6043
24 element	9999.91	39401.5	39373.8	11.7775	41.2023	41.1514
96 element	9850.93	39732.1	39732.1	11.6305	41.5172	41.5069
384 element	9802.40	39808.2	39807.4	11.5745	41.5810	41.5795
1536 element	9788.74	39826.2	39826.0	11.5577	41.5952	41.5950
Exact	9783.48	39831.5	39831.5	11.55	41.60	41.60

Figure 6. Circular plate under uniform loading. Center displacement for various meshes and boundary conditions (DRM plate element)

Conclusions

In the paper we have shown that the condition of vanishing shear strains for the thin plate limit imposes that the coefficients defining the shear strains polinomial field must be a linear function of the nodal rotations and deflections. This explains the sucess of reduced integration techniques and the use of substitute transverse shear strain fields which satisfy that condition. A general expression for the substitute shear strain matrix has been obtained which can be useful for the practical derivation of shear constrained elements based an adequate shear /displacement fields [17]. This methodology has been applied for the derivation of the new linear/quadratic DRM triangular element which shows an excellent behaviour in all the examples analized.

References

1. S. Ahmad, B.M. Irons and O.C. Zienkiewicz, "Analysis of thick and thin shell structures by curved finite elements", *Int. J. Numer. Meth. Eng.*, **2**, 419–451, 1970.
2. K.J. Bathe and E.N. Dvorkin, "A four node plate bending element based on Mindlin/Reissner plate theory and mixed interpolation", *Int. J. Numer. Meth. Eng.*, **21**, 367–383, 1985.
3. M.A. Crisfield, "A four-noded thin-plate bending element using shear constrains-A modified version of Lyon's element", Comp Meth. Appl. Mech. Eng., **38**, 93–120, 1983.
4. E.N. Dvorkin and K.J. Bathe, "A continuum mechanics based four node shell element for general non-linear analysis", *Eng. Comp.*, **1**, 77–88, 1984.
5. E. Hinton and H.C. Huang, "A family of quadrilateral Mindlin plate element with substitute shear strain fiels", *Comp. and Struct.*, **23**, 409–431, 1986.

6. T.J.R. Hughes, R.L. Taylor and W. Kanoknukulchai, "A simple and efficient element for plate bending", *Int. J. Numer Meth. Eng.*, **11**, 1529–1543, 1977.
7. T.J.R. Hughes, L.Taylor and T.E. Tezduyar, "Finite elements based upon Mindlin plate theory with particular reference to the four node bilinear isoparametric element", *J. Appl. Mech.*, **46**, 587–596, 1981.
8. D.S. Malkus and T.J. R. Hughes, "Mixed finite element methods-reduced and selective intergration techniques: A unification of concepts, *Comp. Meth. Appl. Mech. Eng.* , **15** 63–81, 1978.
9. R.D. Mindlin, "Influence of rotatory inertia and in flexural motions of isotropic elastic plates", *J. Appl. Mech.*, **18**, 31–38, 1951.
10. E.D.L. Pugh, E. Hinton and O.C. Zienkiewicz, "A Study of Quadrilateral plate bending elements with reduced integration", *J. Appl. Mech.*, **12**, 1059-1079, 1978.
11. E. Reissner, "The effect of transverse shear deformation on the bending of elastic plates, *J. Appl. Mech.*, **12** 69–76, 1945.
12. O.C. Zienkiewicz,*The Finite Element Method* , 3rd Edition, Mcgraw Hill, London 1977.
13. O.C. Zienkiewicz and D. Lefebvre, "Three field mixed approximation and the plate bending problem", *Comm. Appl. Numer. Meth.*, **3** 301–309, 1987.
14. O.C. Zienkiewicz and D. Lefebvre, " A robust triangular plate bending element of the Reissner-Mindlin type", *Int. J. Numer Meth. Eng.*, **26**, 1169-1184, 1988.
15. O.C. Zienkiewicz , J. Too and R.L. Taylor, "Reduced integration techniques in general analysis of plates and shells", *Int. J. Numer Meth. Eng.*, **3**, 275–290, 1971.
16. O.C. Zienkiewicz, S. Qu, R.L. Taylor and S. Nakazawa, "The Patch test for mixed formulations", *Int. J. Numer Meth. Eng.*, **23**, 1873–1883, 1986.
17. O.C. Zienkiewicz, R.L. Taylor, P. Papadopoulus and E. Oñate, "Plate bending elements with discrete constraints: New Triangular Elements", report NO UCB/SEMM-89/09, Dpt. Civil Engng., Univ. California at Berkeley (to be published in Computer and Structures).

A Nonlinear Shell Theory with Through-The-Thickness Stretch

J. C. Simo, D. D. Fox and M. S. Rifai

Applied Mechanics Division
Stanford University
Stanford, CA 94305, U. S. A.

ABSTRACT

This paper summarizes the important features of a geometrically exact resultant based shell model which incorporates through-the-thickness stretch. The shell is modeled as a single extensible director Cosserat surface where variable thickness and thickness stretch effects are accounted for through the extensibility of the director field. Even though thickness effects are included, the resulting equations are numerically well-conditioned in the thin shell limit and a simple linear elastic constitutive model is given which possesses the correct asymptotic behavior as the thickness tends to zero and identically recovers the plane stress constitutive relations in the thin shell limit. The performance of the theory is demonstrated by two numerical examples.

1. INTRODUCTION

To a large extent, there exists a gap between the classical mathematical analysis of nonlinear shell theories and the corresponding numerical analysis. On the one hand, mathematical approaches advocated by such researchers such as Ericksen & Truesdell [4] or Antman [2] to name but a few, are complex and difficult to implement numerically. On the other hand, computational approaches presented by researchers such as Ahmad, et. al. [1] or Hughes & Liu [6,7] among many others, rely heavily on the degenerated solid concept, with little incorporation of mathematical treatments of shell theories. This work, which is presented entirely in stress resultants and falls within the abstract category of Cosserat continua, is formulated is such a way as to be easily implemented and suitable for large scale computations.

The current presentation is to some extent an extension, but is mostly a summary, of our previous work on geometrically exact shell theories. As in our earlier works, the parametrization of the weak form of the momentum balance equations avoids the explicit appearance of the Riemannian connection of the mid-surface. In particular, objects such as Christoffel symbols or the second fundamental form need not be computed. From a computational standpoint, the present formulation, even though it incorporates thickness stretch, is essentially identical to that presented in our previous work (which did not account for thickness effects) with simply two additional terms in the weak form of the momentum balance equations. Thickness effects are modeled by an added degree of freedom which is coupled (interactively) to the membrane, shear and bending fields. This is at variance with previous attempts in the context of the degenerated solid approach, which treat thickness effects (essentially) as a post-processing procedure. To avoid numerical difficulties for thin shells, a multiplicative decomposition of the director field into a unit vector (rotation part) and a magnitude parameter (thickness part) is

G. Kuhn · H. Mang (Eds.)
Discretization Methods in Structural Mechanis
IUTAM/IACM Symposium Vienna, Austria, 1989

introduced. A simple quadratic isotropic stored energy function is considered which possesses the proper asymptotic limit as the thickness of the shell tends to zero and, in the thin shell limit, this model reduces to the equations obtained by enforcing the plane stress assumption.

2. THE NONLINEAR SHELL THEORY

2.1. Kinematic Description of the Shell

We will consider deformations of a shell which fall within the category of a single extensible director Cosserat surface. In contrast with previous work on stress resultant geometrically exact shell models ([9]-[11]), this formulation admits extension of the director, or more precisely, allows for variable thickness and through-the-thickness stretch.

The kinematic description of the shell is as follows. Points in the shell are parametrized by a coordinate system $(\xi^1, \xi^2, \xi) \in \mathcal{A} \times I$, where $\xi \in I = [h_0^-, h_0^+]$ is the through-the-thickness parameter; $\mathcal{A}$ and I are fixed regions in $\mathbf{R}^2$ and $\mathbf{R}$, respectively. Any Euclidean placement of the shell $\mathcal{S} \subset \mathbf{R}^3$ is given by a mapping $\boldsymbol{\Phi} : \mathcal{A} \times I \to \mathbf{R}^3$, where

$$\mathcal{S} := \{ \, \boldsymbol{x} \in \mathbf{R}^3 \mid \boldsymbol{x} = \boldsymbol{\Phi}(\xi^1, \xi^2, \xi) \text{ for } (\xi^1, \xi^2, \xi) \in \mathcal{A} \times I \, \} . \tag{2.1}$$

According to the single director Cosserat surface kinematic assumption, any point in the shell $\boldsymbol{x} \in \mathcal{S}$ is assumed to be given by a mapping of the form

$$\boxed{\boldsymbol{\Phi}(\xi^1, \xi^2, \xi) := \boldsymbol{\varphi}(\xi^1, \xi^2) + \xi \boldsymbol{d}(\xi^1, \xi^2) ,} \tag{2.2}$$

where $\boldsymbol{\varphi} : \mathcal{A} \to \mathbf{R}^3$ is the mapping that defines the mid-surface of the shell and $\boldsymbol{d} : \mathcal{A} \to \mathbf{R}^3$ is the director field or fiber direction. Note that $\boldsymbol{d} \in \mathbf{R}^3$ is not restricted to have unit length. From a numerical standpoint, in order to avoid ill-conditioning in the thin shell limit, it proves essential to decompose the director field into a magnitude and a unit vector,

$$\boldsymbol{d}(\xi^1, \xi^2) = \lambda(\xi^1, \xi^2) \boldsymbol{t}(\xi^1, \xi^2) , \text{ with } \|\boldsymbol{t}\| = 1 \text{ and } \lambda > 0 . \tag{2.3}$$

The mid-surface convected or natural basis is obtained by particularizing the three-dimensional convected basis, $\boldsymbol{\Phi}_{,i}$ $(i = 1, 2, 3)$, to the mid-surface at $\xi = 0$:

$$\{\boldsymbol{a}_1, \boldsymbol{a}_2, \boldsymbol{a}_3\} = \{\boldsymbol{\varphi}_{,1}, \boldsymbol{\varphi}_{,2}, \boldsymbol{d}\} . \tag{2.4}$$

2.2. Resultant Form of the Local Balance Equations

Let $\boldsymbol{n}^\alpha$, $\tilde{\boldsymbol{m}}^\alpha$ and $\boldsymbol{l}$ represent the resultant stress, resultant stress couple and resultant through the thickness stress. See [9] for the physical interpretation of these resultants and their definition in terms of three-dimensional quantities. The local or strong form of the (static) momentum balance equations in terms of the resultants are given by $(\alpha = 1, 2)$

$$\begin{aligned} &\frac{1}{\bar{j}}(\bar{j}\boldsymbol{n}^\alpha)_{,\alpha} + \bar{\boldsymbol{n}} = \boldsymbol{0} , \\ &\frac{1}{\bar{j}}(\bar{j}\tilde{\boldsymbol{m}}^\alpha)_{,\alpha} - \boldsymbol{l} + \bar{\boldsymbol{m}} = \boldsymbol{0} . \end{aligned} \tag{2.5}$$

A detailed derivation of these equations and the definition of the applied loads $\bar{\boldsymbol{n}}$ and $\bar{\boldsymbol{m}}$ is given in [9]. Equations (2.5) are the resultant balance of linear momentum and balance of director momentum (equivalent to balance of couples around the mid-surface), respectively. The ***balance of angular momentum*** equation (equivalent to $\boldsymbol{\sigma} = \boldsymbol{\sigma}^T$ in the three-dimensional theory) is written as

$$\boldsymbol{n}^{\alpha} \times \boldsymbol{\varphi}_{,\alpha} + \tilde{\boldsymbol{m}}^{\alpha} \times \boldsymbol{d}_{,\alpha} + \boldsymbol{l} \times \boldsymbol{d} = \boldsymbol{0}\,. \tag{2.6}$$

2.3. Effective Stress Resultants and Balance of Angular Momentum

The balance of angular momentum equation (2.6) plays a crucial role in the definition of effective (symmetric) stress resultants conjugate to the desired strain measures. We define the ***effective stress resultant*** by the following rank two tensor:

$$\tilde{\mathbf{n}} := \boldsymbol{n}^{\alpha} \otimes \boldsymbol{\varphi}_{,\alpha} + \boldsymbol{l} \otimes \boldsymbol{d} - \boldsymbol{d}_{,\alpha} \otimes \tilde{\boldsymbol{m}}^{\alpha}\,. \tag{2.7}$$

It can be shown (see [13]) that the following equivalence holds:

$$\textit{Balance of angular momentum} \iff \boxed{\tilde{\mathbf{n}} = \tilde{\mathbf{n}}^T}\,. \tag{2.8}$$

This equivalence between balance of angular momentum and symmetry of the effective stress resultant admits a simple interpretation in component form. Expressing $\tilde{\mathbf{n}}$ in components relative to the convected basis (2.4), the equivalence (2.8) becomes

$$\textit{Balance of angular momentum} \iff \boxed{\tilde{n}^{\beta\alpha} = \tilde{n}^{\alpha\beta}} \text{ and } \boxed{\tilde{n}^{3\alpha} = \tilde{n}^{\alpha 3} \equiv \tilde{q}^{\alpha}}\,. \tag{2.9}$$

The effective stress components (2.9) are crucial to the development of the weak form of the momentum balance equations and the resultant hyper-elastic constitutive equations that follow.

2.4. Strain Measures and Component Expressions for the Weak Form

Before we consider the weak form of the local balance equations (2.5), we define the appropriate shell strain measures. There are two alternative ways to express the shell strain measures and the resulting component expression for the weak form in terms of these strain measures. The two alternative representations correspond to operating with the director $\boldsymbol{d}$ explicitly or operating with the multiplicative decomposition (2.3).

First, we define the strain measures in terms of the director $\boldsymbol{d}$. These definitions correspond to strain measures identified from the Lagrangian strain tensor of the three-dimensional theory, $\boldsymbol{E} = \frac{1}{2}(\boldsymbol{F}^T\boldsymbol{F} - \mathbf{1})$, where $\boldsymbol{F}$ is the deformation gradient $\nabla\boldsymbol{\Phi}\nabla\boldsymbol{\Phi}^{0-1}$. (In what follows, the superscript o refers to any quantity particularized to the reference configuration.) We define

$$\left.\begin{aligned}
\epsilon_{\alpha\beta} &= \tfrac{1}{2}(\boldsymbol{\varphi}_{,\alpha} \cdot \boldsymbol{\varphi}_{,\beta} - \boldsymbol{\varphi}^0_{,\alpha} \cdot \boldsymbol{\varphi}^0_{,\beta}) && \text{(membrane)}\\
\tilde{\rho}_{\alpha\beta} &= \boldsymbol{\varphi}_{,\alpha} \cdot \boldsymbol{d}_{,\beta} - \boldsymbol{\varphi}^0_{,\alpha} \cdot \boldsymbol{d}^0_{,\beta} && \text{(bending)}\\
\tilde{\delta}_{\alpha} &= \boldsymbol{\varphi}_{,\alpha} \cdot \boldsymbol{d} - \boldsymbol{\varphi}^0_{,\alpha} \cdot \boldsymbol{d}^0 && \text{(transverse shear)}\\
\tilde{\chi}_{\alpha} &= \boldsymbol{d}_{,\alpha} \cdot \boldsymbol{d} - \boldsymbol{d}^0_{,\alpha} \cdot \boldsymbol{d}^0 && \text{(couple shear)}\\
\tilde{\chi} &= \tfrac{1}{2}(\boldsymbol{d} \cdot \boldsymbol{d} - \boldsymbol{d}^0 \cdot \boldsymbol{d}^0) && \text{(thickness stretch)}
\end{aligned}\right\}\,. \tag{2.10}$$

The strain measures defined above are denoted by a superposed $\tilde{}$ to differentiate them from those measures used in the numerical implementation, which are defined subsequently.

To construct the variational formulation of the theory, we introduce the space of kinematically admissible variations $\mathcal{V}$, defined as

$$\mathcal{V} = \{ (\delta\varphi, \delta t, \delta\lambda) : \mathcal{A} \to \mathbf{R}^3 \times T_t S^2 \times \mathbf{R} \mid \delta\varphi|_{\partial_\varphi \mathcal{A}} = 0,\ \delta t|_{\partial_t \mathcal{A}} = 0,\ \delta\lambda|_{\partial_\lambda \mathcal{A}} = 0 \}. \tag{2.11}$$

Here $\delta(\bullet)$ denotes the directional derivative of $(\bullet)$. In the notation of (2.11), the directional derivative of the director $\boldsymbol{d}$ is given by $\delta\boldsymbol{d} = \delta\lambda\boldsymbol{t} + \lambda\delta\boldsymbol{t}$. The weak form of the momentum balance equations, obtained by multiplying the local equations (2.5) by variations $(\delta\varphi, \delta\boldsymbol{d})$ and integrating by parts, is expressed in terms of effective stress components (2.9) and strain measures (2.10) as (see [13])

$$\mathrm{G}(\tilde{\mathbf{n}}, \tilde{m}^\alpha, \Phi; \delta\Phi) = \int_{\mathcal{A}} \Big[\tilde{n}^{\alpha\beta}\delta\epsilon_{\alpha\beta} + \tilde{m}^{\alpha\beta}\delta\tilde{\rho}_{\alpha\beta} + \tilde{q}^\alpha\delta\tilde{\delta}_\alpha + \tilde{m}^{3\alpha}\delta\tilde{\chi}_\alpha + \tilde{l}^3\delta\tilde{\chi} \Big] \bar{j}\, d\xi^1 d\xi^2 - \mathrm{G}_{\mathrm{ext}}(\delta\Phi) = 0\,, \tag{2.12}$$

where $\mathrm{G}_{\mathrm{ext}}(\delta\Phi)$ is the external load term given by

$$\mathrm{G}_{\mathrm{ext}}(\delta\Phi) := \int_{\mathcal{A}} \Big[\bar{\boldsymbol{n}} \cdot \delta\varphi + \bar{\boldsymbol{m}} \cdot \delta\boldsymbol{d}\Big] \bar{j}\, d\xi^1 d\xi^2 + \int_{\partial\mathcal{A}} \Big[\boldsymbol{n}^\alpha \cdot \delta\varphi + \tilde{\boldsymbol{m}}^\alpha \cdot \delta\boldsymbol{d}\Big] \nu_\alpha \bar{j}\, d\Gamma\,. \tag{2.13}$$

In (2.13), ν_α are the components of the unit outward normal to the region $\mathcal{A}$ relative to the standard basis in $\mathbf{R}^2$.

Strain measures incorporating the multiplicative decomposition of the director field are now defined. To motivate these definitions, we make the following two observations about the strain measures defined above. First, the through-the-thickness stretch is coupled with the bending and shear measures. A deformation which produces large thickness change can also result in large changes in the bending and shear strain measures $\tilde{\rho}_{\alpha\beta}$ and $\tilde{\delta}_\alpha$. Second, a quadratic stored energy function in the measures (2.10) (the usual model in classical shell theory) possesses an incorrect limit (finite energy) as the thickness of the shell decreases to zero. Taking the three-dimensional theory as a guide, a more appropriate strain measure which produces the proper asymptotic behavior in a quadratic model is the logarithmic strain. With these observations in mind, we define the following set of strain measures:

$$\left.\begin{aligned}
\epsilon_{\alpha\beta} &= \tfrac{1}{2}(\varphi_{,\alpha} \cdot \varphi_{,\beta} - \varphi^0_{,\alpha} \cdot \varphi^0_{,\beta}) && \text{(membrane)}\\
\rho_{\alpha\beta} &= \varphi_{,\alpha} \cdot \boldsymbol{t}_{,\beta} - \varphi^0_{,\alpha} \cdot \boldsymbol{t}^0_{,\beta} && \text{(bending)}\\
\delta_\alpha &= \varphi_{,\alpha} \cdot \boldsymbol{t} - \varphi^0_{,\alpha} \cdot \boldsymbol{t}^0 && \text{(transverse shear)}\\
\chi_\alpha &= (\ln\lambda)_{,\alpha} - (\ln\lambda^0)_{,\alpha} = \ln(\lambda/\lambda^0)_{,\alpha} && \text{(couple shear)}\\
\chi &= \ln\lambda - \ln\lambda^0 = \ln(\lambda/\lambda^0) && \text{(thickness stretch)}
\end{aligned}\right\}\,. \tag{2.14}$$

The component expression (2.12) for the weak form in terms of the strain measures (2.14) becomes

$$\boxed{\mathrm{G}(\tilde{\mathbf{n}}, \tilde{m}^\alpha, \Phi; \delta\Phi) = \int_{\mathcal{A}} \Big[\tilde{n}^{\alpha\beta}\delta\epsilon_{\alpha\beta} + \tilde{m}^{\alpha\beta}\delta\rho_{\alpha\beta} + \tilde{q}^\alpha\delta\delta_\alpha + \tilde{m}^{3\alpha}\delta\chi_\alpha + \tilde{l}^3\delta\chi \Big] \bar{j}\, d\xi^1 d\xi^2 - \mathrm{G}_{\mathrm{ext}}(\delta\Phi) = 0\,,} \tag{2.15}$$

where the *modified* effective stress resultants in (2.15), denoted by a superposed $\check{}$, are defined in terms of the effective stress resultants and the stress couple resultants as

$$\check{n}^{\alpha\beta} = \tilde{n}^{\alpha\beta}, \quad \check{m}^{\alpha\beta} = \lambda\, \tilde{m}^{\alpha\beta}, \quad \check{q}^{\alpha} = \lambda\left[\tilde{q}^{\alpha} + (\ln\lambda)_{,\beta}\, \tilde{m}^{\alpha\beta}\right], \qquad \check{m}^{3\alpha} = \tilde{\boldsymbol{m}}^{\alpha}\cdot \boldsymbol{d}, \quad \check{l}^{3} = \boldsymbol{l}\cdot\boldsymbol{d} + \tilde{\boldsymbol{m}}^{\alpha}\cdot\boldsymbol{d}_{,\alpha}. \tag{2.16}$$

The advantages of expressing the weak form in terms of the strain measures (2.14) becomes apparent in view of the following observations. The through-the-thickness stretch is not coupled with the bending or shear strain measures and a simple quadratic elastic constitutive model possesses the proper asymptotic behavior as the thickness of the shell tends to zero. Furthermore, the numerical treatment of the equations is identical to that for the inextensible director formulation, with simply two additional terms in the weak form corresponding to the added stress resultants $\check{m}^{3\alpha}$ and $\check{l}^{3}$.

2.5. Matrix Formulation and Constitutive Equations

We now summarize the complete set of equations, in component vector/matrix form, for the resultant based geometrically exact shell model with through the thickness effects. For simplicity, we restrict our attention to a simple linear elastic isotropic constitutive model. (Note that $A_{\alpha\beta} = \boldsymbol{\varphi}^0_{,\alpha}\cdot\boldsymbol{\varphi}^0_{,\beta}$ are the components of the reference mid-surface metric tensor and the 3×2 transformation $\bar{\boldsymbol{\Lambda}}$ is defined in [10] and [11].)

- Modified effective stress and stress couple vectors:

$$\check{\boldsymbol{n}} = \bar{J}\begin{Bmatrix}\check{n}^{11}\\ \check{n}^{22}\\ \check{n}^{12}\end{Bmatrix}, \quad \check{\boldsymbol{q}} = \bar{J}\begin{Bmatrix}\check{q}^{1}\\ \check{q}^{2}\end{Bmatrix}, \quad \check{\boldsymbol{m}} = \bar{J}\begin{Bmatrix}\check{m}^{11}\\ \check{m}^{22}\\ \check{m}^{(12)}\end{Bmatrix}, \quad \check{\boldsymbol{l}} = \bar{J}\begin{Bmatrix}\check{m}^{31}\\ \check{m}^{32}\\ \check{l}^{3}\end{Bmatrix}$$

- Incremental displacement and rotation variations:

$$\delta\boldsymbol{\varphi} = \begin{Bmatrix}\delta\varphi^{1}\\ \delta\varphi^{2}\\ \delta\varphi^{3}\end{Bmatrix}, \quad \delta\boldsymbol{T} = \begin{Bmatrix}\delta T^{1}\\ \delta T^{2}\end{Bmatrix}, \quad \delta\boldsymbol{t} = \bar{\boldsymbol{\Lambda}}\,\delta\boldsymbol{T}, \quad \delta\boldsymbol{\Phi} = \begin{Bmatrix}\delta\boldsymbol{\varphi}\\ \delta\boldsymbol{T}\\ \delta\mu\end{Bmatrix}$$

- Matrix differential operators:

$$\mathbf{B}_m = \begin{bmatrix}\boldsymbol{\varphi}^{\mathrm{T}}_{,1}\frac{\partial}{\partial\xi^1}\\ \boldsymbol{\varphi}^{\mathrm{T}}_{,2}\frac{\partial}{\partial\xi^2}\\ \boldsymbol{\varphi}^{\mathrm{T}}_{,1}\frac{\partial}{\partial\xi^2} + \boldsymbol{\varphi}^{\mathrm{T}}_{,2}\frac{\partial}{\partial\xi^1}\end{bmatrix}_{3\times 3} \quad \mathbf{B}_{sm} = \begin{bmatrix}\boldsymbol{t}^{\mathrm{T}}\frac{\partial}{\partial\xi^1}\\ \boldsymbol{t}^{\mathrm{T}}\frac{\partial}{\partial\xi^2}\end{bmatrix}_{2\times 3} \quad \mathbf{B}_{sb} = \begin{bmatrix}\boldsymbol{\varphi}^{\mathrm{T}}_{,1}\\ \boldsymbol{\varphi}^{\mathrm{T}}_{,2}\end{bmatrix}_{2\times 3}\bar{\boldsymbol{\Lambda}}_{3\times 2}$$

$$\mathbf{B}_{bm} = \begin{bmatrix}\boldsymbol{t}^{\mathrm{T}}_{,1}\frac{\partial}{\partial\xi^1}\\ \boldsymbol{t}^{\mathrm{T}}_{,2}\frac{\partial}{\partial\xi^2}\\ \boldsymbol{t}^{\mathrm{T}}_{,1}\frac{\partial}{\partial\xi^2} + \boldsymbol{t}^{\mathrm{T}}_{,2}\frac{\partial}{\partial\xi^1}\end{bmatrix}_{3\times 3} \quad \mathbf{B}_{bb} = \mathbf{B}_m\bar{\boldsymbol{\Lambda}}_{3\times 2} \quad \mathbf{B}_l = \begin{bmatrix}\frac{\partial}{\partial\xi^1}\\ \frac{\partial}{\partial\xi^2}\\ 1\end{bmatrix}_{3\times 1}$$

$$\mathbf{B}_{\text{total}} = \begin{bmatrix}\mathbf{B}_m & 0 & 0\\ \mathbf{B}_{sm} & \mathbf{B}_{sb} & 0\\ \mathbf{B}_{bm} & \mathbf{B}_{bb} & 0\\ 0 & 0 & \mathbf{B}_l\end{bmatrix}_{11\times 6} \qquad \check{\mathbf{r}} = \begin{Bmatrix}\check{\boldsymbol{n}}\\ \check{\boldsymbol{q}}\\ \check{\boldsymbol{m}}\\ \check{\boldsymbol{l}}\end{Bmatrix}$$

• Weak form $\mathrm{G} : \mathcal{C} \times T_{\Phi}\mathcal{C} \to \mathbb{R}$: (Here, $\check{\mathbf{r}} = \check{\mathbf{r}}(\Phi)$)

$$\mathrm{G}(\Phi; \delta\Phi) := \int_{\mathcal{A}} \check{\mathbf{r}}^T\, \mathbf{B}_{\text{total}}\, \delta\Phi\, \bar{j}^0\, d\xi^1 d\xi^2 - \mathrm{G}_{\text{ext}}(\delta\Phi)$$

• Strain component vectors:

$$\epsilon = \begin{Bmatrix} \epsilon_{11} \\ \epsilon_{22} \\ 2\epsilon_{12} \end{Bmatrix}, \quad \delta = \begin{Bmatrix} \delta_1 \\ \delta_2 \end{Bmatrix}, \quad \rho = \begin{Bmatrix} \rho_{11} \\ \rho_{22} \\ 2\rho_{(12)} \end{Bmatrix}, \quad \chi = \begin{Bmatrix} \chi_1 \\ \chi_2 \\ \chi \end{Bmatrix}$$

• Elastic isotropic constitutive equations:

$$\mathbf{C}_m = \begin{bmatrix} (1-\nu)A^{11}A^{11} & [\nu A^{11}A^{22} + (1-2\nu)A^{12}A^{12}] & (1-\nu)A^{11}A^{12} & \nu A^{11} \\ & (1-\nu)A^{22}A^{22} & (1-\nu)A^{22}A^{12} & \nu A^{22} \\ & & [\frac{1-2\nu}{2}A^{11}A^{22} + \frac{1}{2}A^{12}A^{12}] & \nu A^{12} \\ \text{Symmetric} & & & (1-\nu) \end{bmatrix}$$

$$\mathbf{C}_b = \begin{bmatrix} A^{11}A^{11} & [\nu A^{11}A^{22} + (1-\nu)A^{12}A^{12}] & A^{11}A^{12} \\ & A^{22}A^{22} & A^{22}A^{12} \\ \text{Sym.} & & [\frac{1-\nu}{2}A^{11}A^{22} + \frac{1+\nu}{2}A^{12}A^{12}] \end{bmatrix}$$

$$\begin{Bmatrix} \check{n} \\ \bar{J}l^3 \end{Bmatrix} = \frac{Eh}{(1+\nu)(1-2\nu)}\mathbf{C}_m \begin{Bmatrix} \epsilon \\ \chi \end{Bmatrix}; \qquad \check{q} = \kappa G h \begin{bmatrix} A^{11} & A^{12} \\ A^{12} & A^{22} \end{bmatrix} \delta$$

$$\check{m} = \frac{Eh^3}{12(1-\nu^2)}\mathbf{C}_b\rho; \qquad \begin{Bmatrix} \check{m}^{31} \\ \check{m}^{32} \end{Bmatrix} = \frac{7Eh^3}{240(1-\nu^2)} \begin{bmatrix} A^{11} & A^{12} \\ A^{12} & A^{22} \end{bmatrix} \begin{Bmatrix} \chi_1 \\ \chi_2 \end{Bmatrix}$$

As alluded to previously, the simple linear elastic isotropic constitutive model presented above has several desirable features. First, the energy possess the proper asymptotic limit as the thickness goes to zero. Second, the plane stress assumption need not be enforced at the outset. Nevertheless, in the thin shell limit, plane stress constitutive equations are recovered automatically.

3. NUMERICAL SIMULATIONS

Due to space considerations, the numerical implementation of the shell theory outlined above is omitted. We refer to [10], [11] and [13] for details. In order to simply demonstrate the performance of the numerical implementation in large deformation problems, we present two illustrative simulations.

3.1. In-plane bending of an L-shaped plate strip

The stability analysis of an L-shaped flat plate which is clamped on one end and subjected to an in-plane load on the other is investigated, as shown in Figure 3.1(a). This problem has been investigated by Argyris, et. al. [3] and Simo & Quoc [12] to

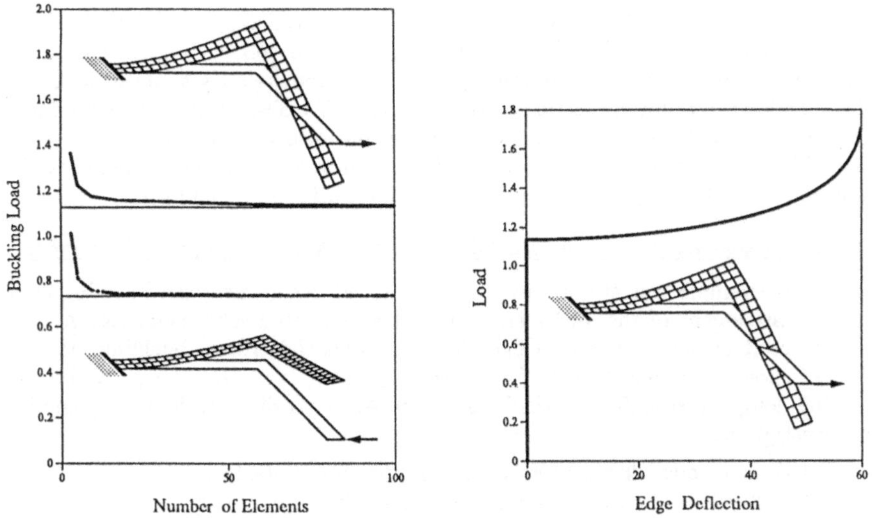

FIGURE 3.1.

(a) Convergence plots for the non-linear buckling load versus number of elements for the analysis of a built-in right angle frame problem. The top curve corresponds to the critical load usually found in the literature. The bottom curve, corresponds to the buckling load when the load direction is reversed, which is of lower magnitude.

(b) Load-deflection curve in the post-buckling regime for the clamped square bracket using a 68-element mesh. The applied load is plotted against the out-of-plane deflection.

evaluate the performance of nonlinear rod elements. The specifications for this problem are: Young's modulus $E = 71240$, Poisson's ratio $\nu = 0.31$, edge length $l = 240$, edge width $w = 30$ and thickness $t = 0.6$. Interestingly, this problem exhibits two different bifurcation points, depending on the load *direction.* The first bifurcation point, shown in the upper part of Figure 3.1(a), is the one reported in the literature. The second bifurcation point, associated with a *lower load* value of *opposite sign*, is shown in the lower part of the figure.

The results for the first bifurcation load, obtained using an extended system of equations as in Wriggers & Simo [14], are shown in Table 3.1. The convergence to the value of the critical loads obtained with this formulation for the two different load directions are shown in Figure 3.1(a).

A 68-element mesh is used to evaluate the post-buckling response and the results

TABLE 3.1. In-plane Buckling Loads for Right-Angle Frame.

Reference	Type	Elmts	Load
Simo & Quoc [12]	Beam	20	1.090
Argyris, et. al. [3]	Beam	20	1.088
Argyris, et. al. [3]	Shell	63	1.155
present	Shell	68	1.137
Converged Solution			1.128

are shown in Figure 3.1(b). The bifurcation point is arrived at by employing an extended system of equations, as described in Wriggers & Simo [14], combined with a mode-switching technique at the bifurcation point. Thus, the stability analysis is performed with no added initial imperfection to the structure. We observe that the load-defection curve remains flat within a wide range of out-of-plane deflections, up to a value of about 50 where a pronounced stiffening of the structure is observed. These results are in agreement with the general response pattern reported in [3] and [12].

3.2. Linearized vs. Non-Linear Buckling Analysis. Axially Compressed Cylinder

This example was studied by Liu, et. al. [8] using linearized buckling analysis in their assessment of the 4-SRI and 4-RSDS degenerated solid elements. A 30 × 14 mesh is used on one-eighth of a cylinder to predict the critical buckling load of the cylinder shown in Figure 3.2(a). The material properties are $E = 567$, $\nu = 0.3$, radius $R = 100$, length $L = 35.95$ and thickness $t = 0.247$. Four different boundary conditions are investigated:

(a) Free ends, with a shortening displacement applied,
(b) Free ends, with a compressive load applied,
(c) Clamped ends, with a shortening displacement applied, and
(d) Clamped ends, with a compressive load applied.

The results by Liu, et. al. [8] are compared with the results obtained in the present investigation (using linearized buckling analysis) in Table 3.2. The critical loads for the free end boundary conditions are approximately half of those of the clamped case. The associated mode shape consists of two sine waves in the circumferential direction and none in the axial direction. These results are in agreement with the classical analytical results of Hoff and Soong [5].

TABLE 3.2. Normalized Critical Load P/P_{cl} for Different B.C.'s

Case	4-RSDS	4-SRI	Present
(a)	—	—	0.5402
(b)	—	—	0.5055
(c)	1.0369	1.1378	1.0837
(d)	1.0379	1.1361	1.0833

The mode shape predicted by the linearized buckling analysis for the clamped cylinder case under the action of a compressive load is axisymmetric, as shown in Figure 3.2(b). However, when the load-deflection curve was traced and the linearized buckling analysis was performed in the vicinity of the bifurcation point, the mode predicted was no longer axisymmetric, but reticulated. Figure 3.2(c) shows the mode shape associated with the critical load when the linearized buckling analysis is performed at the bifurcation point, where $P/P_{cl} = 1.04$. This example illustrates the important fact that results obtained with a linearized buckling analysis should be interpreted with caution–even in a problem like the one at hand which is known to behave linearly up to a neighborhood of the bifurcation point. We further remark that knowledge of the exact mode shape associated with the lowest buckling load often proves to be crucial for a complete imperfection-sensitivity analysis of a thin shell.

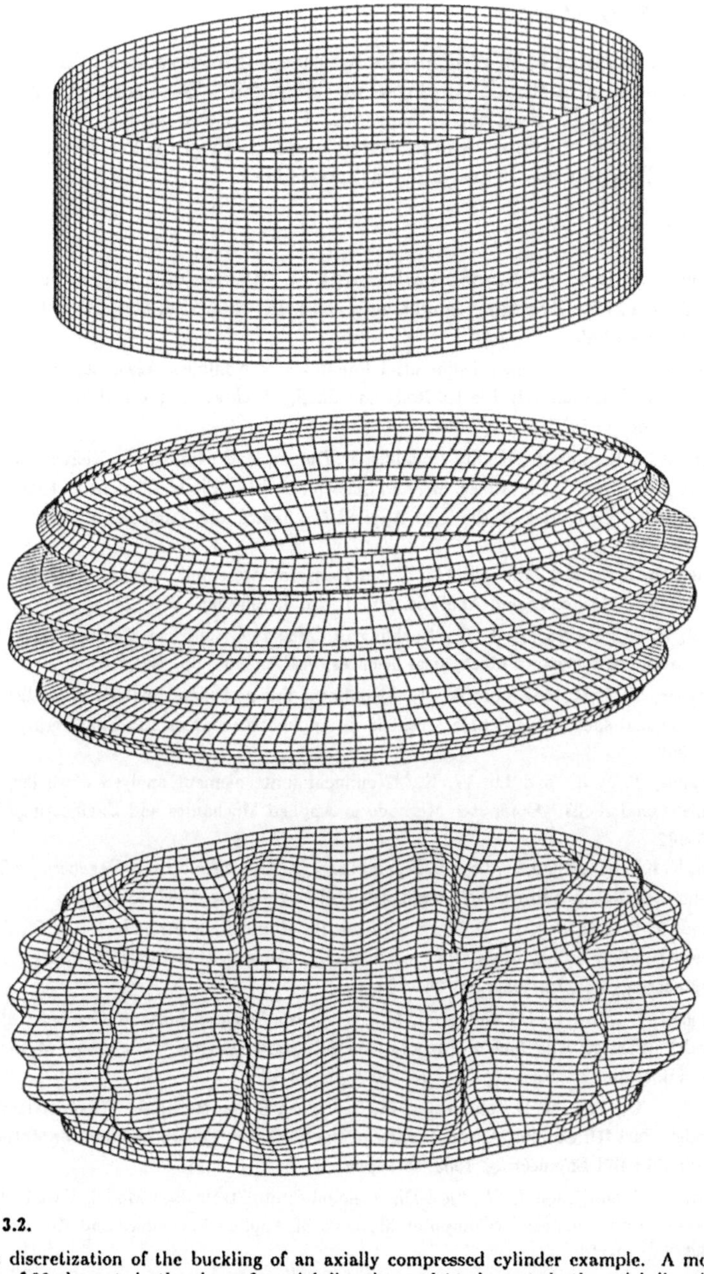

FIGURE 3.2.

(a) Mesh discretization of the buckling of an axially compressed cylinder example. A mesh consisting of 30 elements in the circumferential direction and 14 elements in the axial direction is used to model a one-eighth section of the cylinder using symmetry boundary conditions. This mesh is the same as that used in Liu, et. al. [8].

(b) Axisymmetric buckling mode resulting from *linearized buckling* analysis of the axially compressed clamped cylinder.

(c) Non-axisymmetric buckling mode produced by performing the buckling analysis *at the bifurcation point.* The *bifurcation point* was arrived at by employing an extended system of equations.

ACKNOWLEDGEMENTS

We are indebted to P. Wriggers for his generous involvement in the implementation of the continuation methods employed in this paper. We wish to thank professors S. S. Antman, T. J. R. Hughes, J. Marsden, C. R. Steele and R. L. Taylor for many helpful discussions. Support was provided by N. S. F. grant 2-DJA-491 with Stanford University. This support is gratefully acknowledged.

REFERENCES

1. Ahmad, S., Irons, B. M., and Zienkiewicz, O. C., "Analysis of Thick and Thin Shell Structures by Curved Finite Elements," International Journal of Numerical Methods in Engineering, Vol. 2, 1970, pp. 419-451.
2. Antman, S. S., "Ordinary Differential Equations of Nonlinear Elasticity I: Foundations of the Theories of Non-linearly Elastic Rods and Shell," Archive of Rational Mechanics and Analysis, Vol. 61, no. 4, 1976, pp. 307-351.
3. Argyris, J. H., Balmer, H., St. Doltsinis, J., Dunne, P. C., Hasse, M., Kleiber, M., Malejannakis, G. A., Mlejenek, J. P., Muller, M., and Scharpf, D. W., "Finite Element Method — The Natural Approach", Computer Methods in Applied Mechanics and Engineering, Vol. 17/18, 1979, pp. 1-106.
4. Ericksen, J. L., and Truesdell, C., "Exact theory of stress and strain in rods and shells," Archive of Rational Mechanics and Analysis, Vol. 1, No. 4, 1958, pp. 295-323.
5. Hoff, N. J., and Soong, T. C.,"Buckling of Circular Cylindrical Shells in Axial Compression," International Journal of Mechanical Sciences, Vol. 7, 1965, pp. 489-520.
6. Hughes, T. J. R., and Liu, W. K., "Nonlinear finite element analysis of shells: Part I-Three-dimensional shells," Computer Methods in Applied Mechanics and Engineering, Vol. 26, 1981, 331-362.
7. Hughes, T. J. R., and Liu W. K., "Nonlinear finite element analysis of shells: Part II -Two-dimensional shells," Computer Methods in Applied Mechanics and Engineering, Vol. 27, 1981, 167-182.
8. Liu, K. K., Belytschko, T., Law, S. E., and Lam, D., "Resultant Stress Degenerated Shell Element," Computer Methods in Applied Mechanics and Engineering, 1987, to appear.
9. Simo, J. C. and Fox, D. D., "On a Stress Resultant Geometrically Exact Shell Model. Part I: Formulation and Optimal Parametrization," Computer Methods in Applied Mechanics and Engineering, Vol. 72, 1989, pp. 267-304.
10. Simo, J. C., Fox, D. D., and Rifai, M. S., , "On a Stress Resultant Geometrically Exact Shell Model. Part II: The Linear Theory; Computational Aspects," Computer Methods in Applied Mechanics and Engineering, Vol. 73, 1989, pp. 53-92.
11. Simo, J. C., Fox, D. D., and Rifai, M. S., "On a Stress Resultant Geometrically Exact Shell Model. Part III: Computational Aspects of the Nonlinear Theory," Computer Methods in Applied Mechanics and Engineering, 1989, to appear.
12. Simo, J. C. and Quoc, L. V., "A 3-Dimensional Finite Strain Rod Model. Part II: Geometric and Computational Aspects," Computer Methods in Applied Mechanics and Engineering, Vol. 58, 1986, pp. 79-116.
13. Simo, J. C., Rifai, M. S., and Fox, D. D., "On a Stress Resultant Geometrically Exact Shell Model. Part V: Variable Thickness Shells with Through-the-Thickness Stretching," Computer Methods in Applied Mechanics and Engineering, 1989, to appear.
14. Wriggers, P. and Simo, J. C., "A General Purpose Algorithm for Extended Systems in Continuation Methods," International Journal of Numerical Methods in Engineering, 1989, to appear.

On Symmetrization in Boundary Element Elastic and Elastic-Plastic Analysis

G.Maier*, G.Novati* and S.Sirtori**

(*) Department of Structural Engineering, Technical University of Milan
Piazza Leonardo da Vinci 32, Italy

(**) Department of Mechanics, Technical University of Milan
Piazza Leonardo da Vinci 32, Italy

Summary

Boundary integral equations are formulated by associating in a Betti equation the real elastic state and a fictitious state generated by distributions of static and kinematic discontinuities and of inelastic strains. A suitable discretization is adopted for all variable fields and plastic constitutive laws are enforced in an average sense. Thus a symmetric formulation of the direct BEM is obtained. For the finite-step problem formulated in a backward-difference (stepwise holonomic) sense and solved by an iterative procedure an extremum and a convergence theorem are given. An implementation and some aspects of the computational experience achieved are briefly summarized.

Introduction

Symmetric formulations of the boundary element method (BEM) have been proposed and discussed by various Authors in elasticity [1][2][3], elastoplasticity [4][5][6] and elastodynamics [7].

The purposes of this paper are: (a) to generate a symmetric Galerkin formulation of BEM in quasistatic elastoplasticity by a nonconventional approach; (b) to present two novel contributions (an extremum property and a convergence theorem) concerning the finite-step elastoplastic problem, proved and discussed elsewhere [8]; (c) to briefly outline some aspects of the implementation (probably the first carried out of the symmetric BEM).

The basic notation adopted herein is as follows: Ω = (open, simply connected, bounded) domain occupied by a homogeneous body; Γ_u and Γ_p = constrained and free parts of its (smooth) boundary Γ, respectively; underscored symbols denote matrices (and vectors); $\underline{x} = (x_1 x_2 x_3)$ = Cartesian coordinates; $\underline{u}$, $\underline{p}$ = displacement and traction vectors; $\underline{\sigma}$, $\underline{\varepsilon}$, $\underline{\theta}$ = vectors of stresses, ("engineering") strains and inelastic strains; $\underline{k}$ = constant matrix of elastic moduli; $\Delta(\underline{x} - \underline{\xi})$ = Dirac functions; Δ . = finite increment of . ; exponent t means transpose.

G. Kuhn · H. Mang (Eds.)
Discretization Methods in Structural Mechanis
IUTAM/IACM Symposium Vienna, Austria, 1989

Symmetric formulation and discretization of the integral equations

Ignoring body forces for brevity and marking by starred and unstarred symbols a fictitious and the actual elastic state, respectively, one can easily derive the following expression for Betti's reciprocity theorem of linear elasticity:

$$\int_\Gamma (\underline{p}^t\underline{u}^* - \underline{u}^t\underline{p}^*)d\Gamma + \int_\Omega \underline{\theta}^t\underline{\sigma}^*\, d\Omega = \int_\Omega \underline{\sigma}^t\underline{\theta}^* d\Omega \qquad (1)$$

If the Kelvin state were assumed as fictitious elastic state, eq.(1) would lead to the usual (Somigliana's) integral equation. Let instead the fictitious elastic state in eq.(1) be chosen as follows:

$$\underset{\underline{x}\in\Gamma^-}{\underline{u}^*(\underline{x})} = \int_\Gamma \underline{G}_{uu}(\underline{x},\underline{\xi})\underline{F}^*(\underline{\xi})d\Gamma + \int_\Gamma \underline{G}_{up}(\underline{x},\underline{\xi})\underline{D}^*(\underline{\xi})d\Gamma + \int_\Omega \underline{G}_{u\sigma}(\underline{x},\underline{\xi})\underline{\theta}^*(\underline{\xi})d\Omega \qquad (2a)$$

$$\underset{\underline{x}\in\Gamma^-}{\underline{p}^*(\underline{x})} = \int_\Gamma \underline{G}_{pu}(\underline{x},\underline{\xi})\underline{F}^*(\underline{\xi})d\Gamma + \int_\Gamma \underline{G}_{pp}(\underline{x},\underline{\xi})\underline{D}^*(\underline{\xi})d\Gamma + \int_\Omega \underline{G}_{p\sigma}(\underline{x},\underline{\xi})\underline{\theta}^*(\underline{\xi})d\Omega \qquad (2b)$$

$$\underset{\underline{x}\in\Omega}{\underline{\sigma}^*(\underline{x})} = \int_\Gamma \underline{G}_{\sigma u}(\underline{x},\underline{\xi})\underline{F}^*(\underline{\xi})d\Gamma + \int_\Gamma \underline{G}_{\sigma p}(\underline{x},\underline{\xi})\underline{D}^*(\underline{\xi})d\Gamma + \int_\Omega \underline{G}_{\sigma\sigma}(\underline{x},\underline{\xi})\underline{\theta}^*(\underline{\xi})d\Omega \qquad (2c)$$

Here: $\underline{F}^*(\underline{\xi}) \equiv \underline{p}(\underline{\xi}^+) + \underline{p}(\underline{\xi}^-)$ and $\underline{D}^*(\underline{\xi}) \equiv \underline{u}(\underline{\xi}^+) - \underline{u}(\underline{\xi}^-)$ represent static and kinematic discontinuities distributed on the boundary with superscripts + and - denoting points infinitely close to Γ from outside and inside respectively. The Green's functions $\underline{G}$ concern the elastic space Ω_∞ embedding Ω and can be shown to possess the following reciprocity properties for indeces h,k = u, p, σ:

$$\underline{G}_{hk}(\underline{x}^-,\underline{\xi}) = \underline{G}^t_{kh}(\underline{\xi}^+,\underline{x});\quad \text{but: } \underline{G}_{up}(\underline{x}^-,\underline{\xi}) = \underline{G}^t_{pu}(\underline{\xi}^+,\underline{x}) + \Delta(\underline{\xi}-\underline{x})\underline{I} \qquad (3a,b)$$

Let us discretize both the actual fields and the source distributions:

$$\underline{p}(\underline{x}) = \underline{\Psi}_p(\underline{x})\underline{P}\ ,\quad \underline{u}(\underline{x}) = \underline{\Psi}_u(\underline{x})\underline{U}\ ,\quad \underline{\theta}(\underline{x}) = \underline{\Psi}_\theta(\underline{x})\underline{\Theta} \qquad (4a,b,c)$$

$$\underline{F}^*(\underline{x}) = \underline{\Psi}^*_p(\underline{x})\underline{F}^{**}\ ,\ \underline{D}^*(\underline{x}) = \underline{\Psi}^*_u(\underline{x})\underline{D}^{**}\ ,\ \underline{\theta}^*(\underline{x}) = \underline{\Psi}^*_\theta(\underline{x})\underline{\Theta}^{**} \qquad (5a,b,c)$$

where vectors $\underline{P}$, $\underline{U}$, $\underline{\Theta}$, $\underline{F}^{**}$, $\underline{D}^{**}$, $\underline{\Theta}^{**}$ collect the values assumed by the relevant quantities at the nodes (say n nodes on Γ, m on Ω). The interpolation matrices $\underline{\Psi}$ (constructed over each element or cell but defined for each node over all Γ or Ω) are assumed under the following basic restrictions: (a) the displacements $\underline{u}(\underline{x})$ and the kinematic discontinuities

$\underline{D}^*(x)$ are continuous; (b) the models for corresponding actual and source fields are equal: $\underline{\Psi}_i = \underline{\Psi}_i^*$, i = u, p, θ .

Let us set: $$\underline{\Sigma} \equiv \int_\Omega \underline{\Psi}_\theta^t(\underline{x})\underline{\sigma}(\underline{x})d\Omega \qquad (6)$$

$$\hat{\underline{G}}_{ij} \equiv \iint \underline{\Psi}_{i'}^t(\underline{x})\underline{G}_{ij}(\underline{x},\underline{\xi})\underline{\Psi}_{j'}(\underline{\xi})\, d\underline{x}d\underline{\xi} \ , \ i,j = u,p,\sigma \ , \ i',j' = p,u,\theta \qquad (7)$$

where: i' and j' mark the conjugate of i and j, respectively; $d\underline{x}$ stands for $d\Gamma(\underline{x})$ or $d\Omega(\underline{x})$, $d\underline{\xi}$ for $d\Gamma(\underline{\xi})$ or $d\Omega(\underline{\xi})$ and the integration domains are defined by the indeces (e.g., if i = σ, j = p, and, hence, i' = θ, s' = u: $\underline{x}$ runs over Ω and $\underline{\xi}$ over Γ_p). As a consequence of the reciprocity eqs.(3):

$$\hat{\underline{G}}_{ij} = \hat{\underline{G}}_{ij}^t \quad i,j = u,\ p,\ \sigma \quad ; \quad \hat{\underline{G}}_{up} = \hat{\underline{G}}_{pu}^t + \int_\Gamma \underline{\Psi}_u^t\, \underline{\Psi}_p d\Gamma \qquad (8)$$

Introduce now the discretizations (4) and (5) into eqs.(1) and (2), account taken of the above restriction (b). Substituting (2) into (1), using the definitions (6) and (7) and the reciprocities (8), after some manipulations the Betti equation (1) can be given the form:

$$\begin{Bmatrix} \underline{F}^{**} \\ -\underline{D}^{**} \\ \underline{\Theta}^{**} \end{Bmatrix}^t \left(\begin{bmatrix} \hat{\underline{G}}_{uu} & -\hat{\underline{G}}_{up} & \hat{\underline{G}}_{u\sigma} \\ -\hat{\underline{G}}_{pu} & \hat{\underline{G}}_{pp} & -\hat{\underline{G}}_{p\sigma} \\ \hat{\underline{G}}_{\sigma u} & -\hat{\underline{G}}_{\sigma p} & \hat{\underline{G}}_{\sigma\sigma} \end{bmatrix} \begin{Bmatrix} \underline{P} \\ \underline{U} \\ \underline{\theta} \end{Bmatrix} - \begin{Bmatrix} \underline{0} \\ \underline{0} \\ \underline{\Sigma} \end{Bmatrix} \right) = 0 \qquad (9)$$

Partition the 6n-vectors of boundary variables into subvectors pertaining to Γ_p and Γ_u , marked by subscript p and u, respectively:

$$\underline{P} = \begin{Bmatrix} \bar{\underline{P}}_p \\ \underline{P}_u \end{Bmatrix} , \quad \underline{U} = \begin{Bmatrix} \underline{U}_p \\ \bar{\underline{U}}_u \end{Bmatrix} \quad ; \quad \underline{F}^{**} = \begin{Bmatrix} \underline{F}_p^{**} \\ \underline{F}_u^{**} \end{Bmatrix} , \quad \underline{D}^{**} = \begin{Bmatrix} \underline{D}_p^{**} \\ \underline{D}_u^{**} \end{Bmatrix} \qquad (10)$$

The 3n boundary unknowns will be gathered in the vector $\underline{X}$. A system of 3n independent linear algebraic equations is generated by requiring Betti equation (9) to hold for any $\underline{F}_u^{**}$, $\underline{D}_p^{**}$. This system can be written as:

$$\underline{A}\ \underline{X} + \bar{\underline{B}} + \underline{C}\ \underline{\Theta} = 0 \qquad (11)$$

where $\bar{\underline{B}}$ denotes the resulting vector which contains the boundary data. The coefficient matrix $\underline{A}$ is easily seen to be symmetric.

Alternatively, nonsymmetric systems of 3m independent equations might be generated from eq.(1): e.g., using static discontinuities only is equivalent to enforce the customary Somigliana equation on Γ in a Galerkin weighted-residual sense (instead by collocation). The arbitrariness of vector $\underline{\Theta}^{**}$ in eq.(9) gives rise to the equations ($\underline{B}'$ being a new vector of data):

$$\underline{C}^t\underline{X} + \underline{B}' + \hat{\underline{G}}_{\sigma\sigma}\underline{\Theta} = \underline{\Sigma} \tag{12}$$

Symmetric formulation of plastic laws in cell variables

The inelastic strains $\underline{\theta}$ are now identified with plastic strains $\underline{\varepsilon}^p$ and related to stresses $\underline{\sigma}$ through the following associated perfectly plastic constitutive law:

$$\phi = \frac{\partial\phi}{\partial\,\underline{\sigma}^t}(\underline{\sigma})\underline{\sigma} - y \leq 0 \quad , \quad \dot{\underline{\varepsilon}}^p = \frac{\partial\phi}{\partial\underline{\sigma}}\dot{\lambda} \quad , \quad \dot{\lambda} \geq 0 \; , \; \phi\,\dot{\lambda} = 0 \tag{13}$$

ϕ denotes the yield function (assumed as convex, positively homogeneous of order one), y the yield limit, λ the plastic multiplier.

We derive below from the material model (13) a relationship between generalized strains $\underline{\Theta}$ introduced through eq.(4c) and generalized stresses defined by eq.(6), as proposed first in [8]. The requirement of "generalized" variables in Prager's sense: $\int_\Omega \underline{\sigma}^t\underline{\varepsilon}^p d\Omega = \underline{\Sigma}^t\underline{\Theta}$ for any $\underline{\Sigma}$, $\underline{\Theta}$, is satisfied if the stresses are modeled as:

$$\underline{\sigma}(\underline{x}) = \underline{\Psi}_\sigma(\underline{x})\underline{\Sigma} \text{ , with: } \underline{\Psi}_\sigma(\underline{x}) = \underline{\Psi}_\theta(\underline{x})[\int_\Omega \underline{\Psi}_\theta^t\,\underline{\Psi}_\theta\, d\Omega]^{-1} \tag{14a,b}$$

Similarly, generalized yield functions $\underline{\Phi}$ and plastic multipliers $\dot{\underline{\Lambda}}$ can be introduced to govern the fields of the relevant local quantities:

$$\dot{\lambda}(\underline{x}) = \underline{\Psi}_\lambda(\underline{x})\dot{\underline{\Lambda}} \quad ; \qquad \phi(\underline{x}) = \underline{\Psi}_\phi(\underline{x})\underline{\Phi} \tag{15a,b}$$

and will be required to satisfy the condition $\int_\Omega \phi\,\dot{\lambda}\,d\Omega = \underline{\Phi}^t\dot{\underline{\Lambda}}$ for any $\underline{\Phi},\dot{\underline{\Lambda}}$. This leads to the "inverse" relationships of models (15) and to a dependence between them:

$$\int_{\Omega} \underline{\Psi}_{\phi}^{t}\, \dot{\lambda}\, d\Omega = \dot{\underline{\Lambda}} \quad , \quad \int_{\Omega} \underline{\Psi}_{\lambda}^{t}\, \phi\, d\Omega = \underline{\Phi} \quad , \quad \underline{\Psi}_{\phi} = \underline{\Psi}_{\lambda}\left[\int_{\Omega} \underline{\Psi}_{\lambda}^{t}\, \underline{\Psi}_{\lambda}\, d\Omega\right]^{-1} \qquad (16a,b,c)$$

By combining the material model (13) with the eqs.(4c)(6) and (14)-(16), we obtain the following plastic relationship in generalized variables:

$$\underline{\Phi} = \frac{\partial \underline{\Phi}}{\partial \underline{\Sigma}^{t}} (\underline{\Sigma})\underline{\Sigma} - \underline{Y} \le \underline{0} \quad , \text{ where } \quad \underline{Y} = \int_{\Omega} \underline{\Psi}_{\lambda}^{t}\, y\, d\Omega \qquad (17a,b)$$

$$\dot{\underline{\Theta}} = \frac{\partial \underline{\Phi}^{t}}{\partial \underline{\Sigma}}(\underline{\Sigma})\dot{\underline{\Lambda}} \quad , \quad \dot{\underline{\Lambda}} \ge \underline{0} \quad , \quad \underline{\Phi}^{t}\dot{\underline{\Lambda}} = 0 \qquad (17c)$$

Eqs.(17) enforce the material plastic constitution in a weak, weighted average sense and can be regarded as nonlocal constitutive laws written for each one of all cells simultaneously.

Similarly, elastic laws for cells can be derived from eq.(6) and interpolation (4c), the latter being attributed to total strains $\underline{\varepsilon}(\underline{x})$ and to relevant generalized variables gathered in vector $\underline{E}$:

$$\underline{\Sigma} = \int_{\Omega} \underline{\Psi}_{\theta}^{t}\, \underline{k}(\underline{\varepsilon} - \underline{\varepsilon}^{p})d\Omega = \underline{K}(\underline{E} - \underline{\Theta}), \quad \text{with } \underline{K} \equiv \int_{\Omega} \underline{\Psi}_{\theta}^{t}\, \underline{k}\, \underline{\Psi}_{\theta} d\Omega \qquad (18a,b)$$

Time discretization and solution of the finite step problem

Let the loading history be sampled at suitably chosen instants. Knowing all variables at instant $\bar{t}$ (t being an ordinative time variable), the "finite step problem" consists of finding them at the subsequent instant $\bar{t} + \Delta t$. With reference to the discretized boundary integral equations formulated in what precedes, the determination of the response to the load increments $\Delta\underline{\bar{P}}$, $\Delta\underline{\bar{U}}$ over the interval Δt will be pursued here by an implicit backward-difference (or stepwise holonomic) approach (see e.g. [10]). Namely, marking by bars quantities at $\bar{t}$ and increments by Δ, the path-dependent (nonholonomic, nonlinear differential) elastic-plastic laws (17)(18) are replaced by their backward difference version:

$$\underline{\Phi} = \frac{\partial \underline{\Phi}}{\partial \underline{\Sigma}^{t}} (\underline{\Sigma})\underline{\Sigma} - \underline{Y} \le \underline{0} \quad , \text{ with } \underline{\Sigma} = \bar{\underline{\Sigma}} + \Delta\underline{\Sigma} \qquad (19a)$$

$$\Delta\underline{\Theta} = \frac{\partial \underline{\Phi}^{t}}{\partial \underline{\Sigma}} (\underline{\Sigma})\Delta\underline{\Lambda} \quad , \quad \Delta\underline{\Lambda} \ge \underline{0} \quad , \quad \Delta\underline{\Sigma} = \underline{K}(\Delta\underline{E} - \Delta\underline{\Theta}) \qquad (20b,c,d)$$

These are path-independent (holonomic, nonlinear algebraic) relations for the step. To obtain the full set of relations governing the finite-step, eqs.(20) is associated to the discretized integral equations (11) and (12) written for increments $\Delta\underline{X}$, $\Delta\underline{\Theta}$, $\Delta\underline{\Sigma}$. Alternatively, eq.(11) can be solved with respect to $\Delta\underline{X}$ and substituted into (12), so that we can associate to eq.(20) the single equation:

$$\Delta\underline{\Sigma}^e + \underline{Z}\,\Delta\underline{\Theta} = \Delta\underline{\Sigma} \tag{21}$$

having set: $\Delta\underline{\Sigma}^e \equiv \underline{B} - \underline{C}^t\underline{A}^{-1}\underline{B}$, $\underline{Z} \equiv \hat{\underline{G}}_{\sigma\sigma} - \underline{C}^t\underline{A}^{-1}\underline{C}$ (22a,b)

Vector $\Delta\underline{\Sigma}^e$ defines the linear elastic stress response to the load increments; matrix $\underline{Z}$ transforms (generalized) plastic strain increments into consequent selfstresses $\Delta\underline{\Sigma}^s = \Delta\underline{\Sigma} - \Delta\underline{\Sigma}^e$, exhibits selfevident symmetry and is easily shown to be negative semidefinite (in fact, $-\Delta\underline{\Theta}^t\,\Delta\underline{\Sigma}^s$ represents the strain energy stored in the body due to $\Delta\underline{\Theta}$), see [5].

The finite-step problem (20)(21) can be numerically solved by the following iterative algorithm, which is basically the one adopted in the literature on (nonsymmetric) BEM in plasticity (see e.g. [11]).

1. Generate the coefficient matrices once for all. 2. For the current step compute the linear elastic stress response to the load increments. 3. Prediction: from the iterate r of plastic strain increments compute the new stress increments through eq.(21). 4. Correction: compute through the constitutive laws (20) the new plastic strain increments. 5. Initialization: for r = 1 assume $\Delta\underline{\Theta} = \underline{0}$ or = best guess based on the preceding step. 6. Termination test on the changes of $\Delta\underline{\Theta}$ from iteration r to r+1.

The solution to problem (20)(21), or (20)(11)(12), can be characterized by extremum properties or saddle-point properties as generalizations of those pointed out e.g. in [5][12]. Only the following statement, established in [9] for more general material models, will be mentioned here.

Extremum theorem. The finite-step problem (20)(21) is equivalent to:

$$\min_{\Delta\underline{\Theta},\ \Delta\underline{\Lambda},\ \Delta\underline{E}} \{\Omega \equiv -\frac{1}{2}\,\Delta\underline{\Theta}^t\underline{Z}\Delta\underline{\Theta} + \underline{Y}^t\Delta\underline{\Lambda} - (\bar{\underline{\Sigma}} + \Delta\underline{\Sigma}^e)^t\Delta\underline{\Theta}\ \} \tag{23a}$$

subject to: $\Delta\underline{\Theta} = \dfrac{\partial\underline{\Phi}^t}{\partial\underline{\Sigma}}(\underline{\Sigma})\Delta\underline{\Lambda}$, $\Delta\underline{\Lambda} \geq \underline{0}$, $\underline{\Sigma} = \bar{\underline{\Sigma}} + \underline{K}(\Delta\underline{E} - \Delta\underline{\Theta})$ (23b,c,d)

The above proposition permits to transform the finite step solution to the optimization of a convex quadratic function under generally nonconvex constraints, i.e. to nonconvex nonlinear programming (to convex quadratic programming for piecewiselinearized yield surfaces [12]). The alternative formulation (23) can be shown [9] to give rise to the following statement.

Convergence theorem. The above successive substitution algorithm generates a sequence $\Delta\underline{\Theta}^r$, $\Delta\underline{\Sigma}^r$ which converges to the solution of the finite-step problem (20)(21).

The above two statements are linked, inasmuch the algorithm can be shown to generate, within the feasible domain of the optimization (23), a monotonic sequence $\Omega^r \geq \Omega^{r+1}$, the equality holding at the solution only.

Computational aspects and closing remarks

A first implementation of the theory outlined in what precedes has been carried out on the basis of the following modelling assumptions: straight BEs, with linear displacements and constant tractions; triangular internal cells, with constant plastic strains.

In this case it turns out that the required double integrations along the boundary can be performed analytically in an accurate and economical way; besides, the above assumptions allow to circumvent the evaluation of singular and hyper-singular integrals, arising when the "source element" and the "field element" coincide: in fact, first the integrations are carried out with the two integration segments at a finite distance and afterwards their distance is made to vanish.

As for the integrations over the internal cells, in view of the piecewise-constant model for the plastic strains and of the polygonal cell geometry, they can be reduced to integrations along cell boundaries, in a fashion which is formally equivalent to the one for the boundary elements. An antecedent of this procedure can be found in [14]. The generalized stresses are simply the average cell stresses and, hence, the plastic constitutive law can also be conceived as enforced in the traditional pointwise sense.

As a test problem in plasticity, the classical perforated tension strip of Fig.1a, is considered (the same as in [15]). The material, elastic perfectly-plastic, obeys the von Mises yield criterion and the associated flow rule. Although a three-dimensional "corrector" is used, the plane stress condition $\sigma_z = 0$ is fulfilled at convergence (for each loading step) since it is incorporated in the BE equations of the "predictor". The parameter σ_m/y governs the monotonic loading, where $\sigma_m \equiv \frac{p.a}{(a-r)}$ is the reference nominal stress and y is the uniaxial yield limit.

Two comparative analyses are carried out using the meshes of Fig.1b and 1b which have the same boundary subdivision (29 BEs) but differ in the internal discretization (23 and 68 cells, respectively). The load increments range from the 20% (in the initial elastic-plastic stages) to the 3% (near collapse) of the load at first yield (σ_m/y = 0.51). The spreading of plasticity predicted by the two meshes is displayed in Fig.1d,e for different load levels; the cells marked by dotted areas become plastic only in the load step at which there is convergence failure. On a 80286 PC (equipped with math co-processor) the following CPU times were required to generate all the matrices and to perform the evolutive analysis up to collapse (marked by lack of convergence), respectively: for the coarse mesh, 8.5 min. and 27 min.; for the fine mesh, 37.5 min. and 190 min.

In elasticity, various numerical tests have been carried out comparatively by the present symmmetric BEM and by a standard collocation BEM. The computational experience achieved so far (expounded in [13]), turns out to substatiate the following remarks. (a) The implementation of the symmetric approach is made only slightly more laborious by the double integrations as long as they are performed analytically. (b) Generally more accurate solutions are obtained at equal number of variables. (c) The computational burden is comparable when the integrations are analytical in both cases. (d) The condition number of matrix $\underline{A}$ turns out to be generally lower ("better").

In elasto-plasticity additional favorable features of symmetric formulations appear to be as follows (though only partly dealt with here): extremum and convergence theorems for the finite-step are available; shake-down and bounding theory of classical plasticity can be cast into a BE format.

References

1. Sirtori, S.: General stress analysis method by means of integral equations and boundary elements. Meccanica 14 (1979) 210-218.

2. Costabel, M.: Symmetric methods for the coupling of finite elements and boundary elements, in C.A.Brebbia, W.L.Wendland and G.Kuhn (Eds.), Boundary Elements IX, Vol.1, Springer-Verlag, Berlin (1987) 411-420.

3. Wendland, W.L.: Strongly elliptic boundary integral equations, in A.Iserles and M.Powel (Eds.) The State of the Art in Numerical Analysis, Oxford University Press, Oxford (1987).

4. Maier, G.; Polizzotto, C.: A Galerkin approach to boundary element elastoplastic analysis. Comp. Meth. Appl. Mech. Engng. 60 (1987) 175-194.

5. Polizzotto, C.: An energy approach to the boundary element method. Part I: Elastic solids. Part II: Elastic plastic solids. Comp. Meth. Appl. Mech. Engng. 69 (1988) 167-184, 263-276.

6. Maier, G.; Novati, G.; Sirtori, S.: Symmetric formulation of indirect boundary element method for elastic-plastic analysis and relevant extremum properties. Proc. First Joint Japan-USA Symposium on Boundary Element Methods, Tokyo, October 3-6, 1988.

7. Maier, G.; Diligenti, M.; Carini, A.: A variational formulation of BEM in elastodynamics. Int. Symp. on Boundary Element Methods, East Hartford, Conn., 2-4 Oct. 1989.

8. Maier, G.: On elastoplastic analysis by boundary elements. Mech. Res. Comm. 10 (1983) 45-82.

9. Comi, C.; Maier, G.: Extremum and convergence theorem for the finite-step problem in elastoplastic boundary element analysis. To appear.

10. Maier, G.; Nappi, A.: Backward difference time integration, nonlinear programming and extremum theorems in elastoplastic analysis. Solid Mechanics Archives 14 (1989) 17-36.

11. Brebbia, C.A.; Telles, J.C.F.; Wrobel, L.C.: Boundary element techniques. Springer-Verlag, Berlin, 1984.

12. Maier, G.; Novati, G.: Elastic-plastic boundary element analysis as linear complementarity problem. Appl. Math. Modelling 7 (1983) 74-82.

13. Sirtori, S.; Maier, G.; Miccoli, S.; Novati, G.: Symmetric formulation and implementation of the boundary element method in elastoplasticity. To appear.

14. Eshelby, J.D.: Elastic inclusions and inhomogeneities, in I.N.Sneddon, R.Hill (Eds.), Progress in Solid Mechanics, 2, North Holland, Amsterdam (1961) 89-140.

15. Zienkiewicz, O.C.; Valliappant, S.; King, I.P.: Elasto-plastic solutions of engineering problems 'initial stress', finite element approach. Int. J. Num. Meth. Engn. 1(1969) 75-100.

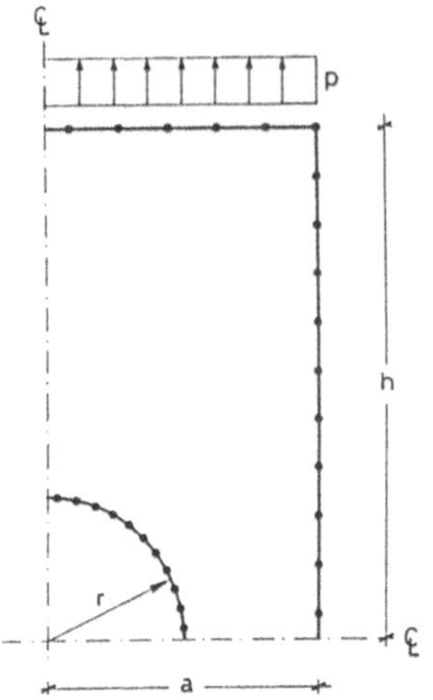

Fig.1a

Fig.1. The perforated tension strip with the two adopted discretizations and the corresponding predicted plastic zones. The material and geometry data are: yield stress y = 240 N/mm^2 , E = 69.000 N/mm^2, ν = 0.2, h = 18 mm, a = 10 mm, r = 5 mm.

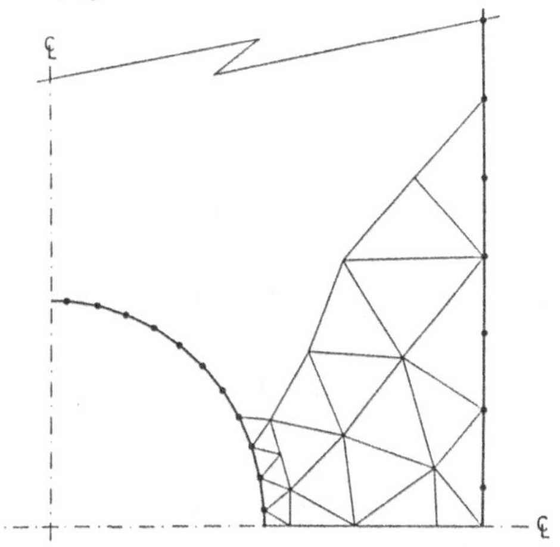

Fig.1b

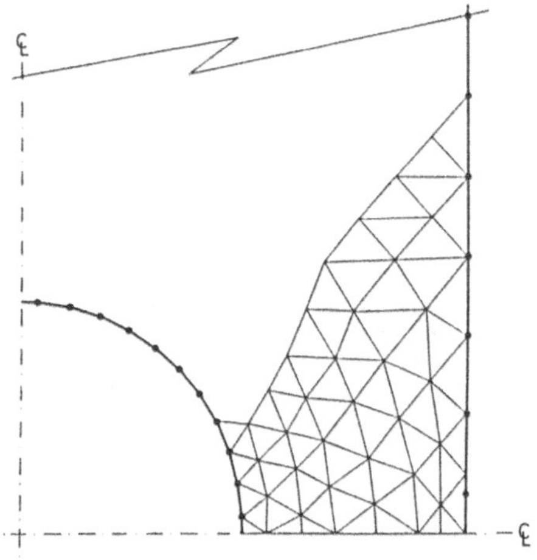

Fig.1c

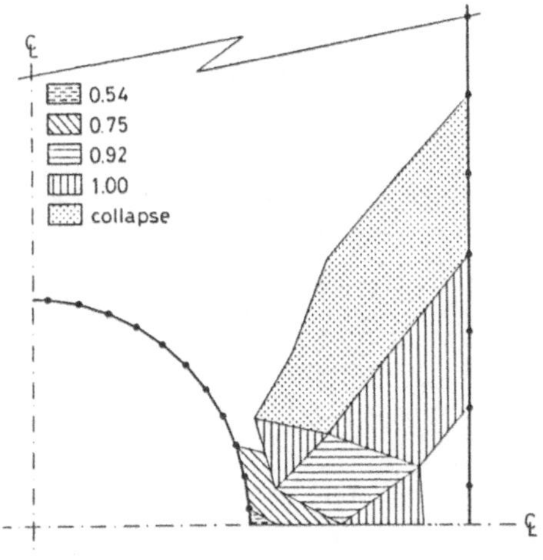

Fig.1d

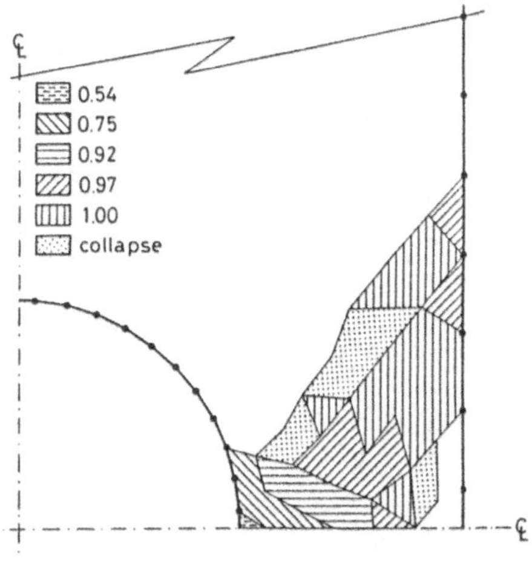

Fig.1e

A Variational Formulation of the BEM for Elastic-Plastic Analysis

C. POLIZZOTTO and M. ZITO

Dipartimento di Ingegneria Strutturale & Geotecnica (DISEG)
Università di Palermo, Palermo, Italy

Summary

The quasi-static elastic perfectly plastic analysis problem is approached by the boundary element method (BEM). To this purpose, a time semidiscretization is first achieved by finite intervals (FI) in order to transform, through a variationally consistent procedure, the evolutive problem into a discrete sequence of inelastic holonomic-type "weighted" problems for each of which a mixed boundary/domain min-max principle is established. This principle is then discretized by means of boundary elements (BE) and cell elements (CE), the latter having the only purpose of suitably interpolating the FI weighted yielding laws within the domain. The algebraic governing equations obtained show symmetry and sign definiteness. Double integrations are needed.

Introduction

The Boundary Element Method (BEM) in its usual direct formulation, though being numerically efficient and competitive with respect to the Finite Element Method (FEM) in many application fields, shows some shortcomings not actually desiderable for a consistent discretization method. Namely, it transforms a self-adjoint problem of continuum mechanics into a discrete no longer self-adjoint one, with few obvious negative consequences, including the difficulties met in error and convergence analysis [1, 2]. Although one may reasonably conjecture that self-adjointness tends to be recovered at the limit as the boundary mesh refinement improves, nevertheless there are enough motivations in order that self-adjointness be a *natural* property of the discretized problem. The procedure often adopted [3, 4] consisting in an a-posteriori symmetrization of a naturally unsymmetric coefficient matrix has certain computational advantages but not a fully rational justification [5], and even it may lead to worse numerical results [6].

The variational approach, most suitable for preserving the problem self-adjointness, was attempted quite recently. The Hellinger-Reissner principle was used by Dumont [5] to derive a " hybrid BEM" showing symmetry, and a similar formulation was given in [7, 8]. The minimum total potential and minimum complementary energy principles were used in [9]. Polizzotto [10–12] employed mixed-type principles, like Hu-Washizu's and Hellinger-Reissner's, as well as the boundary min-max principle [12, 13]. The

* This paper was completed whith the financial support of the (Italian) Ministry of Education.

G. Kuhn · H. Mang (Eds.)
Discretization Methods in Structural Mechanis
IUTAM/IACM Symposium Vienna, Austria, 1989

Hu-Washizu principle was also employed in [14] to approach the dynamic problem. Theorems given by Rieder [15] may also be utilized for BE formulations.

This paper is aimed at presenting an extension of the boundary min-max principle given in [12, 13] to the quasi-static elastic-plastic evolutive analysis problem. This same purpose was pursued in [16] basing on the traditional approach in which the problem is treated in its rate form with the aid of some marching algorithm. Here a different approach will be utilized [17]. Namely the total time interval is subdivided into finite intervals (FI), and for each of them a "weighted" elastic-plastic holonomic-type problem is builded. A mixed boundary/domain min-max principle is then established such as to characterize the solution to the latter weighted inelastic problem.

Notations. The matrix notation will be used, with vectors and matrices denoted by bold face symbols. The symbol $(\cdot)^T$ denotes the transpose of $(\cdot)$. Domain or surface integrals will be denoted by parentheses, i.e.

$$(\mathbf{a},\mathbf{b})_R = \int_R \mathbf{a}^T\mathbf{b}dR, \quad ((\mathbf{a},\mathbf{b}))_{R\times S} = \int_R\int_S \mathbf{a}^T\mathbf{b}dSdR, \quad ((\mathbf{a},\mathbf{b}))_{R^2} = \int_R\int_R \mathbf{a}^T\mathbf{b}dRdR.$$

The unit matrix is denoted by $\mathbf{I}$. Other symbols will be defined where they first appear.

The quasi-static elastic-plastic problem

Let a homogeneous elastic perfectly plastic solid, occupying the (open) region Ω with boundary surface Γ, be referred to the Cartesian orthogonal co-ordinate system $\mathbf{x} = (x_1, x_2, x_3)$ and let it be restrained over the portion Γ_1 of Γ. The body is acted upon by body forces $\bar{\mathbf{b}}$ in Ω, tractions $\bar{\mathbf{t}}$ on $\Gamma_2 = \Gamma - \Gamma_1$, imposed displacements $\bar{\mathbf{u}}$ on Γ_1 and imposed (or initial) strains $\bar{\boldsymbol{\theta}}$ in Ω (e.g. thermal strains). All these loads are variable with some monotonically increasing time-like parameter $t:\ 0 \le t \le t_f$. At every t the mechanical problem is governed by the equations

$$\boldsymbol{\varepsilon} = \mathbf{D}\,\mathbf{u} \text{ in } \Omega, \qquad \mathbf{u} = \bar{\mathbf{u}}(\mathbf{x},t) \text{ on } \Gamma_1 \quad (\textit{compatibility}) \tag{1}$$

$$\mathbf{D}^T\boldsymbol{\sigma} + \bar{\mathbf{b}}(\mathbf{x},t) = 0 \text{ in } \Omega \qquad \mathbf{D}_n^T\boldsymbol{\sigma} \equiv \mathbf{t} = \bar{\mathbf{t}}(\mathbf{x},t) \text{ on } \Gamma_2 \quad (\textit{equilibrium}) \tag{2}$$

$$\boldsymbol{\varepsilon} = \mathbf{C}\,\boldsymbol{\sigma} + \bar{\boldsymbol{\theta}}(\mathbf{x},t) + \boldsymbol{\varepsilon}^p \text{ in } \Omega \quad (\textit{Hooke's law}) \tag{3}$$

where $\boldsymbol{\varepsilon}, \boldsymbol{\sigma}$ are the strain and stress vectors, $\boldsymbol{\varepsilon}^p$ is the plastic strain vector, $\mathbf{D}$ is the differential operator trasforming the displacements $\mathbf{u}$ into the strains $\boldsymbol{\varepsilon}$, $\mathbf{D}^T$ = adjoint of $\mathbf{D}$, $\mathbf{D}_n^T$ is a matrix builded with the unit normal $\mathbf{n}$ to the plane element upon which the traction $\mathbf{t}$ acts. Equations (1)-(3) must be completed by the relevant plasticity laws. The latter, in case of associated plasticity, read:

$$\dot{\boldsymbol{\varepsilon}}^p = \dot{\lambda}\,\frac{\partial\phi}{\partial\boldsymbol{\sigma}}, \qquad \phi(\boldsymbol{\sigma}) \le 0, \qquad \dot{\lambda} \ge 0, \qquad \dot{\lambda}\,\phi(\boldsymbol{\sigma}) = 0 \tag{4}$$

where $\phi(\boldsymbol{\sigma})$ is the (convex, smooth) yield function. Upper dots denote differentiation with respect to t.

In the following, a discretization method for solving eqs.(1)-(4) will be presented. To this purpose, let the body be embedded within the infinite domain Ω_∞, showing the same homogeneous elastic properties as Ω, and let all the loads applied upon Ω be treated as applied upon Ω_∞. The elastic response of Ω_∞ can be represented by the Somigliana formulas [18, 19]

$$\mathbf{u} = (\mathbf{I}, \mathbf{G}_{uu}\mathbf{f})_{\Gamma_1} + (\mathbf{I}, \mathbf{G}_{ut}\mathbf{v})_{\Gamma_2} + (\mathbf{I}, \mathbf{G}_{u\sigma}\boldsymbol{\varepsilon}^p)_{\Omega} + \mathbf{u}_I \tag{5a}$$

$$\mathbf{t} = (\mathbf{I}, \mathbf{G}_{tu}\mathbf{f})_{\Gamma_1} + (\mathbf{I}, \mathbf{G}_{tt}\mathbf{v})_{\Gamma_2} + (\mathbf{I}, \mathbf{G}_{t\sigma}\boldsymbol{\varepsilon}^p)_{\Omega} + \mathbf{t}_I \tag{5b}$$

$$\boldsymbol{\sigma} = (\mathbf{I}, \mathbf{G}_{\sigma u}\mathbf{f})_{\Gamma_1} + (\mathbf{I}, \mathbf{G}_{\sigma t}\mathbf{v})_{\Gamma_2} + (\mathbf{I}, \mathbf{G}_{\sigma\sigma}\boldsymbol{\varepsilon}^p)_{\Omega} + \boldsymbol{\sigma}_I \tag{5c}$$

Here, the matrices $\mathbf{G}_{hk}(\mathbf{x}, \mathbf{y})$ collect different kinds of fundamental solutions. They give effects at $\mathbf{x} \in \Omega_\infty$ (i.e. displacements for $h = u$, tractions for $h = t$ and stresses for $h = \sigma$) due to unit concentrated actions applied at $\mathbf{y} \in \Omega_\infty$ (i.e. forces for $k = u$, relative displacements for $k = t$, and imposed strain for $k = \sigma$) [10, 11, 20], with the symmetry properties

$$\mathbf{G}_{hh}(\mathbf{x}, \mathbf{y}) = \mathbf{G}_{hh}(\mathbf{y}, \mathbf{x}) \qquad \text{for } h = u, t, \sigma \tag{6a}$$

$$\mathbf{G}_{hk}(\mathbf{x}, \mathbf{y}) = \mathbf{G}^T_{kh}(\mathbf{y}, \mathbf{x}) \qquad \text{for } h, k = u, t, \sigma, \quad h \neq k. \tag{6b}$$

Furthermore, the vector quantities $\mathbf{f}, \mathbf{v}$ are unknown layers of force and relative displacement in Γ_1 and Γ_2, respectively, whereas the vectors $(\cdot)_I$ denote the known (elastic) response of Ω_∞ to the assigned loads.

<u>Time modelling of the constitutive equations and FI discretization</u>

Let $t_0 = 0, t_1, t_2, \ldots, t_k, \ldots, t_m = t_f$ be a subdivision of the time interval $(0, t_f)$ into m subintervals of equal length h and let the material obey, within every step (or FI), the ***plastic evolutive model*** [17]:

$$\dot{\boldsymbol{\varepsilon}}^p(t_{k-1} + \tau) = \frac{1}{h}\, g(\tau)\, \mathring{\boldsymbol{\varepsilon}}^p_k, \qquad 0 \leq \tau \leq h \tag{7}$$

where $\mathring{\boldsymbol{\varepsilon}}^p_k$ is an unknown vector and $g(\tau)$ is a shape function satisfying the condition

$$\frac{1}{h^2} \int_0^h \int_0^\tau g(\tau)\, g(\rho)\, d\rho\, d\tau = 1. \tag{8}$$

The representation (7) corresponds to a linear plastic strain path characterized by the least value of the plastic dissipation work with respect to any other strain path—it in fact coincides with the minimum work strain path by Martin and Ponter [21].

Suitable generalized yielding laws for the FI can be derived by a variational principle based on the maximum plastic work theorem [21]. This principle reads [17]

$$\max_{(\boldsymbol{\sigma})} W \equiv \int_0^h \boldsymbol{\sigma}^T \dot{\boldsymbol{\varepsilon}}^p d\tau \qquad \text{s.t. } \phi(\boldsymbol{\sigma}) \leq 0 \text{ in } (0, h) \tag{9}$$

for any fixed plastic strain rate history, e.g. as specified by (7). Accounting for (7) and using the Lagrange multiplier method, the problem (9) can be recast as

$$\min_{(\sigma)} \max_{(\dot\lambda)} -W_L \equiv -\hat{\sigma}_k^T \overset{\circ}{\epsilon}{}_k^p + \int_0^h \dot\lambda\, \phi(\sigma) d\tau \qquad \text{s.t. } \dot\lambda \geq 0 \quad \text{in } (0,h) \tag{10}$$

where $\hat{\sigma}_k$ is the ***weighted stress*** vector associated with the stress history $\sigma(t_{k-1}+\tau)$, i.e.

$$\hat{\sigma}_k = \frac{1}{h}\int_0^h g(\tau)\sigma(t_{k-1}+\tau) d\tau \tag{11}$$

A stress history equally associated with $\hat{\sigma}_k$ is

$$\sigma(t_{k-1}+\tau) = \tilde{g}(\tau)\,\hat{\sigma}_k, \qquad 0 \leq \tau \leq h \tag{12}$$

where $\tilde{g}(\tau)$ is some complementary shape function, suitably chosen as to satisfy

$$\frac{1}{h}\int_0^h g(\tau)\,\tilde{g}(\tau) d\tau = 1, \tag{13}$$

as it can be shown by substituting (12) in (11) and bearing in mind (13). Let the stress history σ in (10) be chosen as in (12) and let the function $\dot\lambda$ be shaped as

$$\dot\lambda = l(\tau)\overset{\circ}{\lambda}_k, \qquad l(\tau) \geq 0 \text{ in } (0,h), \qquad \overset{\circ}{\lambda}_k \geq 0 \tag{14}$$

where $l(\tau)$ is some weight function such that $l(\tau) = 0$ for those τ where $g(\tau) = 0$, but

$$\frac{1}{h}\int_0^h l(\tau) d\tau = 1. \tag{15}$$

The problem (10) thus becomes a discrete one, i.e.

$$\min_{(\hat{\sigma}_k)} \max_{(\overset{\circ}{\lambda}_k)} -\hat{W}_L \equiv -\hat{\sigma}_k^T \overset{\circ}{\epsilon}{}_k^p + \overset{\circ}{\lambda}_k\, \hat{\phi}(\hat{\sigma}_k) \qquad \text{s.t. } \overset{\circ}{\lambda} \geq 0 \tag{16}$$

where $\hat{\phi} = \hat{\phi}(\hat{\sigma}_k)$ is the convex function defined as

$$\hat{\phi}(\hat{\sigma}_k) \equiv \frac{1}{h}\int_0^h l(\tau)\,\phi\,[\tilde{g}(\tau)\hat{\sigma}_k]\, d\tau. \tag{17}$$

The Kuhn-Tucker conditions of (16) are easily shown to be

$$\overset{\circ}{\epsilon}{}_k^p = \overset{\circ}{\lambda}_k \frac{\partial\phi}{\partial\hat{\sigma}_k}, \qquad \hat{\phi}(\hat{\sigma}_k) \leq 0, \qquad \overset{\circ}{\lambda}_k \geq 0, \qquad \overset{\circ}{\lambda}_k\, \hat{\phi}(\hat{\sigma}_k) = 0 \tag{18}$$

which represent the desired yielding laws for the k-th FI, with $\hat{\phi}$ playing the role of yielding function. Through (9) the FI plastic dissipation density turns out to be

$$W_k = \hat{\sigma}_k^T \overset{\circ}{\varepsilon}{}_k^p = W(\overset{\circ}{\varepsilon}{}_k^p), \qquad \text{such that } \hat{\sigma}_k = \frac{\partial W}{\partial \overset{\circ}{\varepsilon}{}_k^p}. \tag{19}$$

The plastic strain increment $\Delta\varepsilon_k^p$ is related to $\overset{\circ}{\varepsilon}{}_k^p$ by

$$\Delta\varepsilon_k^p = \gamma \overset{\circ}{\varepsilon}{}_k^p, \qquad \text{where } \gamma \equiv \frac{1}{h}\int_0^h g(\tau)d\tau. \tag{20}$$

Applying the weighting operation (11) to eqs. (1)-(3) within the k-th FI gives

$$\hat{\varepsilon}_k = \mathbf{D}\,\hat{\mathbf{u}}_k \text{ in } \Omega, \qquad \hat{\mathbf{u}}_k = \hat{\bar{\mathbf{u}}}_k \text{ on } \Gamma_1 \quad (\textit{compatibility}) \tag{21}$$

$$\mathbf{D}^T\hat{\sigma}_k + \hat{\bar{\mathbf{b}}}_k = \mathbf{0} \text{ in } \Omega, \qquad \mathbf{D}_n^T\hat{\sigma}_k = \hat{\bar{\mathbf{t}}}_k \text{ on } \Gamma_2 \quad (\textit{equilibrium}) \tag{22}$$

$$\hat{\varepsilon}_k = \mathbf{C}\,\hat{\sigma}_k + \hat{\bar{\theta}}_k + \gamma\varepsilon_{k-1}^p + \overset{\circ}{\varepsilon}{}_k^p \text{ in } \Omega \quad (\textit{Hooke's law}) \tag{23}$$

The latter equations together with (18) describe some "weighted" elastic-plastic holonomic-type problem $\mathbb{P}_k$ with the load data taken at their weighted values. A number of variational statements can be given for $\mathbb{P}_k$, but here we want to establish a mixed boundary/domain min-max principle.

<u>The min-max principle for the weighted inelastic problem</u>

Two principles of this kind can be established according to whether, besides $\hat{\mathbf{f}}_k$ and $\hat{\mathbf{v}}_k$, both $\hat{\sigma}_k$ and $\overset{\circ}{\varepsilon}{}_k^p$ are unknown, or only $\overset{\circ}{\varepsilon}{}_k^p$ is unknown. For simplicity, here we consider the second case only. Let us consider the functional

$$\Pi_k = \frac{1}{2}a[\hat{\mathbf{f}}_k, \hat{\mathbf{v}}_k, \overset{\circ}{\varepsilon}{}_k^p] - (1, W(\overset{\circ}{\varepsilon}{}_k^p))_\Omega - (\hat{\bar{\mathbf{u}}}_k - \hat{\mathbf{u}}_{Ik}, \hat{\mathbf{f}}_k)_{\Gamma_1^-} - (\hat{\bar{\mathbf{t}}}_k - \hat{\mathbf{t}}_{Ik}, \hat{\mathbf{v}}_k)_{\Gamma_2^-} + (\hat{\sigma}_{Ik}, \overset{\circ}{\varepsilon}{}_k^p)_\Omega \tag{24}$$

where $W(\overset{\circ}{\varepsilon}{}_k^p)$ is the FI plastic dissipation function and

$$\begin{aligned} a[\hat{\mathbf{f}}_k, \hat{\mathbf{v}}_k, \overset{\circ}{\varepsilon}{}_k^p] \equiv {} & ((\hat{\mathbf{f}}_k, \mathbf{G}_{uu}\hat{\mathbf{f}}_k))_{\Gamma_1^2} + 2((\hat{\mathbf{f}}_k, \mathbf{G}_{ut}\hat{\mathbf{v}}_k))_{\Gamma_1\times\Gamma_2} + ((\hat{\mathbf{v}}_k, \mathbf{G}_{tt}\hat{\mathbf{v}}_k))_{\Gamma_2^2} \\ & + 2((\hat{\mathbf{f}}_k, \mathbf{G}_{u\sigma}\overset{\circ}{\varepsilon}{}_k^p))_{\Gamma_1\times\Omega} + 2((\hat{\mathbf{v}}_k, \mathbf{G}_{t\sigma}\overset{\circ}{\varepsilon}{}_k^p))_{\Gamma_2\times\Omega} + ((\overset{\circ}{\varepsilon}{}_k^p, \mathbf{G}_{\sigma\sigma}\overset{\circ}{\varepsilon}{}_k^p))_{\Omega^2} \end{aligned} \tag{25}$$

A mixed boundary/domain min-max principle can be established, expressed as :

$$\Pi_k^* = \min_{(\hat{\mathbf{f}}_k,)} \max_{(\hat{\mathbf{v}}_k, \overset{\circ}{\varepsilon}{}_k^p)} \Pi_k[\hat{\mathbf{f}}_k, \hat{\mathbf{v}}_k, \overset{\circ}{\varepsilon}{}_k^p] \tag{26}$$

under continuity of $\hat{\mathbf{v}}_k$. The principle (26) provides the solution $\hat{\mathbf{f}}_k^*$, $\hat{\mathbf{v}}_k^*$, $\overset{\circ}{\varepsilon}{}_k^{p*}$ which also solves eqs.(21)-(23) and (18) in the sense that eqs. $(5a-c)$ provide, for $\mathbf{f} = \hat{\mathbf{f}}_k^*$, $\mathbf{v} = \hat{\mathbf{v}}_k^*$ and $\varepsilon^p = \varepsilon_{k-1}^p + \gamma \overset{\circ}{\varepsilon}{}_k^{p*}$, the solution to $\mathbf{P}_k$. In fact, the first variation of (24) reads

$$\begin{aligned}\delta\Pi_k = &(\delta\hat{\mathbf{f}}_k, \{(\mathbf{I}, \mathbf{G}_{uu}\hat{\mathbf{f}}_k)_{\Gamma_1} + (\mathbf{I}, \mathbf{G}_{ut}\hat{\mathbf{v}}_k)_{\Gamma_2} + (\mathbf{I}, \mathbf{G}_{u\sigma}\overset{\circ}{\varepsilon}{}_k^p)_\Omega + \hat{\mathbf{u}}_{Ik} - \hat{\bar{\mathbf{u}}}_k\})_{\Gamma_1^-}\\ &+ (\delta\hat{\mathbf{v}}_k, \{(\mathbf{I}, \mathbf{G}_{tu}\hat{\mathbf{f}}_k)_{\Gamma_1} + (\mathbf{I}, \mathbf{G}_{tt}\hat{\mathbf{v}}_k)_{\Gamma_2} + (\mathbf{I}, \mathbf{G}_{t\sigma}\overset{\circ}{\varepsilon}{}_k^p)_\Omega + \hat{\mathbf{t}}_{Ik} - \hat{\bar{\mathbf{t}}}_k\})_{\Gamma_2^-}\\ &+ (\delta\overset{\circ}{\varepsilon}{}_k^p, \{(\mathbf{I}, \mathbf{G}_{\sigma u}\hat{\mathbf{f}}_k)_{\Gamma_1} + (\mathbf{I}, \mathbf{G}_{\sigma t}\hat{\mathbf{v}}_k)_{\Gamma_2} + (\mathbf{I}, \mathbf{G}_{\sigma\sigma}\overset{\circ}{\varepsilon}{}_k^p)_\Omega + \hat{\sigma}_{Ik} - \frac{\partial W}{\partial \overset{\circ}{\varepsilon}{}_k^p}\})_\Omega\end{aligned} \tag{27}$$

from which we readily derive the relevant Euler-Lagrange equations, namely

$$\hat{\mathbf{u}}_k \equiv (\mathbf{I}, \mathbf{G}_{uu}\hat{\mathbf{f}}_k)_{\Gamma_1} + (\mathbf{I}, \mathbf{G}_{ut}\hat{\mathbf{v}}_k)_{\Gamma_2} + (\mathbf{I}, \mathbf{G}_{u\sigma}\overset{\circ}{\varepsilon}{}_k^p)_\Omega + \hat{\mathbf{u}}_{Ik} = \hat{\bar{\mathbf{u}}}_k \quad \text{on } \Gamma_1^- \tag{28a}$$

$$\hat{\mathbf{t}}_k \equiv (\mathbf{I}, \mathbf{G}_{tu}\hat{\mathbf{f}}_k)_{\Gamma_1} + (\mathbf{I}, \mathbf{G}_{tt}\hat{\mathbf{v}}_k)_{\Gamma_2} + (\mathbf{I}, \mathbf{G}_{t\sigma}\overset{\circ}{\varepsilon}{}_k^p)_\Omega + \hat{\mathbf{t}}_{Ik} = \hat{\bar{\mathbf{t}}}_k \quad \text{on } \Gamma_2^- \tag{28b}$$

$$\hat{\sigma}_k \equiv (\mathbf{I}, \mathbf{G}_{\sigma u}\hat{\mathbf{f}}_k)_{\Gamma_1} + (\mathbf{I}, \mathbf{G}_{\sigma t}\hat{\mathbf{v}}_k)_{\Gamma_2} + (\mathbf{I}, \mathbf{G}_{\sigma\sigma}\overset{\circ}{\varepsilon}{}_k^p)_\Omega + \hat{\sigma}_{Ik} = \frac{\partial W}{\partial \overset{\circ}{\varepsilon}{}_k^p} \quad \text{in } \Omega \tag{28c}$$

Bearing in mind eqs.$(5a-c)$, we can state that eqs.$(28a, b)$ represent the classical *boundary integral equations for dispacements and tractions*, respectively, but *taken at their FI weighted values.* Also, eq. $(28c)$ is the Somigliana formula for the weighted stresses $\hat{\sigma}_k$ which, being derivable from the plastic dissipation function $W(\overset{\circ}{\varepsilon}{}_k^p)$, are related to the $\overset{\circ}{\varepsilon}{}_k^p$ through (18). Therefore, the solution to (26) provides a solution to $(28a-c)$ and (18) and thus to (21)-(23) and (18), that is to $\mathbf{P}_k$.

The converse is also true. In fact, let the solution to $\mathbf{P}_k$ be labeled by a star and let us set

$$\Pi_k = \Pi_k^* + \delta\Pi_k + \frac{1}{2}\delta^2\Pi_k \tag{29}$$

where $\delta\Pi_k = 0$ due to $(28a-c)$, and $\delta^2\Pi_k$ is easily shown to be expressed as

$$\delta^2\Pi_k = a[\delta\hat{\mathbf{f}}_k, \delta\hat{\mathbf{v}}_k, \delta\overset{\circ}{\varepsilon}{}_k^p] - (\delta\hat{\sigma}_k, \delta\overset{\circ}{\varepsilon}{}_k^p)_\Omega, \quad \text{where } \delta\hat{\sigma}_k^T\,\delta\overset{\circ}{\varepsilon}{}_k^p \geq 0. \tag{30}$$

The virtual work principle as applied to Ω_∞ gives

$$(\hat{\sigma}'_k, \mathbf{C}\hat{\sigma}'_k + \overset{\circ}{\varepsilon}{}_k^p)_{\Omega_\infty} = (\hat{\mathbf{f}}_k, \hat{\mathbf{u}}'_k)_{\Gamma_1} - (\hat{\mathbf{v}}_k, \hat{\mathbf{t}}'_k)_{\Gamma_2} \tag{31}$$

where the $\hat{\mathbf{u}}'_k$, $\hat{\mathbf{t}}'_k$, $\hat{\sigma}'_k$, the response of Ω_∞ to the loads $\hat{\mathbf{f}}_k$ on Γ_1, $\hat{\mathbf{v}}_k$ on Γ_2 and $\overset{\circ}{\varepsilon}{}_k^p$ in Ω, are given by the l.h. sides of $(28a-c)$ with $\hat{\mathbf{u}}_{Ik}$, $\hat{\mathbf{t}}_{Ik}$ and $\hat{\sigma}_{Ik}$ dropped. Substituting from eqs. $(28a-c)$ in (31) yields

$$\begin{aligned}(\hat{\sigma}'_k, \mathbf{C}\hat{\sigma}'_k) &= ((\hat{\mathbf{f}}_k, \mathbf{G}_{uu}\hat{\mathbf{f}}_k))_{\Gamma_1^2} - ((\hat{\mathbf{v}}_k, \mathbf{G}_{tt}\hat{\mathbf{v}}_k))_{\Gamma_2^2}\\ &\quad - 2((\hat{\mathbf{v}}_k, \mathbf{G}_{t\sigma}\overset{\circ}{\varepsilon}{}_k^p))_{\Gamma_1\times\Omega} - ((\overset{\circ}{\varepsilon}{}_k^p, \mathbf{G}_{\sigma\sigma}\overset{\circ}{\varepsilon}{}_k^p))_{\Omega^2}\\ &= a[\hat{\mathbf{f}}_k, 0, 0] - a[0, \hat{\mathbf{v}}_k, \overset{\circ}{\varepsilon}{}_k^p].\end{aligned} \tag{32}$$

Since the l.h. member of (32) is twice the strain energy stored in Ω_∞, which is finite and essentially positive whater $\hat{\mathbf{f}}_k$, $\hat{\mathbf{v}}_k$, and $\mathring{\varepsilon}^p_k$ not identically vanishing, we write

$$a\,[\hat{\mathbf{f}}_k,\,\mathbf{0},\,\mathbf{0}] > 0, \qquad a\,[\mathbf{0},\,\hat{\mathbf{v}}_k,\,\mathring{\varepsilon}^p_k] < 0\,. \tag{33}$$

Thus (29), with the aid of (30) and (33), provides

$$\Pi_k[\hat{\mathbf{f}}_k,\,\hat{\mathbf{v}}^*_k,\,\mathring{\varepsilon}^{p*}_k] = \Pi^*_k + a\,[\delta\hat{\mathbf{f}}_k,\,\mathbf{0},\,\mathbf{0}] > \Pi^*_k \tag{34a}$$

$$\Pi_k[\hat{\mathbf{f}}^*_k,\,\hat{\mathbf{v}}_k,\,\mathring{\varepsilon}^p_k] = \Pi^*_k + a\,[\mathbf{0},\,\delta\hat{\mathbf{v}}_k,\,\delta\mathring{\varepsilon}^p_k] < \Pi^*_k. \tag{34b}$$

As a consequence the following continued inequality can be written

$$\Pi_k[\hat{\mathbf{f}}_k,\,\hat{\mathbf{v}}^*_k,\,\mathring{\varepsilon}^{p*}_k] \geq \Pi_k[\hat{\mathbf{f}}^*_k,\,\hat{\mathbf{v}}^*_k,\,\mathring{\varepsilon}^{p*}_k] \geq \Pi_k[\hat{\mathbf{f}}^*_k,\,\hat{\mathbf{v}}_k,\,\mathring{\varepsilon}^p_k] \tag{35}$$

which holds for any $\hat{\mathbf{f}}_k$, $\hat{\mathbf{v}}_k$, $\mathring{\varepsilon}^p_k$, but $\hat{\mathbf{v}}_k$ continuous on Γ_2, and the equality sign applies only if $\delta\hat{\mathbf{f}}_k$, $\delta\hat{\mathbf{v}}_k$, $\delta\mathring{\varepsilon}^p_k$ all vanish identically. So the solution to $\mathbf{P}_k$ solves also the min-max principle (26). It can also be proved that the solution is unique. For $\mathring{\varepsilon}^p_k \equiv \mathbf{0}$, the above principle reduces to the boundary min-max principle given in [12, 13].

Coupled BE/CE discretization of the inelastic FI weighted problem $\mathbb{P}_k$

The inelastic FI problem $\mathbb{P}_k$ may be numerically addressed through a suitable discretization by BEs and CEs (cell elements) by writing [11, 12, 16]:

$$\hat{\mathbf{f}}_k(\mathbf{x}) = \mathbf{N}_1(\mathbf{x})\mathbf{X}_{1k} \text{ on } \Gamma_1, \quad \hat{\mathbf{v}}_k(\mathbf{x}) = \mathbf{N}_2(\mathbf{x})\mathbf{X}_{2k} \text{ on } \Gamma_2, \quad \mathring{\varepsilon}^p_k(\mathbf{x}) = \mathbf{N}_p(\mathbf{x})\mathring{\mathbf{p}}_k \text{ in } \Omega \tag{36}$$

where $\mathbf{N}_1(\mathbf{x})$, $\mathbf{N}_2(\mathbf{x})$, and $\mathbf{N}_p(\mathbf{x})$ are matrices of shape functions and $\mathbf{X}_{1k}$, $\mathbf{X}_{2k}$ and $\mathring{\mathbf{p}}_k$ are unknown vector parameters. The cell is a portion of the body volume where the plastic yielding laws (18) are suitably interpolated. This goal can be independently achieved by means of the maximum plastic work theorem, as shown in [11, 12]. The resulting global yielding laws for the individual cell occupying the volume Ω_c read

$$\mathring{\mathbf{p}}_k{}^{(c)} = \mathring{\Lambda}_k{}^{(c)}\frac{\partial\Phi^{(c)}}{\partial\mathbf{P}^{(c)}_k}, \qquad \Phi^{(c)}(\mathbf{P}^{(c)}_k) \leq 0, \qquad \mathring{\Lambda}_k{}^{(c)} \geq 0, \qquad \mathring{\Lambda}_k{}^{(c)}\,\Phi^{(c)}(\mathbf{P}^{(c)}_k) = 0 \tag{37}$$

where $\mathbf{P}^{(c)}_k$ is the CE stress vector, dual of $\mathring{\mathbf{p}}_k{}^{(c)}$, such that the product $\mathbf{P}^{(c)T}_k\,\mathring{\mathbf{p}}_k{}^{(c)} =$ plastic work dissipated within the cell in the $k-th$ FI. The cell yield function $\Phi^{(c)}(\mathbf{P}^{(c)})$, convex in the $\mathbf{P}^{(c)}$ space, is defined as

$$\Phi^{(c)}(\mathbf{P}^{(c)}) = (l_c(\mathbf{x}),\,\phi\,[\tilde{\mathbf{N}}(\mathbf{x})\mathbf{P}^{(c)}])_{\Omega c} \tag{38}$$

where $l_c(\mathbf{x}) \geq 0$ is some weight function and $\tilde{\mathbf{N}}(\mathbf{x})$ is a matrix of shape functions such that $(\mathbf{N}, \tilde{\mathbf{N}})_{\Omega_c} = \mathbf{I}$. The functional Π_k in (24) takes on the algebraic form

$$\Pi_k = \frac{1}{2}\mathbf{X}_{1k}^T\mathbf{A}_{11}\,\mathbf{X}_{1k} + \mathbf{X}_{1k}^T\mathbf{A}_{12}\,\mathbf{X}_{2k} + \frac{1}{2}\mathbf{X}_{2k}^T\mathbf{A}_{22}\,\mathbf{X}_{2k} + \mathbf{X}_{1k}^T\mathbf{B}_1\,\mathring{\mathbf{p}}_k + \mathbf{X}_{2k}^T\mathbf{B}_2\,\mathring{\mathbf{p}}_k$$
$$+\frac{1}{2}\mathring{\mathbf{p}}_k{}^T\mathbf{C}\,\mathring{\mathbf{p}}_k - \mathcal{D}_k(\mathring{\mathbf{p}}_k) - \mathbf{R}_{1k}^T\mathbf{X}_{1k} - \mathbf{R}_{2k}^T\mathbf{X}_{2k} + \mathbf{P}_{Ik}^T\mathring{\mathbf{p}}_k \tag{39}$$

where $\mathring{\mathbf{p}}_k$ collects the plastic strain parameters of the entire set of CEs and furthermore the following definitions hold:

$$\mathbf{A}_{11} = ((\mathbf{N}_1,\, \mathbf{G}_{uu}\mathbf{N}_1))_{\Gamma_1{}^2},\; \mathbf{A}_{12} = ((\mathbf{N}_1,\, \mathbf{G}_{ut}\mathbf{N}_2))_{\Gamma_1\times\Gamma_2},\; \mathbf{A}_{22} = ((\mathbf{N}_2,\, \mathbf{G}_{tt}\mathbf{N}_2))_{\Gamma_2{}^2} \tag{40 a}$$

$$\mathbf{B}_1 = ((\mathbf{N}_1,\, \mathbf{G}_{u\sigma}\mathbf{N}_p))_{\Gamma_1\times\Omega},\; \mathbf{B}_2 = ((\mathbf{N}_2,\, \mathbf{G}_{t\sigma}\mathbf{N}_p))_{\Gamma_2\times\Omega},\; \mathbf{C} = ((\mathbf{N}_p,\, \mathbf{G}_{\sigma\sigma}\mathbf{N}_p))_{\Omega^2} \tag{40 b}$$

$$\mathbf{R}_{1k} = (\mathbf{N}_1,\, \hat{\mathbf{u}} - \hat{\mathbf{u}}_{Ik})_{\Gamma_1^-}, \qquad \mathbf{R}_{2k} = (\mathbf{N}_2,\, \hat{\bar{\mathbf{t}}} - \hat{\mathbf{t}}_{Ik})_{\Gamma_2^-}, \qquad \mathbf{P}_{Ik} = (\mathbf{N}_p,\, \hat{\boldsymbol{\sigma}}_{Ik})_\Omega \tag{40 c}$$

$$\mathcal{D}_k(\mathring{\mathbf{p}}_k) = (1,\, W[\mathbf{N}_p(\mathbf{x})\,\mathring{\mathbf{p}}_k])_\Omega \tag{40 d}$$

Thus the min-max principle (26) trasforms into the following mathematical programming (MP) problem:

$$\min_{(\mathbf{X}_{1k})}\ \max_{(\mathbf{X}_{2k},\mathring{\mathbf{p}}_k)}\ \Pi[\mathbf{X}_{1k},\, \mathbf{X}_{2k},\, \mathring{\mathbf{p}}_k]. \tag{41}$$

The first variation of (39) reads

$$\delta\Pi_k = \delta\mathbf{X}_{1k}^T(\mathbf{A}_{11}\,\mathbf{X}_{1k} + \mathbf{A}_{12}\,\mathbf{X}_{2k} + \mathbf{B}_1\,\mathring{\mathbf{p}}_k - \mathbf{R}_{1k})$$
$$+\,\delta\mathbf{X}_{2k}^T(\mathbf{A}_{21}\,\mathbf{X}_{1k} + \mathbf{A}_{22}\,\mathbf{X}_{2k} + \mathbf{B}_2\,\mathring{\mathbf{p}}_k - \mathbf{R}_{2k}) \tag{42}$$
$$+\,\delta\mathring{\mathbf{p}}_k{}^T(\mathbf{B}_1^T\,\mathbf{X}_{1k} + \mathbf{B}_2^T\,\mathbf{X}_{2k} + \mathbf{C}\,\mathring{\mathbf{p}}_k - \mathbf{P}_{Ik})$$

from which we obtain the Khun-Tucker conditions

$$\mathbf{A}_{11}\,\mathbf{X}_{1k} + \mathbf{A}_{12}\,\mathbf{X}_{2k} + \mathbf{B}_1\,\mathring{\mathbf{p}}_k = \mathbf{R}_{1k}, \qquad \mathbf{A}_{21}\,\mathbf{X}_{1k} + \mathbf{A}_{22}\,\mathbf{X}_{2k} + \mathbf{B}_2\,\mathring{\mathbf{p}}_k = \mathbf{R}_{2k} \tag{43}$$

$$\mathbf{P}_k \equiv \mathbf{B}_1^T\mathbf{X}_{1k} + \mathbf{B}_2^T\mathbf{X}_{2k} + \mathbf{C}\mathring{\mathbf{p}}_k + \mathbf{P}_{Ik} = \frac{\partial\mathcal{D}_k}{\partial\mathring{\mathbf{p}}_k} \tag{44}$$

which are the algebraic equations governing the inelastic FI weighted problem $\mathcal{P}_k$ discretized by BEs and CEs. The $\mathbf{P}_k$ in (44) are the cell internal forces which, being derivable from the plastic dissipation function $\mathcal{D}_k$, turn out to be related to $\mathring{\mathbf{p}}_k$ through eqs. (37). Therefore the solving equation system for the discretized model can finitely be cast in the following form:

$$\mathbf{A}\,\mathbf{X}_k + \mathbf{B}\,\mathring{\mathbf{p}}_k = \mathbf{R}_k, \qquad \mathbf{P}_k = \mathbf{B}^T\mathbf{X}_k + \mathbf{C}\,\mathring{\mathbf{p}}_k + \mathbf{P}_{Ik} \tag{45}$$

$$\mathring{\mathbf{p}}_k = \frac{\partial\boldsymbol{\Phi}_k(\mathbf{P}_k)}{\partial\mathbf{P}_k}\mathring{\boldsymbol{\Lambda}}_k \qquad \boldsymbol{\Phi}_k(\mathbf{P}_k) \leq 0, \qquad \mathring{\boldsymbol{\Lambda}}_k \geq 0, \qquad \mathring{\boldsymbol{\Lambda}}_k{}^T\boldsymbol{\Phi}_k(\mathbf{P}_k) = 0 \tag{46}$$

where, for more compactness, we have set

$$\mathbf{A} = \begin{bmatrix}\mathbf{A}_{11} & \mathbf{A}_{12}\\ \mathbf{A}_{21} & \mathbf{A}_{22}\end{bmatrix}, \qquad \mathbf{B} = \begin{bmatrix}\mathbf{B}_1\\ \mathbf{B}_2\end{bmatrix}, \qquad \mathbf{R} = \begin{bmatrix}\mathbf{R}_{1k}\\ \mathbf{R}_{2k}\end{bmatrix}, \qquad \mathbf{X} = \begin{bmatrix}\mathbf{X}_{1k}\\ \mathbf{X}_{2k}\end{bmatrix}$$

and the vectors $\boldsymbol{\Phi}_k$ and $\mathring{\boldsymbol{\Lambda}}_k$ collect the analogous quantities for all the cells. Eqs.(45) and (46) describe a holonomic-type, or deformation-theory, plasticity problem and a number of computational methods can be adopted for its numerical solution [18, 19, 21-23]. A MP procedure can also be followed, provided that the cell plastic yielding laws are piecewise linearized [11, 12, 20]. It is not the purpose of this paper to examine the above computational methods.

Conclusions

We have addressed the evolutive analysis problem for an elastic perfectly plastic solid body under quasi-static loads. First the evolutive time-continuous analysis problem has been transformed into a set of "weighted" inelastic holonomic-type problems through a time discretization procedure. Then, a mixed boundary/domain min-max principle has been established for such inelastic problem. This principle discretized by means of boundary elements and cell elements provides a system of solving equations showing symmetry and sign definiteness [10-13]. Several question remain still open, primarily for what concernes the the evaluation of hypersingular integrals existing only as finite-part integrals [24, 25].

References

1. Costabel, M.: Boundary Element Methods. Lecture Notes at 1st graduate summer course on "Finite Elements in Physics", Lausanne 1986.
2. Costabel, M.: Symmetric methods for the coupling of finite elements and boundary elements. In: Brebbia, C.A., Wendland, W.L. and Kunh, G. (eds.), Boundary Elements IX, Vol. 2, 414-420. Berlin Heidelberg: Springer-Verlag 1987.
3. Zienkiewicz, O.C., Kelly, D.W. and Bettes, P.: The coupling of the finite element method and boundary solution procedures. Int. Journal Num. Meths. in Engng. 11 (1977) 355-375.
4. Hartmann, F., Katz, C. and Protopsaltis, B.: Boundary elements and symmetry. Ingenieur-Archiv 55 (1985) 440-449.
5. Dumont, N.A.: The hybrid boundary element method. In: Brebbia, C.A., Wendland, W.L. and Kunh, G. (eds.) Boundary Elements IX, Vol. 1, 117-130. Berlin Heidelberg: Springer-Verlag 1987.
6. Mathews, I.C. and Geers, T.L.: A doubly asymptotic nonreflecting boundary for ground-shock analysis. J. Appl. Mechanics 54 (1987) 489-497.
7. De Figueiredo, T.B.G. and Brebbia, C.A.: A hybrid displacement variational formulation of BEM. In: Brebbia, C.A. (ed.), Boundary Elements X, Vol. 1, 33-42. Berlin Heildeberg: Springer-Verlag 1988. Southampton: Computational Mechanics Pubs. 1988.
8. Schnack, E., Becker, I. and Karaosmanoglu, N.: Modern Coupling of FEM and BEM with mixed variational formulations. In: Brebbia, C.A. (ed.), Boundary Elements X, Vol. 1, 59-73. Berlin Heildeberg: Springer- Verlag 1988. Southampton: Computational Mechanics Pubs. 1988.

9. Lo Bianco, M., Panzeca T. and Zito, M.: Variational formulations of the BEM (in Italian). Proc. of the (Italian) National Congress of the AIMETA, Bari, Oct. 4-7, 1988.

10. Polizzotto, C.: An energy approach to the boundary element method. Part I: elastic solids. Comput. Meth. Appl. Mech. Engng. 69 (1988) 167-184.

11. Polizzotto, C.: An energy approach to the boundary element method. Part II: elastic-plastic solids. Comput. Meth. Appl. Mech. Engng. 69 (1988) 263-276.

12. Polizzotto, C.: A consistent formulation of the BEM within elastoplasticity. In: Cruse, T.A. (ed.), Advanced Boundary Element Methods, 315-324. Berlin Heildeberg: Springer-Verlag 1988.

13. Polizzotto, C.: A boundary min-max principle as a tool for boundary element formulations (to appear).

14. Polizzotto, C. and Zito, M.: A variational approach to boundary element methods. In Tanaka, M. and Cruse, T.A. (eds.), Boundary Element Methods in Applied Mechanics, 13-24. Oxford and New-York: Pergamon Press 1988.

15. Rieder, G.: Iterationsverfahren und Operatorgleichungen in der Elastizitätstheorie. Abhandlungen der Braunschweigischen Wissenschaftlichen Gesellschaft. Band XIV, Zweiter Halbband 1962, Verlag Friedr. Vieweg and Sons, Brounschweig.

16. Polizzotto, C.: A symmetric-definite BEM formulation for the elastoplastic rate problem. In: Brebbia, C.A., Wendland, W.L. and Kunh, G. (eds.), Boundary Elements IX, Vol. 2, 315-334. Berlin Heidelberg: Springer-Verlag 1987.

17. Borino, G., Fuschi, P. and Polizzotto, C.: Consistent time modelling for the evolutive analysis of elastc-plastic solids. Complas-II (to appear).

18. Brebbia, C.A., Telles, J.C.F. and Wrobel, L.C.: Boundary Element Tecniques. Berlin Heidelberger: Springer-Verlag 1984.

19. Banarjee, P.K. and Butterfield, R.: Boundary Element Methods in Engineering Science. London: Mc Graws-Hill 1981.

20. Maier, G. and Polizzotto, C.: A Galerkin approach to boundary element elastoplastic analysis. Comput. Meth. Appl. Mech. Engng. 60 (1987) 175-194.

21. Martin, J.B.: Plasticity: Fundamentals and General Results. Cambridge, MA: The MIT Press 1975.

22. Ortiz, M. and Simo, J.C.: An analysis of a new class of integration algorithms for elastoplastic constitutive relations. Int. J. Num. Meth. Engng. 23 (1986) 353-366.

23. Hinton, E. and Owen, D.R.J.: Finite Element in Plasticity, Theory and Practice. Swansea: Pineridge Press 1980.

24. Kutt, H.R.: On the numerical evaluation of finite-part integrals involving an algebraic singularity, SIR Special Report, WISK 179, Pretoria 1975.

25. Rudolphi, J.J., Krishnasamy, G., Schmerr, L.W. and Rizzo, F.J.: on the use of the strengly singular integral equations for crack problems. In: Brebbia, C.A. (ed.), Boundary Elements X, Vol. 3, 249-263. Berlin Heildeberg: Springer- Verlag 1988. Southampton: Computational Mechanics Pubs. 1988.

Discretized Strategies for the Investigation of Instability Problems in Structural Mechanics

W.B. KRÄTZIG

Institute for Statics and Dynamics
Faculty of Civil Engineering
Ruhr-University, Bochum, Germany

Summary

The classical way of tackling instability problems in structural mechanics is their transformation into a space of special functions followed by the detection of instability phenomena as properties of this function space. From the point of view of computational mechanics this procedure is an unnecessary detour: Direct discretized strategies are more efficient and include automatically all instability phenomena. Thus the present paper will outline a general instability theory of discretized structures based on purely computational concepts.

1. Nonlinear Equation of Motion and 1st Variation

Let us consider the response of an arbitrary discontinuum S mapped into a 2n-dimensional normed vector space R^{2n} of time-dependent nodal displacements $\mathbf{V}$ and loads $\mathbf{P}$:

$$\begin{aligned} \mathbf{V} &= \{V_1\ V_2\ \dots\ V_i\ \dots\ V_n\}\ , \\ \mathbf{P} &= \{P_1\ P_2\ \dots\ P_i\ \dots\ P_n\}\ . \end{aligned} \tag{1.1}$$

R shall be structured by the inner product

$$(\mathbf{P},\mathbf{V}) = P_1V_1 + P_2V_2 + \dots P_iV_i + \dots P_nV_n\ . \tag{1.2}$$

Any load-deflection path $\mathbf{V}(\mathbf{P},t)$ of S in R^{2n} shall be described by the **nonlinear equation of motion**

$$\mathbf{M}\cdot\ddot{\mathbf{V}} + \mathbf{G}(\dot{\mathbf{V}},\mathbf{V},t) = \mathbf{P}(t)\ , \qquad (\dot{\dots}) = d\dots/dt \tag{1.3}$$

with the mass matrix $\mathbf{M}$, nodal velocities $\dot{\mathbf{V}}$ and accelerations $\ddot{\mathbf{V}}$, the matrix function $\mathbf{G}$ of internal nodal forces due to viscous damping as well as elasto-plastic restoring mechanisms, and with $\mathbf{P}$ as forcing vector. $\mathbf{G}$ may be subject to arbitrary constraints, e.g. yield conditions; $\mathbf{G}$ and $\mathbf{P}$ shall be assumed

G. Kuhn · H. Mang (Eds.)
Discretization Methods in Structural Mechanis
IUTAM/IACM Symposium Vienna, Austria, 1989

as continuous, Gateaux-differentiable and including their derivatives bounded in R^{2n} .

Nonlinear responses $V(P,t)$ of (1.3) may exhibit various instability phenomena, e.g. bifurcating, flutter, diverging or snapping. In order to construct incremental-iterative procedures for the detection of nonlinear response paths including their instabilities we decompose (1.1)

$$V = \mathring{V} + \bar{V} + \overset{+}{V} \,, \qquad P = \bar{P} + \overset{+}{P} \tag{1.4}$$

into initial imperfections $\mathring{V}$, the fundamental state $\bar{V},\bar{P}$ and infinitesimal perturbations $\overset{+}{V},\overset{+}{P}$ leading to the adjacent state. Analogous decompositions follow for $\dot{V},\ddot{V}$. Forming the 1. variation of (1.3) with respect to the fundamental state

$$M\cdot\delta\ddot{\bar{V}} + \frac{\partial G}{\partial \dot{V}}\bigg|_{\dot{\bar{V}}} \cdot \delta\dot{\bar{V}} + \frac{\partial G}{\partial V}\bigg|_{\bar{V}} \cdot \delta\bar{V} = \delta\bar{P} \,, \tag{1.5}$$

we derive under observation of $\delta\bar{V} = \overset{+}{V}$, $\delta\dot{\bar{V}} = \dot{\overset{+}{V}}$, $\delta\ddot{\bar{V}} = \ddot{\overset{+}{V}}$, $\delta\bar{P} = \overset{+}{P}$ and of mass conservation $(\dot{M} = 0)$ the following linearized equation of motion for $\overset{+}{V}(\overset{+}{P},t)$:

$$M\cdot\ddot{\overset{+}{V}} + \frac{\partial G}{\partial \dot{V}}\bigg|_{\dot{\bar{V}}} \cdot \dot{\overset{+}{V}} + \frac{\partial G}{\partial V}\bigg|_{\bar{V}} \cdot \overset{+}{V} = \overset{+}{P} = P - \bar{P} \;. \tag{1.6}$$

Re-substituting $\bar{P} = F_I = M\cdot\ddot{\bar{V}} + G(\dot{\bar{V}},\mathring{V},\bar{V})$ into (1.6) furnishes the final form of the **tangential equation of motion**

$$M\cdot\ddot{\overset{+}{V}} + C_T\cdot\dot{\overset{+}{V}} + K_T\cdot\overset{+}{V} = P - F_I = P - M\cdot\bar{V} - G(\dot{\bar{V}},\mathring{V},\bar{V}) \tag{1.7}$$

with the tangential damping matrix C_T , the tangential stiffness matrix K_T and F_I as vector of internal d'Alembert's, viscous, elasto-plastic nodal forces.

2. Phase Transformation and Stability Theorems

In order to connect the evaluation of stability properties of nonlinear time-dependent processes (1.3) with the theory of 1st order differential equations, we transform (1.7) into the 2n-dimensional phase space $X = \{\dot{V},V\}$. Assuming regularity for M , $\det M \neq 0$, we receive:

$$\frac{d}{dt}\left[\begin{array}{c}\dot{\overset{+}{V}}\\ \hline \overset{+}{V}\end{array}\right] = \left[\begin{array}{c|c} -M^{-1}\cdot C_T & -M^{-1}\cdot K_T \\ \hline I & 0\end{array}\right]\cdot \left[\begin{array}{c}\dot{\overset{+}{V}}\\ \hline \overset{+}{V}\end{array}\right] + \left[\begin{array}{c} M^{-1}\cdot(P-F_I)\\ \hline 0\end{array}\right]$$

$$\dot{\overset{+}{X}} \quad = \quad A\cdot \overset{+}{X} + B\ , \qquad (2.1)$$

with A of order (2nx2n), non-hermitean, but with (nxn) hermitean submatrices. (2.1) holds in an exact sense only for (infinitesimally small) 1st variations $\delta\bar{X}$, for which the nonlinearity condition [1,2] $\|B\| \to 0$ as $\overset{+}{X} \to 0$ is satisfied.

To decide on the stability of an arbitrary nonlinear response $\bar{X}(\bar{X}_o,t)$ we apply a perturbation $\overset{+}{X}$ with initial conditions:

$$\|\overset{+}{X}\ (\overset{+}{X}_o,\bar{X}_o,t_o,t_o)\| \ < \delta(\varepsilon) > 0\ . \qquad (2.2)$$

Due to **Ljapunow's stability theorems** [3,4] $\bar{X}$ is identified as

- weakly stable, if for $t>t_o$: $\quad \|\overset{+}{X}\ (\overset{+}{X}_o,\bar{X}_o,t_o,t)\| \le \varepsilon$,
- stable in the limit: ε ,
- unstable: $> \varepsilon$, (2.3)
- convergent, if for $t \to \infty$: $\lim \|\overset{+}{X}\ (\overset{+}{X}_o,\bar{X}_o,t_o,t)\| = 0$.

Stable and convergent responses are named asymptotically stable.

The evaluation of time-invariant stability properties will be based directly on the **tangential stiffness equation**

$$K_T\cdot \overset{+}{V} \ = \ P - F_I \ = \ P - G(\overset{\circ}{V},\bar{V}) \quad , \qquad (2.4)$$

an algebraic relation, derived by $\ddot{\overset{+}{V}} = \dot{\overset{+}{V}} = 0$ from (1.7).

3. Basic Equations of Nonlinear Continuum Dynamics

Let us consider an arbitrary continuum, imbedded into the E3, in its initial state with domain $\overset{\circ}{F}$ and boundary $\overset{\circ}{C}$. General convective curvilinear coordinates Θ^i determine points on F For an arbitrary large deformation we define the **vector fields** of given loads $\overset{\circ}{\mathbf{p}}$, displacements $\mathbf{u},\mathbf{r}$, boundary tractions $\mathbf{t}$ and d'Alembert's forces $\mathbf{f} = -\overset{\circ}{\rho}\,\ddot{\mathbf{u}}$ with mass density function $\overset{\circ}{\rho}$, all related to bases of the initial state. We further introduce tensor fields of Piola-Kirchhoff stresses $\boldsymbol{\sigma}$,

Green-Lagrange strains ε and corresponding strain velocities $\dot{\varepsilon}$ leading to the following

field equations: $\forall \{p, u, f, \sigma, \varepsilon\} \in F$

$$-(p+f) = D_e \cdot \sigma = (D_{eL} + D_{eN}(u)) \cdot \sigma , \tag{3.1}$$

$$\varepsilon = D_k \cdot u = (D_{kL} + \tfrac{1}{2} D_{kN}(u)) \cdot u , \tag{3.2}$$

constitutive relation:

$$\sigma = \mathring{\sigma} + E(\varepsilon,\dot{\varepsilon},t)\cdot\varepsilon + D(\varepsilon,\dot{\varepsilon},t)\cdot\dot{\varepsilon} , \tag{3.3}$$

boundary conditions: $\forall \{u,\sigma\} \in F$, $\{r,t\} \in C$

$$t = \mathring{t} = R_t \cdot \sigma = (R_{tL} + R_{tN}(u))\cdot\sigma \quad \text{along } \mathring{C}_t , \tag{3.4}$$

$$r = \mathring{r} = R_r \cdot u \quad \text{along } \mathring{C}_r . \tag{3.5}$$

Both D_e and D_k , the dynamic and kinematic field operators, consist of linear parts D_{eL}, D_{kL} and nonlinear parts $D_{eN}(u)$, $D_{kN}(u)$. R_t and R_r are recognized as dynamic and kinematic trace operators. The proposed material law (3.3) represents a nonlinear Kelvin-Voigt model with residual stresses $\mathring{\sigma}$, the elasticity tensor E and the viscousity tensor D .

In conclusion the **functional of virtual work**

$$\int_{\mathring{F}} \mathring{\rho}\, \ddot{u}^T\cdot\delta u \, d\mathring{F} + \int_{\mathring{F}} \sigma^T\cdot\delta\varepsilon \, d\mathring{F} = \int_{\mathring{F}} \mathring{p}^T\cdot\delta u \, d\mathring{F} + \int_{\mathring{C}_t} \mathring{t}^{\,T}\cdot\delta r \, d\mathring{C}_t \tag{3.6}$$

$$\forall \{\mathring{p},u,\sigma,\varepsilon\} \in F , \quad \{\mathring{t},r\} \in C_t ,$$

a weak formulation of dynamic equilibrium, is constrained by

$$\delta\varepsilon = (D_{kL}+D_{kN}(u))\cdot\delta u \text{ on } F , \quad \delta r = R_r\cdot\delta u \text{ along } C_r . \tag{3.7}$$

Together with (3.3), (3.6, 3.7) are regarded as functional analogue of (1.3).

4. From Incremental Principle of Virtual Work to Tangential Equation of Motion

In order to derive the incremental form of (3.6) we decompose the vector fields of $u,\dot{u},r$

$$u = \mathring{u} + \bar{u} + \overset{+}{u} = \mathring{u} + \bar{u} + \overline{\delta u}\ , \quad \dot{u} = \dot{\bar{u}} + \dot{\overset{+}{u}}\ ,$$
$$r = \mathring{r} + \bar{r} + \overset{+}{r} = \mathring{r} + \bar{r} + \overline{\delta r} \tag{4.1}$$

into imperfections $\mathring{u},\mathring{r}$, finite variables $\bar{u},\dot{\bar{u}},\bar{r}$ of the fundamental motion as well as small increments $\overset{+}{u} = \overline{\delta u},\dot{\overset{+}{u}},\ \overset{+}{r} = \overline{\delta r}$ of the adjacent motion. A similar decomposition of the tensor fields of strains, strain velocities and of $\mathring{p}$, t, f, σ transforms (3.6) into [5]:

$$\int_{\mathring{F}} \mathring{\rho}\ddot{\overset{+}{u}}^{T}\cdot\delta\overset{+}{u}\ d\mathring{F} + \int_{\mathring{F}} [(\mathring{\sigma}+\bar{\sigma})^{T} \cdot \tfrac{1}{2}\delta\overset{++}{\varepsilon} + \overset{+}{\sigma}^{T}\cdot\delta\overset{+}{\varepsilon}\ ]\ d\mathring{F}$$
$$= \delta[\int_{\mathring{F}} \bar{p}^{\circ T}\cdot\overline{\delta u}\ d\mathring{F} - \int_{\mathring{F}} \mathring{\rho}\ddot{\bar{u}}^{T}\cdot\overline{\delta u}\ d\mathring{F} + \int_{\mathring{C}_t} \bar{t}^{\circ T}\cdot\overline{\delta r}\ d\mathring{C} - \int_{\mathring{F}} (\mathring{\sigma}+\bar{\sigma})^{T}\cdot\overline{\delta\varepsilon}\ d\mathring{F}]$$
$$+ \int_{\mathring{F}} \overset{+}{p}^{\circ T}\cdot\delta\overset{+}{u}\ d\mathring{F} + \int_{\mathring{C}_t} \overset{+}{t}^{\circ T}\cdot\delta\overset{+}{r}\ d\mathring{C}\ , \tag{4.2}$$

the **incremental principle of virtual work.** Like its origin it holds independently of the material law applied. Obviously by comparison with (3.6), the centered variation $\delta[\ldots]$ vanishes for fundamental motions in dynamic equilibrium, also the last line for unexcited neighboring motions $\overset{+}{p} = \overset{+}{t} = \mathbf{0}$.

In the course of the discretization process of (4.2) the constitutive relation (3.3) and the strain fields $\varepsilon,\dot{\varepsilon},\delta\varepsilon$ are substituted, whereat powers of arbitrary order of $\bar{u},\dot{\bar{u}}$ are retained and higher than quadratic ones in $\overset{+}{u},\dot{\overset{+}{u}}$ cancelled. Further a standard displacement discretization is carried out leading finally to the **global tangential equation of motion:**

$$\mathbf{M}\cdot\ddot{\overset{+}{\mathbf{V}}} + \mathbf{C}_T\cdot\dot{\overset{+}{\mathbf{V}}} + \mathbf{K}_T\cdot\overset{+}{\mathbf{V}} = \mathbf{P} - \mathbf{F}_I\ : \tag{4.3}$$
$$\mathbf{M}\cdot\ddot{\overset{+}{\mathbf{V}}} + (\mathbf{C}_v + \mathbf{C}_{uL} + \mathbf{C}_{uN})\cdot\dot{\overset{+}{\mathbf{V}}} + (\mathbf{K}_e + \mathbf{K}_{\mathring{\sigma}} + \mathbf{K}_{\sigma L} + \mathbf{K}_{\sigma N} + \mathbf{K}_{\dot{\sigma}L} +$$
$$\mathbf{K}_{\dot{\sigma}N} + \mathbf{K}_{uL} + \mathbf{K}_{uN} + \mathbf{K}_{\dot{u}L} + \mathbf{K}_{\dot{u}N})\cdot\overset{+}{\mathbf{V}} = \mathbf{P} - \mathbf{F}_{\mathring{\sigma}} - \mathbf{F}_{\sigma} - \mathbf{F}_{\dot{\sigma}} - \mathbf{F}_{m}$$

with the following abbreviations [5] (functional indices L: linear in $\bar{\mathbf{V}}$, N: quadratic in $\bar{\mathbf{V}}$ or bilinear in $\bar{\mathbf{V}},\dot{\bar{\mathbf{V}}}$):

$\mathbf{M}$ mass matrix,

$\mathbf{C}_v$ viscous damping matrix,

$\mathbf{C}_u$ initial displacement damping matrix,

$\mathbf{K}_e$ elastic stiffness matrix,

$\mathbf{K}_\sigma^\circ$ residual stress matrix,

$\mathbf{K}_\sigma(\mathbf{K}_{\dot\sigma})$ elastic (viscous) initial stress matrix,

$\mathbf{K}_u(\mathbf{K}_{\dot u})$ elastic (viscous) initial displacement matrix,

$\mathbf{F}_\sigma^\circ(\mathbf{F}_\sigma,\mathbf{F}_{\dot\sigma},\mathbf{F}_m)$ residual (elastic, viscous, D'ALEMBERT's) internal nodal forces.

Because of the assumed symmetry of $\mathbf{K}_T$ flutter instabilities have been ignored.

5. Stability Analysis of Time-Invariant Responses

For time-invariant unstable responses we restrict our attention to single-parametric load systems

$$\mathbf{P} = \lambda \overset{\circ}{\mathbf{P}} \tag{5.1}$$

and load paths in the R^{n+1} of elements $\{V_n, \lambda\}$. Thus the following problems have to be solved:

- evaluation of nonlinear primary, secondary and higher-order responses by path-following procedures [6,7],
- detection of limit points,
- determination of arbitrary bifurcation points and their multiplicity,

where the **tangential stiffness relation**

$$\ddot{\overset{+}{\mathbf{V}}} = \dot{\overset{+}{\mathbf{V}}} = \mathbf{0} : \quad \mathbf{K}_T(\bar{\mathbf{V}}) \cdot \overset{+}{\mathbf{V}} = \mathbf{P} - \mathbf{F}_I = \lambda \mathbf{P}^\circ - \mathbf{F}_I \tag{5.2}$$

is applied as basic process description. In a bifurcation point $\bar{\mathbf{V}}_c, \lambda_c$ the corresponding buckling mode $\overset{+}{\mathbf{V}}_e$ serves as starting vector for the bifurcated branch:

$$\mathbf{V} = \bar{\mathbf{V}}_c + \varepsilon \, \overset{+}{\mathbf{V}}_e , \tag{5.3}$$

with ε as suitable control parameter: $\varepsilon=0$ follows the primary path, $\varepsilon=\pm\varepsilon^* \neq 0$ one bifurcated branch. The iteration of (5.2) via $\mathbf{F}_I$ leads, e.g. for the Riks-Wempner-Wessels scheme, after a few corrector-steps to the first bifurcated point:

$$V = \bar{V}_c + \varepsilon \overset{+}{V}_e + \sum_j \overset{+}{V}_j . \qquad (5.4)$$

Instabilities have to be determined as properties of $\mathbf{K}_T$. Because $\mathbf{K}_T$ is (nxn) hermitean for conservative loads, all eigenvalues λ_n (eigenvectors $\mathbf{U}_n$) of $\mathbf{K}_T$ are real (mutually orthogonal), and the decomposition

$$\mathbf{U}^T \cdot \mathbf{K}_T \cdot \mathbf{U} = \mathbf{\Lambda} \quad \text{with} \quad \mathbf{U} = [\mathbf{U}_1 \ldots \mathbf{U}_n],\ \mathbf{\Lambda} = \text{diag}\,\{\lambda_1 \ldots \lambda_n\} \qquad (5.5)$$

exists, likewise the triangularization

$$\mathbf{K}_T = \mathbf{L}^T \cdot \mathbf{D} \cdot \mathbf{L} \quad \text{with} \quad \mathbf{D} = \text{diag}\,\{D_{11} \ldots D_{nn}\} . \qquad (5.6)$$

For judging the stability of a reached fundamental state $\{\bar{\mathbf{V}}, \lambda\}$, non-trivial states of equilibrium

$$\mathbf{K}_T(\bar{\mathbf{V}}) \cdot \overset{+}{\mathbf{V}} = \lambda \mathbf{P}^{\circ} - \mathbf{F}_I = 0 \rightarrow \overset{+}{\mathbf{V}} = \overset{+}{\mathbf{V}}_e \qquad (5.7)$$

yield the following classification: $\bar{\mathbf{V}}$ is

- stable, if $\overset{+}{\mathbf{V}}$ does not exist, since det $\mathbf{K}_T > 0$: all λ_i, $D_{ii} > 0$;
- unstable, if $\overset{+}{\mathbf{V}}$ does not exist, < 0: at least one λ_i, $D_{ii} < 0$;
- indifferent, if $\overset{+}{\mathbf{V}}$ exists, $= 0$: at least one λ_i, $D_{ii} = 0$.

λ_i, occasionally also D_{ii}, are denoted as **stability coefficients**, positive for stable responses. If negative, the structure responds in unstable modes $\mathbf{U}_i$. The number of negative elements $\{\lambda_i, D_{ii}\}$ is known as **degree of instability**. In **indifferent (neutral) states of equilibrium** the degree of instability may change. From the matrices $\mathbf{K}_T$, $\mathbf{F}_I$ and their derivatives with respect to V_i, λ all initial post-buckling properties, which uniquely determine the response behavior of an arbitrary bifurcation point, can be computed [9].

As an example Fig. 1 demonstrates the unstable nonlinear response of a simple 2-bar structure with different bending stiffnesses. Points of neutral equilibrium were detected by a linearized eigenvalue technique according to [8].

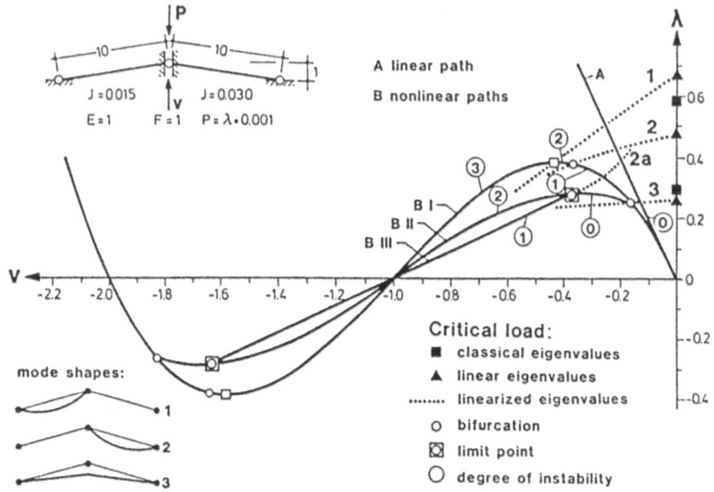

Fig. 1. Load deflection path of 2-bar structure

6. Stability Analysis of Time-Dependent Responses

Time-dependent responses will be treated as load-displacement histories in the R^{2n+2} of elements $\{V_n, \dot{V}_n, \lambda, t\}$. Stability assessments are based on the phase transformation (2.1) of the tangential equation of motion, or directly on (4.3).

Apart from special load conditions **instability evaluations** have to be founded in general on Ljapunow's stability theorems (2.3) applying the following procedure:

- Evaluation of a specific response $\bar{V}(t)$ respectively $\bar{X}(t)$, whose stability is a matter of question, through direct numerical integration of (4.3) or (2.1).
- Performance of small perturbations of $\bar{V}(t)$, $\bar{X}(t)$ at time instants t_o through deviating initial conditions

$$\|\ddagger{X} (\ddagger{X}_o, \bar{X}_o, t_o, t_o)\| < \delta \tag{6.1}$$

  and computation of the future responses $\ddagger{X}(\ddagger{X}_o, \bar{X}_o, t_o, t)$.
- Stability assessment of $\|\ddagger{X}(\ddagger{X}_o, \bar{X}_o, t_o, t)\|$ due to (2.3).

An example demonstrating this procedure is presented in Figs. 2 and 3, tackling a flat cylindrical panel under a constant transverse step loading of increasing intensity p :

$$p^3 = 0 \quad \text{for} \quad t<t_s \ , \qquad p^3 = \max p \quad \text{for} \quad t \geq t_s \ . \qquad (6.2)$$

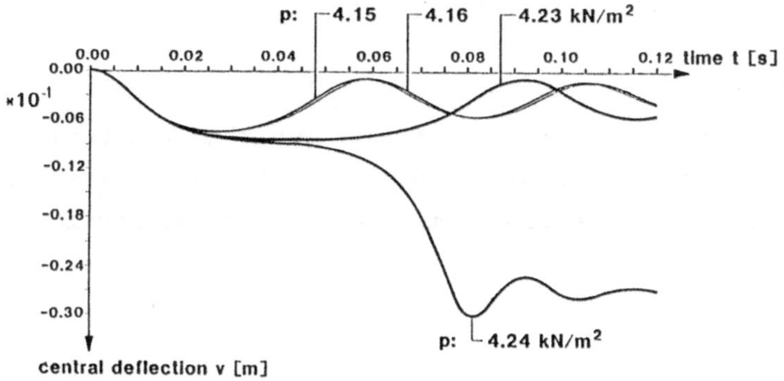

Fig. 2. Cylindrical panel under constant step loading of various intensity p

All responses for $p \leq 4.23$ kN/m² obviously can be classified as asymptotically stable because they reach, after some transient vibrations, the static deflection for $t \to 0$. Solutions from $p = 4.24$ kN/m² upwards are identified as unstable, if the central deflection v on Fig. 2 exceeds a bound $\varepsilon < v_{max}$.

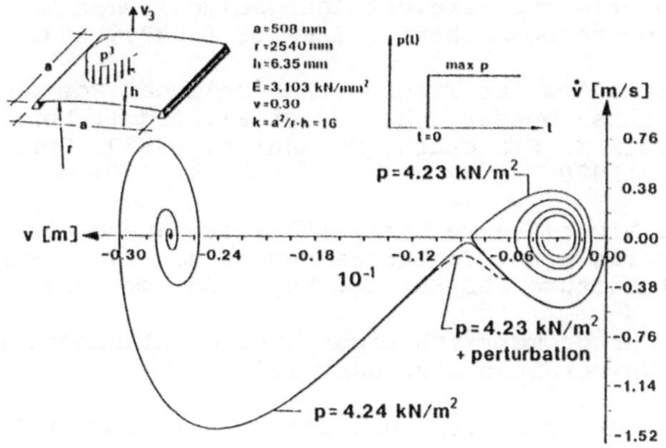

Fig. 3. Phase diagram for 2 responses of Fig. 2

If the response variety of Fig. 2, including its transition to instability, be unknown, every computed path would have to be perturbed (6.1) at different time instants t_o. But finally, only the response for $p = 4.23$ kN/m² would have demonstrated instability, for instance by a perturbation with small, but final velocity at t_s (see the phase diagram on Fig. 3).

Acknowledgement

This research has been carried out partly within the Research Center for Structural Dynamics SFB 151 at the Ruhr-University. Financial support of the DFG-German Science Foundation is gratefully acknowledged.

References

1. La Salle, J.P.: Stability Theory and the Asymptotic Behavior of Dynamical Systems. In: Herrmann, G. (ed.): Dynamic Stability of Structures, 53-63. Pergamon Press, Oxford 1967
2. Bellmann, R.: Stability theory of differential equations. Dover Publications, Inc., New York 1969.
3. Ljapunow, A.M.: Problème général de la stabilité du mouvement. Transl.thesis 1893, Ann. math. studies, Princeton 1949.
4. Willems, J.L.: Stability Theory of Dynamical Systems. Th. Nelson & Sons Ltd, London 1970.
5. Krätzig, W.B.: Eine einheitliche statische und dynamische Stabilitätstheorie für Pfadverfolgungsalgorithmen in der numerischen Festkörpermechanik. ZAMM 69 (1989), 203.
6. Ramm, E.: Strategies for Tracing the Nonlinear Response near Limit Points. Wunderlich, W., et al. (ed.): Nonlinear Elastic Analysis in Structural Mechanics, 63-89. Springer-Verlag, Berlin 1980.
7. Eckstein, U.: Nichtlineare Stabilitätsberechnung elastischer Schalentragwerke. Technical Report No. 83-3, Ruhr-University, Institute for Struct. Eng., Bochum 1983.
8. Brendel, B.: Zur geometrisch nichtlinearen Elastostatik. Dissertation Universität Stuttgart 1979.
9. Jürcke, R.K.: Zur Stabilität und Imperfektionsempfindlichkeit elast. Schalentragwerke - Finite Element Formulierung der Anfangs-Nachbeul-Theorie. Technical Report No.85-5, Ruhr-University, Institute for Struct. Eng., Bochum 1985.

On Static and Dynamic Instability Analysis of Thin-Walled Structures

Bernd Kröplin and Dieter Dinkler - ISD Universität Stuttgart, FRG

Summary

The paper deals with the development of stability estimates for structures under time depending loads. The estimates are based on energy norms and can be used for the investigation of safety against loss stability by buckling. The static analysis is used to define a point on the load path with a certain critical strain energy. Starting from this state the presented procedure allows to decide whether the structure stays for a certain load history within the critical bounds, which separate the motion around the prebuckling state from the motion in a postbuckling region. For an arbitrary load history Galerkin's procedure is applied using a representation of the deformations from static modal analysis. The stability is proved by comparing the load induced energy and a so-called critical strain energy.

1 Introduction

For elastic limit load analysis of thin-walled structures under static loads a number of well-known methods exist. These are for linear structural behaviour mostly based on eigenvalue analysis and for nonlinear structural behaviour on the accompanying eigenvalue calculations along a load path often in conjunction with path tracing methods like constant arc length. For time-dependent loads in structures with linear behaviour load factors are well-known in order to represent the load time dependence. Nonlinear structures require more detailed treatment as far as buckling is concerned. Because of the lack of reliable estimates the equations of motion including all nonlinearities are often integrated in time. However, this effort seems to be often inadequate, particularly as the results are hard to interprete because of the numerous interacting modes.

G. Kuhn · H. Mang (Eds.)
Discretization Methods in Structural Mechanis
IUTAM/IACM Symposium Vienna, Austria, 1989

This paper gives a method to calculate estimates for instationary nonlinear buckling behaviour in a simplified manner without direct integration of the equations of motion. The method consists of three steps:

- identification of limit states between prebuckling and postbuckling and calculation of the energy related to these critical states.
- calculation of the energy related to the time-dependent load influence.
- comparison of the energy related to the critical state and the energy to the external load.

The methods presented here are based on elastic material behaviour the load functions are short time impulse loads or load level changes.

2 Basic assumptions

Fundamental works considering the theory of stability of elastic structures are e. g. Koiter /1/, Leipholz /2/, Huseyin /3/, Thompson/Hunt /4/ and Ziegler /5/. These papers discuss mostly questions of stability of equilibrium states, while the stability of nonlinear dynamic systems is discussed e. g. in Hayashi /6/, Thompson/Stewart /7/ and Haken /8/. Here the following questions are considered:

- How stable is the considered equilibrium state?
- What is the size of additional loads which cause loss of stability?

Basis of the theory of stability of elastic continuous systems is the energy balance in rateform (1). It contains the rates of potential energy, kinetic energy and the external loads. Temperature effects /9/ are neglected.

$$\dot{V}_v = \int \{\dot{\gamma}_{ij}\, e_{ijkl}\, \gamma_{kl} + \rho\, \dot{v}_i\, \ddot{v}_i\}\, dR^3 - \int \dot{v}_i\, \overline{s}_i\, dO = 0 \qquad (1)$$

with the initial condition $V_{v0} = c$ at time $t = t_0$.

$\dot{()}$: derivative with respect to time
e_{ijkl} : elasticity of the material
ρ : material density
$\dot{v}_i$: displacements in direction of the space coordinates x_i, $i = 1, 2, 3$
$\overline{s}_i$: loads

The Green strains γ_{ij} are defined as

$$\gamma_{ij} = \frac{1}{2}\left(v_{i,j} + v_{j,i} + v_{k,i}\, v_{k,j}\right). \tag{2}$$

Alternatively the potential energy can be formulated in stresses and strains as a mixed formulation. Equation (3) contains the nonlinearity one order lower than (1), which leads to some advantages in numerical investigations.

$$\begin{aligned} \dot{V}_{\sigma,v} &= \int \{\dot{\gamma}_{ij}\,\sigma_{ij} + \gamma_{ij}\,\dot{\sigma}_{ij} - \dot{\sigma}_{ij}\,f_{ijkl}\,\sigma_{kl} + \rho\,\dot{v}_i\,\ddot{v}_i\}\; dR^3 \\ &\quad - \int \dot{v}_i\,\overline{s}_i\,dO_\sigma - \int s_i\left(\dot{v}_i - \dot{\overline{v}}_i\right)\,dO_v = 0 \end{aligned} \tag{3}$$

f_{ijkl} : flexibility of the material
σ_{ij} : 2. Piola–Kirchhoff stress–tensor
s_i : stress–vector

Thin–walled shell structures are treated by means of the Kirchhoff–Love–Hypothesis and by introducing a reference surface. The mixed formulations with moderately large rotations describe the rate of the potential dependent on the displacements u_α, u^3, the velocities $\dot{u}_\alpha$, $\dot{u}^3$, the stress resultant axial forces $n^{\alpha\beta}$ and bending moments $m^{\alpha\beta}$ and the loads p_α, p^3 of the reference surface /11/

$$\dot{V} = \dot{V}(u_\alpha, u^3, \dot{u}_\alpha, \dot{u}^3, n^{\alpha\beta}, m^{\alpha\beta}, p_\alpha, p^3). \tag{4}$$

The discretisation of the reference surface with finite elements leads after the integration over the reference surface to the matrix notation (5)

$$\dot{V}_z = \dot{\boldsymbol{z}}^t\,\{A^L(\boldsymbol{z}_0)\,\boldsymbol{z} + A^N(\tfrac{1}{2}\boldsymbol{z})\,\boldsymbol{z} + M\,\ddot{\boldsymbol{z}} - \boldsymbol{p}\} = 0. \tag{5}$$

$\boldsymbol{z}$: vector of the unknowns $\boldsymbol{z}^t = \{\boldsymbol{u}, \boldsymbol{n}, \boldsymbol{m}\}^t$
$A^L(\boldsymbol{z}_0)$: linear matrix, dependent on the initial state $\boldsymbol{z}_0$
$A^N(\frac{1}{2}\boldsymbol{z})$: geometric nonlinear matrix
M : mass matrix
$\boldsymbol{p}$: vector of loads

3 Stability of motions

The fundamental investigations concerning stability of motions with perturbal initial conditions date back to Liapunov /10/. A motion is 'stable', if equations (6) hold. Otherwise the motion is said to be 'unstable'

$$\begin{aligned} [v_{t_0,a_s} - v_{t_0,a_0}] &\leq \delta_1(\varepsilon) \quad , \quad t = t_0 \, , \\ [v_{t,a_s} \; - \; v_{t,a_0}] &< \varepsilon \qquad , \quad t > t_0 \, . \end{aligned} \tag{6}$$

'Asymptotic stability' is given if equation (7) holds

$$\lim_{t \to \infty} [v_{t,a_s} - v_{t,a_0}] = 0 \, . \tag{7}$$

In case of a permanent perturbation R_s, which is not only a perturbation of the initial conditions, the motion is said to be 'totally stable' if also (8) is satisfied

$$[R_{s\,(t,\,v)}] \leq \delta_2(\varepsilon) \, , \; t > t_0 \, . \tag{8}$$

Motions are called 'locally stable', if for the same energy level motions in different regions, e. g. and the pre- and post-buckling range exist. It might happen, that for this case (6) is not satisfied. Motions are called 'orbitally stable', if the trajectories of the perturbed and unperturbed motion in the phase plane are closed to each other. A judgement, which allows to check the stability without direct solution of the equations of the preturbed motions is given by Liapunov (9) and (10). V is a Liapunov–Function now, which is given for the considered case with (1) or (5).

$$V[z_s] > 0 \, . \tag{9}$$

$$\dot{V}[z_s] \cdot \mathrm{sign}(V[z_s]) \leq 0 \, . \tag{10}$$

Both conditions are a priori satisfied. Hence the considered motions are 'orbitally stable'. In case of a finite, not arbitrary small perturbation of the equilibrium state (6) in the sense of 'orbital stability' is necessary to consider. This can be shown with the structure given in figure 1 /2/.

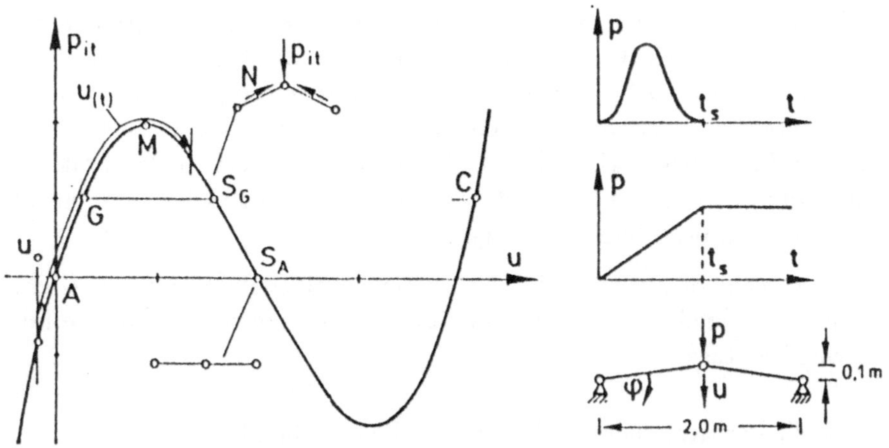

$\sin\varphi \quad \varphi, \quad EA = 5{,}0 \cdot 10^4\ kN, \quad \varrho \cdot A = 1{,}5 \cdot 10^3\ kNs^2/m^2$

Fig. 1 : Internal reactions of a flat truss

The reaction forces p_{iT} of the structure in direction of the displacements of the middle node are considered as control parameters and are plotted with respect to the related displacement u.

First an initial displacement u_0 is given. The system moves around the unloaded state A. If u_0 is large enough, the system may pass the maximum of the inner reaction forces. It returns as long as the flat position S_A is not passed. Hence the flat position is the critical deformation state for the motion around A. If S_A is passed, condition (6) is not satisfied. With respect to the state A the critical deformations are characterized by a vanishing 'secant–stiffness' $A - S_A$. The same is true for a preloaded structure, e. g. the equilibrium state G.

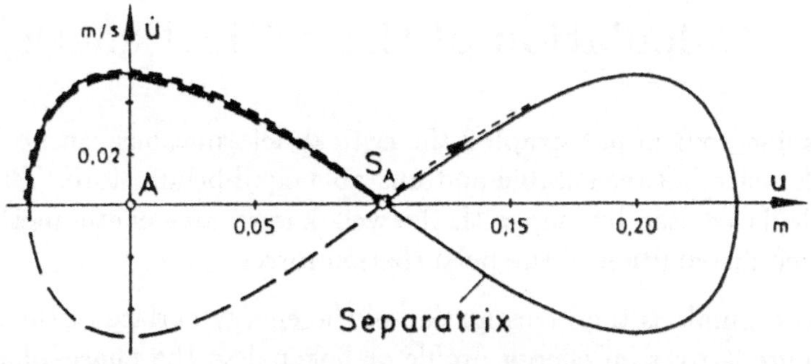

Fig. 2 : Motion of the truss of figure 1

The motion describes a trajectory in the phase-plane of displacements and velocities, figure 2. The distance of the trajectory from the static equilibrium state is a measure for induced energy by the initial conditions. The critical deformations, see figure 1, are here represented on the static axis ($u = 0$) as saddle-point S_A. For investigations to stability the energy level ΔV_{crit} is of interest. This holds for all states of motion on the limit-curve, which is called 'critical trajectory' or 'separatrix'. therefore this curve is the limit-curve for all 'locally stable' motions in the pre- and post-buckling range.

A judgement of the stability of a motion against finite perturbations, which is defined by its initial conditions needs the solution of the following two tasks :

- The critical energy ΔV_{crit} has to be calculated.
- The energy V_0 which is induced by the initial conditions and the energy T_e which is a function of perturbations have to be quantified.

The proof of stability is then possible with (11).

$$\Delta V_{crit} - T_e - V_0 \quad \begin{array}{ll} > 0 : & \text{stable} \\ = 0 : & \text{indifferent} \\ < 0 : & \text{unstable} \end{array} \tag{11}$$

In a normalized version (11) may be written as

$$\eta = \frac{\Delta V_{crit} - T_e - V_0}{\Delta V_{crit}} \, . \tag{12}$$

η is called the 'degree of stability'.

4 Calculation of the critical energy

As lined out in paragraph 3 the critical deformations can be defined as difference between stable and unstable equilibrium state. They can be calculated step by step with the well-known arc-length-methods for a given distribution of the perturbation forces.

More simple is the investigation of the energy surface of the structure. Figure 3 gives an energy profile orthogonal to the phase-plane for the special case of $\dot{u} = 0$.

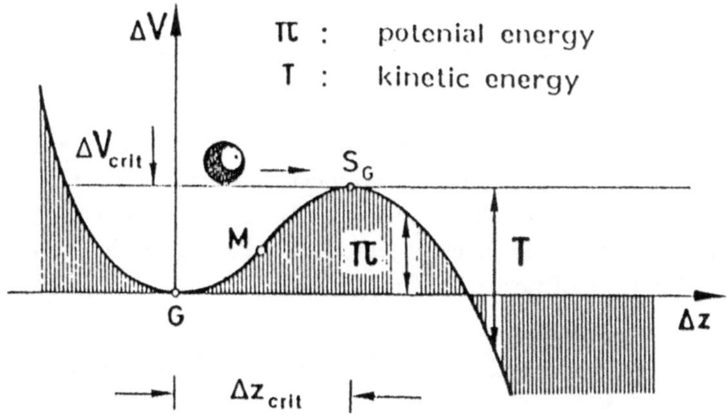

Fig. 3 : Energy profile for $\dot{u} = 0$

The stable equilibrium state is given by the minimum G of the potential energy. A perturbation of this equilibrium state leads to a 'locally stable' motion of the structure with the energy level ΔV. An increase from ΔV to ΔV_{crit} shifts the structure to the saddle-point S_G. S_G gives the critical deformation state of the structure, since two 'locally stable' motions in the pre- and the postbuckling range meet here. A further increase of the energy leads to a motion in the whole pre- and postbuckling state. Hence the decisive factor for the stability of the equilibrium state G is the energy-level, which is given by the maximum of the potential energy. The maximum is related to the unstable equilibrium state.

Therefore the first variation of the potential energy has to vanish in the case of the critical deformations.

$$\delta \Pi = 0 . \tag{13}$$

(13) leads to a nonlinear algebraic system of equations for the critical state vector Δz_{crit}, the non-trivial solution of which describes the unstable equilibrium state.

$$\dot{z}_i^t \{A^L(z_G) + A^N(\tfrac{1}{2}\Delta z_{crit})\} \Delta z_{crit} = 0 . \tag{14}$$

Some possibilities for an approximate calculation of the critical vector Δz_{crit} are given in /12/.

The critical energy, which is necessary for the stability criteria can be calculated now for the equilibrium state G by means of integration of (5) for the static case.

$$\Delta V_{crit} = \Delta z^t_{crit}\{\tfrac{1}{2} A^L(z_G) + \tfrac{1}{6} A^N(\tfrac{1}{2}\Delta z_{crit})\} \Delta z_{crit} . \qquad (15)$$

5 Calculation of the load energy

The aim of the investigations is to compare the critical energy with the load energy which is induced by perturbations, see (11). The external energy depends on the function in time of the perturbation and the related state vector development. Hence the energy has to be integrated over the time of perturbation t_s

$$V_{t_0} = \int_{t_0}^{t^s} \dot{z}^t \{A^L(z_0)\, z + A^N(\tfrac{1}{2}\Delta z)\, z + M\, \ddot{z} - p\}\, dt = 0 . \qquad (16)$$

The load is given by

$$p = \Delta p_i \varphi_{p_i} , \quad t_0 \le t \le t_s \qquad (17)$$

Since the exact function is not known, a Ritz–like method is chosen. The spatial distribution of the unknown state–vector is estimated by a static analysis of a reference load of the same shape as the perturbation . In case of different critical modes every mode has to be investigated with a special perturbation. The approximation of the space–time–function of the state–vectors

$$z = \alpha_{ij}\, \Delta z_{i(\Delta p_i)}\, \varphi_{j(t)} t_0 \le t \le t_s \qquad (18)$$

consists of the spatial pattern $\Delta z_{i(\Delta p_i)}$ and a number of freely chosen shape–functions $\varphi_{j(t)}$ with unknown scalers α_{ij} for the time–dependent part. The shape functions have to satisfy the initial conditions of the perturbed motion for the deformations and the velocities. In our case this is

$$\varphi_{j(t)} = t^{j+1} , \quad j \ge 1 . \qquad (19)$$

Experience has shown that polynomials are well suited since the initial motion oscillates very little.

With the approximation (18) the energy balance (16) is first calculated for spatial coordinates. The calculation of the quadratic forms can be performed with standard finite element programs under static loading. This is shown in (20).

$$\alpha_i \{ \int_{t_0}^{t_s} \dot{\varphi}_i\, \Delta z^t \{A^L(z_0)\, \varphi_j + A^N(\tfrac{1}{2}\Delta z\, \varphi_k\, \alpha_k)\, \varphi_j + M\, \ddot{\varphi}_j\}\, \Delta z\, dt\, \alpha_j$$

$$- \int_{t_0}^{t_s} \dot{\varphi}_i\, \Delta z^t\, \Delta p\, \varphi_p\, dt\} = 0 \qquad (20)$$

are arbitrary and do not influence the accuracy of the result.

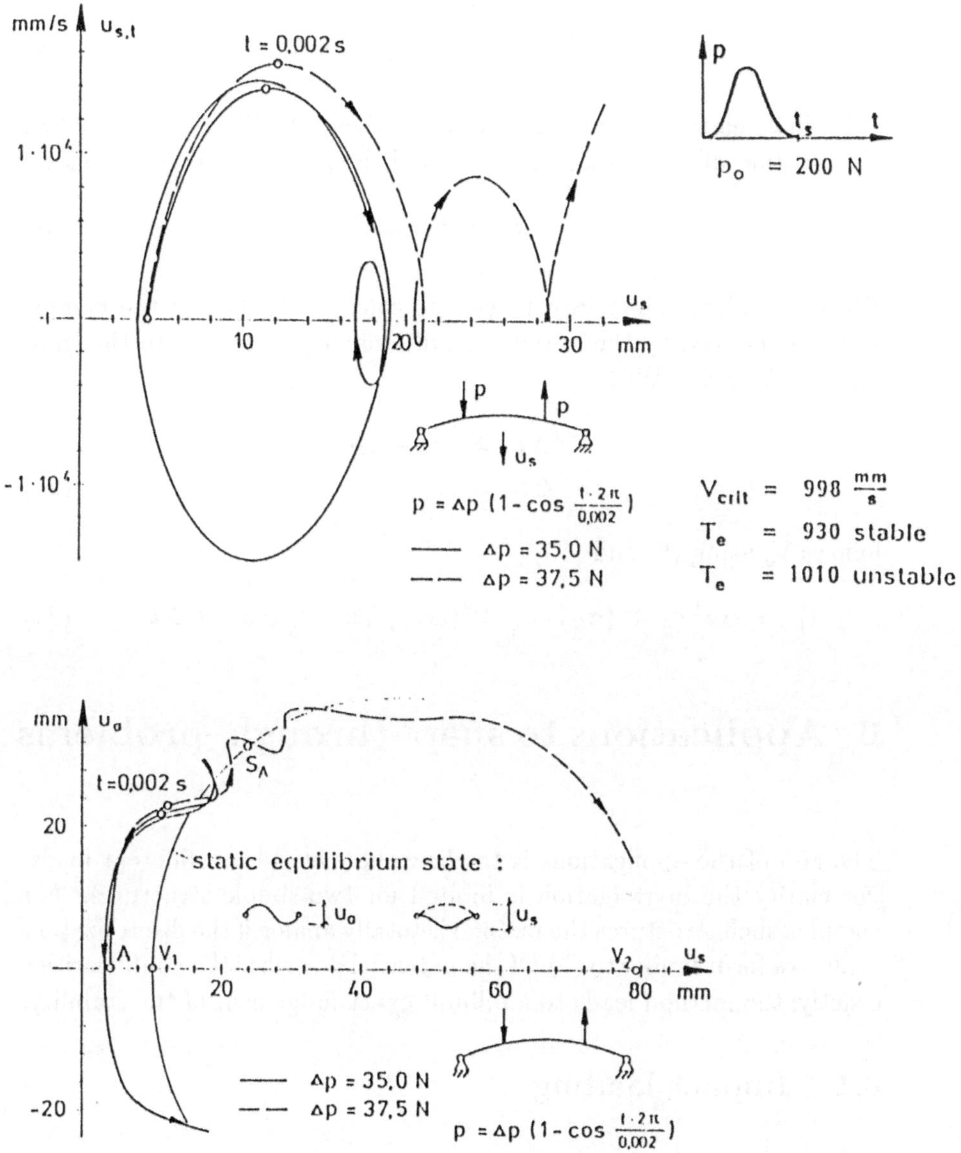

Fig. 4 : Motion of a flat arch under unsymmetric impact loading

The integration in time can be performed analytically or numerically for example with Gauss. After integration a nonlinear algebraic system of equations has to be solved for the unknown factors α_j.

$$\{V_{AL} + V_{AN}(\alpha_k) + V_M\}_{ij}\,\alpha_j - V_{pi} = 0 \tag{21}$$

With little effort (21) can be solved iteratively. With the calculated α_j the function of the motion is known and the induced energy T_e follows

$$T_e = \alpha_i V_{pi}\,. \tag{22}$$

With less effort the energy V_0 can be calculated. This is the energy, which is induced by the initial v conditions z_0, $\dot{z}_0$ related to the equilibrium state z_G. With

$$\begin{aligned} \Delta z &= z_0 - z_G \\ \Delta \dot{z} &= \dot{z}_0 \end{aligned} \tag{23}$$

follows V_0 using (5) and (15)

$$V_0 = \Delta z^t\,\{\tfrac{1}{2}\,A^L(z_G) + \tfrac{1}{6}\,A^N(\Delta z)\}\Delta z) + \tfrac{1}{2}\,\Delta z^t\,M\,\Delta\dot{z}\,. \tag{24}$$

6 Applications to snap–through–problems

The aim of the applications is to show the method for different loads. For clarity the investigation is limited for two simple structures. For complex shell structures the method is totally analog if the discretization leads to a form similar to (5). If the potential describes the nonlinearities exactly, the method leads to an illimit exact judgement of the stability.

6.1 Impact loading

The structure, see figure 4 and its initial equilibrium state G in the prebuckling range is given. For this state first the critical energy ΔV_{crit} is calculated after paragraph 4 for a symmetric constant load. A sufficiently large perturbation of an unsymmetric load pattern, e. g. with the time distribution after figure 4 accelerates the structure beyond the separatrix such that loss of stability occurs. For small perturbations the structure moves in the prebuckling range around the equilibrium state G. The level of the initial load and the perturbation pattern in the time

The approximation of the state-vector (18) is done with the first eigenmode of the equilibrium state. The motion in time caused by the related energy T_e are calculated after paragraph 5. The values T_e is calculated with a polynomial (19) with eight terms. The comparison of critical energy and external energy shows that the error can be kept arbitrarily small. The convergence of the external energy T_e for different approximations in time is given in /12/.

6.2 Transition Loads

Abrupt changes of the load-level cause dynamic transitions. These have to be investigated in a different manner since not the stability with respect to the initial state A has to be proved but the stability of the motion around the equilibrium state G. Dependent on the load-velocity the structure reaches at time $t = t_s$ the state of motion z_0, $\dot{z}_0$ which gives the initial condition for the following motion around the new equilibrium state G.

The calculation of this critical energy ΔV_{crit} can be performed after paragraph 4 with respect to the new equilibrium state G. The state of motion z_0, $\dot{z}_0$ is calculated after paragraph 5. Then the calculation of the energy V_0 with respect to G can be performed with (24). The values in table two show, that the loss of stability can be predicted also for a transition.

$$p_{(t)} = \Delta p \frac{t}{t_s} \quad , \quad t_0 \leq t \leq t_s = \frac{\pi}{2}$$
$$p_{(t)} = \Delta p \quad , \quad t > t_s$$

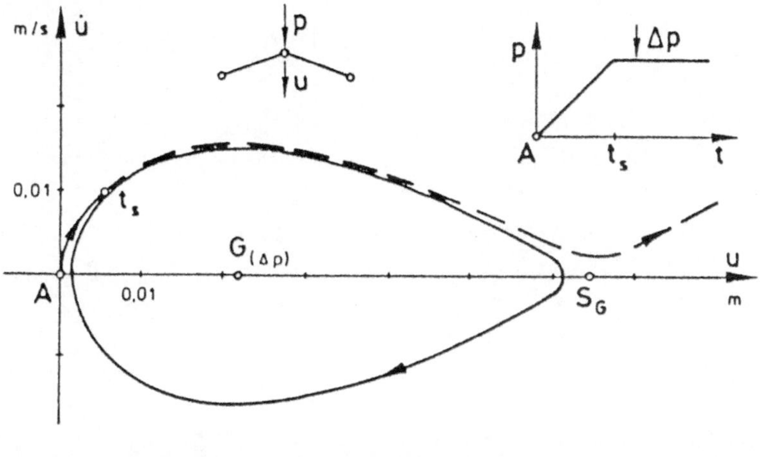

Fig. 5 : Motion under abrupt transition loads

Table 1

	Δp	$10^7 \cdot \delta V_{crit}$	$\leftrightarrow$	$10^7 \cdot T_e$	
--	0,0152	1163	$<$	1198	$\Rightarrow$ unstable
--	0,0151	1206	$>$	1189	$\Rightarrow$ stable

7 Conclusion

A method is given, which allows a simplified stability investigation of snapping and buckling structures under time-dependent loads. The degree of stability is calculated from the comparison of the load-induced energy and the critical energy stored in the structure using energy norms. The deformation state related to the critical energy is calculated from the condition of vanishing secant stiffness with respect to the initial load state of the structure. For different limit criteria it is possible to use other deformation state vectors without changing the method.

The results show that the loss of stability is not related to the induced impulse but to the work of the perturbating loads, in this case the induced energy.

References

1. Koiter, W. T.: General Equations of Elastic Stability for Thin Shells. Proc. Symp. in Honor of Lloyd H. Donnell, D. Muster Ed. Houston: University of Houston Press 1966 187–228

2. Leipholz, H.: Stabilität elastischer Systeme. Karlsruhe: Verlag G. Braun 1980

3. Huseyin, K.: Nonlinear Theory of Elastic Stability. Leyden: Noordhoff Int. Publ. 1975

4. Thompson, J. M. T.; Hunt, G. W.: A General Theory of Elastic Stability. London: John Wiley & Sons 1973

5. Ziegler, H.: Die Stabilitätskriterien der Elastomechanik. Ingenieur Archiv 20 (1952) 49–56

6. Hayashi, C.: Nonlinear Oscillations in Physical Systems. New York: Mc Graw–Hill Book Company 1964

7. Thompson, J. M. T.; Stewart, H. B.: Nonlinear Dynamics and Chaos. Chichester: John Wiley & Sons 1986

8. Haken, H.: Order in Chaos. Computer Methods in Applied Mechanics and Engineering 52 (1984) 635–652

9. Koiter, W. T.: On the Thermodynamic Background of Elastic Stability Theory. Problems of Hydrodynamics and Continuum Mechanics. L. I. Sedov Anniversary Vol. SIAM. Philadelphia: 1969

10. Malkin, J. G.: Theorie der Stabilität einer Bewegung. München: Verlag R. Oldenburg 1959

11. Dinkler, D.: Stabilität dünner Flächentragwerke bei zeitabhängigen Einwirkungen. Bericht Nr. 87 - 52, Institut für Statik – TU Braunschweig, Herausgeber Prof. Dr.–Ing. H. Duddeck. Braunschweig: Eigenverlag 1987

12. Dinkler, D.: Stabilität elastischer Tragwerke mit nichtlinearem Verformungsverhalten bei instationären Einwirkungen. Ingenieur Archiv 59 (1989). Berlin: Springer Verlag

Probabilistic Finite Element Analysis of Unsymmetrical Buckling of Thin Shallow Spherical Shells

S. NAKAGIRI

Institute of Industrial Science
University of Tokyo

Summary

The effect of uncertain initial imperfections of thin shallow spherical shells under external pressure on the buckling is investigated by Monte Carlo simulation. The buckling pressure is estimated by means of nonlinear eigenvalue analysis of the shells discretized by triangular finite elements. The imperfect middle surface of the shells is represented as a two-dimensional spatial stochastic process discretized according to the finite element division. The simulation results show the effect of the input correlation function on the mean and coefficient of variation of the buckling pressure.

Introduction

The finite element discretization of structures has been applied for long to the analyses of large deformation and buckling of thin-walled shells. The effect of initial imperfections on the buckling of shallow spherical shells has been investigated by analytical methods and finite element method, mainly under the assumption that the imperfections are axisymmetric [1]. The uncertainty of the initial imperfections has been a focus of attention since early times [2]. This paper deals with the effect of uncertain middle plane of clamped shallow spherical shells subject to external pressure. The middle plane is discretized into triangular finite elements [3]. Probabilistic variables are assigned to the height coordinates of the nodes. It turns out that the uncertainty is discretized into two-dimensional stochastic process. Monte Carlo simulation is employed to estimate the statistical properties of the shell buckling which is expected highly nonlinear with respect to the probabilistic variables. Attention is paid to the availability of Monte Carlo

G. Kuhn · H. Mang (Eds.)
Discretization Methods in Structural Mechanis
IUTAM/IACM Symposium Vienna, Austria, 1989

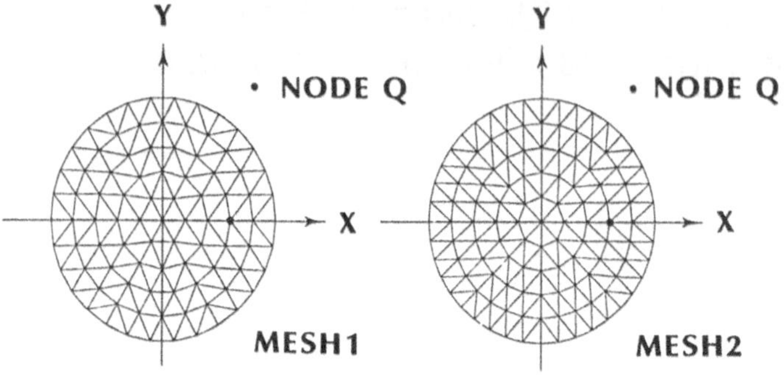

Figure 1 Top view of mesh division
(MESH1: 150 elements, MESH2: 200 elements)

simulation [4] in the presence of many probabilistic variables.

Nonlinear Finite Element Analysis of Shallow Spherical Shell

As a preparatory basis of the subsequent buckling analysis, this section describes the outline of the nonlinear finite element analysis which follows the deterministic formulation developed so far [5],[6]. Figure 1 is the top-view of two types of mesh division used. The element stiffness equation formulated in terms of the convected coordinate system is summarized as given below,

$$([k] + \sum_{n=1}^{3} N_n [k_{Gn}])\{d\} = \{f\} \tag{1}$$

where [k] denotes the linear stiffness matrix, N_n stands for forces per unit length in the middle plane of shell and $[k_{Gn}]$ indicates the matrices derived from the relationship between the middle plane slopes and nodal displacements. They are dependent on the initial imperfections. The stiffness with respect to the rotation in the middle plane is adjusted by adding virtual rotational stiffness [3].

The element stiffness equation in terms of the convected coordinates is transformed and arranged into the nonlinear stiffness equation in terms of the fixed global coordinates by taking the rigid body rotation of the elements into account. It is necessary to solve iteratively the nonlinear stiffness

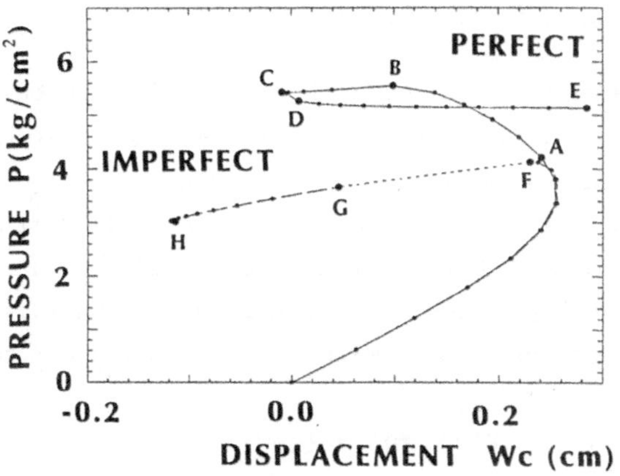

Figure 2 External pressure-center point deflection curve (Clamped shell of λ=7)

equation in the incremental form.

Figure 2 depicts the external pressure-deflection curve at the center of a clamped shell of λ=7 by MESH1, λ being a geometric shell parameter [7]. The shell thickness is 1 mm, the rise 7.401 mm, and the radius of shell curvature 679.3 mm. The deflection contour lines are shown in Fig.3. The pressure-deflection curve agrees well with the one given by Endou et al. [8] up to the maximum pressure points B and F in Fig.2. The contour is not axisymmetric however after the maximum pressure for both the perfect shell and the imperfect shell with a protrusion at node Q, indicating the deficiency of MESH1 and the inaccuracy of the iterative solution in the large deflection domain. The deficiency can be mitigated by the use of MESH2. It can be said thereby that the maximum pressure is estimated rather accurately.

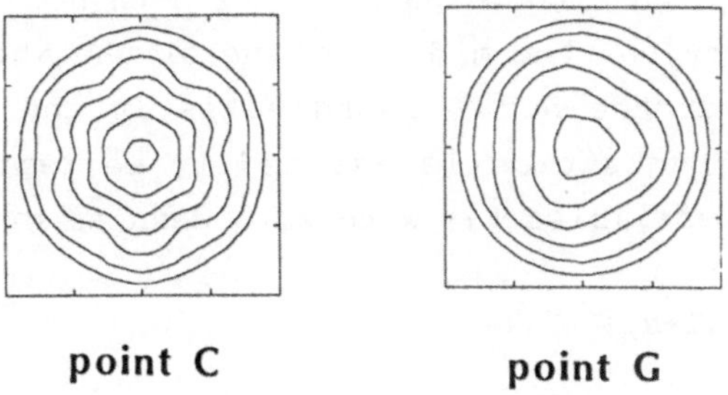

Figure 3 Deflection contours after maxium pressure

Estimation of Spherical Shell Buckling

In this paper the statistical properties of the buckling of imperfect shells are evaluated by Monte Carlo simulation of the maximum pressure which is estimated not by the pressure-deflection curve but by the nonlinear eigenvalue problem in order to diminish the computational effort required for a sample of the imperfections. Equation 2 summarizes the global stiffness equation at a loading level in the incremental form,

$$([K(F)] + [K_g(F)])\{\Delta D\} = \{\Delta F\} \tag{2}$$

where $\{\Delta U\}$ and $\{\Delta F\}$ denote the incremental displacement and force vectors. The matrices $[K(F)]$ and $[K_g(F)]$ are nonlinear depending on the loading level. The nonlinear eigenvalue problem in the form of Eq.3 is employed for estimating the buckling pressure by means of finding the minimum of a loading factor γ equal to 1. In doing so, the incremental stiffness equation is solved iteratively again to evaluate the effect of the nonlinearity dependent on the loading level. The buckling pressure thus estimated by five iterations is 0.565 for a perfect clamped shell of λ=4, which is in fairly good agreement with 0.578 of the solution given by Huang [7].

$$\det([K] + \gamma[K_G]) = 0 \tag{3}$$

Initial Imperfections Expressed as Stochastic Process

The uncertain imperfections dealt with in this paper are the fluctuation of the z coordinate of the n-th node, except those on the boundary, of the middle plane of the shells given as the function of the nodal coordinates in the fixed global system. The fluctuation is assumed to be represented by N probabilistic variables $\{\alpha\}$ with zero mean as given by Eq.4,

$$z_n = \bar{z}_n(1+\alpha_n) \tag{4}$$

where the upper bar indicates the mean. The property of the

fluctuation has not been elucidated yet. Assumption is made thereby that the stochastic process $\alpha(X,Y)$ is stationary, isotropic and Gaussian. By inputting samples of the stochastic process into the nonlinear eigenvalue analysis stated in the preceding section, the statistical properties of the uncertain buckling pressure due to the imperfections can be simulated by Monte Carlo approach. N-dimensional random number $\{p\}$ of Gaussian distribution with zero mean and unit variance can be generated easily by computer.

Two types of the stochastic process are examined, that is, uncorrelated and correlated. The correlation is expressed by a covariance matrix [C] derived from correlation function R(t) of the probabilistic variables based on the Wiener-Khintchine relation. The standard deviation s is taken constant for all the probabilistic variables under the assumption of a stationary process. Assumption is made that the correlation of the imperfections decreases with distance. The following function is input to express the decay,

$$R(t) = s^2 \exp(-t/L) \tag{5}$$

where L denotes correlation distance, a scale of correlation decay, and t is taken equal to the distance between the nodes. It is necessary to transform the Gaussian random number $\{p\}$ generated in computer into the random number $\{\alpha\}$ whose covariance matrix equals the input [C]. The following linear transformation is employed to this end [9].

$$\{\alpha\} = [Q]\{p\} \tag{6}$$

The relationship between the probabilistic variables $\{\alpha\}$ subject to the covariance matrix [C] and the random number $\{p\}$ with unit variance gives rise to the following equation.

$$[C] = [Q]\,[Q]^T \tag{7}$$

There are no established criteria to determine the matrix [Q] uniquely. When [Q] is taken as lower triangular matrix, the ingredients are calculated by the formulae given below.

$$q_{n1} = c_{n1}/\sqrt{c_{11}} \qquad (1 \le n \le N) \tag{8}$$

$$q_{nn} = (c_{nn} - \sum_{i=1}^{n-1} c_{ni}{}^2)^{1/2} \qquad (1 \le n \le N) \tag{9}$$

$$q_{nm} = (c_{nm} - \sum_{i=1}^{m-1} c_{ni}c_{mi})/c_{mm} \qquad (1 < m < n \le N) \tag{10}$$

The samples of the probabilistic variables thus generated are used as the input to Monte Carlo simulation.

Monte Carlo Simulation of Uncertain Buckling

Monte Carlo simulation is a robust technique able to deal with various stochastic processes. It is difficult however to assess how many samples should be taken to afford sufficient simulation. Figure 4 depicts the simulation history of the mean of the buckling pressure, normalized by that of perfect shell, in the case of 61 uncorrelated probabilistic variables of s=1% for a clamped shell of λ=7. It is seen that the simulation result is converged well after 400 samples. Figure 5 is a histogram of the buckling pressure in the same case as above. The figure hints that the histogram is of Gaussian type. Figure 6 illustrates the effect of the correlation distance L on the buckling pressure. The minimum distance between the nodes is 20 mm. Consequently, the probabilstic variables can be regarded uncorrelated for L less than 20 mm.

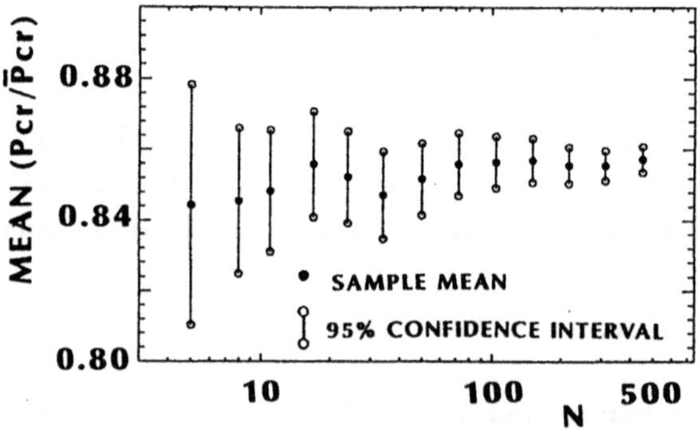

Figure 4 Simulation history of mean buckling pressure

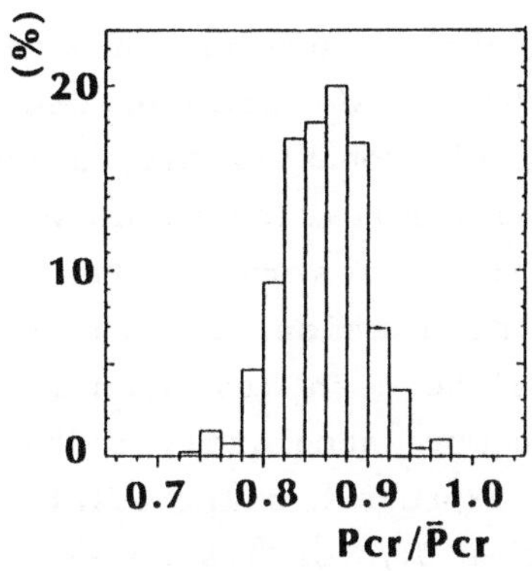

Figure 5 Histogram of buckling pressure

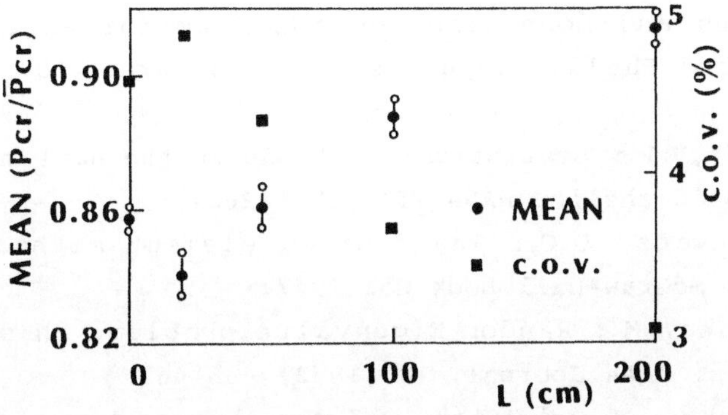

Figure 6 Effect of correlation distance on mean and coefficient of variation of buckling pressure

The mean buckling pressure is increased with L, since the imperfect shell geometry rather nears the perfect shell when the correlation of the probabilistic variables is intensified. This tendency is reflected in the decrease of the coefficient of variation with L in Fig.6.

Conclusion

The effect of the two-dimensional stochastic process on the buckling pressure of shallow spherical shells is investigated by means of Monte Carlo simulation. The mean and coefficient of variation of the buckling pressure are shown in relation to the correlation distance. 400 samples seem to be sufficient for the simulation by 61 probabilistic variables.

To the end of general simulation of uncertain, geometrical imperfections, it is necessary to take into account the fluctuation of the inclination as well as the height at the nodes, and to refine the mesh division in order to estimate the buckling pressure more accurately in the future.

The author wishes to express his appreciation to Mr. I.Nakada, Toa Nenryou Kogyo Co. Ltd., for his constructive support and interest in this study.

References

1. Yamamoto,Y. and Kokubo,K.: Effects of Geometrical Imperfections and Boundaries on the Buckling Strength of a Spherical Shell. Computers and Structures. 19 (1984) 285-290.
2. Boltotin,V.V.: Statistical methods in the nonlinear theory of elastic shells. NASA TT F-85 (1962).
3. Zienkiewicz, O.C.: The finite element method. 3rd ed. London: McGraw-Hill Book Co. 1977.
4. Shinozuka, M.: Random Eigenvalue problems in structural analysis. AIAA Journal, 10 (1972) 456-462.
5. Belytschko, T and Hsieh, B.J.:Non-linear transient finite element analysis with convectedco-ordinates.Int.J. Num. Meth. in Engrg. 7 (1973) 255-271.
6. Kondoh, K. and Atluri, S.N.: A simplified finite element method for large deformation, post-buckling analyses of large frame structures, using explicitly derived tangent stiffness matrices. Int. J. Num. Meth. in Engrg. 23 (1986) 69-90.
7. Huang,N.C.: Unsymmetrical buckling of thin shallow spherical shells. Trans. ASME. Ser.E, 31 (1964) 447-457.
8. Endou,A., Hangai,Y. and Kawamata,S.: Post-buckling analysis of elastic shells of revolution by the finite element method. Report of the Institute of Industrial Science, the University of Tokyo, 26 (1976) 48-81.
9. Elishakoff, I.: Probabilistic methods in the theory of structures. New York: John Wiley & Sons (1983).

A Combined Analytical and Finite Element Method for Non-Collinear Crack Growth in Brittle Solids

Y. Sumi* and Y. Yamamoto†

*Department of Naval Architecture and Ocean Engineering
Yokohama National University
Yokohama 240, Japan
†Department of Mechanical Engineering
Tokyo Denki University
Ishizaka Hatoyama Saitama 350-03, Japan

Summary

Recent new developments in fracture mechanics sometimes require not only stress intensity factors but also new quantities which characterize the crack growth behavior of materials and engineering structures. Computational methods are discussed for the determination of the stress field parameters of higher order at a crack tip, derivatives of stress intensity factors with respect to crack length for collinear crack extension, and stress intensity factors at the virtually extended crack tip for non-collinear crack extension. These methods are applied to the computaional crack path prediction for brittle fracture in welding residual stress fields.

Introduction

Recent fracture mechanics research is on one hand directed to the understanding of micromechanisms of crack growth in non-metallic materials such as rocks, ceramics, and composites [1,2], while it is on the other hand directed to construct a structural design concept such as the fracture control of engineering metallic structures [3]. These new trends in fracture mechanics require not only the conventional crack tip parameters such as stress intensity factors, K, and J-integrals, but also new parameters which are relevant to the fracture mechanisms of the materials and to the structural design concept, respectively.

In case of concrete, rocks, and ceramics the size effect of the fracture process zone is influenced by the stress distribution which corresponds to the coefficients of higher order terms in the Irwin-Williams expansion at the crack tip [4]. In order to determine the energetically favourable growth regime of the system of interacting cracks in brittle solids, the second derivatives of the potential energy with respect to crack lengths, which correspond to the first derivatives of stress intensity factors, have to be calculated [5]. Since fracture control of engineering structures requires precise knowledge of paths of possible crack propagation, along

G. Kuhn · H. Mang (Eds.)
Discretization Methods in Structural Mechanis
IUTAM/IACM Symposium Vienna, Austria, 1989

which the crack resistance such as fracture toughness of materials has to be evaluated, computational crack path prediction may play an essential role in the structural design [6]. In the present paper combined analytical and finite element methods are presented for the determination of a near tip stress field [7,8,9], direct calculation of the derivatives of stress intensity factors [5], and computational crack path prediction [10,11].

Governing Stress Field Parameters of Cracks in Brittle Solids

It has recently been recognized that the concept of the linear fracture mechanics cannot directly be applied to the fracture behavior in non-metallic brittle solids such as concrete, rocks, and ceramics. The fracture toughness, K_{Ic}, of the materials is highly dependent on the geomerty and the size of specimens because of the so-called fracture process zone, where the micro-mechanical fracture behavior such as the nucleation and growth of micro-cracks takes place. Even when the apparent stress intensity factors are the same, different sizes of the process zone are observed ahead of the crack tip [1,2] (see Fig.1). Horii et al. [4] suggested the importance of the elastic non-singular terms in the Irwin-Williams expansion at the elastic crack tip, because non-selfsimilar fracture process zone may be formed due to a variety of the second terms of the following elastic crack tip stress field;

$$\sigma_{22}(x_1,0)=\frac{k_I}{\sqrt{2\pi x_1}}+b_I\sqrt{\frac{x_1}{2\pi}}+O(x_1), \tag{1}$$

where the elastic stress component normal to the crack line is calculated at the crack tip by disregarding the process zone. They roughly estimate the fracture process zone size in terms of k_I and b_I, by using Dugdale-Barenblatt model with the linear tension-softening relation in the fracture process zone.

Tension cracks are also formed in non-metallic brittle solids such as glass, to which the linear fracture mechanics is applicable. Their initiation and propagation are influenced by a variety of mechanisms

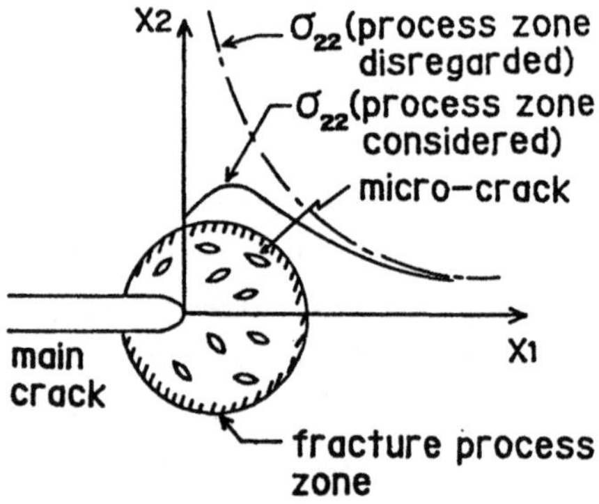

Fig.1 Process zone ahead of a crack tip in a brittle solid.

such as the developments of non-uniform temperature fields, non-uniform volume changes due to phase transformation and others. Since fracture of this kind occurs under a strain-controlled condition, cracks extend quasi-statically with increasing a load parameter. In the case where two or several cracks exist, the interaction effect may induce abrupt changes in their growth pattern. For example, as cracks continuously grow with a continuous supply of elastic energy, a critical state may be reached, where some cracks become stationary, as the others grow at faster rates [5]. For simplicity, let us consider a system of one load parameter, Λ , with two Mode-I interacting cracks (crack 1 and crack 2) of h_1 and h_2 in length, whose stress intensity factors are denoted by K_1 and K_2, respectively (see Fig.2). The changes of the stress intensity factors during the crack growth are represented as

$$dK_1 = \frac{\partial K_1}{\partial h_1} dh_1 + \frac{\partial K_1}{\partial h_2} dh_2 + \frac{\partial K_1}{\partial \Lambda} d\Lambda ,$$

$$dK_2 = \frac{\partial K_2}{\partial h_1} dh_1 + \frac{\partial K_2}{\partial h_2} dh_2 + \frac{\partial K_2}{\partial \Lambda} d\Lambda , \qquad (2)$$

where $d\Lambda$, dh_1, and dh_2 are the increases of the load parameter and the crack lengths. During the quasi-static brittle fracturing process we can assume the conditions $K_1=K_2=K_{Ic}$; i.e. $dK_1=dK_2=0$, where K_{Ic} is the plane strain fracture toughness of the material. Equation (2), therefore, gives the incremental relation between the load parameter and the crack lengths. A strain controlled problem is characterized by the condition,

$$\frac{\partial K_i}{\partial h_i} < 0. \qquad (3)$$

Since these two cracks have an interactive shielding effect, signs of the cross derivatives of the stress intensity factors are negative; i.e.

$$\frac{\partial K_i}{\partial h_j} < 0, \ i \neq j . \qquad (4)$$

In a suitable setting, the interactive effect is small at the initial stage of the crack growth, which can be characterized by the following condition;

$$\frac{\partial K_i}{\partial h_i} < \frac{\partial K_i}{\partial h_j} < 0 , \ i,j=1,2 . \qquad (5)$$

As far as inequality (5) holds, the

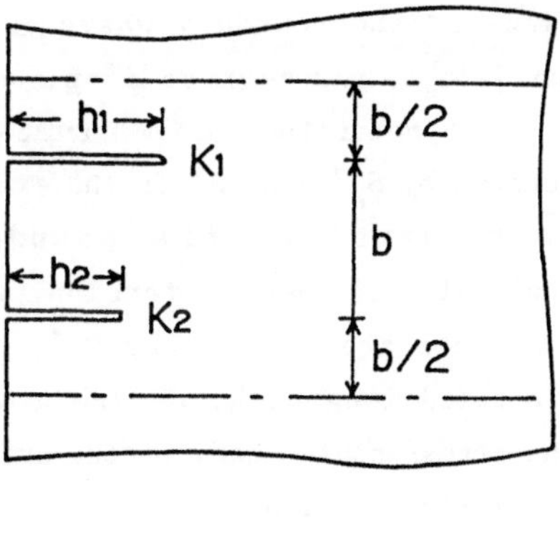

Fig.2 Two-interacting edge cracks in a unit cell of breadth 2b.

matrix $[\partial K_i/\partial h_j]$, $(i,j=1,2)$ is positive definite and dh_i can be determined uniquely for a given increase of the load parameter. As cracks grow, a critical state may be reached, at which the uniqueness of the crack growth regime is lost. It is called a stable bifurcation point, and is characterized by the condition

$$\det[\partial K_i/\partial h_j] = 0. \tag{6}$$

In order to determine the postcritical behavior of the interactive crack growth, it is essential to calculate accurate values of derivatives of stress intensity factors with respect to crack lengths (see details in Ref.[5]).

Non-Collinear Crack Growth

Curved crack paths are sometimes observed for brittle cracks propagating in engineering welded structures, whose curving is mainly caused by the residual stresses and material inhomogeneity along a welded joint. Since brittle crack propagation takes place under a predominantly Mode I loading condition, the local symmetry condition $(K_{II}=0)$ is used as a crack path criterion for the computational prediction [10,11].

In order to predict a crack path based on the above criterion, a first order perturbation solution of a slightly curved crack is utilized. Fig. 3 shows a cracked body, where V denotes the domain of the body, S_t and S_u are the parts of the boundary where surface traction t_i and surface displacement v_i are prescribed. Crack surfaces are denoted by $S_c^{\pm}$, and n_i is the external unit normal of the boundary. Non-collinear crack extension, whose surfaces are denoted by $S_b^{\pm}$, has a projected length, h, on the x_1-axis. The stress state ahead of the original crack tip is given by

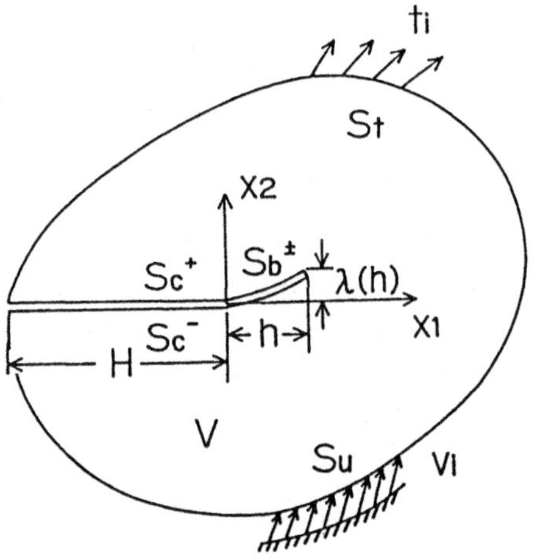

Fig.3 A straight crack with a slightly branched and curved extension.

$$\sigma_{11}(x_1,0)= \frac{k_I}{\sqrt{2\pi x_1}} + T + b_I\sqrt{\frac{x_1}{2\pi}} + O(x_1),$$

$$\sigma_{22}(x_1,0)=\frac{k_I}{\sqrt{2\pi x_1}}+b_I\sqrt{\frac{x_1}{2\pi}}+0(x_1),$$

$$\sigma_{12}(x_1,0)=\frac{k_{II}}{\sqrt{2\pi x_1}}+b_{II}\sqrt{\frac{x_1}{2\pi}}+0(x_1), \tag{7}$$

where k_I and k_{II} are the stress intensity factors at the original crack tip, and T, b_I, and b_{II} are the coefficients corresponding to the higher order terms of the Irwin-Williams expansion. The crack growth profile is assumed to be of the following form;

$$\lambda(x_1) = \alpha x_1 + \beta x_1^{3/2} + \gamma x_1^2 + 0(x_1^{5/2}), \tag{8}$$

where α, β, and γ are the shape parameters. The stress intensity factors at the extended crack tip are calculated as

$$\begin{aligned} K_I(h;\lambda) &= (k_I-\frac{3}{2}\alpha k_{II})-\frac{9}{4}\beta k_{II}h^{1/2}+\{\frac{b_I}{2}-\frac{5}{4}\alpha b_{II}-3\gamma k_{II}+\bar{k}_{II}k_I \\ &\quad -\alpha(\bar{k}_{III}+\frac{3}{2}\bar{k}_{III})k_I+\bar{k}_{III}k_{II}-\alpha(\bar{k}_{II}+\frac{3}{2}\bar{k}_{IIII})k_{II}\}h, \\ K_{II}(h;\lambda) &= (k_{II}+\frac{1}{2}\alpha k_I)+(\frac{3}{4}\beta k_I-2\sqrt{\frac{2}{\pi}}\alpha T)h^{1/2}+\{\frac{b_{II}}{2}-\frac{\alpha}{4}b_I-\frac{3\sqrt{2\pi}}{4}\beta T+\gamma k_I \\ &\quad +\bar{k}_{III}k_I+\alpha(\frac{1}{2}\bar{k}_{II}-\bar{k}_{IIII})k_I+\bar{k}_{IIII}k_{II}+\alpha(\frac{1}{2}\bar{k}_{III}-\bar{k}_{III})k_{II}\}h. \end{aligned} \tag{9}$$

Considering straight crack extension; i.e. $\alpha=\beta=\gamma=0$, the derivatives of stress intensity factors are obtained in the following form;

$$\frac{\partial K_I}{\partial h}=\frac{b_I}{2}+\bar{k}_{II}k_I+\bar{k}_{III}k_{II}$$

$$\frac{\partial K_{II}}{\partial h}=\frac{b_{II}}{2}+\bar{k}_{III}k_I+\bar{k}_{IIII}k_{II} \tag{10}$$

In case of a multi-cracked system considered in the previous section, the cross derivatives can be calculated by disregarding the first terms in equation (10). In equations (9) and (10) the quantities $\bar{k}_{I\mu}$ and $\bar{k}_{II\mu}$ (μ= I, II) represent the effect of the boundary, and they are calculated as the stress intensity factors of the following problems;

$$\begin{aligned} \sigma_{ij,j} &= 0 && \text{in } V, \\ \sigma_{ij}n_j &= -\sigma^f_{\mu ij}n_j && \text{on } S_t, \\ u_i &= -u^f_{\mu i} && \text{on } S_u \qquad \mu = I,\ II. \end{aligned} \tag{11}$$

$\bar{k}_{II}$ and $\bar{k}_{III}$ are the stress intensity factors of Mode I and Mode II for μ=I in equation (11), while $\bar{k}_{III}$ and $\bar{k}_{IIII}$ are those corresponding to μ=II. Stresses σ^f_{Iij} and σ^f_{IIij}, and displacements u^f_{Ii} and u^f_{IIi} represent the so-called fundamental fields of Mode I and Mode II [12].

Assuming the local symmetry condition and using the latter of equation (9), the shape parameters of the predicted crack paths are obtained as

$$\alpha = -2k_{II}/k_I,$$

$$\beta = \frac{8}{3}\sqrt{\frac{2}{\pi}}(T/k_I)\alpha,$$

$$\gamma = -(\bar{k}_{III}k_I + \bar{k}_{IIII}k_{II} + \frac{b_{II}}{2})/k_I$$
$$+[\{(2\bar{k}_{IIII} - \bar{k}_{II})k_I + \frac{b_I}{2}\}/(2k_I) + 4(T/k_I)^2]\alpha \,. \tag{12}$$

The Method of Superposition for the Determination of Stress Field Parameters at Crack Tip

Let us consider a basic boundary value problem (see Fig.4) defined by

$$\sigma_{ij,j}=0 \qquad \text{in } V,$$
$$\sigma_{ij}n_j = t_i \qquad \text{on } S_t \text{ and } S_c^{\pm},$$
$$u_i = v_i \qquad \text{on } S_u. \tag{13}$$

The asymptotic behavior of stresses is expressed by

$$\sigma_{ij}=\sum_{k=1}^{5} a^{(k)}\sigma_{ij}^{(k)} + \sigma_{ij}^{(0)} + O(r), \tag{14}$$

in which r is the distance from the crack tip. $\sigma_{ij}^{(0)}$ is readily obtained by the surface tractions on the crack surfaces at the crack tip. $\sigma_{ij}^{(1)}$, and $\sigma_{ij}^{(2)}$ represent the Mode I and Mode II singular stresses with unit stress intensity factors, respectively. $\sigma_{ij}^{(3)}$ is the unit tensile stress acting parallel to the tangent of the crack at the tip, while $\sigma_{ij}^{(4)}$ and $\sigma_{ij}^{(5)}$ represent the Mode I and Mode II terms of order 1/2 with respect to r of the Irwin-Williams expansion. The stress intensity factors at the crack tip, $a^{(1)}=k_I$ and $a^{(2)}=k_{II}$, and the higher order coefficients $a^{(3)}=T$, $a^{(4)}=b_I$, and $a^{(5)}=b_{II}$ will be determined by the method of superposition of analytical and finite-element solutions proposed by the present authors [7,8,9].

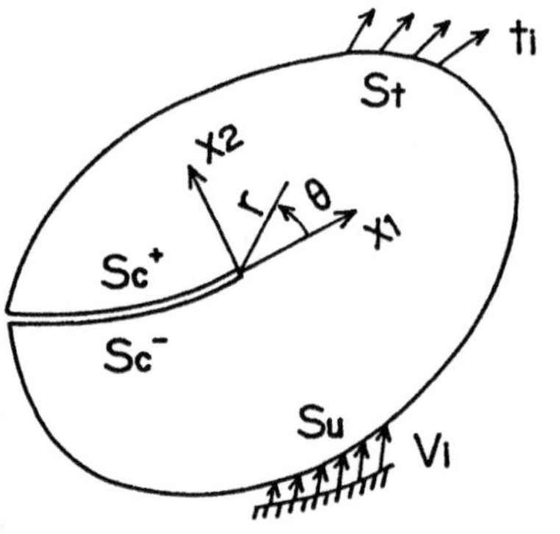

Fig.4 Basic boundary value problem.

According to the procedure of this method, the following problems are defined, and solved by the conventional finite element method;

$$\sigma_{ij,j}=0 \qquad \text{in } V,$$
$$\sigma_{ij}n_j=-\sigma_{ij}^{(k)}n_j \qquad \text{on } S_t \text{ and } S_c^{\pm},$$
$$u_i=-u_i^{(k)} \qquad \text{on } S_u,\ k=1,\cdots,5, \tag{15}$$

in which $u_i^{(k)}$ $(k=1,\cdots,5)$ is the displacement field corresponding to $\sigma_{ij}^{(k)}$. Let the finite-element solution of the original problem (13) and those of problem (15) be denoted, respectively, by (u_i^e, σ_{ij}^e) and $(u_i^{(k)e}, \sigma_{ij}^{(k)e})$ $(k=1,\cdots,5)$. The solution of the original problem is approximated by the following form which combines the analytical and finite-element solutions;

$$\sigma_{ij}=\sum_{k=1}^{5} a^{(k)}(\sigma_{ij}^{(k)}+\sigma_{ij}^{(k)e})+\sigma_{ij}^{e}. \tag{16}$$

Since equation (14) holds near the crack tip, the following condition is obtained at the crack tip;

$$\sum_{k=1}^{5} a^{(k)}\sigma_{ij}^{(k)e}+\sigma_{ij}^{e}-\sigma_{ij}^{(0)}=O(r). \tag{17}$$

Since equation (17) vanishes at the crack tip, the unknown constants $a^{(k)}$'s $(k=1,\cdots,5)$ are determined by satisfying this condition at five collocation points properly chosen near the crack tip. They are determined accurately, and they are not sensitive with regard to the finite element discretization near the crack tip. These and related topics are discussed in Ref.[9].

Finite Element Implementation for Computational Crack Path Prediction

The stress field parameters k_I, k_{II}, T, b_I, b_{II}, and $\bar{k}_{ij}$ $(i,j=I,II)$ are determined for the pre-existing crack tip as described in the previous section. Having predicted a non-collinear crack extension, $\lambda(h)$, by using equations (8) and (12), the crack tip is advanced to a certain point on the predicted path. The stress field parameters are again determined for the updated crack tip. This procedure will be repeated until the final stage of crack growth by taking account of material properties.

In the present analysis a regular finite element mesh pattern shown in Fig.5 is used in the neighbourhood of the crack tip at each step of calculations, while the remaining part of the domain is automatically subdivided into finite elements by the modified quadtree algorithm [13]. Finite element meshes thus generated are examplified in Fig.6. The numerical error of the stress intensity factor is proved to be less than

0.5 percent by using this mesh subdivision. In order to reduce computational errors involved in the curved crack path prediction, the ratio of the shear and opening modes of stress intensity factors is examined after each incremental crack extension. The size of an increment is restrained within the range that the ratio is less than 0.1.

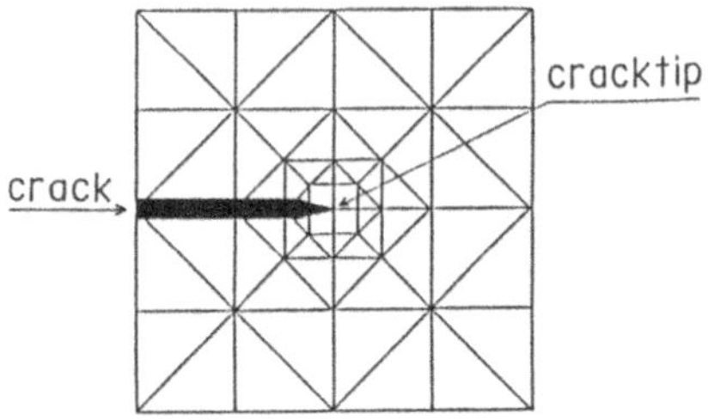

Fig.5 Mesh pattern near the crack tip.

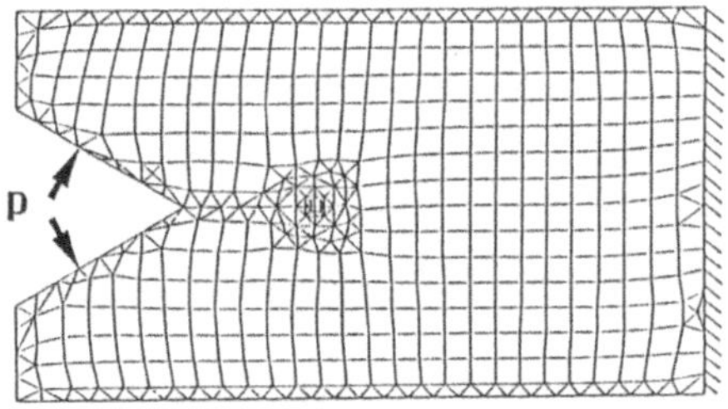

Fig.6 Automated finite element mesh subdivision for straight crack extension.

Brittle Cracks Propagating Along a Welded Joint

Behavior of a brittle crack, which is initiated and propagating along a welded joint, is affected by various conditions such as applied stresses, welding residual stresses, and material degradation in the heat affected zone. Since the decrease of the fracture toughness along the heat affected zone is relatively small in welded structures made of mild steel, it has been observed experimentally that a brittle crack initiated along the welded joint is sharply curved towards the base metal, and propagates in the base metal with certain distance off the weld line [6]. Assuming same fracture toughness throughout the whole welded plate, this problem is investigated by the computational crack path prediction, where an initial edge crack of length 100mm exists along the weld metal of a butt welded square plate of 500mm in length. As shown in Fig.7, uniform stress is applied perpendicularly to the initial crack line, and a self-equilibrating welding residual stress is assumed to act parallel to the weld line. Poisson's ratio and Young's modulus are chosen as 0.3 and 206GPa, respectively. Computer simulations are performed for three levels of the applied stress, σ_a = 59, 88, 118 MPa with a small initial kink of 2.0 degrees directed towards the base metal at the initial crack tip which is considered as an initial imperfection of the problem. Fig.8 shows the crack paths obtained by the simulation, and the final stage of the finite element mesh subdivision is illustrated in Fig.9 for σ_a= 118MPa. In all cases cracks are sharply curved to the base metal, and it is also observed that the cracks extend closer to the welded joint with the increase of the uniform applied stress level.

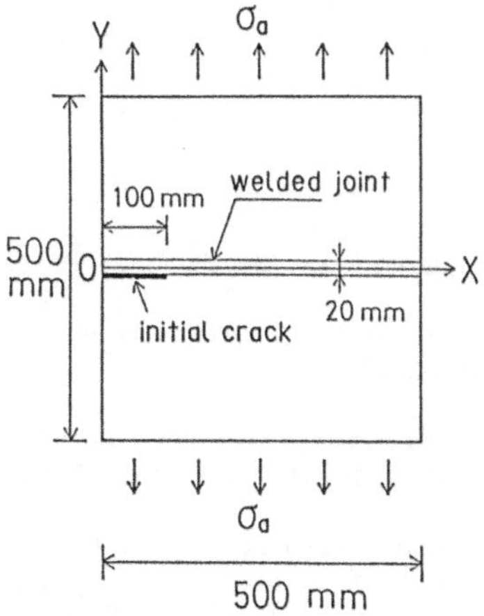

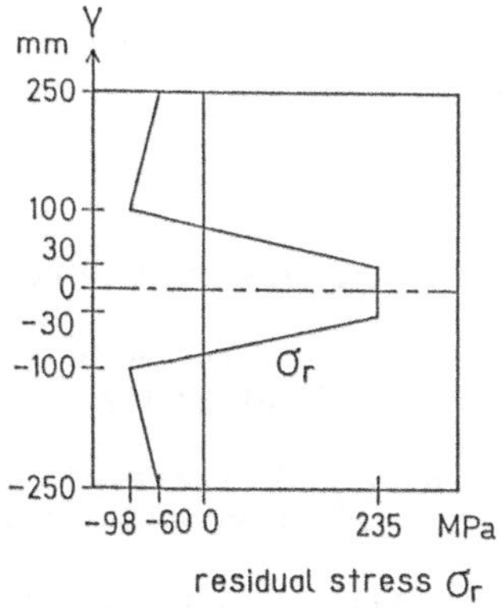

Fig.7 Geometry of a butt-welded specimen with longitudinal residual stress.

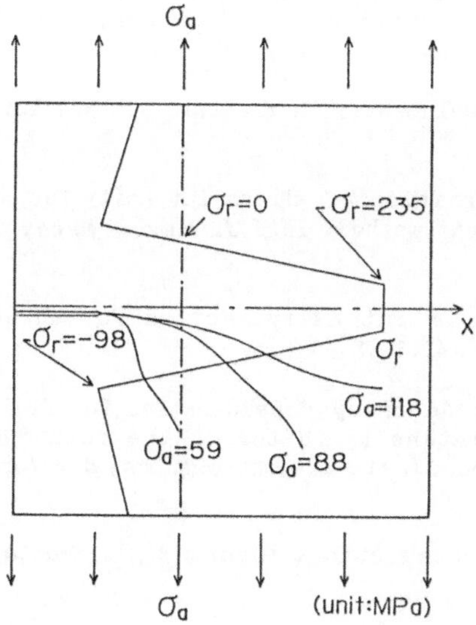

Fig.8 Curved crack paths obtained by computer simulations.

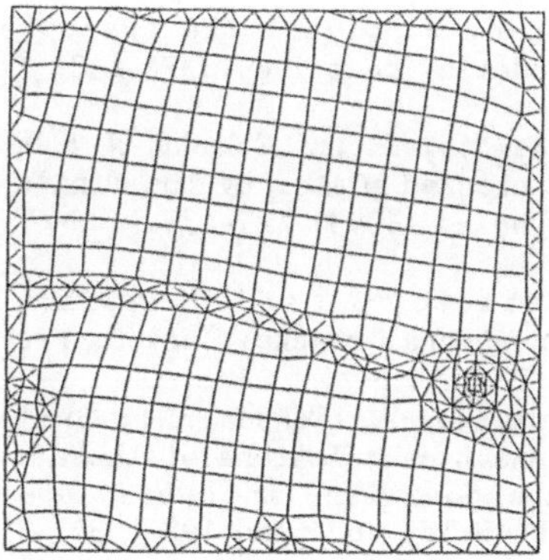

Fig.9 Final stage of mesh subdivision for σ_a=118MPa.

Concluding Remarks

In the present paper a combined analytical and finite element method is proposed and effectively applied to non-collinear crack growth in brittle solids. The stress field parameters near a crack tip, which include the coefficients of higher order terms in the Irwin-Williams expansion, are determined by the method of superposition of analytical and finite element solutions. Incorporating a perturbation analysis and the superposition method, a computational technique is presented for crack path prediction, which is examplified by the analysis of a brittle crack propagating along a welded joint.

References

1. Evans, A.G. (ed.) *Fracture in ceramic materials*. Noyes Pub. 1984.

2. Hutchinson, J.W. *Micro-mechanics of damage in deformation and fracture*. Lyngby: Technical University of Denmark 1987.

3. van Elst, H.C. and Bakker, A. (ed.) *Fracture control of engineering structures*. Warley: Engineering Materials Advisory Services 1986.

4. Horii, H.; Hasegawa, A.; Nishino, F.: Fracture process and bridging zone model and influencing factors in fracture of concrete, in *Fracture of Concrete and Rock* (Proc. SEM-RILEM Int. Conf.), Houston, 1988, 205-219.

5. Sumi, Y.; Nemat-Nasser, S.; Keer, L.M.: A new combined analytical and finite-element solution method for stability analysis of the growth of interacting tension cracks in brittle solids. *Int. J. Engng. Sci. 18* (1980) 211-224.

6. Munse, W.H.: Brittle fracture in weldments, in *Fracture IV*, New York: Academic Press 1969, 371-438.

7. Yamamoto, Y. and Tokuda, N.: Determination of stress intensity factors in cracked plates by the finite element method. *Int. J. Numer. Meth. in Engng. 6* (1973) 427-439.

8. Yamamoto, Y. and Sumi, Y.: Stress intensity factors for three-dimensional cracks. *Int. J. Fracture 14* (1978) 17-38.

9. Yamamoto, Y.; Tokuda, N.; Sumi, Y.: Accuracy consideration for finite element calculations of the stress intensity factor by the method of superposition, in *Hybrid and mixed finite element methods*, New York: John Wiley and Sons 1983, 361-379.

10. Sumi, Y.: Computational crack path prediction. *Theor. Appl. Fracture Mech. 4* (1985) 149-156.

11. Sumi, Y.: Computational crack path prediction for brittle fracture in welding residual stress fields. *Int. J. Fracture* (to appear).

12. Bueckner, H.F.: Field singularities and related integral representations, in *Methods of analysis and solutions of crack problems*, Leyden: Noordhoff 1973, 239-314.

13. Yerry, M.A. and Shephard, M.S.: A modified quadtree approach to finite element mesh generation. *IEEE Computer Graphics and Applications Jan./Feb.* (1983) 39-46.

Numerical Simulation of Crack Propagation in Shells

SHENG-CHUNG LIN and JOHN F. ABEL

School of Civil & Environmental Engrg. and Program of Computer Graphics, Cornell University, Ithaca, New York 14853, USA

Abstract

Numerical simulations of general crack propagation in arbitrary shells are accomplished by the use of automatic finite element remeshing and interactive computer graphics techniques. As part of each stage of this simulation, the curvature as well as the direction of crack propagation are calculated from the rate of energy release rate and the energy release rate, respectively. A fatigue model for mixed-mode fracture of arbitrary shell surface geometry is posed. For brittle fracture, a parameter which can be used to predict crack stability is also suggested.

Introduction

Some numerical crack propagation simulations have been performed by researchers, for example, recently Souma and Zatz [1] and Gerstle et al. [2]. However, relatively a little work has been accomplished for numerical simulation of discrete, curved crack propagation in general shells. The primary difficulty of the efforts to extend 2-D and 3-D crack simulation methodologies to shell fracture propagation is the development of crack propagation models for the highly mixed-mode fracture behavior of arbitrarily loaded, general, doubly curved shells.

Strategies and procedures for the finite element simulation of slow elastic crack propagation similar to the ones employed earlier in 2-D crack propagation programs are adopted here. The predictions of shell fracture initiation and propagation are performed by using the maximum energy release rate theory. The computation of the energy release rate is carried out with

---

[1] This is an extended version of a 4-page abstract first published in the Proceedings of the I Pan American Congress of Applied Mechanics, PUC-Rio, Jan. 1989, pp. 150-153.

G. Kuhn · H. Mang (Eds.)
Discretization Methods in Structural Mechanis
IUTAM/IACM Symposium Vienna, Austria, 1989

the direct integration technique of the virtual crack extension method [3] as extended to shell finite elements [4].

Direct Integration Form of the Stiffness Variation

A direct integration form of the stiffness variation for 2-D problems [3] is extended to a specific doubly curved shell element which is based on classical thin shell theory [4]. Separate contributions of the membrane and bending actions are considered in local, natural coordinates as

$$k_M = \int_A t\, B_M^T E_N B_M \, dA \qquad k_B = \int_A \frac{t^3}{12} B_B^T E_N B_B \, dA \qquad (1)$$

in which B_M is the membrane strain-displacement matrix, B_B is the bending strain-displacement matrix, E_N is the natural elastic constitutive matrix, t is the thickness, and A is the area of the shell element's middle surface. Details are not repeated here, but for example the variation of the membrane stiffness matrix can be derived as

$$\delta k_M = \int_A [\delta B_M^T E_N B_M + B_M^T E_N \delta B_M + B_M^T \delta E_N B_N] t \, dA \qquad (2)$$

The variation of coordinate transformations must also be considered [4]. The derivative of the stiffness matrix and the corresponding energy release rate can be obtained without perturbing the mesh, i.e., without carrying out an actual trial advance of the crack [3]. That is, $\delta k = k'$ for a unit δa, the (sparse) global stiffness derivative K' is assembled from the element stiffness derivatives k', and the energy release rate $G = 1/2\, u^T K' u$ in which u is a vector of nodal displacements.

Since the evaluation of the energy release rate is direct, the rate of the energy release rate can be obtained either from the second order derivative of the stiffness matrix or from the difference of the energy release rates of perturbed and unperturbed meshes:

$$G' = \frac{\delta G}{\delta a} = \lim_{\Delta a \to 0} \left[\frac{G(a+\Delta a) - G(a)}{\Delta a}\right] \tag{3}$$

For 2-D problems, one can obtain the stress intensity factors (SIF's) and the rates of SIF's from G and G', respectively, by mutual energy concepts [3,4].

Curvilinear Crack Propagation

Because the highly mixed SIF's for the various modes of curving through-cracks in thin shells with complicated surface geometries cannot be readily separated, linear elastic fracture initiation and propagation are determined here by the maximum energy release rate. This rate and its direction can be represented in a vectorized form [5] by

$$G_{max} = (G_1^{\,2} + G_2^{\,2})^{1/2} \quad \text{and} \quad \theta_{max} = \tan^{-1}(- G_2/G_1) \tag{4}$$

in which the 1 and 2 directions are straight-on and perpendicular to the current crack, respectively. These two expressions are similar expressions for 2-D fracture problems [6].

The curvature of crack propagation, κ, can be obtained by the derivative of the crack propagation direction with respect to the crack increment

$$\kappa = \frac{d\nu^{max}}{da} = \left[\frac{\partial v_1^{max}}{\partial a_1} v_1^{max} + \frac{\partial v_1^{max}}{\partial a_2} v_2^{max} \right] i + \left[\frac{\partial v_2^{max}}{\partial a_1} v_1^{max} + \frac{\partial v_2^{max}}{\partial a_2} v_2^{max} \right] j \tag{5}$$

in which v_1^{max} and v_2^{max} are the orthogonal direction cosines of the angle of maximum energy release rate obtained by

$$v_1^{max} = G_1(G_1^{\,2} + G_2^{\,2})^{-1/2}; \quad v_2^{max} = G_2(G_1^{\,2} + G_2^{\,2})^{-1/2} \tag{6}$$

Use of κ improves simulation of curved crack propagation with finite crack advancements [4].

Similarly, the change of the maximum energy release rate in the direction of θ_{max} can be derived as

$$\frac{dG_{max}}{da} = (G_{1,1}v_1^{max})\ G_{1,2}v_2^{max})v_1^{max} + \\ + G_{2,1}v_1^{max} + G_{2,2}v_2^{max})v_2^{max} \tag{7}$$

The above equation can be used to predict crack stability for an already propagating brittle fracture based on the following conditions:

$$\frac{dG_{max}}{da} = \begin{cases} \geq 0 & \text{Unstable} \\ < 0 & \text{Stable} \end{cases} \tag{8}$$

An improved fatigue crack growth rate follows from taking da/dN as a function of $\Delta\bar{K}_{Ieq}$, the modified range of the equivalent mode I SIF based upon not only the ranges of the SIF but also the ranges of the SIF rates:

$$\Delta\bar{K}_{Ieq} = \Delta K_{Ieq} + \Delta K'_{Ieq}\Delta a/2 \tag{9}$$

for a finite crack increment Δa, in which $\Delta K'_{Ieq} = d(\Delta K_{Ieq})/da$. Finally, the prediction of fatigue life between crack lengths a_i and a_f can be represented as

$$N = \int_{a_i}^{a_f} \frac{da}{f(\Delta\bar{K}_{Ieq})} \tag{10}$$

in which $da/dN = f(\Delta\bar{K}_{Ieq})$ is some fatigue propagation relation such as a Paris model.

Based on the fatigue models and fracture criteria in 2-D, a direct extension of an SIF-based fatigue model to shells is difficult due to, first, the absence of a generally applicable mode uncoupling technique for arbitrary shell geometry with its highly mixed fracture modes, and second, the absence of an appropriate fatigue propagation law under complex mixed-mode fracture environment. However, one can propose an expression analogous to the equivalent mode I SIF based on the maximum energy release rate criterion which states that the crack will propagate in the direction of maximum energy release rate

$$K_{Ieq}^2 = G_{max} H / t \tag{11}$$

in which H is the equivalent elastic modulus, and t is the thickness. Consequently,

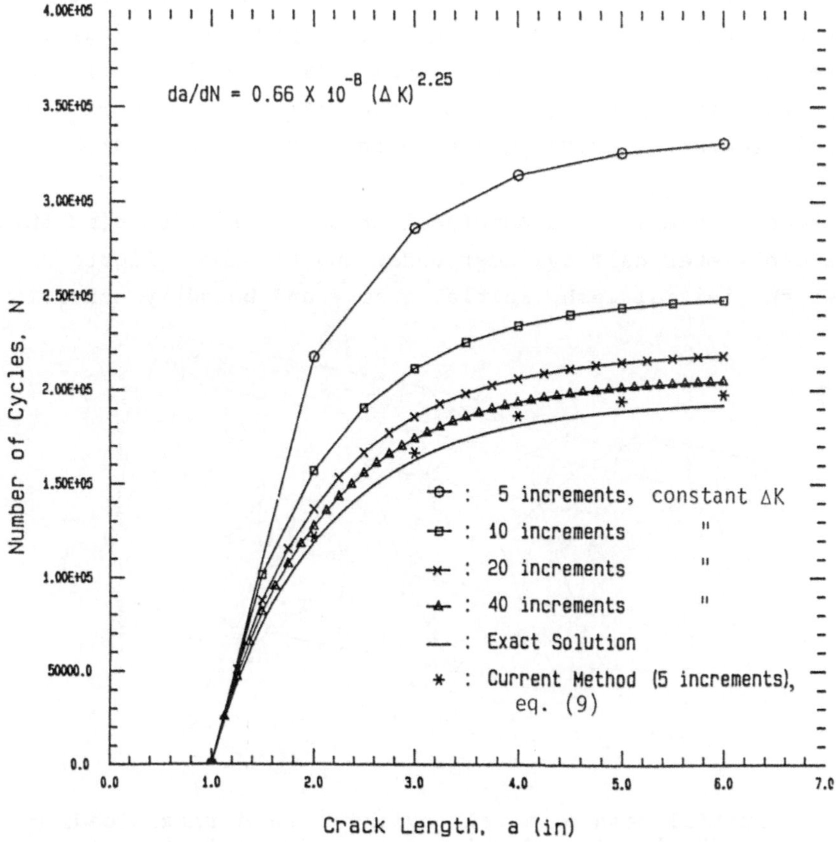

Fig.1. Number of cycles vs. crack length for different total number of crack increments employed in fatigue simulation of a single edge crack tension plate [3].

$$\Delta K_{Ieq} = \frac{1}{2} \frac{\Delta G_{max}}{\sqrt{(G_{max}\ t/H)}} \tag{12}$$

This expression is logical since the maximum energy release rate criterion has been shown [7] to yield almost identical results to the maximum principal stress criterion proposed by Erdogan and Sih [8] for 2-D plane cracks.

<u>Numerical Examples</u>

The first example is a 2-D fatigue problem of a single edge crack tension plate with the properties reported in [3]. Figure 1 indicates that the current model, which includes an estimate of the rate of the mode I stress intensity factor with change of crack length (eqn. 9), approaches rapidly the converged solution obtained by more conventional means; i.e., with crack increments when a constant ΔK_I is assumed within each increment by omitting the second term of eqn. (9).

The second example is a numerical crack propagation simulation of a cantilever half cylinder under end torsion. Figure 2 shows the initial mesh, initial crack, and boundary conditions

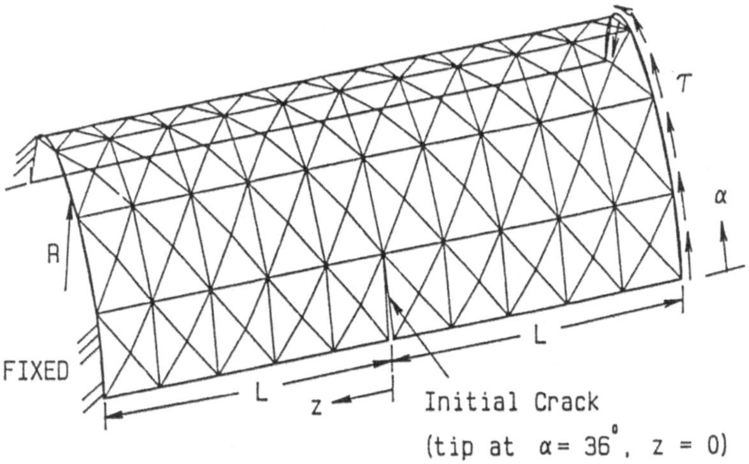

Fig.2. Initial mesh of a half cylinder under shear loading ($R=5''$, $L=10''$, $E=3.0\times10^7$ psi, $\nu=0.3$, $t=0.1''$, $\tau=1.0$ lb/in).

of the half cylinder. Figures 3 and 4 indicate the overall crack trajectory and the history of the maximum energy release rate after several increments. Here the results for four increments arise from use of the mesh shown in Figure 2, while those for eight increments arise from a finer mesh with four times as many elements.

Conclusions

The approach presented for simulation of slow, elastic crack propagation appears promising. However, it will require further testing and verification. Moreover, it is only a first step toward solving a difficult problem, and it needs to be extended to cover a more comprehensive set of theoretical and practical considerations [4].

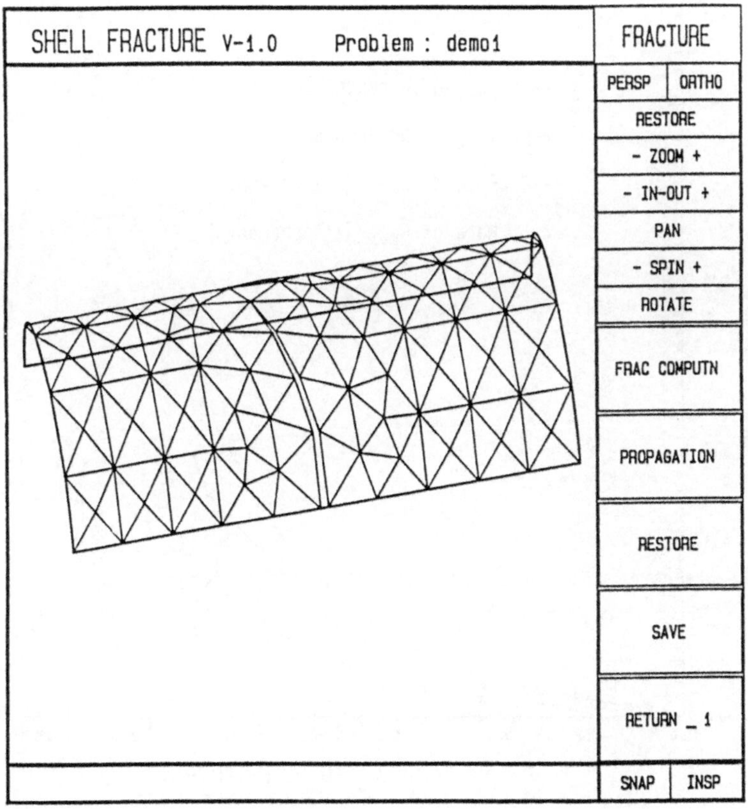

Fig.3. Crack propagation simulation for cantilever half cylinder.

References

1. Saouma, V.E.; Zatz, I.J.: An automated finite element procedure for fatigue crack propagation analysis. Eng. Frac. Mech. 20 (1984) 321-333.

2. Gerstle, W.H.; Martha, L.F.; Ingraffea, A.R.: Finite and boundary element modelling of crack propagation in two and three dimensions. Engrg. with Computers 2 (1987) 167-183.

3. Lin, S.C.; Abel, J.F.: Variational approach for a new direct-integration form of the virtual crack extension method. Int. J. Frac. 38 (1988) 217-235.

4. Lin, S.C.: Numerical simulation of through-crack propagation in general shells by energy integrations. Ph.D. Dissertation, Cornell University, Ithaca, NY (1988).

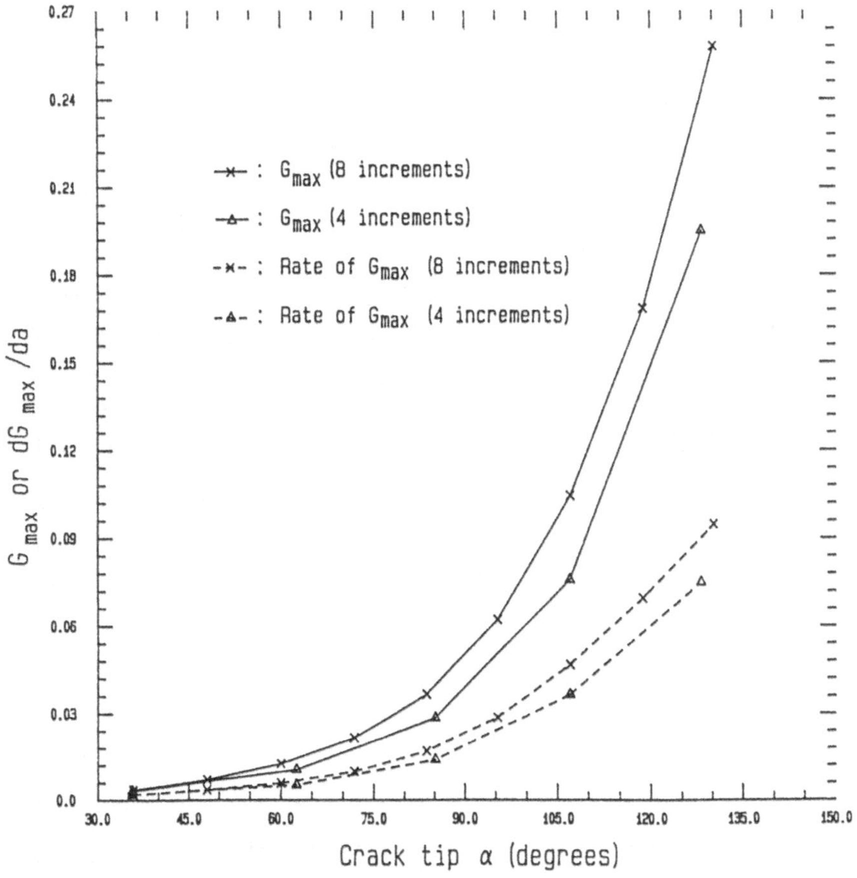

Fig.4. Energy release rate and the rate of energy release rate of the cantilever half cylinder.

5. Lin, S.C.; Abel, J.F.: An energy integral formulation suitable for numerical propagation of through-cracks in general shells. Proc. 4th Int. Conf. Numer. Meth. in Frac. Mech., A.R. Luxmoore et al. eds., Swansea, Wales: Pineridge Press (1987) 727-740.

6. Hellen, T.K.; Blackburn, W.S.: The calculation of stress intensity factors for combined tensile and shear loading. Int. J. Frac. 11 (1975) 605-617.

7. Nuismer, R.J.: An energy release rate criterion for mixed mode fracture. Int. J. Frac. 11 (1975) 245-250.

8. Erdogan, F.; Sih, G.C.: On the crack extension in plates under plane loading and transverse shear. J. Basic Engrg. 85 (1963) 519-527.

Discretization Considerations with Hypersingular Integral Formulas for Crack Problems

G. KRISHNASAMY, T. J. RUDOLPHI, L. W. SCHMERR, JR., F. J. RIZZO

Iowa State University
Department of Engineering Science and Mechanics
Ames, Iowa 50011, USA

Summary

A hypersingular integral equation is derived to solve the cracklike problem of acoustic scattering from a thin rigid screen. Discretization considerations regarding modeling the problem and computing the integrals to a required precision, using the Boundary Element Method (BEM), are discussed. Numerical results are presented.

Introduction

The presence of cracks in structural components is an issue of continuing interest and concern to everyone involved with the safety and performance of such components. The disciplines of fracture mechanics and nondestructive evaluation (NDE) are just two areas of research concerned with the detection of cracks and the analysis of the fields in the vicinity of cracks under static and dynamic loading. Both disciplines have need for analytical/ numerical methods to supplement and interpret data obtained from physical experiments.

Dealing effectively as it does with high stress gradients and surface-only discretizations in many cases, the boundary element method has enjoyed some success in providing numerically obtained information near cracks and other shapes [1] [2], and it continues to offer promise for problems of interest in fracture mechanics and NDE. However, there is a well-understood limitation in modeling a crack in a structure using conventional integral equations as the basis for the boundary element method [3]. Frequently, therefore, new integral equations, obtained by taking the gradient of the conventional (indeterminate) ones, are introduced. Such equations are free of the indeterminacy; but the gradient, operating on the already-singular kernels of the integral equations, introduces a hypersingularity.

In this paper the hypersingular integrals are interpreted carefully in the sense of a Hadamard finite part [4] and are evaluated analytically and computations are done only with regular integrals, cf. [5].

G. Kuhn · H. Mang (Eds.)
Discretization Methods in Structural Mechanis
IUTAM/IACM Symposium Vienna, Austria, 1989

Although we are ultimately interested in scattering of elastic waves by cracks, in this paper we consider the related but simpler problem of scattering of an acoustic wave by an infinitely thin rigid screen. This problem has all of the essential mathematical and computational features, which form the thrust of this paper, without the nonessential complications of the (vector) elastic counterpart.

In [5] a representative scattering problem was solved numerically using simple discretization and a crude integration scheme. The results obtained were reasonably good, which gave us confidence in this method, but the stability and the accuracy of the results were not discussed. In the present work, the discretization considerations for solving hypersingular integrals and a simple integration scheme to compute these integrals are discussed. A comparison of this new integration and the crude one used in [5] is made.

Scattering Problem

Suppose S^+ is one side of the surface of a thin rigid scatterer in an unbounded acoustic medium and suppose $\underline{x}$ is any point on S^+, $n(\underline{x})$, the normal to S^+ at $\underline{x}$, and $\underline{\xi}$ is any point not on the boundary, then the scattered field ϕ^S due to an incident field ϕ^i is

$$\phi^S(\underline{\xi}) = \int_{S^+} \frac{\partial G^D(\underline{x},\underline{\xi})}{\partial n(\underline{x})} \left[\phi^+(\underline{x}) - \phi^-(\underline{x})\right] dS(\underline{x}) \tag{1}$$

where $G^D(\underline{x},\underline{\xi}) = e^{ikR}/4\pi R$, $R = |x-\xi|$, the free space Green's function, k the wave number, $\phi(\underline{x})$ the total field, and the superscripts "+,-" henceforth refer to the two sides of the scatterer surface. On taking the gradient of Eq. (1) and in the limit as $\underline{\xi} \to \underline{\xi}_o$ a point on S^+, we obtain

$$\frac{\partial\phi^S(\underline{\xi}_o)}{\partial\xi_r} = \int\!\!\!\!\!\!=_{S^+} \frac{\partial G^D(\underline{x},\underline{\xi})}{\partial\xi_r\,\partial n(\underline{x})} u(\underline{x})\, dS(\underline{x}) \tag{2}$$

where $u(\underline{x}) = \phi^+(\underline{x}) - \phi^-(\underline{x})$, and the double dash means the Hadamard finite part of the hypersingular integral.

For Eq. (2) to exist, it is necessary for the density function to be at least $C^{1,\alpha}$ at (the collocations points) $\underline{\xi}_o$, cf. [5,6], i.e. $|u(\underline{x}) - u(\underline{\xi}_o) - u,_p(\underline{\xi}_o)(x_p - \xi_{op})| < B\,|x_p - \xi_{op}|^{\alpha+1}$ where $0\leqslant\alpha<1$ and $|B|$ is bounded. If we now consider S^+ to be a flat rigid scatterer in the x_1x_2 plane and $\hat{\underline{t}}(x)$ to be the tangent to C at a point $\underline{x}$ on C, then a process of regularization with proper use of Taylor's theorem and Stokes' theorem on Eq. (1) and limit $\underline{\xi} \to \underline{\xi}_o$, cf. [5], results in an alternate but equivalent expression for Eq. (2) as follows:

$$4\pi \left(\frac{\partial \phi^i}{\partial n}\right) = \int_{S^+} \frac{1}{R^3} \left[e^{ikR} - 1 - ikRe^{ikR}\right] u(\underline{x})ds(\underline{x})$$

$$+ \int_{S^+} \frac{1}{R^3} \left[u(\underline{x}) - u(\underline{\xi}_o) - u,_p(\underline{\xi}_o)(x_p - \xi_{op})\right] ds$$

$$+ u(\underline{\xi}_o) \oint_C \frac{1}{R^3} \left\{(x_1 - \xi_{o1})\hat{t}_2 - (x_2 - \xi_{o2})\hat{t}_1\right\} dl$$

$$+ \oint_C \frac{1}{R} \left\{\frac{\partial u(\underline{\xi}_o)}{\partial x_1} \hat{t}_2 - \frac{\partial u(\underline{\xi}_o)}{\partial x_2} \hat{t}_1\right\} dl. \tag{3}$$

The above equation, which involves regular or weakly singular integrals, is an integral equation where $\partial\phi^i/\partial n$ is known at all points on the surface and we are interested in $u(\underline{x})$. This equation can be solved numerically using the BEM. It is very common in BEM [1] to use conforming elements to describe the geometry and the density function, and this guarantees u(x) to be continuous across interelement boundaries, which is more appealing physically. Such elements are not $C^{1,\alpha}$ at nodes on the edges of elements, and so one cannot collocate at the nodes when computing hypersingular integrals. Hence, we have used nonconforming elements, cf. [5], which have collocation points away from the edges, to approximate the density function and they satisfy the $C^{1,\alpha}$ requirement at the collocation points.

Numerical Integration Scheme

The integrals which need to be computed numerically when using Eq. (3) to solve the scattering problem can be classified as follows:

(a) Line integrals - These integrals are over the element edges and the integrand is of $O(1/R^2)$. Special attention is required when integrating this term and the collocation point (CPT) is close to the edges. This is usually referred to as the "near singularity condition" where the integrand varies rapidly over the domain and so would require higher quadrature order or subdivision of the domain.

(b) Non-singular area integration - The CPT is not on the integration segment, but this could be a near-singularity situation when the CPT is very close to the integration element. This is not uncommon in BEM and the integrand is $O(1/R^3)$. As in the line integration, the near singularity situation requires special care.

(c) Singular area integration - These integrals are area integrals when the CPT is on the integrating element. The integrand is $O(1/R)$ and this integral is usually computed by coordinate transformation to polar coordinates.

When this problem was first solved numerically in [5], a crude integration scheme was used. The line integrals over each side of the element were computed using 10 function evaluations. All non-singular area integrals which include the near-singular situation were computed using 6x6 function evaluations. The singular area integrals were computed by dividing the element into four triangular sub-elements by joining the CPT to the four corners of the element, using polar coordinates and 10x10 function evaluation in each of the sub-elements.

Though the integrals were poorly handled, the preliminary results obtained as in [5] gave confidence in solving crack problems using hypersingular integrals as we have proposed. But there were important questions left to be addressed regarding accuracy and stability of this method. To gain insight into these issues, emphasis was placed on developing an efficient but simple integration scheme to compute these integrals to required accuracy.

When Gaussian quadrature formulas are used to compute these integrals, the accuracy of the result will depend on the quadrature order used. In the case of near-singular line and area integration, the integrand varies rapidly over one part of the domain and therefore requires a high quadrature order to achieve a required accuracy. Thus, the Gaussian quadrature formula is not an efficient method, as equal number of quadrature points are distributed on either side of the center of the domain, i.e. distributed independent of the variation of the integrand over the domain. There is also a restriction on the maximum quadrature order that could be used in our program, which is 10 for line integrals and 10x10 for area integrals. An alternate method of handling these integrals would be to subdivide the domain, thereby isolating the areas where the integrand varies rapidly, and use the optimum quadrature order (OQO) in each to obtain the required accuracy. To implement this integration scheme, it is necessary to find out the OQO to integrate a certain integrand when the variation of the integrand over the domain is known, i.e. when the dimensions of the domain of integration and the location of the collocation point with respect to the domain are known. With this in perspective, three different numerical experiments where performed for the three types of integrals encountered.

The numerical experiment for line integrals involves a straight line ab which is the domain of integration, h being its length, and a set of points P_n which are at a distance l_n from C, the center of the line ab. The line CP_n makes an angle of $\pm\theta_n$ with the normal to the line ab. For each of these points, the integral $\int_a^b dx/R^3$, $R = |\underline{x} - \underline{P}_n|$ is computed for different quadrature orders. When the difference in computed value for two successive quadrature orders is less than a particular value, then the larger of the two quadrature orders is the optimum, OQO, for the required precision. In this work we are demanding a precision of 5 significant figures for all three types of integrals. This exercise is repeated for a large number of points away from the line. Instead of having to repeat this experiment for different lengths of segment ab, it is enough if we take one length and estimate the OQO for different l/h ratio and angle θ. The outcome of this experiment is tabulated as follows for use in the proposed integration scheme.

Table I - OQO as a function of l/h and θ for line integrals.

l/h	θ	OQO
< 1.0	$\theta \leqslant 90°$	> 6
1.0 - 1.5	$\theta \leqslant 45°$	5
	$45° < \theta \leqslant 90°$	6
1.5 - 2.0	$\theta \leqslant 75°$	4
	$75° < \theta \leqslant 90°$	5
2.0 - 2.5	$\theta \leqslant 60°$	3
	$60° < \theta \leqslant 90°$	4
2.5 - 3.5	$\theta \leqslant 90°$	3

The numerical experiment for area integrals involves a square with diagonal D and a set of collocation points P_n outside the square, and S_n is the shortest distance from P_n and the center or edges of each of the sides of the square. This experiment is similar to the one for line integrals but involves an area integral over a square. The integrands are H_i/R^3 and H_i/R

for non-singular and singular area integration, respectively, where H_i, i = 1,2...8, are the shape functions approximating density function and $R = |\underline{x} - \underline{P}_n|$. The OQO for different D/S ratios are given in Table II for non-singular area integrals and in Table III for singular area integrals.

Table II.
OQO as a function of D/S for non-singular area integrals.

D/S	OQO
> 3.633	> 9 x 9
3.633 - 3.535	9 x 9
3.535 - 2.828	8 x 8
2.828 - 2.197	7 x 7
2.197 - 1.480	6 x 6
1.480 - 0.857	5 x 5
0.857 - 0.372	4 x 4
< 0.372	3 x 3

Table III.
OQO as a function of D/S for singular area integrals.

D/S	OQO
> 5.596	> 9 x 9
5.596 - 4.714	9 x 9
4.714 - 3.535	8 x 8
3.535 - 2.695	7 x 7
2.695 - 1.767	6 x 6
1.767 - 0.831	5 x 5
0.831 - 0.150	4 x 4
< 0.150	3 x 3

When solving scattering problems, one has to deal with area integrals which are not only singular or near singular as in static problems but also oscillatory in space and this requires more care when computing the integrals. In the numerical experiment for area integrals the shape functions were included in the integrand and have a spatial variation comparable to a half sine wave. Hence, for a scattering problem it is appropriate to assume that as long as the kernel oscillation over the domain of integration is less than or equal to half the wave length, Tables II and III can be used. In other words, if $OS = k\Delta r/2\pi$ where Δr is the distance between points on the domain furthest and closest to the collocation point, and $OS < 0.5$, then Tables II and III can be used; otherwise, the domain has to be divided into subdomains and the OS must be less than 0.5 before Tables II and III can be used for integration.

The integration scheme devised here is most suited for 8- and 9-node nonconforming elements and would require little modification for other types. Now that the OQO for the different types of integrals is known, a subdivision scheme that would result in the number of function evaluation (NFE) over the domain to be close to minimum is devised. The subdivision process for the line integral works as follows. For a specific collocation point and a line segment, here the side of the 8-node quadratic element, the OQO from Table I is picked out. If this were > 6, then the segment is divided into three subsegments and the OQO, for these subsegments and the CPT, are extracted from Table I. If the OQO is < 6 for a subsegment, then the integral is computed over that subsegment; if not, the subsegment is evenly divided into 2, 3, or 4 subsubsegments until all the subsubsegments have an OQO less than 6 and the integral is computed.

For nonsingular area integrals, the subdivision scheme similar to that of the line integrals involves D/S where D and S are as defined in the numerical experiment for area integrals. If D/S for the whole element is < 2.197, then the integral is computed using OQO as found in Table II. Otherwise, the element is subdivided into 9 subelements (Fig. 1) and the D/S for each of those is computed. If D/S for these subelements is < 3.633, then the integrals are computed once again using the OQO in Table II; otherwise, the subelement is divided into 1^2 or 2^2 or 3^2 subsubelements until D/S < 3.633 on each of the subsubelements and the integral is computed.

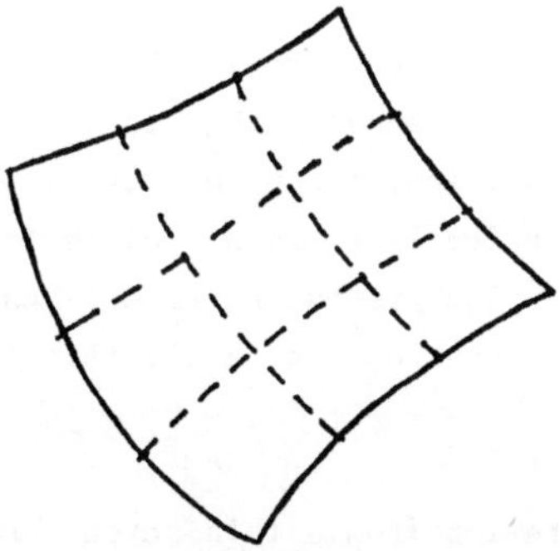

Fig. 1. Subdividing an element into 9 sub-elements.

In the case of singular area integration, the element is first divided into 9 subelements. Then the subelement in which the CPT falls is subdivided into four triangular segments so as to use polar coordinates to compute the singular integral. A Gaussian quadrature order of 5x5 is used [7] after transformation. In the rest of the subelements, the integrals are treated as in the nonsingular area integrals but now using Table III.

Numerical Example

This improved integration scheme which computes the integrals to 5 places of precision is used to solve the previously mentioned scattering problem through BEM. The zero value of the solution at the edges and the square root behavior of the solution close to the crack edges are built into the shape functions in the elements along the edges of the crack. The model is discretized by 33 elements as in [5] and a significant improvement in the result is observed as expected, but the improvement in the results at nodes close to the center is not significant.

On further experimentation, it was observed that the solution at nodes close to inter-element boundaries which were not conforming geometrically, as in the 33-element model used, was poor as the irregularities in the geometry, i.e. gaps and overlaps of adjacent elements, are magnified by the strong singularity associated with that node. Hence, a geometrically conforming model with 20 elements, which was similar to the 33-element model in radial division but had 8 elements in each of the two outer rings and four in the center, was used to solve this scattering problem again. The comparison of solution from [5] and this 20-element model with all improvements suggested in this work is as in Fig. 2.

One numerical problem not discussed here is the order of singularity of the integrand of the second area integral in Eq. (3). From the smoothness requirement on the density function, the numerator is $O(R^2)$ theoretically and often this is not true numerically at points close to the CPT due to the numerical round-off error. The present numerical scheme helps alleviate this problem and work is in progress to handle this term better.

Conclusion

When formulating the scattering from a thin screen problem using integral equations using our proposed regularization technique, a certain smoothness requirement on the density function is demanded. The regular BEM which uses conforming elements to approximate density function cannot be used. Either nonconforming elements or splines are appropriate and the former has an advantage of being simple to implement. The strong singularity associated with this integral equation demands care when performing the integration. An efficient integration routine is developed and the results thus obtained are close to the exact. Though this integration scheme used is suited for 8- and 9-node nonconforming elements, it can be used for other elements with little change. The influence of geometrically nonconforming elements on solution accuracy was discovered and properly handled.

Acknowledgment

Partial support for this work was provided by the U.S. Office of Naval Research under Contract N0014-86-K-0551.

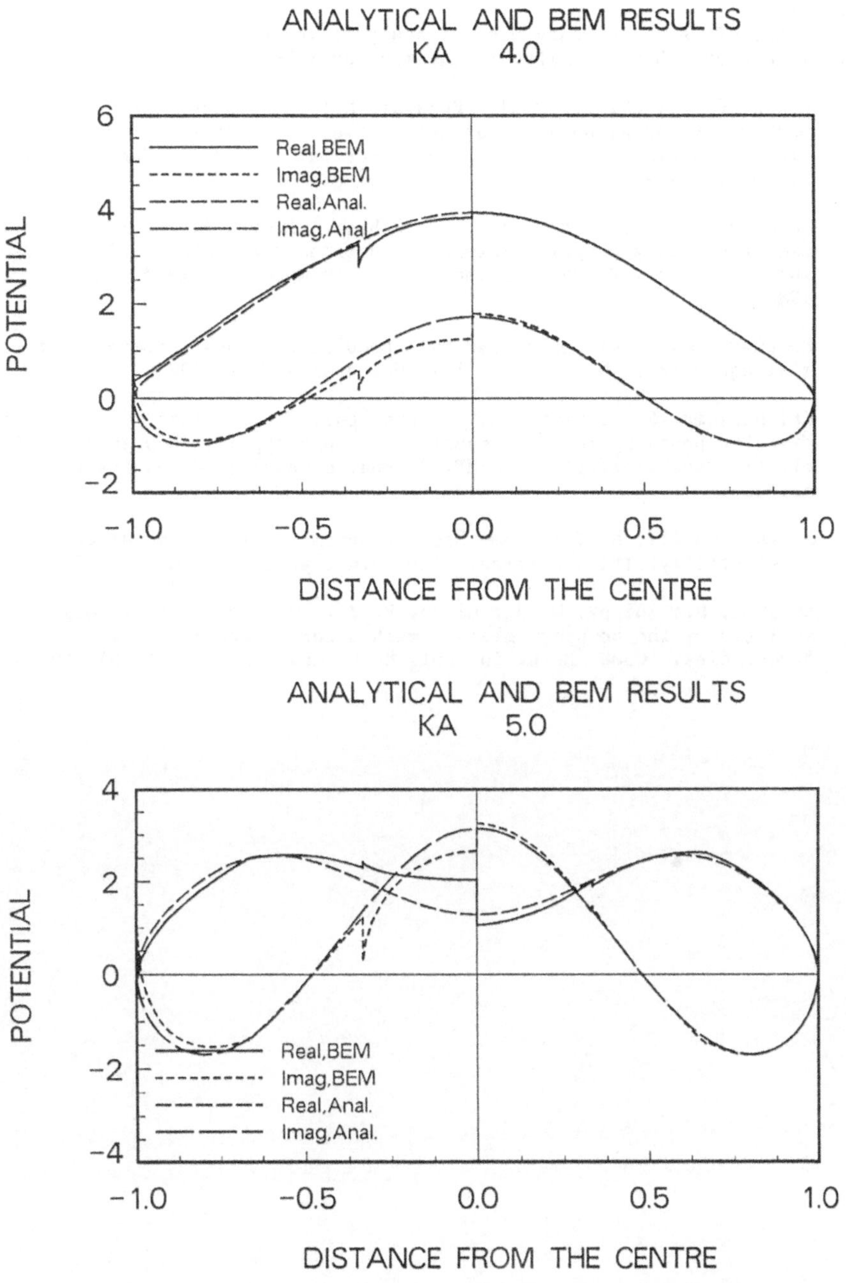

Fig. 2. The results from [5] are to the left of the origin and the results from all improvements suggested are to the right of the origin.

References

1. Cruse, T. A.: Boundary element analysis in computational fracture mechanics. Boston: Kluwer Academic Publishers 1988.

2. Rizzo, F. J.; Shippy, D. J.; Rezayat, M.: A boundary integral equation method for radiation and scattering of elastic waves in three dimensions. International Journal for Numerical Methods in Engineering 21 (1985) 115-129.

3. Rudolphi, T. J.; Krishnasamy, G.; Schmerr, L. W.; Rizzo, F. J.; On the use of strongly singular integral equations for crack problems. Proc. 10th Int. Conf. Boundary Elements. Southampton, England: September 1988.

4. Hadamard, J.; Lectures on Cauchy's problem in linear partial differential equations. New Haven: Yale University Press 1923.

5. Krishnasamy, G.; Schmerr, L. W.; Rudolphi, T. J.; Rizzo, F. J.: Hypersingular boundary integral equations: some applications in acoustic and elastic wave scattering. ASME, Journal of Applied Mechanics (to appear).

6. Muskhelishvili, N. I.: Some basic problems of the mathematical theory of elasticity, third edition. The Netherlands: Nordhoff 1953.

7. Rezayat, M.; Shippy, D. J.; Rizzo, F. J.: On time-harmonic elastic-wave analysis by the boundary element method for moderate to high frequencies. Comp. Meth. in Appl. Mech. and Engr. 55 (1986) 349-367.

FRANSYS: A Software System for the Simulation of Crack Propagation in Three Dimensions

Paul Wawrzynek, Luiz Martha & Anthony Ingraffea
Department of Structural Engineering and
Program of Computer Graphics Fracture Group
Cornell University, Ithaca, New York 14850

Summary

The FRacture ANalysis SYStem (FRANSYS) is a collection of programs developed and/or integrated into one cohesive package. The purpose of this package is to allow a person knowledgeable about fracture mechanics and numerical stress analysis to perform simulations of the behavior of structures containing cracks. This system is suitable for the simulation of structures which are three-dimensional in nature and which may exhibit changes in geometry and topology during the course of the simulation. These simulations may account for time dependent and/or nonlinear material responses.

This document provides a broad but concise overview of the system, and presents some preliminary results.

Introduction

The FRacture ANalaysis SYStem (FRANSYS) is a new software system developed for numerical investigations into three-dimensional fracture mechanics. The primary motivation for the development of the FRANSYS system arises out of the authors' frustration with using existing programs for research in the area of computational fracture mechanics.

In general, existing codes have been found to be insufficient for this task for two reasons. Firstly, the simulation of fracture processes involves geometries and topologies which evolve as a result of crack initiation and propagation. Many available codes will not allow the geometry of the problem to change during the course of the simulation. Crack motion may require a modification to the topology (mesh) as well as the geometry of the object being analyzed. In lieu of routines to update automatically the topology of an object, the analyst is often forced to perform tedious manual updates to the model for fracture analysis. A fracture simulation code must allow the topology as well as the geometry of an object to evolve during the

G. Kuhn · H. Mang (Eds.)
Discretization Methods in Structural Mechanis
IUTAM/IACM Symposium Vienna, Austria, 1989

course of an analysis and must provide automatic routines to assist in these updates.

A second insufficiency of existing programs is that they tend to have a closed software architecture. That is, the program may not be easily modified to incorporate new features and algorithms. In the case of proprietary codes, often the source code is not available for direct modification. If the source code for the program is available, it is often not a trivial task to determine where a program must be modified and what effects the modifications will have on the rest of the program. In order for a program to be a good tool for prototyping computational fracture techniques, it must be easy for one to identify where the program is to be modified and it must be modular enough so that modifications have only local impact.

Scope

The system has been designed to incorporate both finite and boundary element methods for the solution of stress analysis problems. Other techniques may fit into the FRANSYS framework but there are no current plans to support other techniques.

FRANSYS always maintains a consistent underlying solid representation of any object being modeled. FRANSYS, however, is not a solid modeler. Many solid-model-like features and algorithms are available within FRANSYS. These are necessary to support the underlying representation, but the system is not designed to be a tool for creating the initial geometric description of an object to be modeled.

FRANSYS requires as input a solid model boundary representation. This representation consists of a collection of planar and curved polygonal patches which describe the geometry of the object. This representation may include internal features such as interfaces and cracks. The boundary representation is converted to a representation internal to FRANSYS. Internally, the geometry of the model is allowed to change during the course of the simulation, but no attempt is currently made to reincorporate the updated geometry with the representation in the solid modeler. The integration of solid modelers and numerical simulators is an open and active area of research with many unanswered questions. It is anticipated that advances in this area will be incorporated into FRANSYS as they emerge.

Hardware Environment

FRANSYS requires three conceptual hardware components for operations: a workstation, a 3D graphics processor, and a compute node. These three components are defined below:

Workstation - for the present purposes, a workstation is defined as a single user computer with a 32-bit word length, at least a 1k x 1k color bitmapped terminal, a window manager, and an operating system that supports multi-tasking, virtual memory, and network communication.

3D Graphics Processor - The graphics processor is a hardware, microcode, and software package which is responsible for display transformations, clipping, culling, polygonal filling, and polygonal shading computations. It is assumed that performance requirements will dictate that the graphics processor be tightly coupled with the workstation architecture. It is also assumed that the display processor memory will either be directly mapped into the workstation virtual memory or can be quickly accessed by the workstation CPU.

Compute Node - The compute node is the most variable element of the system. It is assumed that there is available, via a network, a computer system with sufficient computational resources for the analysis tasks at hand. Three interfaces are defined between the workstation and the compute node. In a batch simulation, data and control information are sent to the compute node and simulation results are sent back to the workstation at the end of the simulation. In a listener interface, the compute node periodically downloads intermediate results to the workstation. The workstation may then display the intermediate results. In a dialogue interface, there is continuous two-way communication between the workstation and the compute node. With this interface the workstation can actually interactively guide the simulation. No restriction is placed on the compute node's size or architecture. For small simulations the compute node may be the workstation CPU.

The FRANSYS Model

Figure 1. shows the FRANSYS model of fracture simulation. Geometry is generated via an external modeling package. The boundary representation of the model is then transformed to the FRANSYS internal representation. Attributes are then assigned to the model. At this point the surface of the model is discretized. If a finite element analysis is to be performed, the volume of the model must also be discretized. An analysis is performed using either the finite element or boundary element method (in the future these may be coupled). The results of this analysis are used to update the model geometry and possibly the model topology. At this point remeshing may be required and an analysis can be performed. At any point in this process the analyst can view problem response via a suite of postprocessing tools. Underlying this system is a common internal representation for the problem geometry, topology, and attributes. Interaction with the system is via one common user interface.

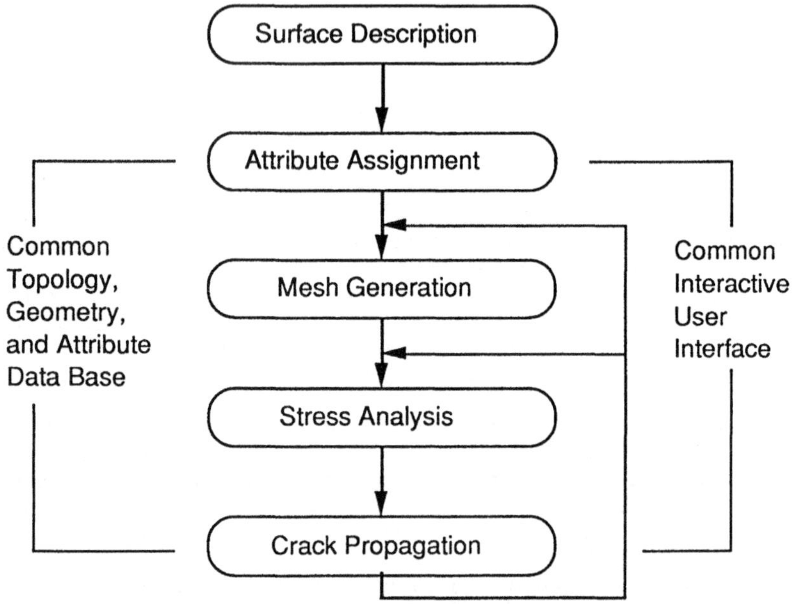

Figure 1. FRANSYS system organization

Current Results

Although the implementation of the FRANSYS system has not been completed, large segments of the system are currently functionally operable. Some of these are highlighted here.

User Interface

Viable computational mechanics software must be both useful and *useable*. That is, someone other than the author of the program should be able to use the program to perform useful calculations. In other words, the user interface should be sufficiently well developed so that a knowledgeable user can perform useful work while interacting with the computer in an intuitive and natural way. In addition, the user interface should shield the user from the superfluous idiosyncrasies and details which underlay the computational techniques.

The user interface for FRANSYS uses the multiple window/popup menu/popup dialog box paradigm which has recently become popular for most personal computer and workstation based software.

The menu system has a hierarchical structure. The user, however, may have any number and/or collection of menus on the screen at any one time. The user can freely move between program options which reside on different menus. The user can position any menu at any arbitrary location on the screen.

Three-dimensional graphics appear within a movable resizable window. Any number of three-dimensional windows can appear on the screen simultaneously (subject to hardware specific constraints). The view orientations for the three-dimensional windows are specified interactively using the toggle-pad (or using a hardware dial-box when available).

Two-dimensional graphics also appear in movable resizable windows. These window are currently used for the display of two axis graphs and contour values on planar cross-sections.

Modeling and Meshing

FRANSYS is not a solid modeler; however, many solid modeling features are provided to aid the analyst in creating a model for analysis and for the explicit geometric representation of cracks. The underlying data structure for all geometric and topological manipulation is the radial-edge data structure, developed by Weiler [1,2].

Meshing is performed by a technique of progressive hierarchical refinement. That is, the user interactively subdivides an object into progressively simpler objects until a standard meshing algorithm can be used to mesh these simpler objects. The method is hierarchical because it is carried out within the framework of a hierarchy of topological representations of the object. The topology and geometry of each level in the hierarchy is constrained by the levels above it [3].

More specifically, the true geometric representation of the object is the highest model in the hierarchy. The next level in the hierarchy is the volume subdivision level. Within this model the user can interactively decompose the object into a collection of distinct three-dimensional regions. The geometry of these regions, however, will be constrained by the geometry of the original object. The next level is a surface subdivision level. At this level the faces of the regions can be interactively decomposed into collections of distinct surface regions. The next lower level is an edge subdivision level.

At this level the edges which bound the faces of the regions can be interactively decomposed into a number of linear segments.

The mesh model is the lowest model in the hierarchy and is constrained by all the levels above it. The first step in meshing is to create a surface mesh. Both mapped meshing and arbitrary triangulation algorithms are provided. Mapping techniques are preferred because of their relative speed and their tendency to generate well formed elements. These algorithms, however, have rather stringent input requirements. When these requirements cannot be met the arbitrary techniques can be used. For boundary element analyses, surface meshes only are required. For finite elements, the surfaces of regions are meshed and then three-dimensional versions of the mapped or arbitrary algorithms can be used to generate volumetric meshes. The requirement that surfaces (including internal surfaces) be meshed first, forces the continuity of the mesh between independent regions.

Visualization

At this point the visualization capabilities of FRANSYS are incomplete. However, some significant capabilities do exist. The capabilities which do exist fall into three categories: deformations, surface quantity plots, and color surface contours.

The deformations of an object are displayed as either a deformed solid, wire-frame mesh, or wire-frame geometric outline. These representations are displayed in a three-dimensional window so the user can interactively manipulate the view of the deformed object. The object can be displayed either alone, with a wire-frame of the undeformed object, or with a solid representation of the undeformed object.

Various response quantities (stresses, displacements, tractions, principal stresses, etc...) can be plotted on a two axis graph (and x-y plot). The responses are evaluated along a line specified by the user and lying on the surface of the object.

Color surface contours allow the user to plot surface quantities as discrete contour ranges displayed on the geometry of either the original or the deformed geometry of the object. The contoured image appears in its own three-dimensional window which can be manipulated independently. In addition, while only a relatively small number of colors are used for contour ranges, a large number of shades of these colors are displayed so that the object maintains its three-dimensional appearance.

Fracture

By definition the fracture capabilities of FRANSYS are incomplete. This is because a significant object of the FRANSYS system was to create a testbed for new fracture theories and techniques. Three-dimensional computational fracture mechanics, and the area of crack propagation in particular, is in its infancy. The fracture literature contains only a small number of references to crack propagation prediction and simulation for arbitrary crack surfaces.

FRANSYS currently contains the following fracture capabilities:

- Arbitrary fracture surfaces can be defined and represented within the FRANSYS system. These fracture surfaces can be either internal cracks or surface cracks. The fracture surfaces can be specified either initially as part of the geometric specification of a new problem, or can be created interactively.

- Stress-intensity-factors can be computed and plotted for a discrete number of points along the crack front.

- The direction of crack propagation can be computed at a number of points along the crack front. The prediction of the direction is based on the distribution of the circumferential stress in the region around the crack front.

- For internal cracks, a new boundary element method based on a hyper-singular integrals has been implemented [4,5]. This technique is effecient and avoids the necessity of multi-domain analysis.

Current activity includes the incorporation of capabilities to predict the relative amount of propagation for a discrete number of points along the crack front. These techniques are based on a generalization of the Griffith fracture criteria. It also includes the generalization of the hyper-singular approach for surface cracks.

Illustrative Example

As an illustrative example, consider the aircraft component shown in Figure 2. This is a door-latch fitting which has been observed to exhibit failure due to fatigue cracking. Figure 2a shows a shaded view of the initial geometry of the object. Figure 2b shows the initial boundary element mesh used to analyze the problem.

Figure 3 shows the object after a number of steps of crack propagation. The initial flaw was semi-circular in shape. As the crack propagates, it evolves into a through-the-thickness crack. The three-dimen-

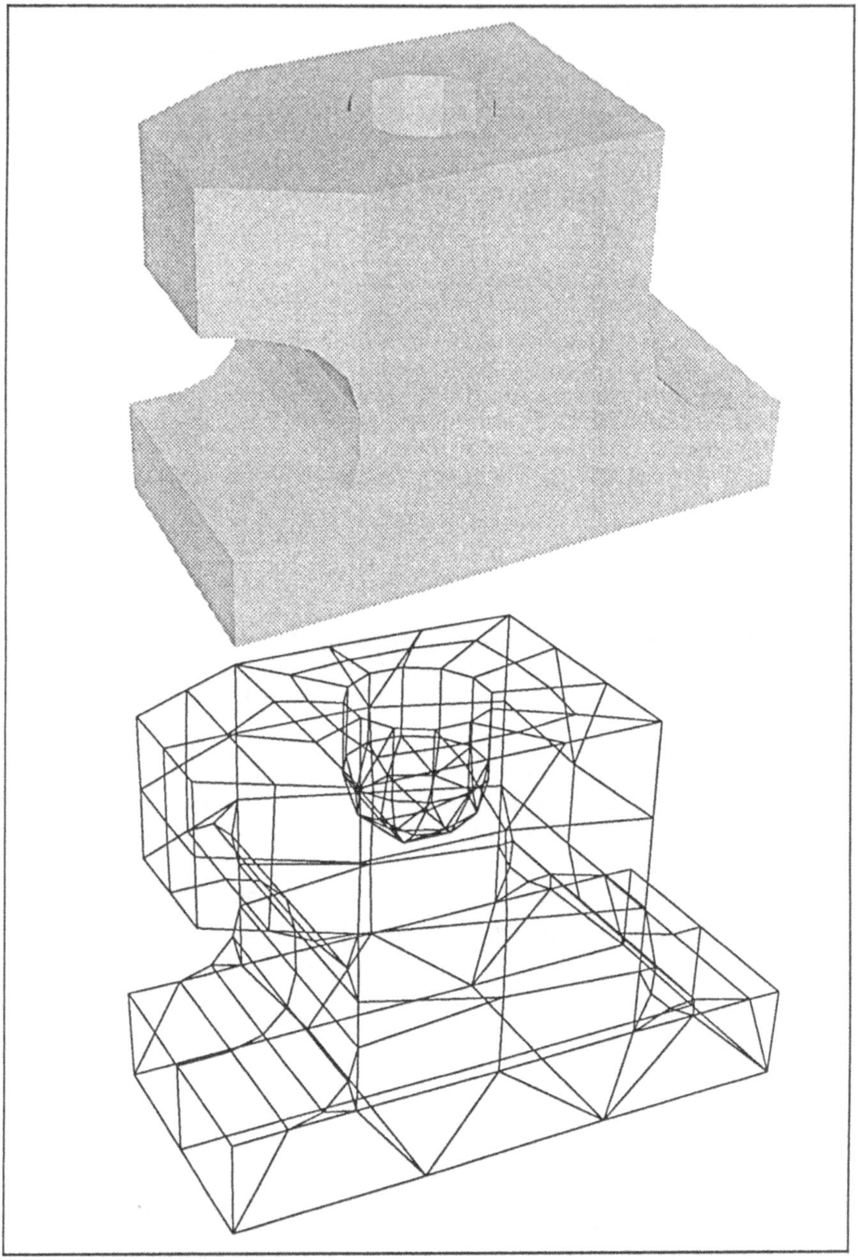

Figure 2. Shaded view and initial mesh configuration of an aircraft fitting.

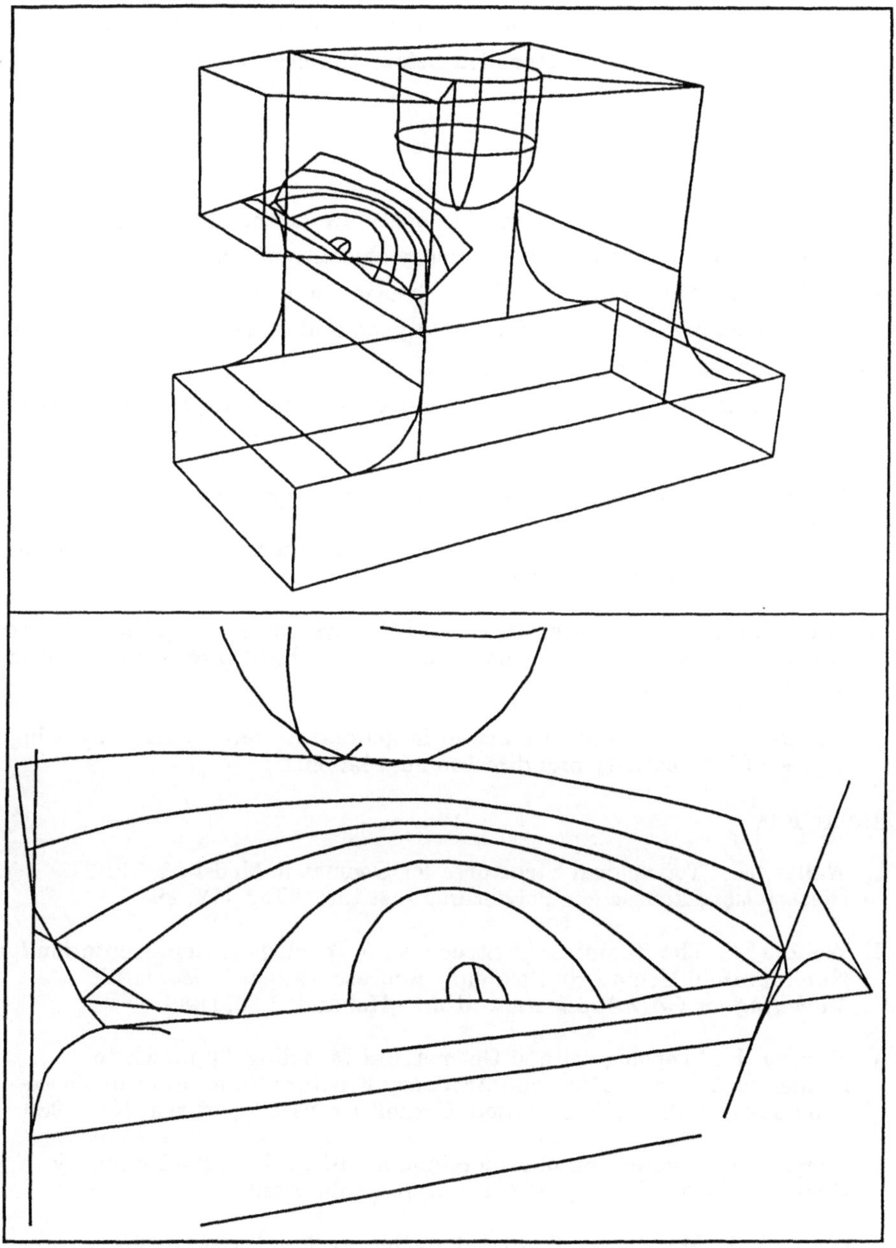

Figure 3. Crack geometry evolution of the aircraft fitting.

sionality of the problem is important, because a significant amount of the fatigue life of the part is expended before the crack is amenable to two-dimensional analysis.

Summary

This paper provides a concise overview of the FRANSYS fracture simulation software system. The primary objective of FRANSYS is to create a test bed in which new algorithms, theories, and ideas in computational fracture mechanics can be quickly prototyped and checked with both simple and complex three-dimensional models.

Some of the system's salient features, outlined in previous sections, are:

- The system incorporates both finite and boundary element methods.
- The system always maintains a consistent underlying solid representation of any object being modeled.
- The vehicle for user interaction with the system is a high-performance engineering workstation which incorporates hardware assisted three-dimensional graphics.
- The user interface and interaction in general is characterized by a high degree of interactivity and direct manipulation.

References

1. Weiler, K., "Topological Structures for Geometric Modeling," Ph.D. Dissertation, Rensselaer Polytechnic Institute, Troy, NY, 1986.

2. Weiler, K., "The Radial-Edge Structure: A Topological Representation for Non-Manifold Geometric Boundary Representations," *Geometric Modelling for CAD Applications*, North Holland, 3-36, 1988.

3. Martha, L., "Topological and Geometrical Modeling Approach to Numerical Discretization and Arbitrary Fracture Simulation in Three-Dimensions," Ph.D. Dissertation, Cornell University, Ithaca, NY, 1989.

4. Gray, L. J., "Evaluation of Hypersingular Integrals in the Boundary Element Method," *Comput. Math. Appl.*, submitted.

5. Gray, L. J., Martha, L. F., and Ingraffea, A. R., "Hypersingular Integrals in Boundary Element Fracture Analysis," *Int. J. Num. Meth. Engng.*, submitted.

Design Sensitivity Coefficients for Linear Elasticity Problems by the Derivative Boundary Element Method

QING ZHANG AND SUBRATA MUKHERJEE
Department of Theoretical and Applied Mechanics
Cornell University
Ithaca, New York 14853, U.S.A.

Summary

This paper discusses the determination of design sensitivity coefficients (DSC's), for planar elasticity problems, by the derivative boundary element method (DBEM). It is shown that this is an efficient and accurate method for the determination of DSC's – especially of stresses on the boundary of a body. Numerical results are presented for two illustrative problems for which analytical solutions are available. These numerical solutions compare very well with the analytical solutions.

Introduction

The subject of this paper is the accurate and efficient determination of design sensitivity coefficients (DSC's), for 2–D linear elasticity problems, by the boundary element method (BEM). DSC's are the rates of quantities, such as displacements, tractions or stresses, in a boundary value problem, with respect to a design variable. A design variable being considered here is a shape parameter that controls the shape of part or whole of the boundary of a body. The DSC's, which are therefore gradients of variables of interest with respect to the shape of a body, are then used in nonlinear programming algorithms which typically iterate on the shape of the object until an optimal shape is obtained.

The approach being used here is the direct analytical differentiation (DDA) of the governing boundary element method (BEM) formulation of the problem. The exact differentiation eliminates errors that might occur from finite difference methods and leads to closed form integral equations for the desired sensitivities. These equations are then solved by numerical discretization. This approach is very accurate and efficient.

There are some papers in the literature that determine DSC's by DDA of the BEM formulation of a problem. Barone and Yang [1] have used this approach for 2D linear elasticity and have solved a simple example of an ellipse without corners. Rice and Mukherjee [2] have solved for axisymmetric elasticity DSC's while Kane and Saigal [3] Saigal, Borggaard and Kane [4] and Aithal, Saigal and Mukherjee [5] have solved planar axisymmetric and some 3D problems, respectively. In refs. [3–5] the authors first discretize the BEM equations and then differentiate them analytically with respect to

G. Kuhn · H. Mang (Eds.)
Discretization Methods in Structural Mechanis
IUTAM/IACM Symposium Vienna, Austria, 1989

shape variables, while the authors in refs. [1] and [2] follow the opposite approach. It seems more appealing, intuitively, to start with differentiation of the relevant BEM equations and follow this by discretization. The two approaches do, as expected, eventually lead to the same equations.

Two difficult problems in this area of research are the accurate determination of sensitivities of boundary stresses and the determination of displacements, tractions, stresses and their sensitivities at corners on the boundary of a body. The finite element method (FEM) has problems with the determination of accurate boundary stresses while the approach of Barone and Yang [1] requires the integration of strongly singular kernels in order to obtain these quantities. The problem of DSC's for stresses, at a point on a corner of the boundary of a body, has not been solved in an elegant manner before this work.

Ghosh et al. [6,7], in two recent papers, have presented a BEM formulation in which the tractions and displacement derivatives are primary boundary variables. This formulation only has logarithmically singular kernels for 2–D problems. *The boundary stresses can be obtained from the boundary values of tractions and displacement derivatives by purely algebraic calculations.* While the original work [6] used nonconforming elements at corners, the present work uses conforming elements, so that a source point can be placed directly at a corner.

The present work, which is based on DDA of the derivative BEM formulation [6], makes two important contributions to the literature in this field. The first is the elegant and accurate determination of stress sensitivities at a regular point on the boundary of a body. The BEM formulation [6] is analytically differentiated with respect to a shape parameter to yield an integral equation for the sensitivities of tractions and tangential derivatives of displacements on the boundary of the body. The new differentiated kernels are completely regular for 2D problems! Then the boundary stress sensitivities are obtained directly as linear combinations of the sensitivities of tractions and tangential derivatives of displacements. *The accuracy of stress sensitivities at a boundary point, therefore, is of the same order as that of the sensitivities of tractions and displacement derivatives.*

The second important contribution is in the treatment of corners. New equations for stress jumps at corners, based on the theory of elasticity, have been recently derived by the present authors [8]. *These equations, together with their sensitivities, permit the accurate and efficient evaluation of stresses and their sensitivities at corners.*

The above ideas have been implemented in a computer program for the determination of DSC's for planar elasticity problems. Numerical results are presented in this paper for DSC's for planar bodies without and with corners. A very interesting example is the determination of the sensitivity of stresses, at the tip of a wedge, with respect to the wedge angle. The numerical results reported here are generally very accurate.

A DBEM Formulation for Planar Elasticity

Ghosh et al. have recently proposed a derivative boundary element method (DBEM) formulation for linear elasticity in which the tractions and tangential derivatives of displacements ([6] for 2D problems) or tractions and displacement gradients ([7] for 3D problems) are the primary variables on the boundary of a body. A similar formulation has been presented also by Okada, Rajiyah and Atluri [9].

Following ref. [6], the BEM equations for two-dimensional linear elasticity for a simply connected region B can be written as:

$$\int_{\partial B} [U_{ij}(P,Q)\tau_i(Q) - W_{ij}(P,Q)\Delta_i(Q)]ds_Q = 0 \tag{1}$$

where U_{ij} is available in many references (eg. [10]) and

$$W_{ij} = \frac{1}{4\pi(1-\nu)} \left[2(1-\nu)\psi\delta_{ij} + \epsilon_{ik} r_{,j} r_{,k} + (1-2\nu)\epsilon_{ji} \ell n r \right]$$

in terms of the source and field points P and Q (with capital letters denoting points on ∂B and lower case letters denoting points inside B), r the Eucledian distance between P and Q and ψ the angle between the vector r(P,Q) and a reference direction. Also, τ_i and Δ_i are the components of the traction and tangential derivative of the displacement $\left[\frac{\partial u_i}{\partial s}\right]$, respectively, with s the curvilinear coordinate measured along the boundary ∂B of the planar body. Finally, $\epsilon_{11} = \epsilon_{22} = 0$, $\epsilon_{12} = -\epsilon_{21} = 1$, ν is the Poisson's ratio and δ_{ij} is the Kronecker delta. A comma following r denotes a derivative with respect to a field point coordinate. It is very important to note that W_{ij} has only a logarithmic singularity (same as U_{ij}) as $r \rightarrow 0$. A constraint equation

$$\int_{\partial B_1} \Delta_i ds = u_i(2) - u_i(1)$$

(where ∂B_1 is a suitable part of ∂B with $u_i(1)$ and $u_i(2)$ the values of u_i at the beginning and at the end of ∂B_1) must be included for certain problems.

As can be seen from equation (1), the traction and tangential displacement derivative vectors are the primary unknowns on ∂B in this formulation. It has been shown by Sladek and Sladek [11] and Cruse and Vanburen [12] that the stress components at a regular point on ∂B, for plane strain, can be written in terms of the components of τ and Δ as

$$\sigma_{ij} = A_{ijk}\tau_k + B_{ijk}\Delta_k \tag{2}$$

where

$$A_{ijk} = (n_i n_j + C_1 t_i t_j)n_k + (n_i t_j + n_j t_i)t_k$$

$$B_{ijk} = C_2 t_i t_j t_k$$

with $C_1 = \nu/(1-\nu)$, $C_2 = 2G/(1-\nu)$ and G the shear modulus of the material. Also, n_i and t_i are the components of the unit (outward) normal and (anticlockwise) tangent vectors at a point on ∂B. Thus, if τ and Δ are primitive variables on ∂B in a BEM formulation, then these quantities, as well as σ_{ij}, can be obtained on ∂B with very high accuracy.

The corresponding DBEM equations for the sensitivities are obtained by differentiating equation (1) with respect to a shape design variable 'b' (see [1–4], [5], [13])

$$\int_{\partial B} \left[U_{ij}(b,P,Q)\overset{*}{\tau}_i(b,Q) - W_{ij}(b,P,Q)\overset{*}{\Delta}_i(b,Q)\right] ds(b,Q)$$

$$+ \int_{\partial B} \left[\overset{*}{U}_{ij}(b,P,Q)\tau_i(b,Q) - \overset{*}{W}_{ij}(b,P,Q)\Delta_i(b,Q)\right] ds(b,Q)$$

$$+ \int_{\partial B} \left[U_{ij}(b,P,Q)\tau_i(b,Q) - W_{ij}(b,P,Q))\Delta_i(b,Q) \right] d\overset{*}{s}(b,Q) = 0 \qquad (3)$$

where a superscribed * denotes a derivative with respect to 'b'. As shown by Barone and Yang [1],

$$\overset{*}{U}_{ij}(b,P,Q) = U_{ij,k}(b,P,Q)[\overset{*}{x}_k(Q)-\overset{*}{x}_k(P)] \qquad (4)$$

where, by virtue of the fact that

$$\overset{*}{x}_k(Q)-\overset{*}{x}_k(P) \approx O(r)$$

$\overset{*}{U}_{ij}$ *is completely regular*! A similar argument is used to show that $\overset{*}{W}_{ij}$ *is also regular*. Finally,

$$d\overset{*}{s}(b,Q) = [\overset{*}{x}_{k,k}(Q)-n_i n_j \overset{*}{x}_{i,j}(Q)]ds(b,Q) \qquad (5)$$

A very interesting feature of equation (3) is that its first line is identical to equation (1) with the sensitivities replacing the tractions and displacement derivatives. Analogous to the usual BEM problem, half of the sensitivities on ∂B must be prescribed and the rest can then be determined from equation (5). *Thus, the sensitivity problem has the same coefficient matrices as the original BEM problem with a known right hand side* (since τ and Δ on ∂B are known at this stage). This known right hand side involves the evaluation of regular integrals which is very easy to perform accurately.

The equation for the sensitivity of stress at a regular point on ∂B is obtained by differentiating equation (2) with respect to the design variable 'b'

$$\overset{*}{\sigma}_{ij} = A_{ijk}\overset{*}{\tau}_k + B_{ijk}\overset{*}{\Delta}_k + \overset{*}{A}_{ijk}\tau_k + \overset{*}{B}_{ijk}\Delta_k \qquad (6)$$

The above equation expresses $\overset{*}{\sigma}_{ij}$ as a linear combination of Δ_k, τ_k and their sensitivities. Hence, one expects $\overset{*}{\sigma}_{ij}$ to be obtained as accurately as $\overset{*}{\tau}_i$ and $\overset{*}{\Delta}_i$.

Corners

A point on a corner on ∂B is one across which there is the jump in the unit vectors n and ι which are normal and tangential to ∂B. Consequently, the quantities of interest, τ, Δ and σ, in general, experience jump discontinuities across a corner point. Therefore, as seen in Figure 1, eight quantities are of interest at a corner, only four of which are prescribed from the boundary conditions. If a source point P is placed exactly at a corner (conforming boundary elements) one obtains two BEM equations at P, but two more independent equations are still necessary. Fortunately, these equations can be obtained by considering the behavior of stress components at a corner point.

The corner issue is addressed in detail in a recent paper by the same authors [8]. The important ideas are presented below.

The first step is to write equation (2) in terms of local variables, taking account of the jump discontinuities of the variables. Next, the elasticity problem of a wedge, loaded by uniform tractions on its faces, is solved in order to determine, explicitly, the stress jumps at the tip of the wedge [14,8]. These jumps are obtained in terms of the geometry (wedge angle) and the tractions τ_n^+, τ_n^-, τ_s^+ and τ_s^- on the wedge faces (Figure 1).

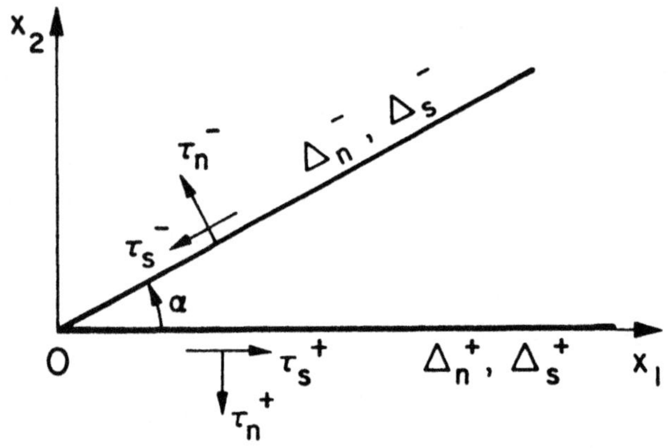

Fig. 1. The elasticity problem in a wedge

The resulting three equations are

$$\mathbf{t}^+ \cdot [\sigma^+ - \sigma^-] \cdot \mathbf{t}^+ = C_1 \tau_n^+ + C_2 \Delta_s^+$$

$$- (\sin^2\alpha + C_1 \cos^2\alpha) \tau_n^-$$

$$- 2(\sin\alpha \cos\alpha) \tau_s^- - C_2 (\cos^2\alpha) \Delta s^-$$

$$= \frac{\alpha + \sin\alpha \cos\alpha}{\alpha - \tan\alpha} f(\tau, \alpha) \tag{7}$$

$$\mathbf{n}^+ \cdot [\sigma^+ - \sigma^-] \cdot \mathbf{n}^+ = \tau_n^+ - (\cos^2\alpha + C_1 \sin^2\alpha) \tau_n^-$$

$$+ 2(\sin\alpha \cos\alpha)\ \tau_s^- - C_2 (\sin^2\alpha) \Delta_s^-$$

$$= \frac{\alpha - \sin\alpha \cos\alpha}{\alpha - \tan\alpha} f(\tau, \alpha) \tag{8}$$

$$-\mathbf{t}^+ \cdot [\sigma^+ - \sigma^-] \cdot \mathbf{n}^+ = -\tau_s^+ - (C_1 - 1)(\sin\alpha \cos\alpha) \tau_n^-$$

$$- (\sin^2\alpha - \cos^2\alpha) \tau_s^- - (C_2 \sin\alpha \cos\alpha) \Delta_s^-$$

$$= \frac{\sin^2\alpha}{\alpha - \tan\alpha} f(\tau, \alpha)$$

where $f(\tau,\alpha) = (\tau_n^+ - \tau_n^-) + (\tau_s^+ + \tau_s^-)\tan\alpha$ (9)

In the actual BEM modelling, of course, only the expressions on the right hand sides of the first equal signs (i.e. relationships between the variables of interest in Figure 1) of equations (7–9) are used. When these equations are added on to the BEM equations at a corner point P on ∂B, an overdetermined system of global equations result. This system, however, is consistent and has been solved in a least squares sense [15] in order to obtain the numerical results that follow.

For $\alpha = \pi/2$, invoking of the principle of complementary shearing stresses, $\tau_s^+ + \tau_s^- = 0$, leads to continuity of stresses at a corner point. In this case, the equations (7–9) simplify to

$$C_1\tau_n^+ + C_2\Delta_s^+ - \tau_n^- = 0 \tag{10}$$

$$\tau_n^+ - C_1\tau_n^- - C_2\Delta_s^- = 0 \tag{11}$$

Now one gets an equal number of equations and unknowns.

The corresponding sensitivity equations are obtained by differentiating equations (7–9) with respect to the design variable 'b'. This would lead to three equations relating sensitivities to the components of τ and Δ and the geometrical quantities in the problem.

These expressions are somewhat lengthy. For illustrative purposes, the simpler expression for the sensitivity of the jump of the first invariant $I = \mathrm{trace}(\sigma)$ is given below. This is

$$\begin{aligned}\Delta\overset{*}{I} &= (C_1+1)\overset{*}{\tau}{}_n^+ + C_2\overset{*}{\Delta}{}_s^+ - (C_1+1)\overset{*}{\tau}{}_n^- - C_2\overset{*}{\Delta}{}_s^- \\ &= \frac{2\overset{*}{\alpha}}{(\alpha\cos\alpha-\sin\alpha)^2}\Big[(\alpha-\sin\alpha\cos\alpha)\Delta\tau_n + (\alpha^2-\sin^2\alpha)(\tau_s^+ + \tau_s^-)\Big] \\ &\quad + \frac{2\alpha}{\alpha-\tan\alpha}\Big[\Delta\overset{*}{\tau}{}_n + (\overset{*}{\tau}{}_s^+ + \overset{*}{\tau}{}_s^-)\tan\alpha\Big]\end{aligned} \tag{12}$$

where $\Delta\tau_n = \tau_n^+ - \tau_n^-$ and $\Delta\overset{*}{\tau}{}_n = \overset{*}{\tau}{}_n^+ - \overset{*}{\tau}{}_n^-$

Discretization of Equations

The BEM equations (1) (for τ and Δ) and (3) (for their sensitivities), are discretized in the usual way. The boundary ∂B is subdivided into N_b piecewise quadratic, conforming boundary elements. The variables τ_i and Δ_i are assumed to be piecewise quadratic on these boundary elements. The logarithmically singular kernels are integrated by using log weighted Gaussian integration. When the corner equations (if any) are added to the usual BEM equations, and all the equations are assembled together, the resulting systems are of the form

$$[A]\{\tau\} + [B]\{\Delta\} = \{C_1\} \tag{13}$$

$$[A]\{\overset{*}{\tau}\} + B\{\overset{*}{\Delta}\} = \{C_2\} \tag{14}$$

and after switching appropriate columns one obtains, for the unknowns $\{x\}$ and $\{\overset{*}{x}\}$ on ∂B

$$[K]\{x\} = \{f_1\} \tag{15}$$

$$[K]\{\overset{*}{x}\} = \{f_2\} \tag{16}$$

Two points deserve mention here. First, the equations (15) and (16) have the same stiffness matrix [K]. The vector $\{f_2\}$, of course, contains the contributions from the second and third lines of equation (3). Second, equations (15) and (16) are overdetermined but consistent. They have been solved by the method of least squares [15] in the numerical examples that follow.

Numerical Results

All the numerical results discussed below are for plane strain with Poisson's ratio ν = 0.3.

The first problem considered is that of an ellipse subjected to the boundary tractions (Figure 2)

$$\tau_1 = 0,\ \tau_2 = n_2$$

which result from removing the ellipse from a rectangular body subjected to unit stress σ_{22}. The idea here is to consider a problem with a closed form solution. Here, of course, the uniform state of stress is preserved if an ellipse with a larger 'b' is cut out of the plate, so that

$$\overset{*}{\sigma}_{22} = \frac{\partial \sigma_{22}}{\partial b} = 0.$$

The sensitivities $\overset{*}{\Delta}_1$ and $\overset{*}{\Delta}_2$ on the surface of the ellipse can be shown to be

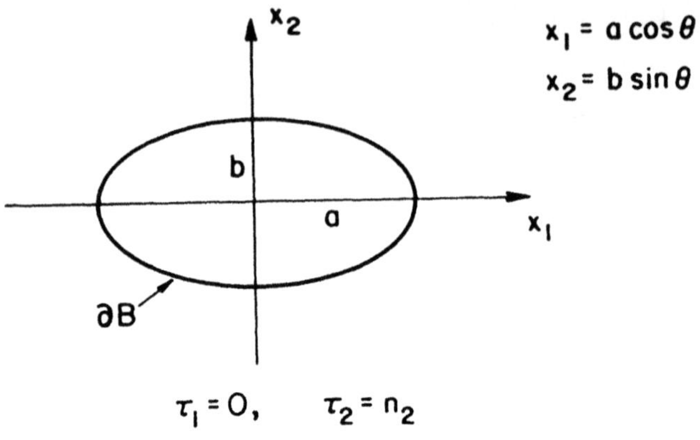

Fig. 2. Ellipse with applied boundary tractions.

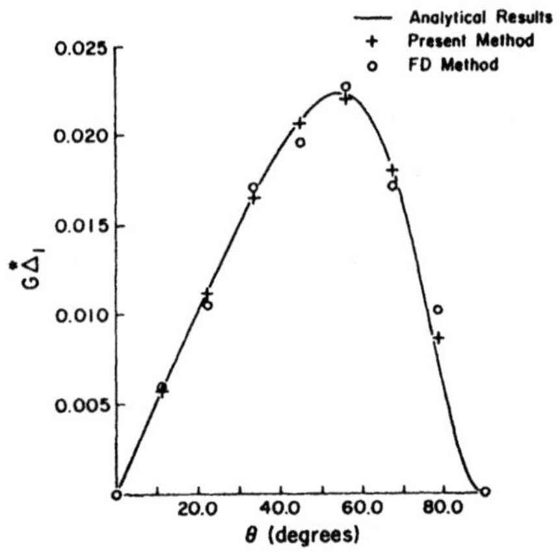

Fig. 3. Angular variation of $G\overset{*}{\Delta}_1$ around the ellipse

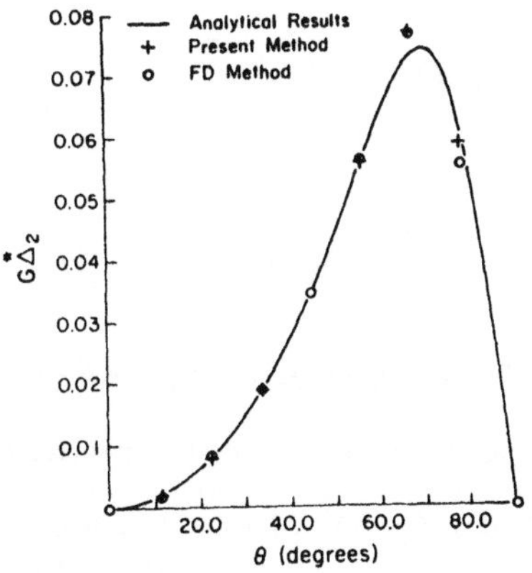

Fig. 4. Angular variation of $G\overset{*}{\Delta}_2$ around the ellipse

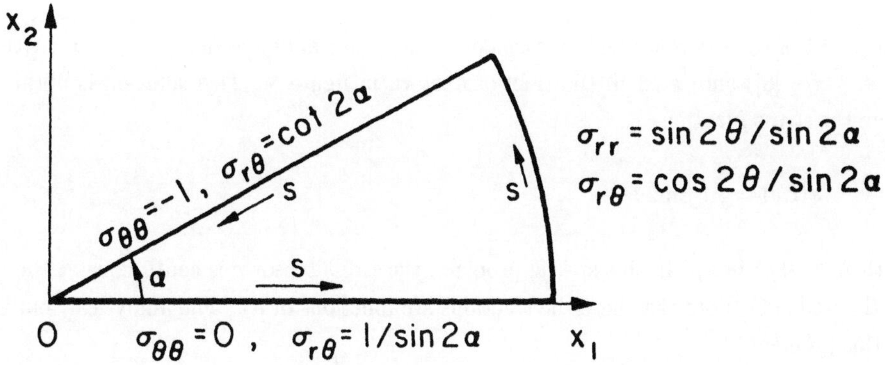

Fig. 5. A special elasticity problem

$$\overset{*}{\Delta}_1 = -\frac{\nu ab \sin\theta \cos^2\theta}{E(a^2\sin^2\theta+b^2\cos^2\theta)^{3/2}}$$

$$\overset{*}{\Delta}_2 = -\frac{1}{E}\left[\frac{b^2\cos^3\theta}{(a^2\sin^2\theta+b^2\cos^2\theta)^{3/2}} - \frac{\cos\theta}{(a^2\sin^2\theta+b^2\cos^2\theta)^{1/2}}\right]$$

A comparison of analytical, finite difference (FD) and the present method, for $G\overset{*}{\Delta}_1$ and $G\overset{*}{\Delta}_2$ (with G the shear modulus) as functions of θ, over a quarter of the ellipse, appear in figures 3 and 4. The FD results are obtained from two neighboring BEM analyses with $\Delta b = 0.001$. (The base problem has a = 2, b = 1). Other FD calculations with $\Delta b = 0.01$ and 0.0001 gave poorer results. The present method uses 16 quadratic boundary elements over the full ellipse. The results from the present method are seen to be very accurate over the entire region.

The sensitivities $\overset{*}{\sigma}_{22}$, which should be zero, are in the range of 10^{-3} to 10^{-4} from the present method.

The effect of corners is studied by considering the problem of a wedge of angle α, (with $\alpha < \pi$, $\neq \pi/2$ [8]) subjected to the tractions shown in figure 5. This solution is obtained from the stress function

$$\phi(r,\theta) = Ar^2\sin2\theta$$

with $A = -1/2\sin2\alpha$. In this special problem, the stress tensor σ is continuous at the tip of the wedge O (note that here the tractions are functions of α). The analytical solution of this problem is

$$\sigma_{11} = \sigma_{22} = 0, \ \sigma_{12} = 1/\sin2\alpha$$

so that

$$\overset{*}{\sigma}_{12} = \frac{\partial\sigma_{12}}{\partial\alpha} = -\frac{2\cos2\alpha}{\sin^2 2\alpha}, \ \overset{*}{\sigma}_{11} = \overset{*}{\sigma}_{22} = 0.$$

This example provides an opportunity to test the present method for the determination of the stress at the tip of a wedge, and its sensitivity with respect to the wedge angle.

In solving this example, it is assumed that the curved boundary of ∂B is stretched uniformly when α is increased, i.e. $\frac{d\theta}{d\alpha} = \frac{\theta}{\alpha}$. This gives

$$\frac{d\overset{*}{s}}{ds} = \frac{1}{\alpha}$$

on the curved surface. Of course, $d\overset{*}{s} = 0$ on the straight faces which undergo rigid body rotation.

The numerical results for the stress components and their sensitivities at O are given in Table 1. It is quite remarkable that *the numerical result for* $\overset{*}{\sigma}_{12}$ *captures 5 significant digits of the analytical solution* with only three quadratic boundary elements used to model the boundary ∂B!

	σ_{11}	σ_{22}	σ_{12}
Analytical σ_{ij}	0.0000000	0.0000000	1.1547005
Numerical σ_{ij}	-0.2115977E-5	0.1675500E-6	1.1546976
% Error	/	/	0.00025
Analytical σ^*_{ij}	0.0000000	0.0000000	-1.3333333
Numerical σ^*_{ij}	0.2448821E-4	0.1550109E-4	-1.3333032
% Error	/	/	0.0030

Table 1. Stress components and their sensitivities at the tip of the wedge.

Conclusions

This paper demonstrates the usefulness of the derivative boundary element method (DBEM) to obtain design sensitivity coefficients (DSC's) for linear elasticity problems. Some of the important advantages of this approach are listed below:

1. Analytical differentiation of the BEM equations avoids the need for finite differencing.

2. The BEM equation for sensitivities has the same stiffness matrix as the original equations for the regular variables. The order of the singularities of the kernels is not increased by differentiation with respect to the shape design parameter.

3. The DBEM is particularly well suited for accurate determination of boundary stress components and their sensitivities.

4. Corners are modelled accurately through the use of stress jump equations. It is particularly encouraging that the stress sensitivity at a corner point is obtained almost perfectly in an example.

Work is in progress on extending this work to materially [13] as well as geometrically [16] nonlinear problems. Also, a DBEM formulation for 3D problems exists [7], so this work can be extended to 3–D problems as well.

Acknowledgements

This research has been supported by grant number MSM–8609391, of the National Science Foundation, to Cornell University. Partial support of the U.S. Army Research Office, through the Mathematical Sciences Institute at Cornell University, is also acknowledged.

References

1. Barone, M.R.; Yang, R.–J.: Boundary integral equations for recovery of design sensitivities in shape optimization. AIAA Journal, 26 (1988) 589–594
2. Rice, J.R.; Mukherjee, S.: Design sensitivity coefficients for axisymmetric elasticity problems by boundary element methods. Submitted for publication.

3. Kane, J.H.; Saigal, S.: Design sensitivity analysis of solids using BEM. ASCE Journal of Engineering Mechanics 14 (1988) 1703–1722.

4. Saigal, S.; Borggaard, J.T.; Kane, J.H.: Boundary element implicit differentiation equations for design sensitivities of axisymmetric structures. International Journal of Solids and Structures. (in press).

5. Aithal, R.; Saigal, S.; Mukherjee, S.: Three–dimensional boundary element implicit differentiation formulation for design sensitivity analysis. Computing and Mathematics with Applications (in press).

6. Ghosh, N.; Rajiyah, H.; Ghosh, S.; Mukherjee, S.: A new boundary element method formulation for linear elasticity. ASME Journal of Applied Mechanics 53 (1986) 69–76.

7. Ghosh, N.; Mukherjee, S.: A new boundary element method formulation for three–dimensional problems in linear elasticity. Acta Mechanics 67 (1987) 107–119.

8. Mukherjee, S.; Zhang, Q.: The modelling on corners for regular and sensitivity analysis and its relationship to the derivative boundary element method. Submitted for publication.

9. Okada, H.; Rajiyah, H.; Atluri, S.N.: A novel displacement gradient boundary element method for elastic stress analysis with high accuracy. ASME Journal of Applied Mechanics 55 (1988) 786–794.

10. Mukherjee, S.: Boundary Element Methods in Creep and Fracture. London. Elsevier Applied Science 1982.

11. Sladek, J.; Sladek, V.: Computation of stresses by BEM in 2–D Elastostatics. Acta Techanica CSAV31 (1986) 523–531.

12. Cruse, T.A.; Vanburen, W.: Three–dimensional elastic stress analysis of a fracture specimen with an edge crack. International Journal of Fracture Mechanics 7 (1971) 1–15.

13. Mukherjee, S.; Chandra, A.: A boundary element formulation for design sensitivities in materially nonlinear problems.ACTA Mechanica (in press).

14. Timoshenko, S.P.; Goodier, J.N. Theory of Elasticity. 3rd ed. New York, McGraw–Hill 1970.

15. Dongarra, J.J.; Moler, C.B.; Bunch, J.R.; Stewart, G.W.: Linpack Users' Guide, SIAM, Philadelphia 1979.

16 Mukherjee, S.; Chandra, A. A boundary element formulation for design sensitivities in problems involving both geometric and material nonlinearities. Computing and Mathematics with Applications (in press).

The Panel Clustering Method for BEM

W. Hackbusch

Institut für Informatik und Praktische Mathematik
Christian-Albrechts-Universität zu Kiel
Olshausenstr. 40
D-2300 Kiel, FRG

Summary

The boundary element method (BEM) leads to a system of linear equations with a full matrix, while the FEM yields sparse matrices. This fact seems to require much computational work for the definition of the matrix, for the solution of the system, and, in particular, for the matrix-vector multiplication, which always occurs as an elementary operation. In this paper a method for the approximate matrix-vector multiplication is described which requires much less arithmetical work. In addition, the storage requirements are strongly reduced.

1. Introduction

We consider discrete analogues of general integral operators

(1.1) $$(Ku)(x) = \int_{\Gamma} k(x,y)u(y)do_y,$$

which may arise in Fredholm equations

(1.2) $$\lambda u = Ku + f$$

of the first ($\lambda=0$) or second kind ($\lambda \neq 0$). Since the results of this paper are important in particular for boundary element methods (BEM), we shall assume that Γ is a $(d-1)$-dimensional manifold in $\mathbb{R}^d$:

(1.3) $$\Gamma = \partial\Omega, \quad \Omega \subset \mathbb{R}^d.$$

Given n boundary elements (panels) we shall discretise the integral operator K by

(1.6) $$(K_n u_n)(x) = \int_{\Gamma_n} k_n(x,y)u_n(y)do_y,$$

where u_n is restricted to a q-dimensional subspace of functions ($q=q_n$). The collocation method reduces the integral equation (1.2) to a linear system

(1.7) $$Ba = b$$

with vectors $a, b \in \mathbb{R}^q$ and a $q \times q$-matrix B. In order to simplify the notation, we restrict ourselves to the collocation method.

The matrix B in (1.7) is a *full* matrix and therefore the usual matrix multiplication $a \mapsto Ba$ requires $O(q^2)$ arithmetical operations. The matrix multiplication arises e.g. in the iterative solution of Eq. (1.7) (cf. [2]) or in the evaluation of the integral. Often this fact is put forward as a drawback of the BEM compared with field methods leading to sparse matrices. In this paper we show that a vector sufficiently close to Ba *can be computed by only* $O(q \log^{d+2} q)$ *operations* (cf. Theorem 6.2).

G. Kuhn · H. Mang (Eds.)
Discretization Methods in Structural Mechanis
IUTAM/IACM Symposium Vienna, Austria, 1989

It may be even more important that computational work and storage is saved also in the definition phase, since it is not necessary to compute and store all q^2 entries of B. It suffices to evaluate and store $O(q \log^{d+1} q)$ quantities (cf. Theorem 6.3).

2. The Discrete Problem

2.1 The Panel Method (BEM)

Instead of the integral operator K from (1.1) we consider

(2.1) $$(K_n u_n)(x) = \int_{\Gamma_n} k(x,y) u_n(y) do_y .$$

where the surface Γ is replaced by Γ_n. Γ_n is the union of so-called *panels* or *boundary elements*. If $d=3$, these panels may be plane or curved triangles, quadrilaterals etc.

In the following the panels are denoted by $\pi_1,\dots,\pi_n$. The set of panels is called P. The number n of its elements is used as discretisation parameter:

(2.2) $$P := \{\pi_1,\dots,\pi_n\}, \quad n := \# P .$$

The integral in (2.1) is taken over

(2.3) $$\Gamma_n := \bigcup_{j=1}^{n} \pi_j = \bigcup_{\pi\in P} \pi .$$

The next step in the discretisation is the restriction of the functions u in (1.1) to functions $u_n \in V_n$ on Γ_n, where V_n is a linear space with finite dimension

(2.4) $$q := q_n := \dim(V_n).$$

V_n may e.g. consist of piecewise constant functions or continuous functions being linear, quadratic etc. on each $\pi\in P$. The quantities q_n and n are proportional, i.e. there is a constant C such that

(2.5) $$n/C \leqslant q = q_n \leqslant Cn .$$

Given q different collocation points

(2.6a) $$\Xi = \{\xi^1,\dots,\xi^q\} \subset S_n, \quad q = \# \Xi ,$$

we may replace the integral equation (1.2) by the *collocation problem* (cf. [4]):

(2.6b) find $u_n \in V_n$ with $\lambda u_n(\xi) = (K_n u_n)(\xi) + f(\xi)$ for all $\xi \in \Xi$.

Using a basis $\{b_1,\dots,b_q\}$, we can formulate the problem (2.6b) as the *linear system*

(2.7a) $$Ba = b, \qquad a = \begin{bmatrix}\alpha_1\\ \alpha_2\\ \vdots\\ \alpha_q\end{bmatrix}, \quad b = \begin{bmatrix}\beta_1\\ \beta_2\\ \vdots\\ \beta_q\end{bmatrix} \in \mathbb{R}^q, \quad B = (\beta_{ij})_{i,j=1,\dots,q} \quad q\times q \text{ matrix}.$$

The coefficients β_{ij} of the matrix B are defined by

(2.7b) $$\beta_{ij} = (K_n b_j)(\xi^i) = \int_{\Gamma_n} k(\xi^i,y) b_j(y) do_y .$$

The coefficients α_j of the vector a describe the discrete surface distribution by

(2.7c) $$u_n = \sum_{j=1}^{q} \alpha_j b_j \qquad (b_j\text{: basis functions in } V_n).$$

We shall see how the information contained in the matrix B can be stored by a smaller number of quantities and how it can be used for the matrix-vector multiplication.

2.2 Assumptions on the Kernel Funktion

In the far field (i.e. if $|\mathbf{x}-\mathbf{y}|$ is sufficiently large) we shall replace $k(\mathbf{x},\mathbf{y})$ by a piecewise polynomial with respect to $\mathbf{y}$, which may be chosen as expansion of order m around a suitable centre $\mathbf{y}_0 \in \mathbb{R}^d$:

(2.9) $$\tilde{k}(\mathbf{x},\mathbf{y}) = \sum_{\iota \in I_m} \varkappa_\iota(\mathbf{x};\mathbf{y}_0)\, \Phi_\iota(\mathbf{y}),$$

where I_m is an *index set* of size

(2.10) $$\# I_m \leq C_I\, m^d.$$

The integer m may be regarded as the *order of the expansion*. The functions $\Phi_\iota(\mathbf{y})$ must be independent of x and the centre $\mathbf{y}_0$. Furthermore, integrals of $\Phi_\iota(\mathbf{y}) u_n(\mathbf{y})$ over single panels $\pi \in P$ must easily be computable for $u_n \in V_n$ (cf. (2.12b)). In the standard applications $\Phi_\iota(\mathbf{y})$ is a piecewise polynomial over each $\pi \in P$.

The error of the expansion depends on the order m and on the distance $|\mathbf{y}-\mathbf{y}_0|$ from the centre. The exact requirements on $\tilde{k}$ are as follows. Let $\eta_0 \in (0,1)$ be given.

(2.11a) There are constants C_1, C_2 such that for all $\mathbf{x},\mathbf{y}_0 \in \mathbb{R}^d$, all $0<\eta \leq \eta_0 < 1$, and all $m \in \mathbb{N}$ there are expansions $\tilde{k}_n$ of the form (2.9) satisfying

(2.11b) $$|k(\mathbf{x},\mathbf{y}) - \tilde{k}(\mathbf{x},\mathbf{y})| \leq C_1 (C_2 \eta)^m\, |k(\mathbf{x},\mathbf{y})| \qquad \text{for all } |\mathbf{y}-\mathbf{y}_0| \leq \eta\, |\mathbf{x}-\mathbf{y}_0|.$$

Inequality (2.11b) provides an estimation of the *relative error* of $\tilde{k}$. In [8] we prove inequality (2.11b) for a standard example and determine the values of C_1, C_2, and η_0.

For the estimation of the complexity we need additional assumptions of the cost for the integration of $k_n(\mathbf{x},\mathbf{y}) u_n(\mathbf{y})$ and $\Phi_\iota(\mathbf{y}) u_n(\mathbf{y})$ over each panel $\pi \in P$:

(2.12a) for all $\pi \in P$, $1 \leq j \leq q$, $\mathbf{x} \in \mathbb{R}^d$ the integral $\int_\pi k(\mathbf{x},\mathbf{y}) b_j(\mathbf{y})\, do_y$ can be evaluated by at most C_Q arithmetical operations.

(2.12b) for all $\pi \in P$, $u_n \in V_n$, $\iota \in I_m$, $m \geq 1$, $\mathbf{x},\mathbf{y}_0 \in \mathbb{R}^d$ the evaluation of $\int_\pi \Phi_\iota(\mathbf{y}) b_j(\mathbf{y})\, do_y$ and $\varkappa_\iota(\mathbf{x};\mathbf{y}_0)$ requires at most $C_Q m$ operations.

Often, $k(\mathbf{x},\mathbf{y})$ is a finite sum of terms like $\mathbf{x}^\nu \mathbf{y}^\mu / |\mathbf{x}-\mathbf{y}|^\gamma$, while $u_n(\mathbf{y})$ is piecewise polynomial. Hence the product $k(\mathbf{x},\mathbf{y}) u_n(\mathbf{y})$ has the same structure. There are exact formulae for the integrals $\int \mathbf{y}^\mu / |\mathbf{x}-\mathbf{y}|^\gamma do_y$ over plane triangles (cf. Nowak [12], Hackbusch [4]). Therefore, condition (2.12a) can be satisfied in practice. Also (2.12b) can be shown to be valid in standard applications. Note that in practice m is a small number (cf. (6.3)). Therefore, the distinction between $O(1)$ and $O(m)$ is only of theoretical interest.

3. Panel Clustering

3.1 The Tree of Clusters

By a ***cluster*** $\tau \in T$ we mean a certain *union of panels*: $\tau = \pi_1 \cup \ldots \cup \pi_k$ $(\pi_j \in P)$. The cluster set T is required to have a *tree structure*:

(3.1a) The cluster set T has a connected tree structure: $\Gamma_n \in T$ is the *root* of T.

(3.1b) The panels of P belong to T; they are the *terminal knots* of T, i.e. there is no $\tau' \in T$ with $\tau \supset \tau'$, $\tau \neq \tau'$.

(3.1c) If $\tau \in T$ is an inner knot, there exist sons $\tau_1, \ldots, \tau_k$ (k depending on τ: $k = k(\tau)$) with

i) $\tau = \tau_1 \cup \ldots \cup \tau_k$,
ii) for any $\tau' \in T$ with $\tau' \subset \tau$, $\tau' \neq \tau$ there is an index $j \in \{1, \ldots, k\}$ such that $\tau' \subset \tau_j$.

A simple consequence of the tree structure is

Remark 3.1 Let T satisfy (3.1a-c). Then the number of clusters $\tau \in T$ is at most $2n-1$, where n is the number of panels:

(3.2) $$\# T \leqslant 2 \# P - 1.$$

The number of clusters not being panels is bounded by $n-1$.

3.2 Coverings of Γ_n by Admissible Clusters

A sphere around $\mathbf{z}$ with radius r is denoted by

(3.3) $$K_r(\mathbf{z}) := \{\mathbf{x} \in \mathbb{R}^d : |\mathbf{x} - \mathbf{z}| < r\}.$$

Each cluster $\tau \in T$ has a closed sphere $K = \overline{K(\tau)}$ with minimum radius containing τ. Therefore, we may define

(3.4) $\mathbf{z}_\tau =$ centre of the sphere $K(\tau)$, $\rho_\tau =$ radius of the sphere $K(\tau)$,

i.e.

(3.4') $$K(\tau) = K_{\rho_\tau}(\mathbf{z}_\tau).$$

Let η be a fixed number satisfying the inequality

(3.5a) $0 < \eta \leqslant \eta_0 < 1$ with η_0 from (2.11a)

and define

(3.5b) $$\overline{\eta} := \eta / \sqrt{1 + \eta^2}.$$

η is a parameter controlling the error of the expansion (2.11b) and will be chosen in chapter 5. We call $\tau \in T$ an ***admissible cluster*** w.r.t. a control point $\mathbf{x} \in \mathbb{R}^d$, if

(3.6) $$\rho_\tau \leqslant \overline{\eta}\, \mathrm{dist}(\mathbf{x}, \tau).$$

Here, $\mathrm{dist}(\mathbf{x}, \tau)$ denotes the distance of $\mathbf{x}$ from τ. The reason for inequality (3.6) is

Lemma 3.2 Inequality (3.6) implies

(3.6') $$\rho_\tau \leqslant \eta\, |\mathbf{x} - \mathbf{z}_\tau|.$$

If $\tau \in T$ is admissible w.r.t. $\mathbf{x}$, inequality (2.11b) holds for all $\mathbf{y} \in \tau$, if we choose $\mathbf{y}_0 := \mathbf{z}_\tau$ as centre of the expansion (2.9).

The proof of this and the following statements can be found in Hackbusch-Nowak [8].

The somewhat complicated definition of the centre $\mathbf{z}_\tau$ and the quantity $\overline{\eta}$ is chosen to have the following properties of admissible clusters:

Lemma 3.3 (a) $\mathbf{x} \notin \tau$ holds, if τ is admissible w.r.t. $\mathbf{x}$.
(b) Let $\tau', \tau \in T$, $\tau' \subset \tau$. If τ is admissible w.r.t. $\mathbf{x}$, then τ' is also admissible w.r.t. $\mathbf{x}$.

Remark 3.4 Condition (3.6) is somewhat inconvenient to implement. In practice one may replace (3.6) by (3.6') and in addition the centre $\mathbf{z}_\tau$ from (3.4) may be replaced e.g. by the mass centre of τ. In any case ρ_τ is to be defined by $\rho_\tau := \sup\{|\mathbf{x} - \mathbf{z}_\tau| : \mathbf{x} \in \tau\}$ or an upper bound.

Definition 3.5 We call $\mathcal{C}=\{\tau_1,\ldots,\tau_\sigma\}\subset T$ an ***admissible covering*** (of Γ_n) w.r.t. $\mathbf{x}$ if

(3.7a) $\qquad \Gamma_n=\bigcup_{j=1}^{\sigma}\tau_j,$

(3.7b) $\qquad$ either $\tau_j\in P$ or τ_j is admissible w.r.t. $\mathbf{x}$.

If $\mathbf{x}\in\Gamma_n$, there is no covering (3.7a) of Γ_n consisting only of admissible clusters. Condition (3.7b) states that the only non-admissible clusters can be the panels. The number of clusters in a covering should be as small as possible. The optimal choice is discussed in

Proposition 3.6 For each $\mathbf{x}\in\mathbb{R}^d$ there is a *unique* admissible covering $\mathcal{C}=\mathcal{C}(\mathbf{x})$ w.r.t. $\mathbf{x}$ with minimum number $\sigma=\sigma(\mathbf{x}):=\#\mathcal{C}(\mathbf{x})$ of clusters. $\mathcal{C}(\mathbf{x})$ is called the ***minimum admissible covering*** w.r.t. $\mathbf{x}$.

The minimum admissible covering $\mathcal{C}(\mathbf{x})$ w.r.t. $\mathbf{x}$ can easily be computed by

(3.8a) $\qquad \mathcal{C}:=\emptyset;\ Divide(S_n,\mathcal{C});$ **comment** result is $\mathcal{C}=\mathcal{C}(\mathbf{x})$;

where $Divide$ is the recursive procedure

(3.8b)
```
procedure Divide(τ,𝒞);
comment τ is a cluster, 𝒞 is a subset of T;
begin integer i,s;
    if (τ is admissible w.r.t. x) then 𝒞:=𝒞 ∪ {τ} else
    if τ∈P then 𝒞:=𝒞 ∪ {τ} else
    begin determine the sons τ_1,...,τ_s of τ;
        for i:=1 step 1 until s do Divide(τ_i,𝒞)
end    end;
```

Remark 3.7 Algorithm (3.8a,b) requires at most $2\sigma(\mathbf{x})-1$ calls of $Divide$, where $\sigma(\mathbf{x}):=\#\mathcal{C}(\mathbf{x})$. Hence, the computation of $\mathcal{C}(\mathbf{x})$ needs $O(\sigma(\mathbf{x}))$ operations.

The complexity of the fast multiplication algorithm depends mainly on the number $\sigma(\mathbf{x})$ of clusters in the covering $\mathcal{C}(\mathbf{x})$. The desired result will follow from the condition that there is a constant C_σ with

(3.9) $\qquad \sigma(\mathbf{x})\leq\sigma(\eta,n):=C_\sigma(\frac{1}{\eta})^{d-1}\log(2+\eta^{d-1}n) \qquad$ for all $\mathbf{x}\in\mathbb{R}^d$

with d: dimension of $\mathbb{R}^d$, η: from (3.6'), $n=\#P$. In [8] we show that T satisfies (3.9) under very natural assumptions. The logarithmic factor can even be omitted if $\mathbf{x}\notin S_n$.

By a suitable choice of η and m explained below we obtain the estimate

(3.10) $\qquad \sigma(\mathbf{x})\leq C\log n.$

Therefore, we have to deal with $O(\log n)$ clusters instead of n panels.

4. The Algorithm for the Matrix Multiplication

4.1 The Near Field Part

Instead of computing the matrix entries we compute the *far field coefficients* and *influence coefficients* in *Phase I*. In *Phase II* we evaluate an approximation $\tilde{K}_n u_n$ of $K_n u_n$ at $\mathbf{x}\in\mathbb{R}^d$ (e.g. at the collocation points $\xi\in\Xi$). We recall that the desired result of the matrix-vector multiplication $b=Ba$ is the sum

(4.1) $\qquad \beta_i=\sum_{j=1}^{q}\beta_{ij}\alpha_j=\sum_{j=1}^{q}\alpha_j\int_{\Gamma_n}k(\xi^i,y)b_j(y)do_y.$

Let $\mathscr{C}(\xi^i)$ be the minimum admissible covering determined in (3.8b). We split $\mathscr{C}(\xi^i)$ into a near and far field part by

$$\mathscr{C}_{near}(\xi^i) := \{\tau\in\mathscr{C}(\xi^i):\ \tau \text{ is a panel}\},$$
$$\mathscr{C}_{far}(\xi^i) := \{\tau\in\mathscr{C}(\xi^i):\ \tau \text{ is not a panel}\}.$$

The integral in (4.1) can be written as

$$\int_{\Gamma_n} \dots = \sum_{\tau\in\mathscr{C}_{near}(\xi^i)} \int_\tau \dots + \sum_{\tau\in\mathscr{C}_{far}(\xi^i)} \int_\tau \dots .$$

This induces the analogous splitting of β_i from (4.1) into

$$\beta_i = \beta_i^{near} + \beta_i^{far}.$$

We shall approximate β_i^{far} by $\tilde\beta_i^{far}$. The near field part of β_i is computed by

(4.2) $$\beta_i^{near} = \sum_{\pi\in\mathscr{C}_{near}(\xi^i)} \sum_j \alpha_j \int_\pi k(\xi^i,y)\, b_j(y)\, do_y .$$

Since the support of the basis functions b_j is small, there is only a constant number of indizes j such that the panel π intersects with $\operatorname{supp}(b_j)$. Hence the sum $\sum_j$ has only $O(1)$ terms. The number of panels in $\mathscr{C}_{near}(\xi^i)$ is bounded by $\sigma(\xi^i)=O(\log n)$ (cf. (3.10)). This proves

Remark 4.1 Under the assumptions (2.12a) and (3.10), the evaluation of β_i^{near} for all $1\leqslant i\leqslant q$ requires $O(n\log n)$ arithmetical operations. Storage of size $O(n\log n)$ is needed to store $\mathscr{C}_{near}(\xi^i)$ and the non-zero integrals $\int_\pi k(\xi^i,y)\,b_j(y)\,do_y$ from (4.2).

4.2 The Far Field Part

Replacing the exakt kernel k in the definition of β_i^{far} by $\tilde k(x,y)$ from (2.9), we obtain the approximation $\tilde\beta_i^{far}$:

(4.3) $$\tilde\beta_i^{far} := \sum_{\tau\in\mathscr{C}_{far}(\xi^i)} \sum_j \alpha_j \int_\tau \tilde k(\xi^i,y)\, b_j(y)\, do_y =$$
$$= \sum_{\tau\in\mathscr{C}_{far}(\xi^i)} \sum_j \alpha_j \int_\tau \sum_{\iota\in I_m} \varkappa_\iota(\xi^i;z_\tau)\, \Phi_\iota(y)\, b_j(y)\, do_y .$$

Summation and integration can be interchanged:

(4.3') $$\tilde\beta_i^{far} = \sum_{\tau\in\mathscr{C}_{far}(\xi^i)} \sum_{\iota\in I_m} \varkappa_\iota(\xi^i;z_\tau) \sum_j \alpha_j \int_\tau \Phi_\iota(y)\, b_j(y)\, do_y .$$

The integrals

(4.4) $$J_\tau^\iota(b_j) := \int_\tau \Phi_\iota(y)\, b_j(y)\, do_y \qquad (\iota\in I_m,\ \tau\in T,\ 1\leqslant j\leqslant q)$$

will be called *far field coefficients*. Of particular interest are the far field coefficients corresponding to panels:

(4.4') $$J_\pi^\iota(b_j) := \int_\pi \Phi_\iota(y)\, b_j(y)\, do_y \qquad (\iota\in I_m,\ \pi\in P,\ 1\leqslant j\leqslant q).$$

They are evaluated in a first phase, since they are independent of the special vector a in the matrix-vector multiplication $b=Ba$. The total number of far field coefficients in (4.4') is $O(n*I_m)=O(n\log^d n)$ (d: dimension, $d=3$ for three-dimensional problems. Note that there is only a fixed number of panels π intersecting with the support of b_j). By assumption (2.12b) the necessary number of arithmetical operations is $O(n\log^{d+1} n)$ using $m=O(\log n)$ (cf. (6.3)).

As soon as the far field coefficients (4.4') are computed, the quantities

$$(4.5)\qquad J_\pi^\iota := \sum_j \alpha_j\, J_\pi^\iota(b_j) \qquad (\iota\in I_m,\ \pi\in P)$$

can be summed up in $O(n*I_m)=O(n\log^d n)$ additions. The same amount of work is needed for the summation over the tree T:

$$(4.6)\qquad J_\tau^\iota := \sum_{\tau' \text{ son of } \tau} J_{\tau'}^\iota \qquad \text{for } \tau\in T\setminus P.$$

The coefficients J_τ^ι represent the sum

$$J_\tau^\iota = \sum_j \alpha_j\, J_\tau^\iota(b_j).$$

Hence the quantities $\tilde\beta_i^{far}$ can be computed from the simple sum

$$(4.7)\qquad \tilde\beta_i^{far} = \sum_{\tau\in\mathscr{C}_{far}(\xi^i)} \sum_{\iota\in I_m} \varkappa_\iota(\xi^i;z_\tau)\, J_\tau^\iota \qquad (1\le i\le q)$$

by $O(n*I_m)=O(n\log^d n)$ operations. Since the evaluation of the coefficients $\varkappa_\iota(\xi^i;z_\tau)$ for all ι, i and centres z_τ requires $O(n\log^{d+1}n)$ operations, we obtain the

Remark 4.2 Under the assumptions (2.12b) and (3.10), the evaluation of $\tilde\beta_i^{far}$ for all $1\le i\le q$ requires $O(n\log^{d+1}n)$ arithmetical operations. Storage of size $O(n\log^d n)$ is needed to store $\mathscr{C}_{far}(\xi^i)$ and J_π^ι.

The given work counts depend on the special choice of η and m leading to (3.10). In general the result is as follows.

Remark 4.3 The computation of J_π^ι from (4.4') requires $O(nm^{d+1})$ operations. The computation of J_τ^ι for all $\iota\in I_m$, $\tau\in T\setminus P$ by (4.5-6) requires $O(nm^d)$ additions. The evaluation of $\tilde\beta_i^{far}$ in (4.7) requires $O(nm^{d+1}\sigma(\eta,n))$ operations with $\sigma(\eta,n)$ from (3.9).

5. Influence of the Additional Quadrature Error

The replacement of $K_n u_n$ by $\tilde K_n u_n$ can be regarded as an additional quadrature error. By (2.11a,b) the error is of order $O(\eta^m)$. In order to make the error equal to the consistency error $O(h^\varkappa)$ (h: panel size) we choose η in (3.5/6) by

$$(5.1)\qquad \eta = \eta(n) := O((n^{-\varkappa/(d-1)})^{1/m}) \le \eta_0, \quad \varkappa\le m,$$

Then $O(\eta^m)$ equals $O(h^\varkappa)$.

6. Complexity of the Algorithm

Inserting (5.1) into (3.9) we obtain the estimate (6.1) for the number $\sigma(\mathbf{x})$ of clusters in $\mathscr{C}(\mathbf{x})$:

$$(6.1)\qquad \sigma(\mathbf{x})\le\sigma(\eta,n)=C_c n^{\varkappa/m}\log(2+C_d n^{1-\varkappa/m}) \qquad \text{for all } \mathbf{x}\in\mathbb{R}^d.$$

According to Remark 4.3 the number of operations is

$$(6.2)\qquad N_n \le C\, n^{1+\varkappa/m}\, m^{d+1}\log(2+C_d n^{1-\varkappa/m})$$

with some constant C independent of n and m. The order m of the expansion (2.9) is now chosen depending on the number n of panels:

$$(6.3)\qquad m := \left\lfloor \tfrac{\varkappa}{d+1}\log n \right\rfloor \qquad (\lfloor x \rfloor := \text{largest integer } i \text{ with } i\le x).$$

Remark 6.1 Assume $\varkappa\in\mathbb{N}$, $n\ge e^{d+1}$ and define m by (6.3). Then $m\ge\varkappa$ (cf. (5.1)) and $e^{d+1}\le n^{\varkappa/m}\le e^{(d+1)(1+1/\varkappa)}$ hold. The upper bound of $n^{\varkappa/m}$ tends to e^{d+1} when $n\to\infty$.

With this choice of m, the inequalities (6.2a,b) become

(6.4) $$N_n \leq C_N\, n \log^{d+2} n$$

with a new constant C_N.

Summarising the preceding results, we obtain

Theorem 6.2 Choose η and m according to (5.1) and (6.3). Then $O(n \log^{d+2} n)$ operations are sufficient to perform the matrix-vector multiplication $a \mapsto Ba$ with an error estimated in Section 5. By (2.5) the complexity order equals $O(q \log^{d+2} q)$.

By the Remarks 4.1, 4.2 and $O(nm^d \sigma(\eta, n)) = O(n \log^{d+1} n)$ we prove

Theorem 6.3 The matrix-vector multiplication algorithm of Section 4 requires a *storage* of size $O(n \log^{d+1} n)$.

References.

[1] Ballmann, J., Eppler, R. and Hackbusch, W. (eds.) Panel Methods in Mechanics with Emphasis on Aerodynamics. Notes on Numerical Fluid Mechanics, vol. 21. Vieweg, Braunschweig 1988

[2] Hackbusch, W. *Multi-Grid Methods and Applications.* Springer Heidelberg, 1985

[3] Hackbusch, W.: *Theorie und Numerik elliptischer Differentialgleichungen.* Teubner, Stuttgart 1986

[4] Hackbusch, W.: *Integralgleichungen. Theorie und Numerik.* Teubner, Stuttgart 1989

[5] Hackbusch, W. and Nowak, Z.P.: Multigrid method for calculating the lifting potential incompressible flows around three-dimensional bodies. In [9] 135-148

[6] Hackbusch, W. and Nowak, Z.P.: A multi-level discretisation and solution method for potential flow problems in three dimensions. In [10]

[7] Hackbusch, W. and Nowak, Z.P.: On the complexity of the panel method (in Russ.). To appear in the proceedings of the conference «Modern Problems in Numerical Analysis», Moscow, Sept. 1986. Also in Report 8608, Universität Kiel, Sept 1986

[8] Hackbusch, W. and Nowak, Z.P.: On the fast matrix multiplication in the boundary element method by panel clustering. Numerische Mathematik 54 (1989) 463-491

[9] Hackbusch, W. and Trottenberg, U. (eds.): Multi-grid methods II. Proceedings, Köln, Oct. 1985. Lecture Notes in Mathematics 1228. Springer Berlin, 1986

[10] Hirschel, E.H. (ed.): Finite approximation in fluid mechanics. Notes on Numerical Fluid Mechanics, vol. 14. Vieweg, Braunschweig 1986

[11] Kral, J. and Wendland, W.: On the applicability of the Fredholm-Radon method in potential theory and the panel method. In [1] 120-136

[12] Nowak, Z.P.: Efficient multi-level panel methods of higher order accuracy. Report Universität Kiel, 1989

Exact Finite Elements for the Linear Buckling Analysis of Unidimensional Structural Problems

Z. WASZCZYSZYN, J. PIECZARA

Institute of Structural Mechanics

Cracow University of Technology

Poland

Summary

Finite difference scheme with the midpoint can be used to integrate the canonic set of ordinary differential equations. The equations describe the 2-point boundary value problem for unidimensional structural problems. The matrices of the exact finite element are obtained as a result of solution of the appropriate set of linear algebraic equations. Algorithm for the analysis of a modified eigenproblems presented. Buckling of Euler's column, a cylindrical shell under external pressure and of a cooling tower structure are considered. These numerical examples confirm accuracy and effectiveness of the proposed exact finite elements application.

Introductory Remarks

An appropriate approximation of field functions is the basis of the Finite Element Method (FEM). Corresponding FE can be called Approximate Finite Element (AFE).

If the basic functions (shape functions) fulfill all field equations then FE can be called as Exact Finite Element (EFE) Such EFE are well known in the simple problems of the stiffenes method. As a good example beam functions can be quoted, which are the basis for the formulation of FE matrices in the linear analysis.

EFE can be formulated not only for framed structures but also for plates and shells. There is such a possibility in a class of Unidimensional Structural Problems (USP) which are governed by ordinary differential equations. Plates and shells of revolution are good examples of such structures after the separation of variables has been made and the ortogonalization procedure has been applied. If the ordinary differential

G. Kuhn · H. Mang (Eds.)
Discretization Methods in Structural Mechanis
IUTAM/IACM Symposium Vienna, Austria, 1989

equations can be analytically solved then 'explicit' formulae for the FE matrices are formulated. This way was followed by Meyer and Harmon [1], Popov, Penzien, Zung-Au Lu [2], Mateja and Pela [3] for conical shells, by Lakis [4] for cylindrical shells, by Pardoen [5] for circular plates.

Analytical solutions are available only for a narrow class of unidimensional problems. That is why numerical approach has been preferred to compute EFE matrices. The approach bases on formulations of the Boundary Value Problem (BVP) in the form of canonic set of ordinary differential equations completed by boundary conditions. In papers [6] this set of equations was solved by means of forward numerical integration. Simple matrix manipulations led to the components of FE matrices.

A more efficient approach was sugested in [7], where finite differences have been used for integrating the canonic set of equations. In such a way all components of EFE matrices can be computed during one solution procedure of a set of algebraic equations with multiple right hand sides. This approach enables us to extend the length of FE significantly and consider variable geometrical and material properties along EFE.

In the paper the concept of EFE is presented as well as the algorithms with respect to the prebuckling state and the eigenvalue analysis. Linear elastic material, small strains and moderate displacements are assumed, as it is usually done in the linear buckling analysis.

Finite-Difference Analysis of the 2-Point BVP

In case of USP the full set of equations (geometrical, equilibrium and constitutive equations) can be transformed into the canonic set of ordinary differential equations:

$$\underline{y}' = \underline{A}(\underline{y},\xi)\underline{y} + \underline{p}(\xi) = \underline{f}(\underline{y},\xi) \;, \tag{1}$$

completed by boundary conditions:

$$\underline{B}_b\, \underline{y}_b = \underline{g}_b \qquad \text{for} \quad b = 0, L \;. \tag{2}$$

In Eqs (1) and (2) the state vector is used:

$$\underline{y}(\xi) = \{\underline{q}, \underline{r}\} , \tag{3}$$

where $\underline{q}$, $\underline{r}$ are the vectors of generalized displacements and forces. The boundary conditions (2) are formulated at the end points ξ_b for b=0,L and $\underline{y}_b \in \mathbb{R}^n$, $\underline{g}_b \in \mathbb{R}^{n/2}$, $\underline{B}_b \in \mathbb{R}^{n/2} \times \mathbb{R}^{n/2}$. In what follows we shall assume that $\underline{f}(\underline{y},\xi)$ is continuous and obeys Lipschitz conditions in the domain $(\xi, \underline{y}(\xi)) \subset D$ for $\xi \in [\xi_0, \xi_L]$.

Eqs (1) and (2) formulate the symmetric 2-Point BVP of the range r = n/2 . Various methods for solving of such a problem were suggested in [8] but we use the finite-difference scheme with the midpoint, which was recommended in [9]:

$$\underline{y}'_{l+1/2} \approx \frac{1}{\Delta\xi} (\underline{y}_{l+1} - \underline{y}_l) = \frac{1}{2} (\underline{f}_{l+1} + \underline{f}_l) + \\ + \underline{O} (\frac{1}{24} \underline{y}'''_{l+1/2} \Delta\xi_l^2) , \tag{4}$$

where l = 0,1,...,L enumerate the FD stations. Applying (4) to (1) and adding (2) the following set of algebraic equations is obtained:

$$\begin{aligned} &\underline{B}_0 \, \underline{y}_0 = \underline{g}_0 , \\ &\underline{C}_l \, \underline{y}_l + \underline{D}_{l+1} \, \underline{y}_{l+1} = \underline{g}_{l+1/2} \qquad \text{for } l=0,1,\dots,L-1 \\ &\underline{B}_l \, \underline{y}_l = \underline{g}_l , \end{aligned} \tag{5}$$

where:

$$\underline{C}_l = -\underline{I} - \frac{\Delta\xi_l}{2} \underline{A}_{l+1/2} , \quad \underline{D}_l = \underline{I} - \frac{\Delta\xi_l}{2} \underline{A}_{l+1/2} , \tag{6}$$

$$\underline{g}_{l+1/2} = \frac{\Delta\xi_l}{2} (\underline{p}_l + \underline{p}_{l+1}) .$$

Eqs (5) can be written in the following canonic form:

$$\underline{H} \, \underline{y} = \underline{g} , \tag{7}$$

where now the vector $\underline{y}$ is composed of subvectors $\underline{y}_l$ for l = 0,1,...,L as it is shown in Fig.1.

In the linear buckling analysis the fundamental and adjacent equilibrium states are considered using the state vectors:

$$\underline{y} = \overset{o}{\underline{y}} + \overset{j}{\underline{y}} , \tag{8}$$

which correspond to the following equations:

$$\underline{H}_o \overset{o}{\underline{y}} = \overset{o}{\underline{g}} \,, \tag{9}$$

$$\underline{H}_T \underline{y} \equiv [\, \underline{H}_o + \underline{H}_\sigma(\Lambda \overset{o}{\underline{g}}{}^{*}) \,] \overset{j}{\underline{y}} = \underline{0} \,. \tag{10}$$

The matrix $\underline{H}_o$ is to be formulated for linear prebuckling state, i.e. the matrix $\underline{A}(\xi)$ in Eq.(1) is independent of the state vector $\overset{o}{\underline{y}}$. The vector $\overset{j}{\underline{y}}$ in the generalized eigenproblem (10) is associated with the j-th buckling mode and Λ is the load factor in the relation $\overset{o}{\underline{g}} = \Lambda \overset{o}{\underline{g}}{}^{*}$.

Matrices of Exact Finite Elements

Let us consider the matrix form of Eq.(9):

$$\underline{H}_o \underline{Y} = \underline{G} \,, \tag{11}$$

where the matrices $\underline{Y}$ and $\underline{G}$ are split into blocks:

$$\underline{Y} = [\underline{Y}_I \,\vdots\, \underline{Y}_p] \,, \qquad \underline{G} = [\underline{G}_I \,\vdots\, \underline{G}_p] \,. \tag{12}$$

The subscript I is related to unit displacements of the FE ends and p to the homogeneous boundary conditions but non-zero loads. In Fig.1 the structure of Eq.(11) is shown with reference to notation (3) and for m load configurations.

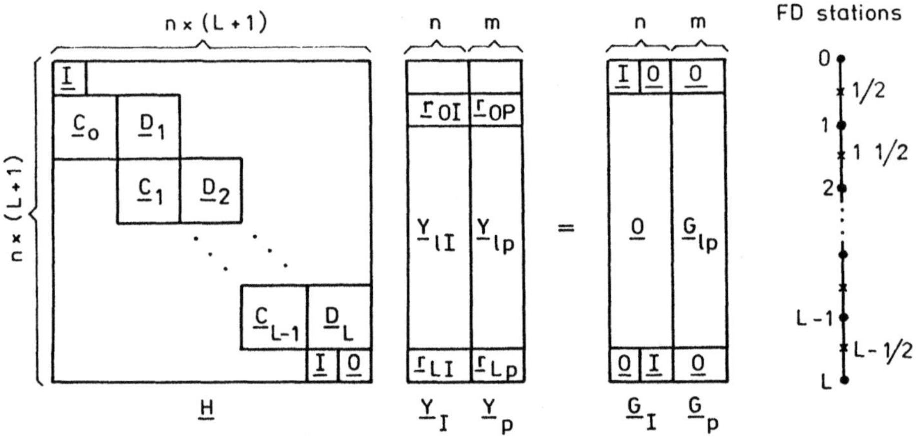

Fig.1. Structure of the matrix equation (11).

Reactions $\underline{r}_{bI}$, $\underline{r}_{bp}$ are obtained as a result of solutions of Eq.(11) - cf.Fig.1. They compose the blocks of the linear stiffness matrix $\underline{k}_o^e$ and the vector of equivalent nodal forces $\underline{f}^e$:

$$\underline{k}_o^e = \begin{bmatrix} -\underline{r}_{OI} \\ \underline{r}_{LI} \end{bmatrix} , \qquad \underline{f}^e = \begin{bmatrix} -\underline{r}_{Op} \\ \underline{r}_{Lp} \end{bmatrix} . \tag{13}$$

The sign minus is added in (13) because of relation of the reactions $\underline{r}_{OI}$, $\underline{r}_{Op}$ to the FE system of coordinates.

The solution of the matrix equation

$$\underline{H}_T \, \underline{Y} = \underline{G}_I \tag{14}$$

is used to compute the geometrical stiffness matrix $\underline{k}_G$:

$$\underline{k}_G^e = \underline{k}_T^e - \underline{k}_o^e . \tag{15}$$

In case of plates and shells of revolution all functions are extended into the Fourier series

$$\underline{y}(\xi,\theta) = \overset{j}{\underline{y}}(\xi) f_j \equiv \{\underline{q},\underline{r}\}^j f_j , \tag{16}$$

where $f_j = \cos j\theta$ or $\sin j\theta$. Instead of the intensities $\overset{j}{\underline{r}}$, the resultant forces $\overset{j}{\underline{R}}$ are used , which satisfy the principle of virtual work:

$$\delta \overset{j}{\underline{q}}{}^T \overset{j}{\underline{R}} = \int_0^{2\pi} \delta \overset{j}{\underline{q}}{}^T \overset{j}{\underline{r}} (f_j)^2 \rho \, d\theta = \delta \overset{j}{\underline{q}}{}^T (c \rho \overset{j}{\underline{r}}) , \tag{17}$$

where ρ is the radius of circumferential circle and parameter c equals:

$$c = \begin{cases} 2\pi & \text{for } j=0 , \\ \pi & \text{for } j\geq 1 . \end{cases} \tag{18}$$

In such a way the symmetric stiffness matrices $\underline{k}^e$ and the vector of nodal forces $\underline{f}^e$ are:

$$\underline{k}^e = c \begin{bmatrix} -\rho_o \, \underline{r}_{OI} \\ \rho_L \, \underline{r}_{LI} \end{bmatrix} , \qquad \underline{f}^e = c \begin{bmatrix} -\rho_o \underline{r}_{Op} \\ \rho_L \underline{r}_{Lp} \end{bmatrix} . \tag{19}$$

The formula $(19)_1$ is valid for both the linear stiffness matrix $\underline{k}_o^e$ and the geometric stiffness matrix $\underline{k}_G^e$. The matrix $\underline{k}_G^e$ is computed by means of the formula (15) using $\overset{j}{\underline{H}}_o$ and $\overset{j}{\underline{H}}_T$.

Algorithm of the Eigenanalysis

Let us assume that nodal systems of coordinates are adopted. After the transformation , assembling and fulfilling the kinematic boundary conditions,the vector of nodal displacements and stiffness matrices are $\underline{Q}$, $\underline{K}_o$, $\underline{K}_\sigma$ respectively. The analysis is carried out in two stages. The linear set of equations is firstly solved for the load factor $\Lambda = 1$ and the reference load vector $\overset{o}{\underline{F}}{}^*$:

$$\overset{o}{\underline{K}}_o \, \overset{o}{\underline{Q}}{}^* = \overset{o}{\underline{F}}{}^* \quad , \tag{20}$$

where overscript o corresponds to the fundamental (pre-buckling) state.

A modified eigenvalue problem

$$[\underline{K}_o + \lambda_{(i)} \underline{K}_\sigma(\Lambda_{(i)})] \, \underline{v}_{(i)} = \underline{O} \tag{21}$$

is solved in an iterative way by means of the following formula

$$\Lambda_{(i)} = \Lambda_{(i-1)}\lambda_{(i-1)} \qquad \text{and} \quad \Lambda_{(1)} = 1 \quad . \tag{22}$$

The predictor $\Lambda_{(1)} = 1$ is associated with the value of the load factor. The iteration is ended if the following convergence condition is satisfied:

$$|\lambda_n - 1| \leq \varepsilon_\lambda \ . \tag{23}$$

Numerical tests have shown that $n = 3\text{-}4$ for $\varepsilon_\lambda = 0.001$.

In case of axisymmetric plates and shells the matrices $\underline{K}_o$ and $\underline{K}_\sigma$ are computed for the j-th harmonics (number of circumferential buckling waves) and the critical value of load factor is defined as:

$$\Lambda_{cr} = \min_j (\Lambda^j_{cr}) \quad , \tag{24}$$

and the corresponding buckling mode is $\overset{j}{\underline{v}}$.

Buckling of Euler's Column

The value of buckling load for the axially compressed column shown in Fig.2 is $P^{an}_{cr} = \pi^2 EI/(4a^2)$, where a is the length of the column. The state vector for the plane buckling is:

$$\underset{(6x1)}{\underline{y}_l} = \{ u,v,w \,\vdots\, N,T,M \}_{\xi_l} \quad , \tag{25}$$

and the canonic set of ordinary differential equations can be easily deduced from the equations given in Appendix.

The data for computation are shown in Fig.2. The results are completed in Tabl.1 for different number of exact finite elements EFE and for the number of finite difference stations FDS. It is clear that even small number of FE and FD stations

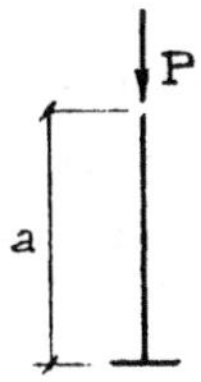

a = 10
A = 1
r = 0.01
E = 1000

Fig.2. Data for Euler's column.

FDS \ EFE	1	2	4
3	4.712	1.127	0.283
4	2.508	0.620	0.112
6	1.130	0.226	0.053
8	0.622	0.127	0.035
10	0.390	0.088	0.022

Tabl.1. Load parameter $\alpha = (P_{cr} - P^{an}_{cr})/P_{cr} * 100\%$

leads to quite accurate results. Convergence of the iteration procedure (22) is rapid, i.e. 3 steps are enough to fulfill the condition (23) for $\varepsilon_{\lambda} = 0.001$.

Buckling of a Cylindrical Shell under External Pressure

The canonic set of equations (A2) is used. The data for computation, taken from [11], p.164, are shown in Fig.3. Numerical results for the fixed number of circumferential waves j=8 are listed in Tabl.2 for different number of exact finite elements EFE and finite different stations FDS. In Tabl.3

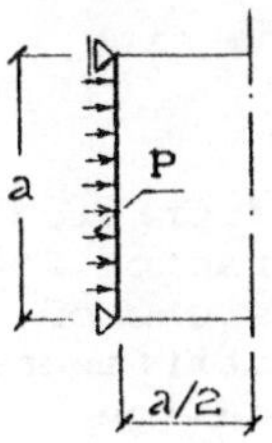

a=20in.
h=0.2in.
$E=1*10^7$psi
$\nu=0.3$

Fig.3. Data for a cylindrical shell.

EFE	FDS	P_{cr} [psi]
1	4	113.95
	10	107.10
	20	106.26
4	5	106.45
3	10	106.35
6	5	106.16

Tabl.2 Critical pressure for j=8.

j	P_{cr}	
	FD	[10]
7	122.61	117
8	106.26	106
9	108.11	108

Tabl.3. Critical pressure for different wave number j .

the results obtained by means of EFE=1, i.e. in fact using FD method are compared with results from [10], p.164.

Stability of a Natural Draught Cooling Tower

This example has been taken from [11]. The cooling tower is composed of a shell of variable thickness, different number of ring-stiffeners and supporting columns. An equivalent shell EFE is considered smearing the supporting columns as it was suggested in [12]. The main data are shown in Fig. 4. the ring-stiffener cross-section is assumed to be rectangular 1.20X0.4m^2 the colummns are circular of diameter 1.26m. Dead load g, internal succion s and equivalent axisymmetric wind pressure 0.6w_o are considered in the frame of single parameter load Λ(g+s+0.6w_o). The distribution of loads s and w_o along the meridian was given in [11].

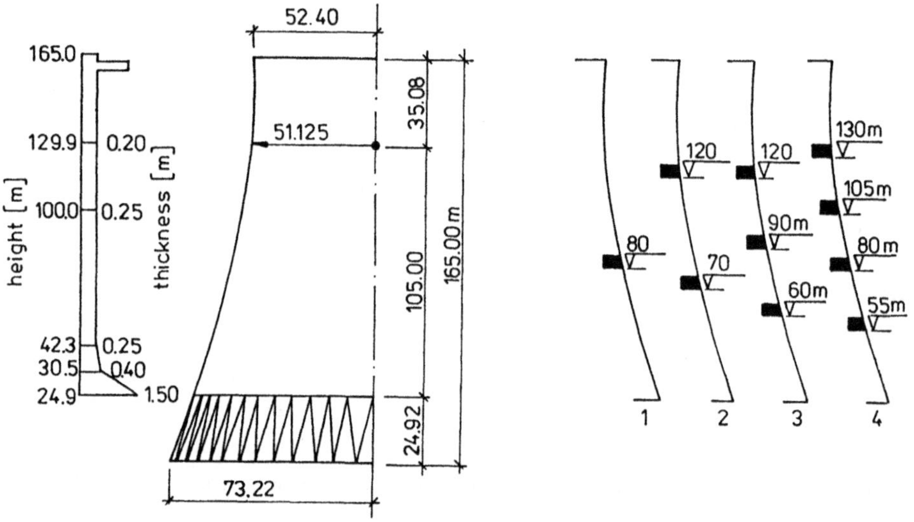

Fig. 4. Cooling tower structure and versions of the ring stiffener application.

Version	0	1	2	3	4
Λ_{cr} EFE (j)	4.72 (7)	5.36 (9)	6.43 (8)	8.36 (11)	9.74 (6)
Λ_{cr} [11] (j)	4.13 (7)	5.96 (9)	6.79 (8)	8.56 (10)	11.11 (5)

Tabl. 4. Critical load factor for 4 versions of ring-stiffeners application (0-only top ring-stiffener

Eqs (A2) are used to derive the stiffness matrices for shell EFE. Formulae for the ring-stiffener matrices have been taken from [12]. Results of the critical load parameter computation for different number of stiffners (shown in Fig. 5) are given in Tabl. 4. They are compared with the results from [11] where 35 standard, conical AFE were used (versus 9 EFE). Other examples of application of EFE to the analysis of cooling towers have been given in [13].

References

1. Meyer, G. H.; Harmon, M. B.: Conical segment method for analysing open crown shells of revolution for edge loading. AIAA J. 4(1963) 886-891.

2. Popov, E. P.; Penzien, J.; Zung-Au Lu: Finite element solution for axisymmetrical shells. Proc. ASCE. EM5. 90 (1964) 119-145.

3. Mateja, O.; Pela, J.: Exact finite element for conical shells of revolution (in Polish). Proc. V Polish Conference on Computational Mechanics. Wroclaw (1989). Vol.2. 45-52.

4. Lakis, A. A.: Cylindrical finite element for analysis of cylindrical structures. Dept. Mech. Engng. Mc Gill Univ. R.No 70-2. Montreal (1969).

5. Pardoen, G. C.: Static, vibration and buckling analysis of axisymmetric circular plates using finite elements. Int. J. Comp. and Stru. 2 (1973) 355-373.

6. Waszczyszyn, Z.: Exact finite elements for thin elastic shells of revolution. Theor. and Appl. Mech. Bulgarian Acad. Sci. 9 (1978) 101-104.

7. Waszczyszyn, Z.; Pieczara, J.: Exact finite elements in the analysis of shells of revolution. Proc. 3rd Conf. Space Stru. Guildford (1984) 349-354.

8. Meyer, G. H.: Initial value methods for boundary value problems. New York, Toronto: Academic Press (1973).

9. Beafait, W.; Reddien, G. W.: Midpoint difference method for analysing beam structures. Int. J. Comp. and Stru. 8 (1978) 745-751.

10. Brush, D. O.; Almroth, B. O.: Buckling of bars, plates and shells. Mc Graw Hill. Book Co. (1975).

11. Eckstein, U. et al.: Improvement of the structural behavior of cooling tower shells by ring-stiffeners. Natural Draught Cooling Towers. Gould, P. L. et al. (Eds.). Springer-Verlag (1984) 61-76.

12. Pieczara, J.: Statics and buckling of elastic shells of revolution by means of exact finite elements(in Polish). Dr. Th. Cracow Univ. of Technology (1987).

13. Waszczyszyn, Z.; Pieczara, J.: Application of exact finite elements to the buckling analysis of cooling towers. Proc. 3-rd Intern. Symp. on Natural Draught Cooling Towers, Paris (1989) 197-206.

Appendix

Canonic set of Equations for Axially Symmetric Shells

The state vector at the point l is as follows:

$$\underset{(8\times1)}{Y_l} = \{ u, v, w, \varphi_1 ; \bar{n}_{11}, \bar{n}_{12}, \bar{t}_1, \bar{m}_{11} \}_l \quad , \qquad (A1)$$

where the components are shown in Fig. A1. The bars over the symbols of internal forces emphasise that edge forces are used, e.g $\bar{t}_1 = t_1 + \partial m_{12}/(r\partial\theta)$ if the Kirchhoff-Love hypothesis is assumed

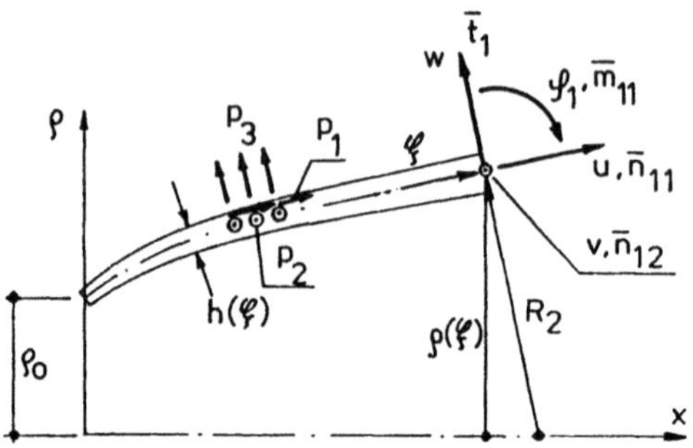

Fig. A1. Generalized displacements and internal forces in a shell of revolution.

The Donnell-Mushtari-Vlasov theory of quasi-shallow elastic shells (cf. [11], p. 195-197), gives the following canonic set of ordinary differential equations:

$$
\begin{aligned}
u' &= -\frac{A_1}{R} w + \frac{A_1}{Eh}(\bar{n}_{11} - \nu n_{22}) - \Lambda A_1 \overset{o*}{\varphi}_1 \varphi_1 , \\
v' &= \frac{jA_1}{\rho} u + \frac{\rho'}{\rho} v + \frac{2(1+\nu)A_1}{Eh} \bar{n}_{12} - \Lambda \frac{jA_1}{\rho} \overset{o*}{\varphi}_1 w , \\
w' &= -A \varphi_1 , \\
\varphi_1' &= \frac{12A_1}{Eh^3}(\bar{m}_{11} - \nu m_{22}) , \qquad \text{(A2)} \\
\bar{n}'_{11} &= -\frac{\rho'}{\rho}(\bar{n}_{11} - n_{22}) - \frac{jA_1}{\rho} \bar{n}_{12} - A_1 p_1 , \\
\bar{n}'_{12} &= -\frac{2\rho'}{\rho} \bar{n}_{12} + \frac{jA_1}{\rho} n_{22} - A_1 p_2 + \Lambda \frac{jA_1}{R_2}\left(\frac{j}{\rho} \overset{o*}{n}_2 w + \overset{o}{\varphi}_1 \bar{n}_{12}\right) , \\
\bar{t}'_1 &= -\frac{\rho'}{\rho} \bar{t}_1 + A_1\left(\frac{\bar{n}_{11}}{R_1} + \frac{n_{22}}{R_2}\right) - A_1 p_3 + \Lambda \frac{jA_1}{\rho}\left(\overset{o*}{\varphi}_1 \bar{n}_{12} + \frac{j}{\rho} \overset{o*}{n}_2 w\right) , \\
\bar{m}'_{11} &= \frac{\rho'}{\rho}(\bar{m}_{11} - m_{22}) - \frac{2jA_1}{\rho} \bar{m}_{12} + A_1 \bar{t}_1 + \Lambda A_1\left(\overset{o*}{\varphi}_1 \bar{n}_{11} + \overset{o*}{n}_1 \varphi_1\right) ,
\end{aligned}
$$

where the indices j are ommitted at the components of the state vector (A1) and $(\ )' = d(\ )/d\xi$. The variables $\overset{o*}{\varphi}_1$, $\overset{o*}{n}_1$, $\overset{o*}{n}_2$ are associated with the axially symmetrical fundamental state for $\Lambda = 1$. The other variables are:

$$
\begin{aligned}
n_{22} &= Eh\left(\frac{\rho'}{\rho A_1} u + \frac{j}{\rho} v + \frac{1}{R_2} w\right) + \nu \bar{n}_{11} , \\
m_{22} &= \frac{Eh^3}{12}\left(\frac{j^2}{\rho^2} w + \frac{\rho'}{\rho A_1} \varphi_1\right) + \nu \bar{m}_{11} .
\end{aligned} \qquad \text{(A3)},
$$

Coupled Numeric-Symbolic Computations in Structural Mechanics

M. KLEIBER

Polish Academy of Science
Institute of Fundamental Technological Research
Swiętokrzyska 21, 00–049 Warsaw, Poland

Summary

A new emerging area of computer-assisted research within the structural mechanics is disscussed. Rather than attempting to formulate a general theory, simple examples illustrate the currently available concepts and software tools which can be employed to 'humanize' structural mechanics computations.

Introductory remarks

In structural mechanics, as well as in many other branches of science and engineering, there still exist areas in which the performance of an experienced researcher is far superior to that of even the most powerful computer. Such areas may generally be identified as those which are hardly susceptible to algorithmization and involve issues difficult to formalize in the precise language of mathematics. As examples from the field of Discretization Metods in Structural Mechanics we may quote:

(a) On the part of general problem modelling:

- selection of the region in which to formulate the boundary-value problem,
- idealization of boundary conditions,
- idealization of loads,
- idealization of material behaviour (selection of the constitutive model),
- decision on whether small or large deformations should be considered,
- decision on the way the b.v. problem is posed (step-by step analysis, safety factors at failure, selection of a particular shell theory, etc.)
- derivation of governing equations.

G. Kuhn · H. Mang (Eds.)
Discretization Methods in Structural Mechanis
IUTAM/IACM Symposium Vienna, Austria, 1989

(b) On the part of the solution methods:

- decision on which basic solution philosophy to employ (analytic vs numerical).
- selection of 'optimal' discretizations,
- selection of numerical algorithms,
- control of algorithm parameters,
- re-computation philosophy,
- output data selection.

The possible sources of human superiority over computers in such situations may be:

(a) existence of enormous experience available to experts, which is not easily transferable to computers,

(b) imprecise, fuzzy and thus more 'general' character of existing informations and the phenomenal human capability to process such data,

(c) human learning capabilities.

The above remarks can be summarized by saying that:

> In order to make computers more effective in supporting the creative work of the researcher and engineer, we have to provide them with a sufficient amount of coded pieces of self-modifying knowledge, and to endow them with the ability to use some approximate reasoning techniques.

Below we shall say more about the issues identified above as (a), (b), (c) by taking reference to some simple examples worked out at the author's home institute in the framework of the AIASM (AI-Aided Solid Mechanics) project. This article is a compiled review of some other recent papers co-authored by this writer, [1-5].

CONMOD - a consultation system in the field of CONstitutive MODelling.

The issue of employing computers for storing, processing and using knowlege in a 'human-like' manner is ilustrated by briefly describing the program CONMOD, [3, 4]. CONMOD is an expert system based on the blackboard architecture; the primary means of deductive inference are provided by the resolution rule, [6].

The term 'constitutive modelling' has in solid mechanics the conventional meaning of the task of constructing the best possible model representing the thermo-mechanical behaviour of a given material. Here, a slightly different meaning of this term is adopted

and constitutive modelling is understood as a selection from the existing variety of constitutive models of an 'optimal' one under specified constraints.
The constraints are grouped in three categories, each of them representing, respectively:

(a) data concerning the overall engineering situation to be studied,

(b) data concerning those features of material behaviour which are presumed to be of significance,

(c) data concerning existing software and hardware resources.

According to the above scheme in the course of an interactive session with CONMOD the user is encouraged to answer questions related to specific topics, and each of the selected options splits into a number of more specific choices and so on, thus forming a problem description hierarchy. A number of heuristic constraints imposed upon different options guide the consultation and allow the program to take 'short cuts' depending on informations inferred from the previous answers. Other important aspects of the program are its capability to accept uncertain answers modelled by means of the so-called certainty factors and the explanation option which can be invoked by the user in case of doubts as to the relevance of a question displayed.
To illustrate the functioning of the program an examplary consultation with CONMOD is reproduced below; options selected by the hypothetical user are indicated by arrows, some options are left out for compactness sake, coments are given only to better ilustrate the program behaviour.

Desired level of analysis sophistication:

→ 'standard' analysis (traditional design needs)
'improved' assessment of the process
sophisticated assessment of the process

Comment: all 'advanced' constitutive models are thus excluded

General field of application:

→ structures
machine elements
geotechnical engineering
....

Problem type:

purely mechanical
purely thermal
→ coupled
→ thermo-mechanical
fluid-structure interactions
....

Philosophy of analysis:

→ complete analysis
safety factors at failure

....

Problem dimensiality:

1D
→ 2D
  bending
  → no bending
    → thin plate loaded in-plane
    axisymmetric

3D

Type of material:

→ metal
polymer
timber
concrete
....

Comment: Material isotropy is inferred but can be denied by the user.

Expected deformations:

→ small displacements
large displacements

Result of the consultation:

Your problem should be posed as
| Plane stress isotropic uncoupled thermo-elasticity |
with the constitutive equations of the form....
Remarks: ....
(material constants apearing in the model, etc.)
References: ....

Fuzzy sets and natural language computations.

A fuzzy subset of some universe U is a collection of objects from U (the set part) such that with each object is associated a degree of membership (the fuzzy part), [7]. The degree of membership is always a real number between 0.0 and 1.0 and it measures the plausibility of an element being in a particular set.

We recall that the characteristic function for an ordinary subset A of an ordinary set U is defined as the binary choice

$$\mu_A(x) : U \rightarrow \{0\ 1\}$$

which determines whether the element $x \epsilon U$ is in $(\mu_A(x) = 1)$ or out $(\mu_A(x) = 0)$ of the subset A. Similarly,but more generally, we assume for a fuzzy subset A of the universe U

$$\mu_A(x) : U \rightarrow [0.0\ 1.0]$$

where the degree of membership function is the characteristic function with the range $[0.0\ 1.0]$ allowing a continuum of possible choices.

One of the main motivations for making such an extension to the ordinary set theory is the need to develop a methodology for the formulation and solution of problems which are too complex or ill-defined to be susceptible to analysis by conventional techniques. In this context we would like now to briefly describe what is known in the literature as the natural language computations, [8]. We first note that, almost at the very core of our traditional view of scientific knowlege, there is a tendency towards precise quantification, towards numbers and the precision they apparently imply. However, sacrificing the 'precision' of the numeric estimate may sometimes mean gaining the believability and confidence of an 'inexact' fuzzy estimate that is both realistic and easier to interpret. To further elaborate on this, let us for example have a closer look at the philosophy underlying the analysis of structural sensitivity to imperfections of design parameters. If we intend to account for imperfections which are either precisely known or have a clearly statistical nature then the appropriate numerical considerations will lead to a rational determination of the structural sensivity. However, some real-world events (negligent checking on quality of welding, unpredicable transportation damages, etc.) will be ignored in this 'precise' analysis because it is unrealistic to calculate the probability of some highly unusual accidents and hazards - there is just not sufficient data for such a mathematically precise estimate. Realizing this inherent lack of precise and complete data, it would seem that rather than estimating the probability of, say, some transportation component damage as 0.0123 it is really more accurate to say that the possibility of this to happen is 'extremely low'.

The above considerations will make sense only if we are able to develop methods of approximate processing of expressions like 'extremely low', 'somewhat high', etc. In other words we need to work out computer utilities which in the process of symbolic computations would assist one in considering simultaneously a very large number of 'fuzzy' factors contributing to the analysis in question.

An ideal model for natural language computations is offered by the theory of fuzzy sets which may be thought of as 'an attempt' to remove linguistic barriers between humans, who think in fuzzy terms and machines, that accept only precise instructions', [9].

Consider the stress analysis at a point within a structure. Let the allowable stress be σ_a; assume that the expression 'high' (stress) has the fuzzy set representation shown in Fig. 1a (positive stresses are only considered for simplicity).

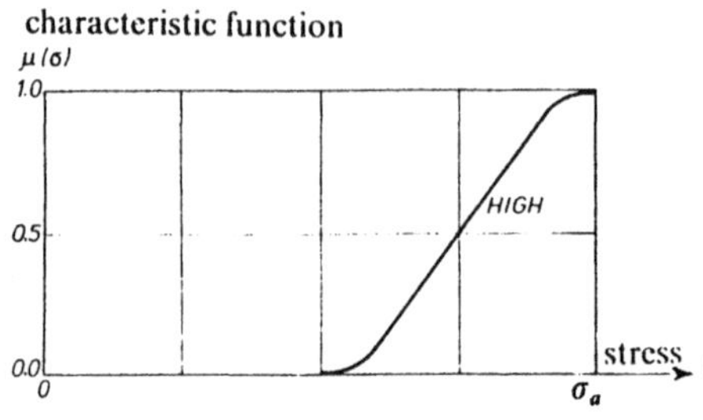

Fig. 1a

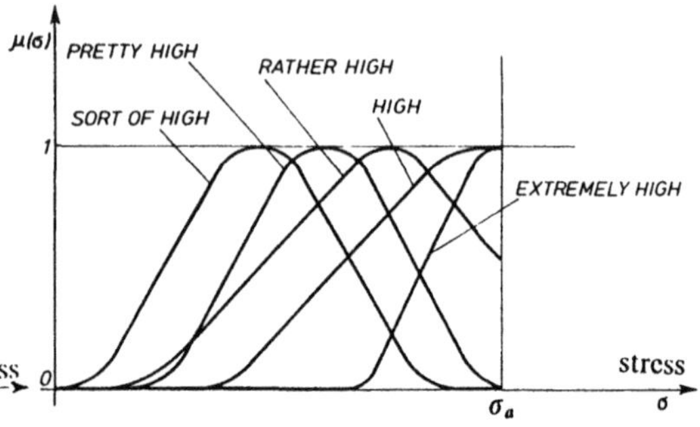

Fig. 1b

Some approximate analysis related to such diverse disciplines like psychology, linguistics and computer science makes it possible to represent terms like 'very high', 'pretty high', etc. in the way shown in Fig. 1b. Even if such definitions are to a great extent biased decisions of the analyst, results of natural language computations turn out to be relatively insensitive to details of particular fuzzy set representations for given expressions.

Every fuzzy set representation can easily be given an analytical description, which can then be used for performing operations (additon, multiplication, etc.) on natural language expressions. Without going into details of such a formalism, [5, 8], we shall just consider one simple example to illustrate the approach.

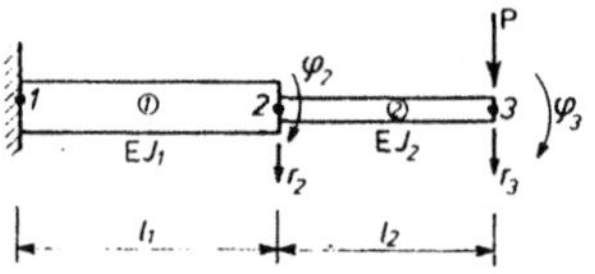

Fig. 2

Consider a cantilever shown in Fig. 2 loaded by the end force P. The expressions for the tip vertical displacement is easily obtained as

$$r_3 = \frac{P}{3EJ_1}(l_1^3 + 3l_1^2 l_2 + 3l_1 l_2^2) + \frac{Pl_2^3}{3EJ_2}$$

The sensitivity of r_3 with respect to imperfections in the element lengths l_1 and l_2 may be expressed as

$$\frac{\delta r_3}{r_3} = \frac{l_1}{r_3} \frac{P}{EJ_1} (l_1 + l_2)^2 \frac{\delta l_1}{l_1} + \frac{l_2}{r_3} \left[\frac{P}{EJ_1} (l_1^2 + 2l_1 l_2) + \frac{Pl_2^2}{EJ_2} \right] \frac{\delta l_2}{l_2}$$

Let $l_1 = l_2 = 100$cm, $J_1 = 2 \cdot 10^3 cm^4$, $J_2 = 10^3 cm^4$, $E = 2 \cdot 10^7 N/cm^2$, $P = 10^4 N$. Assuming that $\frac{\delta l_1}{l_1} ='\ very\ low'$, $\frac{\delta l_2}{l_2} ='\ medium'$ the sequence of natural language transformations (for a particular set of natural expression representations and rules for manipulating them as discussed in [1, 5]) is performed as follows:

$$\frac{\delta r_3}{r_3} = sort\ of\ (very\ low) + slightly\ greater\ than\ (medium)$$
$$= very\ low + slighly\ more\ than\ medium = medium$$

Similar natural language transformations may be applied to assess sensitivity of large (linear and nonlinear) structural systems by using traditional FEM programs implemented with routines performing fuzzy operations, [5].

Inductive knowledge acquisition for expert systems

In this section we shall briefly address the fascinating issue of expert system learning capabilities. A major 'bottle neck' in developing any expert system is the problem of effective knowlege acquisition (usually from one or more experts). $ID3$ induction algorithm described in [10] offers an alternative: it can be used to automatically generate 'if...then' rules from a set of solutions to a sequence of benchmark problems. The algorithm is based on the information-theoretic measure of entropy (i.e. surprise) which is used for assessing the discriminatory power of particular variables appearing in the set of benchmark solutions. The data is partitioned into two subsets on the basis of the most discriminatory variable, and each subset is subsequently partitioned in a similar way until it contains examples of only one class(exception driven filtering of data).

Practical application of the induction algorithm to a FEM-related expert system will be illustrated below on a simple example of generating rules for modelling 2D cantilever structures under plane stress with different kinds and arrangements of elements, Fig. 3.

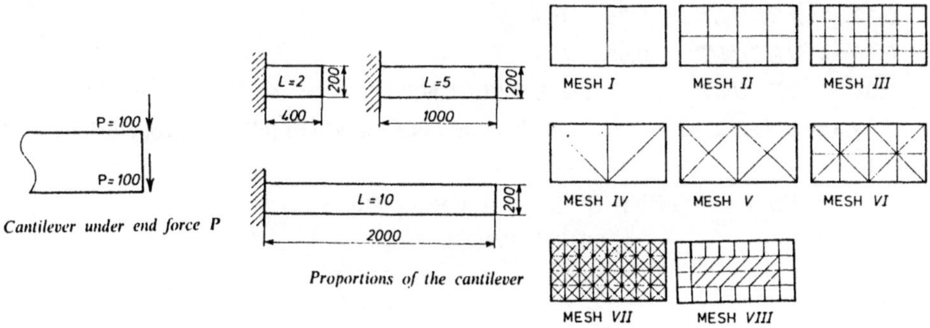

Fig. 3

The randomly selected problems are specified by the following attributes:

$L(=2,5,10)$ - aspects ratios, Fig.3 ,

$MESH(=I,II,...,VIII)$ - mesh type, Fig.3 ,

DOF - number of degrees of freedom in the mesh,

EL - element type selected out of the following list:

A - rectangle with linear displacement functions,

B - rectangle with linear displacement functions and constant shear stress,

C - isoparametric quadrilateral,

D - isoparametric quadrilateral with constant shear stress,

E -triangle with linear displacement functions,

F -triangle with in-plane rotational stiffness.

The results (attribute $RESULT$) of computations are expressed as the ratio of the calculated end displacement r_{cal} to the reference value r:

$$RESULT = \frac{r_{cal}}{r} \quad \text{where} \quad r = \frac{P\,l^3}{3\,E\,J} + \frac{1.5\,P\,l}{A\,G}$$

and P=200., E=2.1E6, G=8.1E5, A=0.05, J=4.167E-3.

Results of 10 (out of 58) solved tests arranged in two columns are given as

((EL	L	MESH	DOF	RESULT)					
(A	2	I	8	0.64)	(C	10	III	400	0.96)
(A	2	II	24	0.81)	(D	10	I	40	0.90)
(A	2	III	80	0.90)	(D	10	II	120	0.97)
(B	2	I	8	0.82)	(D	10	III	400	0.99)
(B	2	II	24	0.88)	(E	10	IV	40	0.29)

By partitioning the results as follows:

$$\begin{aligned} RESULT &< 0.50 : BAD \\ 0.50 \le RESULT &< 0.75 : POOR \\ 0.75 \le RESULT &< 0.95 : GOOD \\ 0.95 \le RESULT & \quad\;\; : EXCELLENT \end{aligned}$$

our own implementation of the $ID3$ algorithm generated the following conclusions:

Rule 1:

IF $MESH = I$ THEN IF $EL = A$ OR C OR F
THEN $POOR$
ELSE IF $EL = B$ OR D
THEN $GOOD$.

Rule 2:

IF $MESH = II$ THEN IF $EL = A$ OR C OR F
THEN $GOOD$
ELSE IF $EL = B$ OR D
THEN IF $L = 2$ OR 3
THEN $GOOD$
ELSE IF $L = 10$
THEN $EXCELLENT$

Rule 3:

IF $MESH = III$ THEN IF $EL = A$ OR F
THEN $GOOD$
ELSE IF $EL = B$ OR C OR D
THEN IF $L = 2$
THEN $GOOD$
ELSE IF $L = 3$ OR 10
THEN $EXCELLENT$.

Rule 4:

IF $MESH = IV$ THEN BAD.

Rule 5:

IF $MESH = V$ OR VI THEN $POOR$.

Rule 6:

IF $MESH = VII$ THEN $GOOD$.

The commonsense interpretation of the above rule would be:

'For cantilever structures the accuracy of FEM results is primarily determined by the type of mesh used whereas for some types of elements results become better with increasing length to height parameter'

It is interesting that the number of degrees of freedom as well as the use of distorted elements (to the extent shown as $MESHVIII$!) are of secondary importance and some meshes are disqualified for all types of elements. We should also note that sensitivity of these results to the number of random tests is clearly an important issue which remains to be investigated.

Bibliography:

1. Breitkopf, P.; Kleiber, M.: Knowlege engineering enhancement of finite element analysis, Comm. Appl. Num.. Meths. 3 (1987) 359-366.

2. Breitkopf, P.; Kleiber, M.: Non-numerical computational techniques in structural mechanics, Adv. in Mechanics, 11 (1988) 109-139.

3. Ambroziak, J. R.; Kleiber, M.: Qualitative reasoning about constitutive models, in: Proc. Inf. Conf. on Computational Engineering Science, Atlanta 1988. New York: Springer 1988.

4. Ambroziak, J. R.,Kleiber, M.: A blackboard consultation system for constitutive modelling in solid mechanics in: AI in Computational Engineering , Kleiber, M.: editor, Ellis Horwood 1989.

5. Kleiber, M.: Natural language estimates of nonlinear response structural sensivity, Comp. Mech. 4 (1989).

6. Genesereth, M. R.; Nilsson, N. J.: Logical foundations of artificial intelligence, Los Altos: Morgan Kaufmann 1987.

7. Zadeh, L. A.: Fuzzy sets, Information and Control, 8 (1965) 338-353.

8. Schmucker, K. J.: Fuzzy sets, natural language computations, and risk analysis, Computer Science Press 1984.

9. Gupta, M. M.: 'Fuzzy-ism' - The first decade, Fuzzy Automata and Decision Processes, Elsevier 1977.

10. Ross Quinlan, J.: Learning efficient classification procedures and their application to chess and games, Machine Learning, Tioga 1983.

The Implementation of Finite Element Computations in Transputer Based Systems

D. R. J. Owen
University College of Swansea - UK

J. S. R. Alves Filho*
Instituto de Pesquisas Espaciais – Brasil

G. P. Mitchell
University College of Swansea - UK

K. Kasper
University of Stuttgart – West Germany

Abstract

The next generation of computers will rely heavily on parallel architectures to achieve better performances. In the pursuit of faster and more powerful finite element programs, the consideration of parallel processing is today a necessity. Transputer based systems are one of the first multi-processor systems available which provides the necessary flexibility, ease of use and low cost. A parallel finite element program is presented in this paper for the solution of elastic-plastic problems. It employs a completely distributed data structure and a parallel Cholesky solver. The results are very encouraging showing that large speed-up factors can be obtained in comparison to popular sequential machines.

1 Introduction

The Finite Element method is well established today as a reliable and general method for solving engineering problems and its development as a numerical method followed closely the advances in computer technology. Better computers produce better analyses in the sense that larger problems can be solved faster and with improved accuracy. The need for greater memory capacity and faster processors led finite element programmers to use the latest generation of computers available. Large mainframes were replaced by *supercomputers* which, in time, will incorporate more and more parallelism in their architectures.

The rapid development of microprocessor technology today allows the production of very fast chips but on the other hand leaves little physical possibilites of further improvements. Consequently, new computer architectures are now having to consider parallelism as a mean of obtaining better performance. There are several ways in which a computer can take advantage of a parallel architecture and

*presently at University College Swansea

G. Kuhn · H. Mang (Eds.)
Discretization Methods in Structural Mechanis
IUTAM/IACM Symposium Vienna, Austria, 1989

the simplest is the exploitation of concurrency inside the processor itself. The next instruction, for example, can be set up while the current instruction is being processed. Techniques of this type are already being used in the design of most processors and its existance is completely transparent to the user. Another form of parallelism is present in vector-processor computers. These machines have the hardware capability of performing the same operation on different components of a vector at the same time. The final performance depends to what extent the code can be *vectorized* and good programming may help in achieving better performances. The ultimate stage in using parallelism is the consideration of multi-processor systems where several processors are joined together to solve the same problem. In this case the software has to be much more specialized to the particular hardware than for previous systems. Transputer based machines fall into this last category.

The transputer is normally defined as a computer on a chip and contains a 32-bit central processing unit, a 64-bit floating-point unit, 4 Kbyte of fast memory and four independent serial links. Up to 4 Gbyte of external memory can be addressed by the processor and an infinite number of units can be connected together to form complex networks. A new parallel language called OCCAM was developed with the transputer which facilitates the implementation of codes. One of the main features of the transputer is its capability of supporting a parallel environment in hardware . This means that the same algorithm coded in OCCAM can run in one or in several transputers with no change to the logical structure of the code. The performance/cost ratio of transputer systems makes them very appealing when compared to other computers with the same nominal performance.

The use of transputer systems in finite element computations is therefore completely justified by the above considerations. Apart from utilizing a faster and less costly system, the development of parallel techniques in finite elements will undoubtedly create a new generation of programs more adapted to the new generation of computers. This paper presents an overview of parallel processing in Section 2 and some considerations on how finite element programs can benefit from such techniques in Section 3. A program called PLAST2 is considered in the rest of the work as an example of implementation of a parallel finite element program. PLAST2 was written in OCCAM and performs elastic-plastic analysis for plane-stress, plane-strain and axisymmetric problems. A summary of elastic-plastic finite element theory is presented in Section 4 and Section 5 introduces the parallel algorithm for the Cholesky solver used in the program. The parallel structure of PLAST2 is fully described in Section 6. Finally a discussion of the results and of future developments is presented in Section 7.

2 Issues in Parallel Processing

The idea behind parallel processing is very simple: concurrent tasks can be simultaneously performed with a consequent reduction of the total processing time. However, the implementation of parallel programs in multi-processor machines is not straightforward. Since, the exploitation of concurrency in the algorithm is dependent on the existance of concurrency in the first place. A strictly sequential process cannot benefit from a parallel system. Even when concurrency is present, the final performance can be strongly compromised by, among other reasons, inter-processor communication, synchronization of processes and unequal processing load between processors. This section discusses some general aspects of parallel processing.

The performance of a parallel machine is usually measured by means of a *speed-*

up factor relating the sequential and the parallel total times for the same application as:

$$S = \frac{T_{seq}}{T(P)} \tag{1}$$

where T_{seq} is the sequential time and $T(P)$ is the time taken by the parallel machine with P processors. The speed-up factor can also be used to measure the performance of a parallel algorithm for an application. In this case, the sequential time is normally considered to be the time taken by the parallel algorithm running with only one processor and therefore $T_{seq} = T(1)$. The efficiency of the algorithm can be defined using the speed-up factor as:

$$E = \frac{S}{P} \tag{2}$$

This efficiency parameter (which is less than or equal to unity) gives an indication of how well all processors are being used by the implementation.

Load balancing is an important consideration when designing a parallel algorithm. If some processors have less work to do than others, it is clear that they will have to wait for the more busy processors to complete their tasks. Poor load balancing can cause a low value of efficiency, even on an ideal parallel machine in which no overheads are present. If M independent processes are distributed among the P processors, the problem grain size given by $\beta = M/P$ can indicate how good the load balancing will be. If P is a factor of M, perfect load balancing is achieved. If it is not, the load balancing will be worse, the smaller the grain size β. Amdahl's Law should also be mentioned: this law states that the efficiency of a parallel application is limited by the amount of sequential code that cannot be parallalised [7].

Inter-processor communication (which does not arise in sequential machines) is also an important issue in parallel systems. The speed of communication between processors is normally much slower than the speed of the processor itself and on a transputer, this difference is approximately one order of magnitude at present. A reasonable ratio between communication and processing must exist in the algorithm for it to provide good performance. The more computer intensive the application and the smaller the need to exchange data between processors, the larger the speed-up that can be extracted from the system.

The way in which the various processors are linked together can also strongly affect the performance of a parallel program. In theory, the hardware configuration should map the software configuration in well designed systems. In practice, however, the limitations imposed by hardware and software can make the situation rather different. The transputer, for instance, can only be connected to four other transputers which limits the number of ways in which transputers can be linked together. On the software side, the need to route messages through the network for cases when the sender and the receiver processors are not directly linked may complicate the code and at least make it very specific to a particular machine or configuration.

A system called *Parallel Harness* (PH) was used to implement the program described in Section 6. It uses a fixed processor configuration and provides simple routines to distribute messages across the network. The messages can simply be forwarded to a target processor or broadcast to all other processors in the network. The system consists of a *master* processor, connected to the host machine, and a number of *worker* processors. Every processor can perform I/O operations which

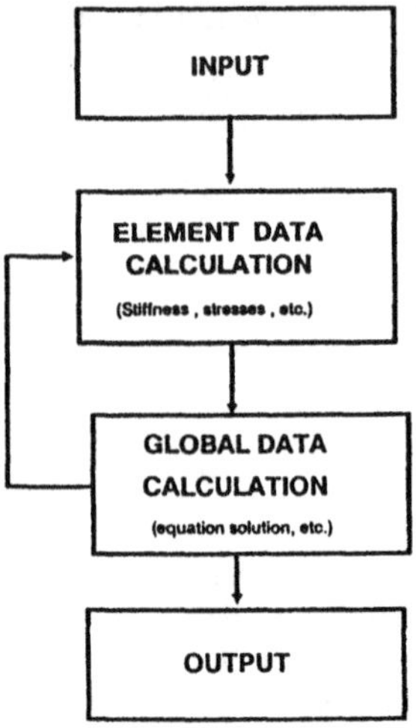

Figure 1: Basic scheme for a general finite element program.

are handled by the host via the master processor. A full description of the system is given in [3].

Designing a parallel program involves the organisation and distribution of data among the various processors. Because the processors are independent, the message-passing must be well managed in order to ensure that the relevant data arrives at the right place at the right time. The data structure must be designed in such way that good load balancing is achieved as well as ensuring that communication is minimised. A parallel finite element program must, therefore, start with a new and distributed data structure.

3 Concurrency in Finite Element Calculations

The data processed in a typical finite element program is summarised by the scheme shown in Figure 1. Three main types of data can be identified. The first is the input data which consists mainly of control parameters, the mesh geometry, element connectivities and material properties. The second, called element data, consists of data processed during the analysis for each element (eg. stresses) and is completely independent of other elements. The third and final group, called global data, usually consists of data associated with the nodes (eg. assembled stiffness matrix).

The input data is not changed during the analysis and should not be communicated between processors once it has been loaded into the appropriate processors for solution. In parallel systems, memory is not (or should not be) as big a limitation as in sequential systems. Optimizing communication usually takes a higher priority than saving memory space. All input data needed for a particular element should be kept local to the processor to which that element was assigned (even though some duplication may occur).

An obvious source of concurrency in a finite element program are the elements themselves. Operations performed on an element level are completely concurrent and different elements can be processed in different processors without the need for any communication whatsoever. Load balancing can be achieved by simply dividing the elements equally among the processors (noting the observations about grain size in the previous section). If a non-homogeneous mesh is considered, an equal distribution may not be so obvious but it is still a fairly straightforward task.

The same level of concurrency cannot be provided for operations involving the global data of finite element problems. In most cases (such as elastic-plastic analysis), a equation solver is used to obtain the solution vector. The Cholesky solver described in Section 5 shows that a great deal of communication is involved and that the way in which data are distributed among the processors affects the final performance. Therefore, the way in which the nodes are distributed among the processors must follow a well-thoughtout strategy. Load balancing is not just achieved by distributing the nodes equally but also by the way in which this distribution takes place.

As can be observed in Figure 1, the finite element program alternates between element level operations and global level operations. The switching between the two different types of data structures must also be considered as it involves interprocessor communication. One important point to note here is that the distribution of elements and nodes among the processors will only affect the performance of the program. Whatever distribution is chosen, the algorithm for the analysis can be considered to remain logically the same. There will, however, always be an optimum data distribution for any particular algorithm.

Substructuring can be used as a way of distributing the data [2,5]. This is not considered here due to the complexity in producing a general algorithm able to deal with all sorts of mesh geometries. A more general approach is proposed in this work and implemented in program PLAST2 in which the elements are freely distributed among the processors and the distribution of nodes follows the strategy required by the solver algorithm. In this way, the overall performance of the program is dependent on the amount of communication needed for the assemblage phase and on the efficiency of the solver itself. Performance tests with PLAST2 will show the validity of these assumptions.

4 Elastic-Plastic Finite Element Analysis

A summary of the basic numerical procedure used in elastic-plastic finite element analysis is presented in this section. A more detailed explanation of the finite element method and of the theory of plasticity can be found in the literature [4,8]. The general finite element equations to be solved in the elastic-plastic problem can be written as:

$$\mathbf{K}(\sigma, q)\Delta q + \Delta f = 0 \tag{3}$$

where $\mathbf{K}$ is the stiffness matrix, σ the stress field, q the nodal displacement vector and f the load vector. The total load is applied in a sequence of increments and a Newton-Raphson iterative procedure is used to solve for the incremental displacements.

The basic steps to solve for the displacements for an increment n are as follows:

1. The stiffness matrix of each element is calculated using the consistent tangent modulus $\hat{\mathbf{D}}_i$ associated with the relevant stress calculation [6] as:

$$\mathbf{K^e}_i = \int \mathbf{B}^T \hat{\mathbf{D}}_i \mathbf{B} dV \tag{4}$$

where subscript i indicates the iteration number and **B** is the strain-displacement matrix where $\Delta\epsilon = \mathbf{B}\Delta q$.

2. The global stiffness matrix is assembled from the element contributions and a displacement increment for the first load increment is obtained as:

$$\Delta q_{ni} = \mathbf{K}^{-1}\Delta f_n \quad (5)$$

3. An approximation to the corresponding stress level in every element is given by:

$$\sigma^e_{ni} = \sigma_{n-1} + \mathbf{DB}\Delta q_n \quad (6)$$

where σ^e is the *elastic predictor* for the stresses and Δq_n is the cumulative displacement increment for this load increment.

4. The yield function φ is calculated using the elastic predictor and if $\varphi > 0$ a plastic relaxation procedure is performed to return the stress point to the yield surface ($\varphi = 0$). For a Von Mises yield function this plastic relaxation can be done in one step [6] but other surfaces, such as Mohr-Coulomb, require an iterative procedure. A new approximation for the element stresses σ_{ni} and the plastic strains ϵ^p_{ni} are obtained.

5. The stresses σ_{ni} are used to calculate the internal force vector for each element and a global residual vector is calculated as;

$$\Psi_i = \sum\left(\int_e \mathbf{B}^T\sigma_{ni}dV\right) - f_n \quad (7)$$

where f_n is the total current applied load.

Convergence of the iterative process is assumed if an appropriate norm (eg. force or displacement norm) of the residual is smaller than a preselected tolerance. In this case the same procedure is followed for the next load increment until the total load is applied. If convergence did not occur, Steps 1-5 are repeated with the residual Ψ_i replacing the load increment Δf_n in Step 2.

5 A Parallel Equation Solver

A parallel Cholesky algorithm was used in the factorisation of matrix **K** for the solution of the set of linear equations (5). This algorithm was developed from the active column equation solver presented by Farhat and Wilson [1] in which the upper triangle of the symmetric matrix **K** is stored by columns using a sky-line storage scheme. The columns are distributed among the processors on a round-robin basis as shown in Figure 2. In this way, load balancing is enforced until every processor has only one equation. The efficiency of this algorithm is largely dependent on the grain size β of the problem. The reason for this is that the elimination of every equation in the factorization process involves the broadcasting of the active column. This broadcasting must be counterbalanced by a minimum amount of processing. If there are not enough columns per processor (low value of β), the efficiency is compromised. It will be seen that for the banded matrix **K**, the semiband width B will dictate the grain size instead of the total number of equations N.

Matrix **K** is factorized by the Cholesky algorithm into matrix **L** so that:

$$\mathbf{K} = \mathbf{L} * \mathbf{L}^T \quad (8)$$

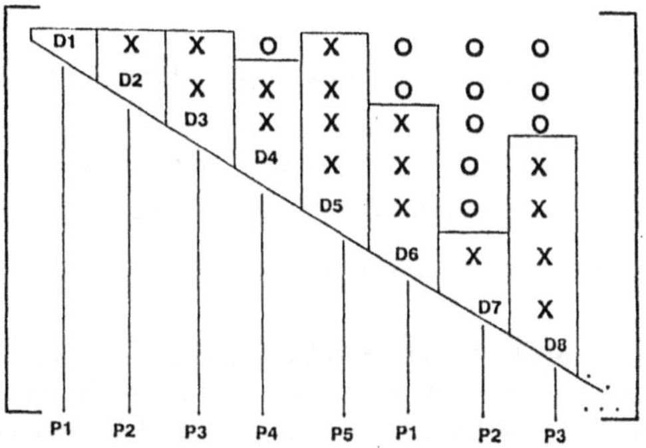

Figure 2: Distribution of matrix columns for the parallel Cholesky algorithm

where

$$L_{ii}^T = \sqrt{K_{ii} - \sum_{k=1}^{i-1} L_{ki}^2} \tag{9}$$

$$L_{ij}^T = (A_{ij} - \sum_{k=1}^{i-1} L_{ki}^T L_{kj}^T)/L_{ii}^T \tag{10}$$

$$i < j \leq N \quad , \quad i = 1, \ldots, N \tag{11}$$

After factorization, equation (5) can be written in the more general form:

$$\mathbf{L} * y = f \tag{12}$$

where

$$y = \mathbf{L}^T * q \tag{13}$$

The solution for vector q can be obtained by performing a forward substitution in equation (11) to calculate vector y which can then be used in equation (12) to obtain q by means of a backward substitution algorithm.

The factorization is performed by calculating the rows of matrix $\mathbf{L}^\mathrm{T}$ starting with the first one and sequentially advancing down until the Nth column is complete. The calculation of the elements of row i requires column $j = i$ which is called the active column at that instance. Therefore, the processor containing column i must broadcast that column to all other processors to enable them to calculate the elements of row i for all the local columns. Due to the banded nature of the finite element matrix $\mathbf{K}$, the number of elements calculated for every row is approximately the same and hence the processing load is constant during the factorization. Consequentely, the processing load will be given by the matrix semiband width and not by the total number of columns N.

The forward substitution can be done at the same time as the factorization as described below. Together with the active column, every processor can also broadcast the respective value of vector y which can be calculated as $y_i = f_i/L_{ii}$. After calculation of the elements of row i in every processor, the right-hand-side vector can be updated as $f_j = f_j - L_{ij}y_i$. This procedure improves the overall efficiency because a separate broadcasting for the forward substitution is no longer necessary.

Equations (9),(10) and (11) can be rewritten for the parallel algorithm outlined above as:

DO sequentially FOR $i = 1 \rightarrow N$

1. Process containing column i calculates diagonal term L_{ii} from equation (9) and $y_i = f_i/L_{ii}$.
2. The active column i plus y_i are broadcast to the network.
3. All processes calculate in parallel the elements of row i using equation (10).
4. All processes update the right-hand-side vector using: $f_k = f_k - y_i * L^T_{ik}$.

The implementation of the backward substitution is complicated by distribution of the matrix by columns which is dictated by the factorization process. This data distribution forces the backward substitution to be done sequentially causing a speed-down factor due to the communication required. Fortunately, the percentage of the total time spent on this process is very small and therefore the overall efficiency is not seriously compromised. Nevertheless, the efficiency for the entire solution procedure is decreased by the backward substitution. The final vector Δq is collected by the master processor.

Figure 3 shows the efficiency values obtained by the algorithm described above written in OCCAM and running with 4,8,12 and 16 processors. The effect of grain size on the overall efficiency can be observed in this figure where semibandwidths of 100%, 50% and 10% are considered. The sequential times used to calculate these efficiency figures were measured using a sequential Cholesky algorithm, written in OCCAM. Due to memory limitations, the maximum value of N considered for the sequential algorithm was 2200. Figure 3.a shows the efficiency of the factorization plus forward substitution phase whereas Figure 3.b shows the overall values including the backward substitution. The effect of the poor performance of the backward substitution algorithm can be seen comparing the two figures. It can be observed that although the efficiency decreases with the number of processors, the speed-up is always increased. It means that for a fixed value of memory size, the distribution of the memory over a larger number of processors may lead to a better final performance than concentrating the memory over a smaller number of processors. While in the latter case the utilisation of the processors would be more efficient, the former presents higher values of speed-up.

The algorithm implemented in program PLAST2 is slightly different from the one described above. This was necessary due to the fact that the program assembles the total stiffness matrix **K** which includes the prescribed degrees-of-freedom. These equations are eliminated from the system solution simply by skipping over them in the algorithm. The inclusion of this modification led to a decrease in the solver efficiency of approximately 15%.

6 Program PLAST2

PLAST2 was developed using a general parallel data structure for finite element problems. All data are distributed among the processors and for inter-processor communication the parallel harness (PH) described in Section 2 was used. The PH system carries out all message-passing and buffering routines so that the code is much clearer and easier to develop.

The program consists of two similar algorithms: the *master* and the *worker*. The *master* algorithm runs on the MASTER processor which is connected to the host. It is not heavily involved in the computations and is mainly responsible for

I/O and control operations. The *worker* algorithm runs on all the other processors available in the network and is responsible for the bulk of the calculation.

The memory is dynamically assigned in both *master* and *worker* using a global storage vector. Pointers to this vector are calculated for all variables which are re-typed whenever required. All data relating to an element are kept as a self-contained set of data in a vector called *element record.* Each *worker* is assigned a number of element records and no element data need to be communicated between processors unless for updating purposes (which is normally done through the *master*). Global data is organized by nodes and the nodes are distributed among the *workers* in a similar way to the elements. All global data relating to a particular node are kept local to the processor to which that node was assigned. Some data is broadcast to all processors to avoid communication. These are: control parameters, material properties and nodal fixity conditions.

The assignment of elements and nodes to the various *workers* is controlled in the program by two vectors: elementmap and nodemap. These vectors give for a particular element or node its logical address (i.e. the *worker* identification number). Referring back to the discussion in Section 3, these vectors are a key factor in achieving two important objectives: load balancing among processors and minimization of inter-processor communications. For the current implementation, elements and nodes are distributed on a round-robin basis which follows the scheme considered for the Cholesky algorithm in Section 5.

Figure 4 shows the basic algorithm for both the *master* and the *workers*. The al-

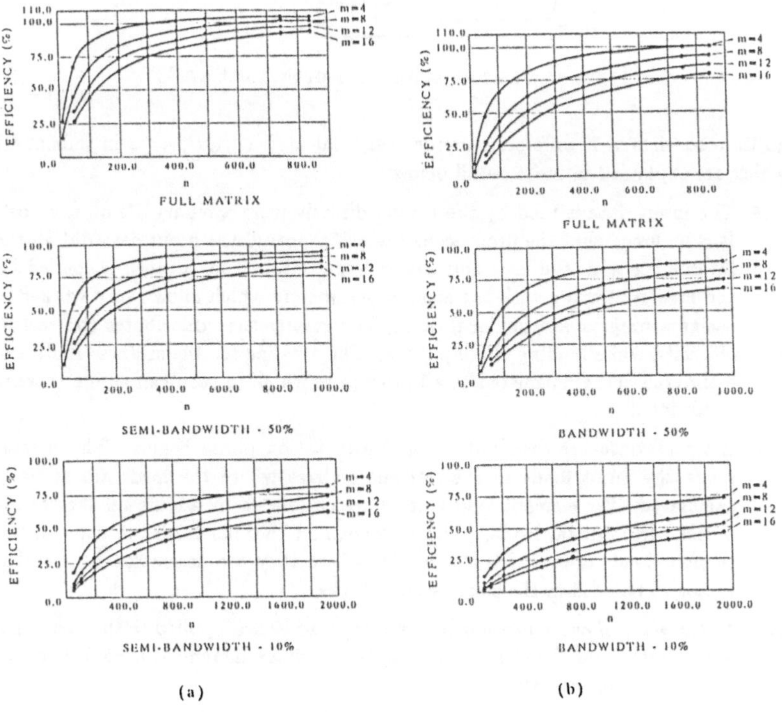

Figure 3: Efficiency of the parallel equation solver.

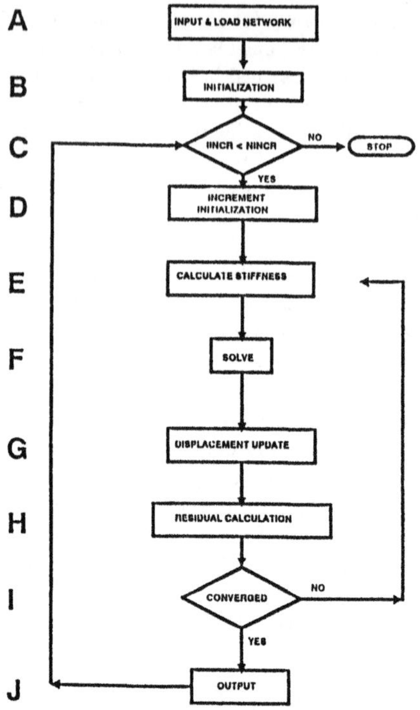

Figure 4: Basic algorithm for program PLAST2

gorithm consists of 10 different subprocesses (indicated as $A, B, \ldots, J$ in the figure) which are explained in more detail below:

A The input data is read by the *master* directly from a binary file on the host. It is assumed that the preprocessor which prepared the input data did all the interfacing with the user and checked the data. The network is loaded by the *master* starting with the control parameters which allow every worker to assign storage space for local data. The *master* then distributes the rest of the data while reading the input file. The sky-line for the stiffness matrix is estimated and transmitted to all processors which in turn can assign storage space for it.

B After all data are distributed, the initialization phase begins. The *master* sends the input loads to the element address where the loads are actually calculated. The element load vectors are calculated by adding up the contributions from nodal loads, pressure loads and thermal loads. A copy of the element loads are sent to the *master* where they are assembled in a global vector to be used in the convergence checks.

C,D At the start of every increment, the total load vector applied to the structure is calculated and the element data of the previous increment is copied to the converged element data.

E A signal is broadcast to all processors which initiates the calculation of the stiffness matrix and load vector for all elements. Simultaneously, the global

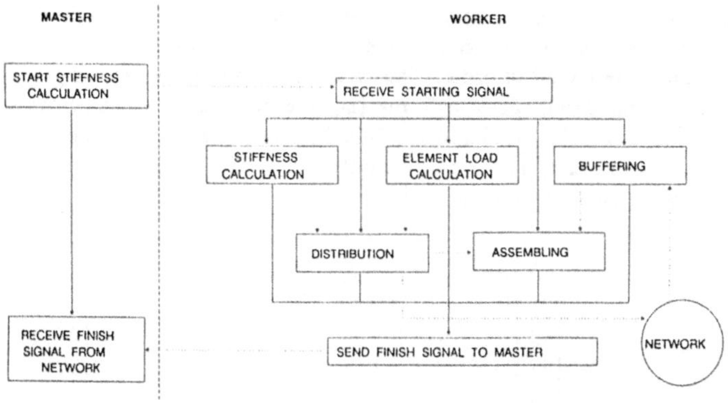

Figure 5: Independent processes involved in the stiffness matrix calculation and assembly.

stiffness matrix and load vector is assembled. Figure 5 illustrates the various processes active at this stage and the channels of communication between them. The first one calculates the stiffness matrix for all elements in this processor, sending them to the distribution process. A second calculates the load vector for all elements, also sending them to the distribution process. The distribution process receives the stiffness matrix and the load vector for every element and distributes them through the network according to the addresses of its various degrees of freedom (dof). The assembling process receives the data sent by all distribution processes in the network and assembles the global stiffness matrix and load vector for the nodes assigned to that particular processor. Each processor sends an acknowledgement to the *master* to signal it is finished. This acknowledgement is necessary to synchronize all processors and ensure that no messages from different phases are mixed up.

F The solution of the system of equations is commenced by a signal from the *master*. The Cholesky algorithm described in Section 5 is used for the factorization of the global stiffness matrix after which forward and backward substitution phases are performed. When the solver is finished, the displacement increment vector is returned to the *master* processor.

G The *master* processor sends to each element an update of its nodal displacement increments.

H,I These displacement increments are used to calculate the internal stresses for all elements. This calculation is done with no communication needed but for the residuals and reactions which are sent to the *master* where a convergence check is performed. The result of this check is broadcast to all processors. If the solution has converged, data is output if requested and a new load increment is applied. If the solution hasn't converged, a new iteration is performed. Every cycle is similar to that previously described with the difference that depending on the solution strategy employed the stiffness matrix is not calculated in every iteration. When this happens, only the load vector is updated, distributed and assembled and the solution procedure uses the previously factorized stiffness matrix to obtain the new displacement vector.

J The output is written directly by the processors into a binary file in the host. A post-processing operation is needed afterwards to organize the data that invariably will be out of order. This dumping of output is necessary to avoid to much time being spent in formatting the output. If a post-processor is conceived to run in parallel in the HOST or in a neighbouring network, the post-processing can be done concurrently and partial results can be visualised before the analysis is finished.

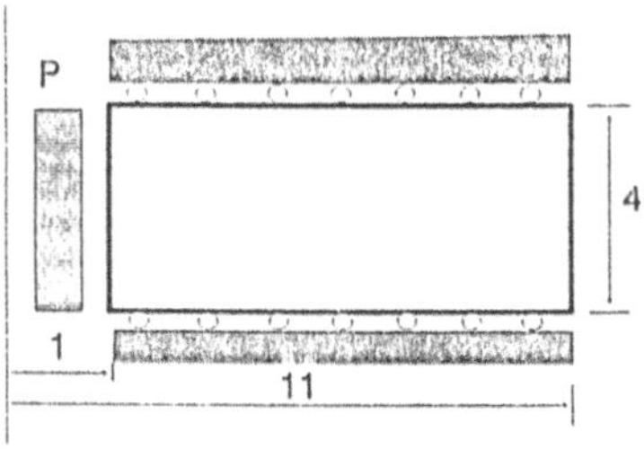

* 8 - noded elements
* Von-Mises Yield Criteria

MODEL	No. of Elements	No. of Nodes
C	8	37
D	16	69
E	32	121
F	64	233
G	128	433
H	256	849

Figure 6: Geometry and mesh sizes for the test model.

7 Discussion

Tests were carried out on a simple problem to study the effect of increasing the number of degrees of freedom of the model and the number of transputers in use. Figure 6 shows the geometry and the different mesh sizes considered for the test. The model consists of a thick cylinder under internal pressure for which an axisymmetric analysis was performed. The loading condition was chosen so as to ensure a plastic state in part of the mesh. The results agree with the similar sequential program μPLAST written in FORTRAN. The performance results can be seen in Figure 7.

Figure 7.a shows the speed-up factors obtained for PLAST2 when compared to program μPLAST running on a μVAX. The effect of the model's grain size can be clearly seen in that figure. For the largest model H, the speed-up with 17 transputers can be as high as 18 which is an encouraging figure. It has to be pointed out that program μPLAST performs some input operations for checking the data which are not done in PLAST2. The effect of these operations on the total time of the program is, however, small and are believed not to affect the speed-up factors shown considerably. Figure 7.b presents the speed-up factors calculated with the time of PLAST2 with 2 processors as the sequential time. The same effect of grain

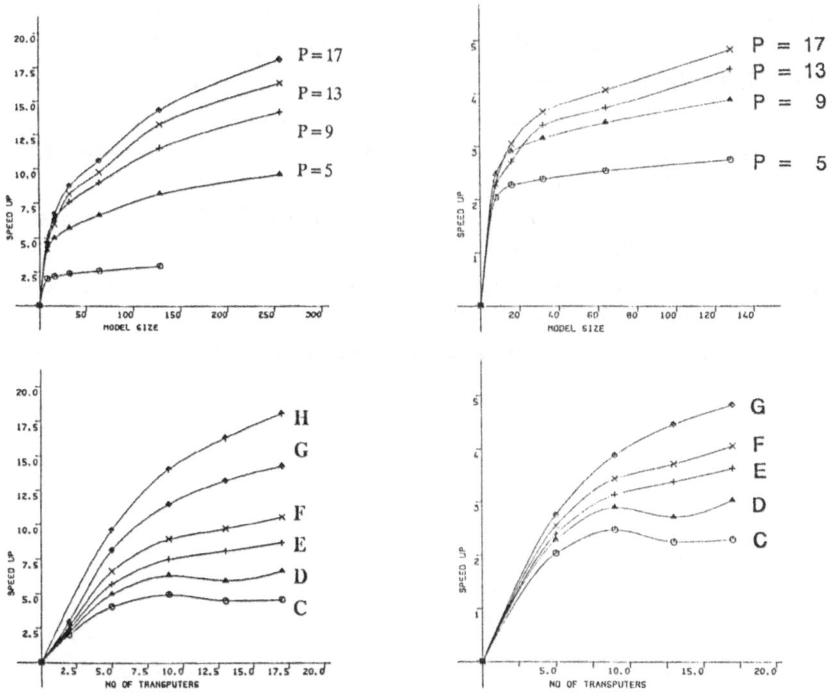

(a) (b)

Figure 7: Speed-up obtained for program PLAST2.

size on performance can be observed. However, the speed-up values are now smaller and can give a better idea of the performance of PLAST2 with different number of processors.

The results obtained show that the use of transputer systems can improve the performance of currently available finite element programs. Comparison with a similar program running on a μVAX illustrates also the high performance/cost ratios of transputer systems. It has been shown that the 17-transputer system presents a performance of up to 18 times that of a μVAX.

Acknowledgements: This work was funded in part by the Science and Engineering Research Council, UK, and this support is gratefully acknowledged.

References

[1] C. Farhat and E. Wilson. A parallel active column equation solver. *Computers and Structures*, 28(2):289–304, 1988.

[2] Charbel Farhat. A simple and efficient automatic fem domain decomposer. *Computers and Structures*, 28(5):579–602, 1988.

[3] J. S. R. Alves Filho. *The Use of Transputer Based Computers in Finite Element Calculations.* PhD thesis, Univerity of Wales, Swansea, 1989. to be submitted.

[4] R. Hill. *The Mathematical Theory of Plasticity.* Oxford University Press, 1950.

[5] R.G. Miles and S.P. Harvard. Multifronts and transputer networks for solving fluid mechanical finite element systems. 1988. Department of Mathematics and Computing of Polytechnic of Wales.

[6] G. P. Mitchell and D. R. J. Owen. Numerical solutions for elastic-plastic problems. *Engineering Computations*, 5(4):274–284, 1988.

[7] A. K. Noor. Parallel processing in finite element structural analysis. *Engineering with Computers*, 3:255–241, 1988.

[8] D.R.J. Owen and E. Hinton. *Finite Elements in Plasticity.* Pineridge Press, Swansea, UK, 1986.

A BEM-Formulation of Nonlinear Plate Vibrations

R. Heuer, H. Irschik, F. Ziegler

Institut für Allgemeine Mechanik
Technical University of Vienna, Austria A-1040 - E201

Summary

Using Mindlin's plate theory along with Berger's hypothesis, large amplitude forced vibrations of moderately thick plates on two-parameter foundations are considered. Thereby influence of shear and rotatory inertia are taken into account and the inplane forces caused due to large deflections are assumed to be constant within the plate domain. Considering time-harmonic force excitation a Kantorovich-Galerkin procedure provides the formulation of this problem in frequency domain. An analogy to thin plates without foundations is given, which is complete in the case of polygonal planforms with "hard" hinged supports. In that case the sixth-order Mindlin-problem is reduced to three (second-order) Helmholtz-Klein-Gordon boundary value problems. The first two are solved by an advanced BEM using Green's functions of rectangular domains and Berger's normal force is evaluated from membran deflections by a simple domain integral. This problem oriented strategy provides the nonlinear frequency response functions for the deflection of the undamped Mindlin plate.

1. Introduction

Dynamic problems of moderately thick plates are efficiently dealt with according to Mindlin's plate theory /1/ which yields precise approximations in comparison with solutions deduced from the three-dimensional equations of elasticity within a wide thickness-span range. Hutchinson /2/ confirms this accuracy for free circular plates, and a comparative study of the dispersion relations is presented in /3, p.282/. In the following large flexural vibrations of polygonal plates with "hard" hinged supports are determined, because for rectangular plates with such boundary conditions excellent agreements with three-dimensional results are known, see e.g. /3, p.290/. (Thomas /4/ demonstrates even numerical instabilities which can occur in case of boundary conditions of "soft" hinged supports and in relation to Reissner's theory). Furthermore, it is pointed out in /5/ that the Mindlin theory provides also the shear stress distribution, which may be used to determine the wrapping of the plate cross-section. Knowing the wrapping allows

G. Kuhn · H. Mang (Eds.)
Discretization Methods in Structural Mechanis
IUTAM/IACM Symposium Vienna, Austria, 1989

further refinement in the determination of the nonlinear inplane normal stress distribution, which in turn can be used to find the variation of the transverse displacement and so on. Therefore, from an engineering point of view the use of Mindlin`s theory seems to be appropriate and preferable with respect to the use of numerical three-dimensional routines.

The present paper shows an extension of an advanced boundary element method (BEM) for Mindlin plates /6/ by including an elastic foundation of the plate domain and admitting large deflections. We assume a two parameter Pasternak-type foundation model (see e.g. /7/), which contains the widely used Winkler`s model as a special case. (Following Vlasov, Leont`ev /8/, that model corresponds to a consistent Ritz-type discretization of a linear elastic soil strip.) Considering also large flexural vibrations we make use of Berger`s approximation: Berger /9/ derived a decomposition of v. Karman`s system of two coupled nonlinear differential equations for plate deflection and stress function by neglecting the strain energy density due to the second invariant of the plate`s middle-surface strain. Since the resulting equation is uncoupled and quasi-linear it has been extensively applied in place of the complete v. Karman`s equations; the influence of the v. Karman`s inplane forces is replaced by tensile forces of hydrostatic type being constant within the whole plate domain. Berger`s equation has frequently been discussed for static isothermal and thermoelastic problems, /10/, /11/, /12/, /13/, and it is also applied in dynamic problems, /14/, /15/, /16/. BEMs for thermally loaded Berger-type plates and shells are derived in /17/. It is now widely accepted that Berger`s hypothesis gives a fairly good approximation to the problem provided that the inplane displacements are restrained on the boundary.

At first, Mindlin`s coupled equations of motion are given in their extended form due to the foundation and Berger`s hydrostatic inplane force. Being interested only in the plate deflection, the cross-sectional rotations are eliminated and a single fourth-order equation for deflection is obtained. Harmonical force excitation is assumed and applying a Kantorovich-Galerkin procedure (see /18/) yields the boundary value problem of a fictitious Lagrange-Kirchhoff plate for the undamped frequency response function. As shown in earlier works /19/, /20/, /21/, this problem can be decomposed into two Helmholtz equations with homogeneous boundary conditions, which finally are solved by an ad-

vanced BEM iteratively along with the computation of Berger's deflection-dependent inplane force. The above mentioned BEM makes use of Green's functions of rectangular domains in the kernels of the integrals: embedding the given polygonal domain into a proper rectangular one, the boundary conditions may be satisfied exactly at the coinciding boundaries, and the boundary integrals have to be applied at the remaining parts only. An extensive review of this BEM including numerous references is presented in /21/.
In a numerical example the nonlinear frequency response functions of a trapezoidal plate are worked out in detail.

2. Mindlin's equations and Berger's approximation for nonlinear vibrations of moderately thick plates

According to Mindlin's theory of shear deformable plates with rotatory inertia /1/ plate motions may be expressed by a sixth-order system of differential equations in terms of the plate deflection $\overline{w}$ and the cross-sectional rotations $\overline{\psi}_x$, $\overline{\psi}_y$. Taking also into account hydrostatic inplane forces $\tilde{n}$ and a two-parameter Pasternak foundation /8/ renders (compare /19/):

$$\frac{K}{2}\left[(1-\nu)\Delta\overline{\psi}_x+(1+\nu)\Phi_{,x}\right]-\frac{1}{s}(\overline{\psi}_x+\overline{w}_{,x})=r\ddot{\overline{\psi}}_x\,, \tag{1.1}$$

$$\frac{K}{2}\left[(1-\nu)\Delta\overline{\psi}_y+(1+\nu)\Phi_{,y}\right]-\frac{1}{s}(\overline{\psi}_y+\overline{w}_{,y})=r\ddot{\overline{\psi}}_y\,, \tag{1.2}$$

$$\frac{1}{s}\phi+\left(\frac{1}{s}+e+\tilde{n}\right)\Delta\overline{w}-d\overline{w}+\overline{p}=\rho h\ddot{\overline{w}}\,, \tag{1.3}$$

where $\phi=\overline{\psi}_{x,x}+\overline{\psi}_{y,y}$, Δ is Laplace's operator and a dot stands for time derivative. The influence of shear is characterized by $s = 1/\kappa^2 Gh$ with $\kappa^2=5/6$, and the coefficient of rotatory inertia is defined by $r=\rho h^3/12$; $K = Gh^3/6(1-\nu)$ is the bending stiffness, $\overline{p}$ denotes the lateral loading and d and e are the elastic moduli of the foundation. Following /19/, in case of free vibrations, the problem under consideration is completely determined by the three potentials, w_1, w_2 and $H=\psi_{x,y}-\psi_{y,x}$ ($w=w_1+w_2$ and ψ_x,ψ_y stand for the amplitudes of plate deflection and rotations), which are solutions of the homogeneous Helmholtz equation and they are coupled by means of the boundary conditions. w_1 is called flexural motion, w_2 denotes the thickness-shear motion, whereas the deformation associated with H, which contains no deflection, is called thickness-twist motion, see /1/. In the following only w_1 and w_2 of laterally loaded plates are to be determined.

First $\overline{\psi}_x$ and $\overline{\psi}_y$ are eleminated from eqs. (1) and a single fourth-order differential equation for $\overline{w}$ is obtained:

$$D\{\overline{w}\} = K[1+(\tilde{n}+e)s]\,\Delta\Delta\overline{w} - [\tilde{n}+e+Ksd]\,\Delta\overline{w} + d\overline{w}$$

$$-[K\rho hs + r + (\tilde{n}+e)rs]\,\Delta\ddot{\overline{w}} + \rho hrs\ddot{\ddot{\overline{w}}} + (\rho h + rsd)\ddot{\overline{w}}$$

$$-\ \overline{p} - rs\ddot{\overline{p}} + Ks\Delta\overline{p} = 0 \tag{2}$$

Furthermore, considering also large deflections we make use of Berger's approximation /9/: Neglecting the strain energy density due to the second invariant of the plate's middle-surface strain, Berger derived a decomposition of v. Karman's system of equations of motion. As a result the inplane forces are approximately replaced by constant hydrostatic tensile forces $\tilde{n}$ and the equation obtained is uncoupled and quasi-linear; in case of moderately thick plates on foundation this equation is of the form of eq. (2). Following /15/ Berger's normal force is not effected by the influence of shear explicitly. If the plate edges are prevented from in-plane motions, $\tilde{n}$ can be computed (see e.g. /17/) from

$$\tilde{n}(t) = -\frac{Gh}{(1-\nu)}\frac{1}{A}\int \overline{w}\,\Delta\,\overline{w}\,dA\,, \tag{3}$$

with the area of the plate domain A.

3. Frequency domain formulation of nonlinear vibrations

Large deflections of plates which are loaded by a time-harmonic lateral force, $\overline{p}(x,y;t) = p(x,y)\sin\omega t$, are investigated. Assuming that the response of the plate is periodic we choose a generalized Ritz-Ansatz, following Kantorovich, compare /18/, $\overline{w}^*(x,y;t) = w(x,y)\sin\omega t$, which enters the differential operator of eq. (2). Finally the "error" is forced to be zero in the average sense of Galerkin's procedure,

$$\frac{1}{\omega}\int_0^{2\pi} D\{\overline{w}^*\}\sin\omega t\ d(\omega t) = 0\,. \tag{4}$$

The resulting equation corresponds to the equation of a fictitious Lagrange-Kirchhoff plate with bending stiffness $\bar{K}$, hydrostatic inplane force $\bar{n}$, and inertia forces $\bar{\mu}$:

$$\bar{K}\Delta\Delta w - \bar{n}\,\Delta w - \bar{\mu} w = \bar{q} - \bar{K}(1+\nu)\Delta\bar{\kappa} \tag{5}$$

with

$$\bar{K} = K[1+(0.75n+e)s] \ , \tag{6.1}$$

$$\bar{n} = (0.75n+e) + Ksd - \omega^2[K\rho hs+r+(0.75n+e)rs] \ , \tag{6.2}$$

$$\bar{\mu} = (\rho h\omega^2 - d)\,(1-sr\omega^2) \ , \tag{6.3}$$

$$\bar{q} = p(1-sr\omega^2) \ , \tag{6.4}$$

$$\bar{\kappa} = ps/\{[1+(0.75n+e)s]\,(1+\nu)\} \ , \tag{6.5}$$

$$n(\omega) = -\frac{Gh}{(1-\nu)}\frac{1}{A}\int_A w\Delta w \, dA \ . \tag{6.6}$$

For vanishing imposed loading, Eq. (5) provides the "nonlinear normal modes" of free vibrations. The boundary conditions of shear-deformable plates with hinged (hard) supports can be modelled in the form, see /1/:

$$C: w=0, \ \psi_s=0, \ m_n=K(\psi_{n,n}+\nu\psi_{s,s})=0 \ , \tag{7}$$

with the bending moment m and a local (n,s)-coordinate system at boundary C with normal (n). Restriction to polygonal contours C, due to $\psi_s=0$, simplifies the last condition to $\psi_{s,s}=\psi_{n,n}=0$. Inserting these conditions into the frequency domain formulation of eq. (1.3) finally leads to two boundary conditions in w:

$$C: w = 0, \ \Delta w = -\bar{\kappa}(1+\nu) \ . \tag{8}$$

The complete boundary value problem (5), (8) can efficiently be solved by a membrane-analogy, which has been prepared for linear free /19/ and forced vibrations /6/, /20/ of Mindlin plates. Thereby the differential equation (5) of the total deflection is decomposed into a set of two second order problems corresponding to deflections of prestressed linear elastic membranes:

$$w = w_1 + w_2 , \tag{9}$$

$$\Delta w_j + \alpha_j w_j = - \alpha_j \vartheta_j p \tag{10}$$

with

$$\alpha_j = -\left[\bar{n} \mp (\bar{n}^2 + 4\bar{K}\bar{\mu})^{1/2}\right] / 2\bar{K} , \; j = 1,2 . \tag{11}$$

Inserting eqs. (9), (10) into eq. (5) and comparing the coefficients of p and Δp, respectively, renders two equations for the unknown coefficients ϑ_j:

$$\alpha_1^2\vartheta_1 + \alpha_2^2\vartheta_2 + \bar{n}\left[\alpha_1\vartheta_1 + \alpha_2\vartheta_2\right]/\bar{K} = (1 - sr\,\omega^2)/\bar{K} , \tag{12}$$

$$\alpha_1\vartheta_1 + \alpha_2\vartheta_2 = s/\left[1 + (0.75n + e)s\right] . \tag{13}$$

Their solution gives

$$\alpha_j\,\vartheta_j = \pm\left\{(1 - sr\,\omega^2)/\bar{K} + \alpha_j\, s/\left[1 + (0.75n + e)s\right]\right\} / (\alpha_1 - \alpha_2) . \tag{14}$$

Inserting eqs. (9), (10) into the boundary conditions (8) together with eq. (13) leads to homogeneous conditions at C:

$$C : w_j = 0, \quad j = 1,2. \tag{15}$$

Berger's integral eq. (6.6), which bears the nonlinearity, is reformulated due to eqs. (9), (10):

$$n(\omega) = -\frac{Gh}{(1-\nu)}\frac{1}{A}\int_A (w_1 + w_2)\left[\left(-\alpha_1\vartheta_1 p - \alpha_1 w_1\right) + \left(-\alpha_2\vartheta_2 p - \alpha_2 w_2\right)\right] dA . \tag{16}$$

Hence, the Mindlin-Pasternak problem (5) with inhomogeneous boundary conditions (8) is, with respect to deflections, governed by the two second-order Helmholtz-type equations (10) with Dirichlet's conditions (15).

4. Computation of frequency response function by BEM

Passing over to a non-dimensional forcing frequency in the form

$$\lambda = \omega\, a_1/c \quad , \quad c = \sqrt{G/\rho} \quad , \tag{17}$$

with a characteristic length a_1, varying λ in steps, and solving the boundary value problems eqs. (10), (15) iteratively along with the integral (16), the undamped frequency response function $w(\lambda) = w_1(\lambda)+w_2(\lambda)$ is obtained pointwise.

The actual polygonal domain is embedded into a rectangular one by satisfying boundary conditions exactly where possible and the solutions of eqs. (10), (15) are found by superposition: Using Green's function G for the basic rectangular domain $0 \leq x \leq a$, $0 \leq y \leq b$ (see /6/), a particular solution to eq. (10) is given by

$$w_{pj}(x,y) = -\,\alpha_j \vartheta_j \int_0^b \int_0^a G_j(x,y;\,\xi,\eta)\; p(\xi,\eta)\; d\xi d\eta \quad . \tag{18}$$

Total deflections w_j are found by superposing homogeneous solutions which are chosen to be due to source functions μ_j^F acting in the rectangular domain at those parts Γ of the polygonal boundary, which do not coincide with the edges of the rectangle:

$$w_j(x,y) = w_{pj}(x,y) + \int_\Gamma G_j(x,y;\sigma)\, \mu_j^F(\sigma) d\sigma \quad . \tag{19}$$

Distribution of the line load μ_j^F is chosen such that deflection w_j vanishes along Γ, which leads to Fredholm's integral equation of the first kind:

$$\Gamma: w_j(s) = w_{pj}(s) + \int_\Gamma G_j(s,\sigma)\, \mu_j^F(\sigma) d\sigma = 0 \quad . \tag{20}$$

Solution of this indirect boundary integral equation is performed numerically by discretizing Γ into m equidistant boundary elements of length ϑ and using e.g. the rectangular formula for integration except for the singular interval, where the singularity of Green's function is analytically integrated. Thus, the system of linear equations

$$w_{pj}(s_i) + \vartheta \sum_{\substack{k=1 \\ k \neq i}}^{m} G_j(s_i, s_j)\, \mu_{kj}^F + \mu_{ij}^F \int_{s_i - \vartheta/2}^{s_i + \vartheta/2} G_j(s_i, n; \sigma)\, d\sigma \Big|_{n \to 0} = 0 \, ; \qquad (21)$$

$$i = 1, \dots, m$$

yields the step-line loading on Γ. Integration over the singular intervals analogously follows the static version of this BEM with Green`s functions of finite domains /22/: G_j is splitted into a logarithmically approximated singular part G_s which when integrated gives

$$\int_{s_i - \vartheta/2}^{s_i + \vartheta/2} G_s(s_i, n; \sigma)\, d\sigma \Big|_{n \to 0} = \frac{\vartheta}{2\pi} \left(1 - \ln \frac{\vartheta}{2} \right) , \qquad (22)$$

and a regular part G_{jR}, which accounts for the boundary conditions. The value of the latter part is found by interpolation of $G_{jR} = G_j - G_s$, calculated in the neighbouring intervals of the singular one and it is integrated again by rectangular formula. Berger`s normal force eq. (16) is evaluated numerically by dividing the actual plate domain into triangular cells and using three-point Gauß integration.

5. Numerical results and computational aspects

As numerical example, the linear and nonlinear frequency response function (FRF) $\tilde{w}(\lambda) = w(\lambda)/h$ is calculated for a trapezoidal plate with right angles, see Fig. 1. The geometry is specified through the aspect ratio $a_1/b = \sqrt{3}$ and skewness angle $\gamma = 45^0$ of the trapezoidal plate and by $a/b = \sqrt{3} + 0.2$ of the rectangular basic domain; thickness ratio is chosen $\tilde{h} = h/a_1 = 0.15$. Non-dimensional soil parameter are $\tilde{d} = da_1/G = 0.2$ and $\tilde{e} = e/a_1G = 0.1$. A constant (harmonic) pressure loading p_o is applied and the FRF due to the load factor $\tilde{p}_o = p_o\, 6(1-\nu)/G\tilde{h}^3$ is evaluated at three specified plate points A(1.3b, 0.5b), B(1.4b, 0.8b) and C(0.8b, 0.2b). Thereby the skew boundary is discretized into 19 equidistant intervals and the BEM presented in Section 4 is applied iteratively along with the computation of Berger`s inplane force n according to eq. (16). The latter is determined numerically by dividing the actual plate domain into 40 triangular cells and using three-point Gauß integration. In the low fre-

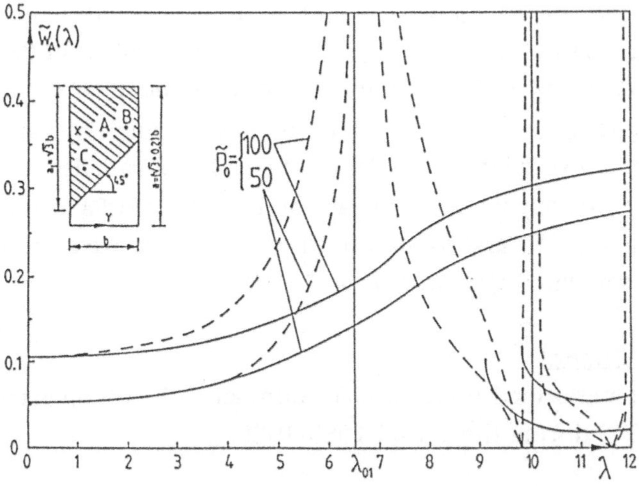

Fig. 1. Linear (- - - -) and nonlinear (———) non-dimensional undamped frequency response function $\tilde{w}$ at a central point A(1.3b, 0.5b) as a function of $\lambda = \omega a_1 \sqrt{\rho/G}$; $\tilde{h} = 0.15$, $\tilde{d} = 0.2$, $\tilde{e} = 0.1$.

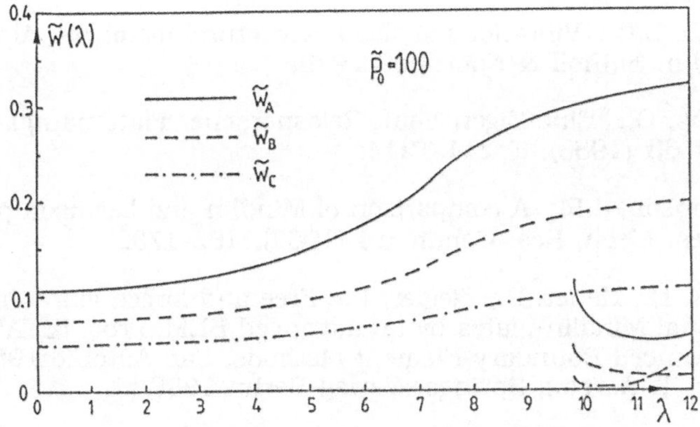

Fig. 2. Nonlinear frequency response function $\tilde{w}$ at the plate points A(1.3b, 0.5b), B(1.4b, 0.8b) and C(0.8b, 0.2b).

quency range, $\lambda \leq 0.5\, \lambda_{01}$ (λ_{01} stands for the fundamental frequency of the linear problem), when starting from the linear solution w(n=0), only up to four iterations provide sufficient accuracy (three digits of n). However, in the range $\lambda > 0.5\,\lambda_{01}$ it is more efficient to initiate the procedure with $n^{(o)}(\lambda_j) \equiv n_j^{(o)} = n_{j-1} + \Delta n$, where Δn is found by extrapolation from previous n_{j-1}, n_{j-2}, and the iteration is continued by varying the normal force in constant steps Δn until eq. (16) is satisfied.

Fig. 1 shows the opposition of linear and nonlinear FRFs at a central point A due to $\tilde{p} = 50$ and $\tilde{p} = 100$. In Fig. 2 nonlinear FRFs at three plate points are compared graphically for $\tilde{p} = 100$.

Acknowledgement

Research is performed under project number S-30/03 sponsored by the Austrian FWF during the period 1985-1989.

References

1. Mindlin, R.D.: Influence of rotatory inertia and shear on flexural motions of isotropic, elastic plates. J. Appl. Mech. 18 (1951), 31-38.

2. Hutchinson, J.R.: Vibrations of thick free circular plates, exact versus approxinate solutions. J. Appl. Mech. 51 (1984), 581-585.

3. Magrab, E.B.: Vibrations of elastic structural members. Alphen aan den Rijn: Sijthoff & Noordhoff 1979.

4. Thomas, G.: Eine Eigenschaft Reissnerscher Plattenaufgaben (II). ZAMM 63 (1983), T 211-T214.

5. Hutchinson, J.R.: A comparison of Mindlin and Levinson plate theories. Mech. Res. Comm. 14 (1987), 165-170.

6. Irschik, H., Heuer, R., Ziegler, F.: Free and forced vibrations of polygonal Mindlin-plates by an advanced BEM. Proc. IUTAM-Symp. on Advanced Boundary Element Methods, San Antonio 1987 (Cruse, T. A., ed.), Springer-Verlag Berlin 1988.

7. Kerr, A.D.: Elastic and viscoelastic foundation models. J. Appl. Mech. 31 (1964), 491-498.

8. Vlasow, V.Z., Leont`ev, U.N.: Beams, plates and shells on elastic foundations. Israel Progr. for Sci. Transl., Jerusalem 1966.

9. Berger, H.M.: A new approach to the analysis of large deflections of plates. J. Appl. Mech. 22 (1955), 465-472.

10. Nowinski, J.L., Ohnabe, H.: On certain inconsistencies in Berger equations for large deflections of elastic plates. Int. J. Mech. Sci. 14 (1972), 165-170.

11. Schmidt, R.: On Berger's method in the nonlinear theory of plates. J. Appl. Mech. 41 (1974), 521-523.

12. Irschik, H.: Large thermoelastic deflections and stability of simply supported polygonal panels. Acta Mechanica 59 (1986), 31-46.

13. Ziegler, F., Rammerstorfer, F.G.: Thermoelastic stability. In: Thermal stresses 3 (Hetnarski, R., ed.), North-Holland Publ., in press.

14. Wah, T.: Large amplitude flexural vibration of rectangular plates. Int. J. Mech. Sci. 5 (1963), 425-438.

15. Wu, C.-I., Vinson, J.R.: Influences of large amplitudes, transverse shear deformation, and rotatory inertia on lateral vibrations of transversely isotropic plates. J. Appl. Mech. 36 (1969), 254-260.

16. Irschik, H.: Influence of large amplitudes on free flexural vibrations of polygonal sheardeformable plates - a unifying dimensionless formulation. Int. J. Solids Structures, in press.

17. Kamiya, N., Sawaki, Y.: Finite deflections of plates. In: Topics in boundary element research 1 (Brebbia, C.A., ed.) Springer-Verlag Berlin 1984, 204-224.

18. Huang, C.-L., Sandman, B.E.: Large amplitude vibrations of a rigidly clamped circular plate. Int. J. Non-Linear Mech. 6 (1971), 451-468.

19. Irschik, H.: Membrane-type eigenmotions of Mindlin plates. Acta Mechanica 55 (1985), 1-20.

20. Irschik, H., Heuer, R., Ziegler, F.: Dynamic analysis of polygonal Mindlin plates on two-parameter foundations using classical plate theory and an advanced BEM. Comp. Mech. 4 (1989), 293-300.

21. Irschik, H., Heuer, R., Ziegler, F.: BEM using Green's functions of rectangular domains: static and dynamic problems of bending of plates. In: Boundary Elements IX, Proc. BEM-Conf. Stuttgart 1987 (Brebbia, C.A., Wendland, W.L., Kuhn, G., eds.), Springer Verlag Berlin-New York 1987, Vol. 2, 35-49.

22. Irschik, H., Ziegler, F.: Application of the Green's function method to thin elastic polygonal plates. Acta Mechanica 39 (1981), 155-169.

A Unified Approach to the FE and Generalized Variational FD Methods in Nonlinear Mechanics, Concepts and Numerical Approach

J. Krok, J. Orkisz

Institute of Structural Mechanics

Cracow University of Technology, Poland

Summary

Presented is a concept of combination and formal unification of the Finite Element Method and Generalized Finite Difference Method at arbitrary irregular grids. Higher order FD approximation and a formulation on differential manifold may be also applied. Symbolic operations were used in order to provide a fully automatic approach in both the FE and FD methods. Results of numerical tests are discussed.

Introduction

A combination approach to both the Finite Element and the Finite Difference Methods is considered here, as opposed to a confrontation which is the usual way these methods are treated. Although by the FEM we understand the classical version of that method, as well very recent FEM versions might be taken into account. When considering the FDM we assume a generalized, fully automatic, variational formulation of that method (GFDM). It is defined on arbitrary irregular meshes [9,11,18,19,22]. Included here are also some recent developments of the GFDM, like formulation on a differential magnifold [15,27], and use of arbitrarily type generalized degrees of freedom (d.o.f.) [6,11,17].

Four particular combinations of the both methods are considered, and followed by a formal unification of their notations and algorithms. That approach is applied to boundary value problems posed in the global formulation, namely either as the extremum of a given functional, eg.

$$\min_{u \in U} J(u) \ , \quad J(u) \equiv B(u,u) - L(u), \tag{1}$$

or in an incremental form as a corresponding principle of the virtual work

$$\delta B(u,\delta u) = L(\delta u) \ , \ \bigwedge_{\delta u \in U} \ . \tag{2}$$

G. Kuhn · H. Mang (Eds.)
Discretization Methods in Structural Mechanis
IUTAM/IACM Symposium Vienna, Austria, 1989

Here u is the required field function (e.g. displacements, stresses,...); U is the space of admissible solutions; the bilinear form $B(u,u)$ presents, in mechanics, the internal energy of the whole body, while the linear form $L(u)$ is the work of all external forces.

The present paper is a continuation and extension of some earlier authors efforts [8-11] spent on combination of the FE and FD methods. Proposed here is a unified algorithm, compatible for both methods. It is aided by use of symbolic programming in order to provide fully automatic generation of the corresponding discrete FE/FD model.

Generalized FDM Approximation

The idea of the FEM is well known, however, basic concepts of the GFDM at irregular grids require some explanation. Although the foundations of that method may be traced to the early fifties [20], its real development took place mostly in the last decade. Various variants of the GFDM were presented either in the local [4,5,13,14,18,20,22,24-26] or in the global [1,2,3,7-11,16,19,22,23] formulations. In this paper we use and develop the most advanced concepts presented in papers [17-19,22] and also in [6-11].

The classical FDM uses regular meshes. Mesh irregularity leads to a number of new problems that must be addressed [18,22]. Some of the more important are: mesh generation, area assignment to a node, selection of stars, generation of the FD formulas, discretization of boundary conditions, generation of FD equations in the local and global (numerical integration, assembling) formulations, effective solution of simultaneous algebraic equations (may be nonsymmetric), approximation approach applied in order to obtain the final results, automation of the procedure, computer implementation.

We discuss here only some recent developments in the generalized FD approximation. The basic concept of the FD approximation in question has been given in papers [17,18] and extended in [6,11]. We now present further developments of that approach oriented also on possible FE/FD combinations. There follows a need for better fulfillment of various types of boundary conditions, and improvement of the order of approximation and convergence rate.

Besides the nodal values of the unknown function u we now additionally introduce also the other types of degrees of freedom like various derivatives of u or values Au of any given operators A acting on u, e.g. the Laplace operator $\nabla^2 u$. Using the Taylor expansion of the uknown function u(x,y) we are to determine the surface that is the local "best fit" for that function. Let us consider for example the situation shown in Fig.1.

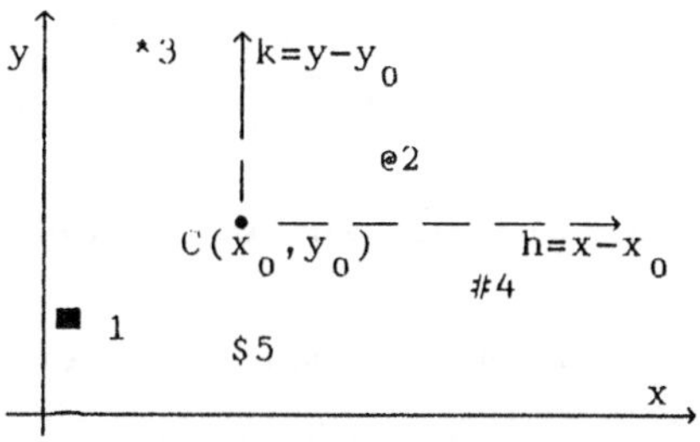

point		number d.o.f	kind d.o.f.
1		− 1 d.o.f.	u function value;
3	*	− 2 d.o.f.	u, $\partial u/\partial n$;
2	@	− 3 d.o.f.	u, $\frac{\partial u}{\partial x}$, $\frac{\partial u}{\partial y}$;
4	#	− 2 d.o.f.	u, $\partial^2 u/\partial n^2$;
5	$	− 1 d.o.f.	$\nabla^2 u = \frac{\partial^2 u}{\partial x^2} + \frac{\partial^2 u}{\partial y^2}$.

Fig. 1. FD star

Vector of all d.o.f. in the considered star is of the form

$$Q = \{u_1 \; u_2 \; \frac{\partial u_1}{\partial x} \; \frac{\partial u_2}{\partial y} \; u_3 \; \frac{\partial u_3}{\partial n} \; u_4 \; \frac{\partial^2 u_4}{\partial n^2} \; \nabla^2 u_5\} \; . \tag{1}$$

All these d.o.f. will be used now in order to determine surface Π_p of the p order, locally strictly tangent in the central node C to the unknown surface z=u(x,y). The required surface Π_p is uniquely determined by the values $\frac{\partial^{(r+s)} u_0}{\partial x^r \, \partial y^s}$ of its derivatives in the central node C. Denoting by

$$Du \equiv \{u \; \frac{\partial u_0}{\partial x} \; \frac{\partial u_0}{\partial y} \; \frac{\partial^2 u_0}{\partial x^2} \; \frac{\partial^2 u_0}{\partial x \, \partial y} \; \frac{\partial^2 u_0}{\partial x^2} \; \ldots \; \frac{\partial^p u_0}{\partial y^p} \} \, , \tag{2}$$

the vector of these derivatives as well as by

$$S \equiv \{1 \;\; h \;\; k \;\; \tfrac{1}{2} h^2 \;\; \tfrac{1}{2} k^2 \;\; h\,k \;\; k^p \}, \quad h \equiv x - x_0, \; k \equiv y - y_0, \tag{3}$$

the vector of basic functions, and expanding the unknown function

$$u = u_0 + h \frac{\partial u_0}{\partial x} + k \frac{\partial u_0}{\partial y} + \tfrac{1}{2} h^2 \frac{\partial^2 u_0}{\partial x^2} + hk \frac{\partial^2 u_0}{\partial x \, \partial y} + \tfrac{1}{2} k^2 \frac{\partial^2 u_0}{\partial x^2} + \ldots + \tfrac{1}{2} h^p \frac{\partial^p u_0}{\partial y^p} , \tag{4}$$

in an arbitrary point (x,y) with the respect to the central

point $C(x_o,y_o)$, we obtain the following local approximation of that function

$$u = S^t\, Du + O(h) \approx S^t\, Du \;,\; h \equiv \{h,k\} \;. \tag{5}$$

We may determine now all elements of the vector Q, e.g.

$$\frac{\partial u}{\partial x} = S_x^t\, Du \;,\; \frac{\partial u}{\partial n} = S_n^t\, Du \;,\; \nabla^2 u = {}^t S_{\nabla^2}\; Du, \ldots, \tag{6}$$

where

$$S_x = \frac{\partial\, S}{\partial x} \;,\; S_n = \frac{\partial\, S}{\partial n} \;,\qquad S_{\nabla^2} = \frac{\partial^2 S}{\partial x^2} + \frac{\partial^2 S}{\partial y^2}, \ldots . \tag{7}$$

In this way we may form the basic functions for the FD approximation of any operator appearing in the vector Q. Let us consider e.g. the normal derivative

$$\frac{\partial u}{\partial n} = l\, \frac{\partial u}{\partial x} + m\, \frac{\partial u}{\partial y} = S_n^t\, Du \;, \tag{9}$$

on the boundary of the domain (Fig.2) defined in a local coordinate system (n,s). Here

$$S_n = \{0 \quad l \quad m \quad lh \quad lk+mh \quad mk \quad \ldots \;\} \;,\quad n = \{l,m\}. \tag{10}$$

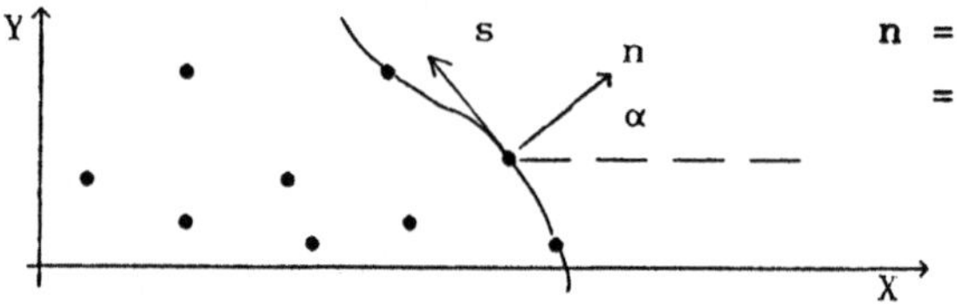

$n = n\,(l\;,\;m) = $
$= (\cos\alpha, \sin\alpha)$

Fig. 2

Finally for the star presented in Fig. 1 we obtain the following relation

$$Q = C\, Du + R \approx C\, Du \;,$$

where R denotes the vector of residuals, resulting from cutting off higher order terms in the considered FD approximation,

$$C = \begin{bmatrix} 1 & h_1 & k_1 & \frac{1}{2}h_1^2 & h_1k_1 & \frac{1}{2}k_1^2 \ldots & \frac{1}{p!} & k_1^p \\ 1 & h_2 & k_2 & \frac{1}{2}h_2^2 & h_2k_2 & \frac{1}{2}k_2^2 & \frac{1}{p!} & k_2^p \\ 0 & 1 & 0 & h_2 & k_2 & 0 \;\ldots & 0 & \\ 0 & 0 & 1 & 0 & h_2 & k_2 \;\ldots & \frac{1}{(p-1)!}k_2^{p-1} & \\ 1 & h_3 & k_3 & \frac{1}{2}h_3^2 & h_3k_3 & \frac{1}{2}k_3^2 \ldots & \frac{1}{p!} & k_3^p \\ 0 & l & m & lh_3 & (lk_3+mh_3) & mk_3 \ldots m & \frac{1}{(p-1)!}k_3^{p-1} & \\ \cdot & \cdot & \cdot & \cdot & \cdot & \cdot & \cdot & \end{bmatrix}, \tag{11}$$

$Q = \{q_j\}$, and q_i are uknown nodal values of quantities corresponding to the assumed d.o.f. in the vector (1).

Instead of the unknown true surface $z=f(x,y)$ we now consider its aproximation given in each point (x,y) by a local "best fit" surface Π_p. This way we replace required nodal values q_j by their approximations $\tilde{q}_j$ and the vector Q by $\tilde{Q} \equiv \{ \tilde{q} \}$. Thus, we obtain for the unknown vector **Du** the following simultaneous algebraic equations

$$\tilde{Q} = C\ \mathbf{Du} \tag{12}$$

Equations (12) usually constitute an overdetermined system. It is solved by a weighted least squares approach [17,18]:find

$$\min_{\mathbf{Du}} I \tag{13}$$

of the weighted error function

$$I \equiv (Q-\tilde{Q})^T\ W^2\ (Q-\tilde{Q})\ , \tag{14}$$

where

$W = \lceil w_j \rfloor$ is a diagonal matrix;

$w_j = w_j(r)$ is a weighting factor taking into account the distance $r_j = (h_j^2 + k_j^2)^{1/2}$ from the origin of the local cordinate system (h,k) to the node C_j, e.g. for the derivative of the order s and the approximation order p $w_j = \alpha_j + r_j^{s-p-1}$, α_j=const.

Finally, we obtain this way the following simultaneous equations

$$C^T\ W^2\ C\ \mathbf{Du} = C^T\ W^2\ Q \tag{15}$$

yielding the unknown vector

$$\mathbf{Du} = \mathbf{B}\ Q\ . \tag{16}$$

Here

$$\mathbf{B} \equiv (C^T\ W^2\ C)^{-1}\ C^T\ W^2 \tag{17}$$

is a matrix providing the generalized FD formulas for all the required derivatives **Du** in the point (x_0,y_0). If that point is identified with the central node C of a considered star, the value u_0 is prescribed. Otherwise u_0 is considered as one of these unknown parameters that define the local "best fit" surface Π_p, and is then found from the solution of the minimization problem (13).

Using representation (4) and formulas (3),(16),(17) we may

also express the local FD approximation in a similar way to the FEM

$$u(x) \approx S^t \, Du = S^T(h) \, B \, Q = N \, Q \, . \quad (18)$$

Here

$$N \equiv S^T(h) \, B \quad (19)$$

are generalized shape functions locally defined in a FD star.

Integration

In any global discrete formulation integration is required. Two approaches may be applied in the variational FDM, namely

- splitting the considered domain into subdomains, each one surrounding a node (Fig 3a), followed by numerical integration performed over these subdomains;
- division of the whole domain into arbitrary subdomains (elements, e.g. triangles using FD nodes), and numerical integration cerried out as in the FEM (Fig.3b).

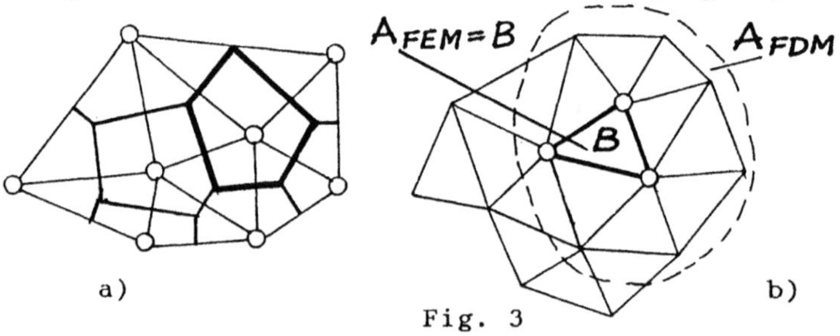

Fig. 3

There is, however, an essential difference between the FD and FE approaches in the last case. In the FEM (Fig. 3b) the same domain (element) is used for both approximation (domain A) and integration (domain B) purposes i.e. $A \equiv B$. In the FDM (Fig 3a) these domains are usually different ($A \neq B$), since different sets of nodes are used for approximation of the field function $u = N \, Q_{star}$ (FD star) and geometry $x = N' X_{elem}$ (integration element) Here N' are the FE shape functions and $X_{elem} = \{x_1, y_1 \ldots y_k\}$ are corresponding nodal coordinates.

It is worth stressing that such an approach, i.e. $A \neq B$, $N \neq N'$, generalizes the classical concept of superparametric and subparametric elements where all approximations are defined within the same element, namely $A \equiv B$, $N \neq N$. For instance, a variety of so called transition elements may be automatically generated in that way.

Formal unification of the FE and FD approach

Though formally similiar, the FD and FE approximations are, from the mathematical point of view, very different from each other. However,we use these approximations only locally, mainly for numerical integration, and apply them both in the same way. Thus, adopting the same notation for both approximations

$$u = N\,Q\ ,\qquad x = N'\,X_{elem}, \tag{20}$$

in order to describe a field function **u** and geometry **x**, we may obtain a formal unification of the FE and FD methods.

FEM/FDM combination considered

Using the above mentioned formalizm, considered are the following four combinations of both methods, when applied to solution of boundary value problems:

- use of both discretizations, at the same time, each one in a different subregion of the whole domain;
- use of each method in a different stage of the solution procedure, e.g. processing by the FEM, postprocessing by the FDM [10];
- use of the FDM for generation of the FEM characteristics;
- use of the FE approximation for generation of the FD formulas by means of differentation.

Automation and computer implementation

Computer implementation of that FE/FD model is done in the form of a system of programs called NAFDEM - "Nonlinear Analysis by the Finite Difference and Element Methods". NAFDEM is aided by a preprocessor JKJK providing the automatic use of all symbolic operations (Fig.4) like formal differentation , derivation of various formulas or generation of source code subroutines for stiffness matrices. Consequently the user has only to define the form of the functional or the virtual work principle applied, to specify boundary conditions, and to describe the domain of the problem considered.

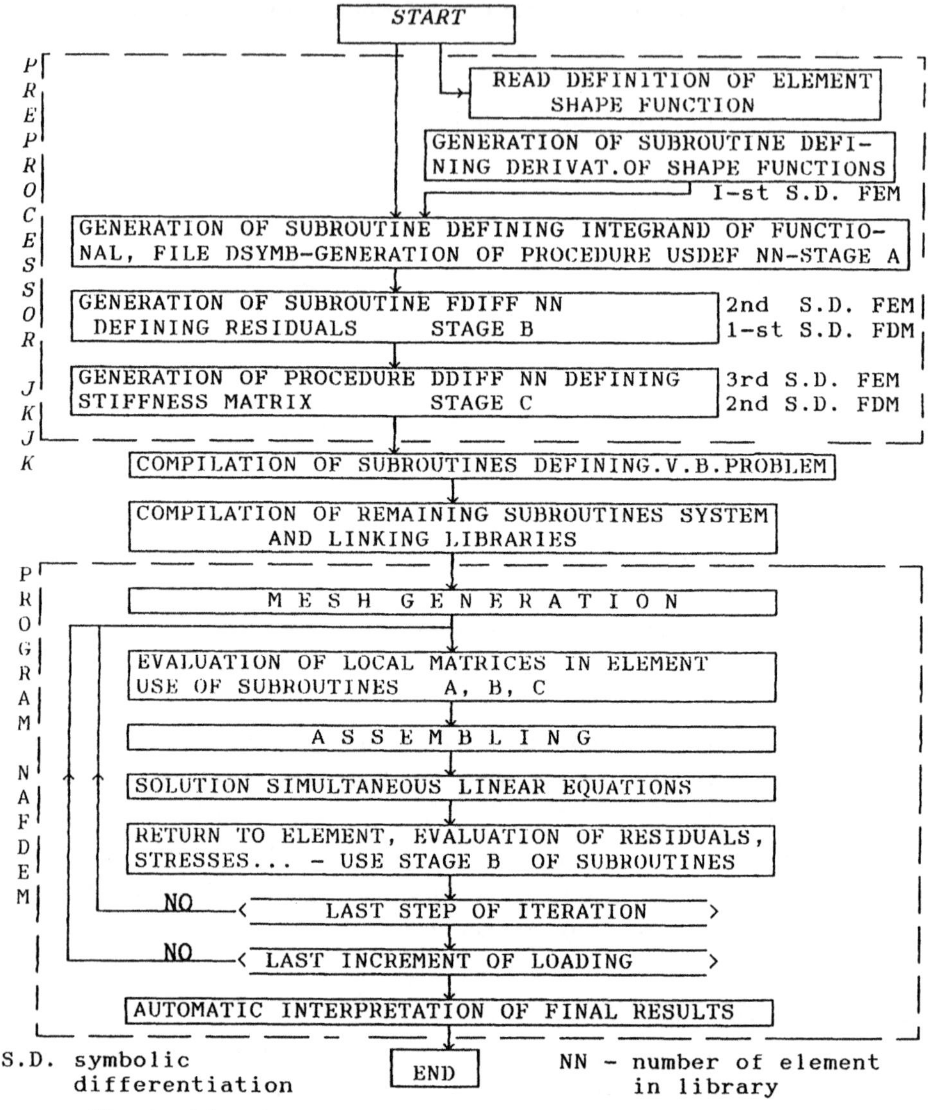

Fig.4. Flowchart of Combined FEM/FDM System NAFDEM

The core of the NAFDEM system is a special element called DUMMY. It uses subroutines automatically generated by the preprocessor JKJK, and provides implementation of all considered FE/FD combination in one computer program for a wide class of boundary value problems, given in the variational formulation. That way both the discretization of variational principles, and linearization of nonlinear problems considered are fully automatic. All programs were implemented in mainframe computer, and IBM PC AT versions.

The discrete approach, presented here, and the software developed have been successfully tested on various linear and nonlinear particular problems of applied mechanics.

References

1. Barve,V.D., and Dey, S.S., Isoparametric Finite Difference Energy Method for Plate Bending Problems, Comput. Struct. 17, pp.459-465 (1983).

2. Bushnell,D., Finite Difference Energy Method Versus Finite Elements Models: Two Variational Approach in one Computer Program, in "Numerical and Computer Methods in Structural Mechanics" , Fenves, S., Perrone, N., Robinson, A.R., Schnobrich, W.C. Academic Press, N.Y. pp.291-336 (1973).

3. Chi-Mou-Ni, A Quadrilateral Finite Difference Plate Element for Nonlinear Transient Analysis of Panels, Comput. Struct. 15, pp.1-10 (1982).

4. Dekker, K., Semi-Discretization Methods for Partial Differential Equations on Non-Rectangular Grids, Int. J. Num. Met. in Engng, 4, pp.405-419 (1980).

5. Godoy,L.A., Ill-Conditioned Stars in the Finite Difference Method for Arbitrary Meshes, Comput. Struct., 22, pp. 469-473 (1986).

6. Kawalec, J., Krok, J., Liszka, T., Some Concepts of Improved Approximation in the FDM with Arbitrary Irregular Meshes, VIII (In Polish), Proc. Conf. Comput. Meth. in Mech., Jadwisin, Poland , 3, pp.395-401 (1987).

7. Krok, J., Kucwaj, J., Orkisz, J., The Application of Symbolic Manipulation to hte FEM and Variational FDM in Nonlinear Mechanics (in Polish), Proc 9th Conf. Comput. Meth. in Mech., Rytro, Poland, pp.547-554 (1989).

8. Krok, J., Orkisz, J., On Unified Approach to the FE and FD Methods, First World Congress on Comp. Mech., Austin Texas, USA,September 22-26, 1986.

9. Krok. J., Orkisz, J., On Unified Approach to the FEM and Variational FDM in Nonlinear Mechanics (in Polish), Proc. 9th Conf. Komput. Meth. in Mech., Jadwisin, Poland, pp.479-487 (1987).

10. Krok, J., Orkisz, J., Application of the Generalized FD Approach to Stress Evaluation in the FE Solution, Proc. Int. Conf. on Comput. Mech., Tokyo, Japan, 12, pp. 32-36 (1986).

11. Krok, J., Orkisz, J., Finite Element and Finite Difference Analysis in Mechanics, Combination and Automation of both Methods (in Polish), Proc. 9th Conf. Komput. Meth. in Mech., Rytro, Poland, 2, pp.555-564 (1989).

12.Kucwaj J., Orkisz, J., Symbolic Differentation (in Polish), Mechanika i Komputer, pp.151-158 (1983).

13.Kucwaj, J., Orkisz, J., Tworzydło, W., Application of the Symbolic Programming to the Solutibn of Boundary Problems by the Finite Difference Methods (in Polish), Arch. Inż. Ląd., 4, pp.595-606 (1984).

14.Kwok, S., K., Numerical Computations of Tensor Quantities on a Curved Surface by the Curvilinear Finite Difference (CFD) Methods, Comp. Struct., 18, pp.1087-1114 (1984).

15.Kwok, S., K., An Improved Curvilinear Finite Difference (CFD) Methods for Arbitrary Mesh System, Comp. Struct., 18, pp.719-731 (1984).

16.Kwok, S., K., Geometrically Nonlinear Analysis of General Thin Shells Using a Curvilinear Finite Difference (CFD) Energy Approach, Comp. Struct., 20, pp.683-697 (1985).

17.Liszka, T., An Interpolation Method for an Irregular Net of Nodes, Int. J. Num. Meth. in Engng, 20, pp.1599-1612 (1984).

18.Liszka, T., Orkisz, J., The Finite Difference Methods at Arbitrary Irregular Grids and Its Application in Applied Mechanics, Comp. Struct., 11, pp.83-95 (1980).

19.Liszka, T., Orkisz, J., The Finite Difference Method for Arbitrary Irregular Meshes - a Variational Approach to Applied Mechanics Problems, Proc. 2-nd Int. Congress Num. Meth. for Engng, GAMNI, Ed. Dunod, Paris, pp.277-288 (1980).

20 MacNeal, R.H., An Asymmetrical Finite Difference Network, Quoterly of Applied Mechanics, 11, pp. 295-310, 1953.

21.Noor, A., K., Andersen, C., M., Computerized Symbolic Maniplulation in Structural Mechanics, Progress and Potential, Comp. Struct., 10, pp.95-118 (1979).

22.Orkisz, J., The Finite Difference Method in Fatigue Design Handbook, AE-10, R.C.Rice ed., Soc. of Automative Engineers, Warrendale, PA, USA, pp. 192-203 (1989).

23.Pavlin, V., Perrone, N., Finite Difference Energy Techniques for Arbitrary Meshes Applied to Linear Plate Problems, Int. J. Num. Meth. Engng, 14, pp.647-664 (1979).

24.Snell, C., Vesey, D.G., and Mullord, P., The Application of a General Finite Difference Method to Some Boundary Value Problems, Comp. Struct., 13, pp.547-552 (1981).

25.Tseng, A.A. A Generalized Finite Difference Scheme for Convention Dominated Metal Forming Problems, Int. J. Num. Meth. Engng, 20, pp.1855-1900 (1980).

26.Tworzydło, W., Analysis of Large Deformations of Membrane Shells by the Generalized Finite Difference Method , Comp. Struct., 1,39-59 (1987).

27.Tworzydło, W., The FDM in Arbitrary Curvilinear Coordinates - Formulation, Numerical Approach and Application, Int. J. Num. Meth. Engng, 28, 261-277 (1989).

Stress Analysis of Laminated Composites by Hybrid Finite Elements

Theodore H.H. Pian and Mingsang Li

Department of Aeronautics and Astronautics
Massachusetts Institute of Technology
Cambridge, MA 02139

Summary

Hybrid stress elements for laminated plate analysis are developed based on a mixed form of Hellinger-Reissner principle which is a function of three displacement components and three transverse stresses. Such elements are computationally much more efficient than the conventional hybrid stress elements. Also, the use of conventional hybrid stress elements and a post-processing smoothing scheme for maintaining the continuity of stress and strain along the laminar interface is not as effective as the use of the present mixed form element.

Introduction

In the analysis of laminated plates of fibrous composite material, the transverse shear effect on the behavior of the plate is more pronounced than that in isotropic plates due to the very large ratios of elastic modulus to shear modulus. Failure modes in such materials often consist of delamination resulting from excessive interlaminar stresses. Thus, efficient analytical and numerical methods are necessary for accurate estimation of these interlaminar stresses.

Across the interface between laminae, the inplane strains ε_x, ε_y and γ_{xy} and the transverse stresses τ_{xz}, τ_{yz} and σ_z are continuous. On the other hand, the continuity conditions cannot be imposed on the remaining stresses σ_x, σ_y and τ_{xy} and strains γ_{xz}, γ_{yz} and ε_z, in general. The present paper is concerning two approaches in analyzing the interlaminar stresses. One is the use of hybrid type finite element which is based on a mixed variational functional of the three displacement components and only the three transverse stresses [1-4]. The other is the use of conventional hybrid stress elements with the supplement of a post-processing smoothing scheme for stresses and strains [5,6].

G. Kuhn · H. Mang (Eds.)
Discretization Methods in Structural Mechanis
IUTAM/IACM Symposium Vienna, Austria, 1989

Mixed Form Variational Principle

The original version of the Hellinger-Reissner Principle which involves all the displacements and stresses, in the form

$$\pi_R = \sum_n \{ \int_{V_n} [-\frac{1}{2}\sigma^T \mathbf{S}\,\sigma + \sigma^T(\mathbf{Du})]dV \} - W = \text{stationary} \qquad (1)$$

where σ are the stresses, $\mathbf{S}$ is the compliance matrix, and the strains ε are expressed in terms of displacements $\mathbf{u}$, by $\mathbf{Du}$. Here W represents the work due to applied loads.

We divide the stresses σ, the strains ε into the inplane and transverse parts,

$$\mathbf{t}_p = \{\sigma_x, \sigma_y, \tau_{xy}\}^T, \text{ and } \mathbf{t}_t = \{\tau_{yz}, \tau_{xz}, \sigma_z\}^T \qquad (2.a)$$

$$\mathbf{e}_p = \{\varepsilon_x, \varepsilon_y, \gamma_{xy}\}^T = \mathbf{D}_p\mathbf{u}, \text{ and } \mathbf{e}_t = \{\gamma_{yz}, \gamma_{xz}, \varepsilon_z\}^T = \mathbf{D}_t\mathbf{u} \qquad (2.b)$$

The stress-strain relation becomes

$$\begin{bmatrix} \mathbf{e}_p \\ \mathbf{e}_t \end{bmatrix} = \begin{bmatrix} \mathbf{S}_p & \mathbf{S}_{pt} \\ \mathbf{S}_{pt}^T & \mathbf{S}_t \end{bmatrix} \begin{bmatrix} \mathbf{t}_p \\ \mathbf{t}_t \end{bmatrix} \qquad (3)$$

By substituting Eqs. (2.a,b) into Eq. (1) but eliminating the inplane stresses $\mathbf{t}_p$ by using Eq. (3), we obtain

$$\pi_{mR} = \sum_n \{ \int_{V_n} [\frac{1}{2}\mathbf{e}_p^T\mathbf{C}_p\mathbf{e}_p + \mathbf{t}_t^T\mathbf{C}_{tp}\mathbf{e}_p - \frac{1}{2}\mathbf{t}_t^T\mathbf{S}_t^*\mathbf{t}_t + \mathbf{t}_t^T\mathbf{e}_t]dV \} - W \qquad (4)$$

where $\mathbf{C}_p = \mathbf{S}_p^{-1}$; $\mathbf{C}_{pt} = -\mathbf{C}_p\mathbf{S}_{pt} = \mathbf{C}_{tp}^T$; and $\mathbf{S}_t^* = \mathbf{S}_t - \mathbf{S}_{pt}^T\mathbf{C}_p\mathbf{S}_{pt}$

The present mixed variation principle was first suggested by Reissner [1,2] and was derived independently by Moriya [3] through the Hu-Washizu principle. Huang [4] also derived the functional by means of a weighted residual scheme.

Mixed Form Hybrid Stress Element

The transverse stresses $\mathbf{t}_t$ and displacements $\mathbf{u}$ are in terms of stress parameters β and nodal displacements $\mathbf{q}$, respectively, as

$$\mathbf{t}_t = \mathbf{P}\,\beta, \text{ and } \mathbf{u} = \mathbf{N}\,\mathbf{q} \qquad (5)$$

where $\mathbf{P}$ and $\mathbf{N}$ are assumed functions, β are stress parameters and $\mathbf{q}$ are nodal displacements.

Substituting Eqs. (5) into Eq. (4), and eliminating β by applying the stationary condition with respect to β, we obtain the element stiffness matrix $\mathbf{k}$, in the form

$$\mathbf{k} = \mathbf{k}_p + \mathbf{G}^T \mathbf{H}^{-1} \mathbf{G} \tag{6}$$

where $\mathbf{k}_p = \int_{V_n} (\mathbf{D}_p\mathbf{N})^T \mathbf{C}_p (\mathbf{D}_p\mathbf{N})\, dV$; $\mathbf{G} = \int_{V_n} \mathbf{P}^T [\mathbf{C}_{tp}(\mathbf{D}_p\mathbf{N}) + \mathbf{D}_t\mathbf{N}]\, dV$;

and $\mathbf{H} = \int_{V_n} \mathbf{P}^T \mathbf{S}_t^* \mathbf{P}\, dV$

A criterion for suppressing the zero energy modes of the present mixed form hybrid stress elements is established in the following [7]. Displacements $\mathbf{u}$ can be expressed in terms of rigid body modes $\mathbf{R}$ and deformation modes α. Kinematic deformation modes can appear in those modes at ($\mathbf{u} = \mathbf{N}_\alpha\, \alpha_t$) which are related only to transverse strains $\mathbf{e}_t$. The corresponding deformation energy is

$$2U_t = \int_{V_n} \mathbf{t}_t^T \mathbf{e}_t\, dV = \int_{V_n} \mathbf{t}_t^T (\mathbf{D}_t\mathbf{N}_\alpha)\, dV = \beta^T \mathbf{G}_\alpha \alpha_t \tag{7}$$

$$\text{and} \quad \mathbf{G}_\alpha = \int_{V_n} \mathbf{P}^T (\mathbf{D}_t\mathbf{N}_\alpha)\, dV \tag{8}$$

Then, the necessary and sufficient condition for suppression of kinematic deformation modes (LBB condition) is to choosing $\mathbf{P}$ such that rank of $\mathbf{G}_\alpha$ is equal to dimension of α_t.

Two families of elements are constructed. One is to assumed that the displacement field and transverse stresses are continuous functions through the entire laminate thickness. For the other, these are piecewise continuous functions. For the two dimensional cases these are illustrated, respectively, in Figs. 1 and 2. Similar 3D elements of quadrilateral plane form and 4-node stations have also been constructed. The traction free conditions at the top and bottom surfaces of the laminate are satisfied. For type II elements, the stress continuity condition along the laminar interface is satisfied. For 2D elements and for 3D elements with rectangular geometry, the assumed transverse stresses do satisfy the equilibrium equation. For 3D elements of irregular plan form, the tensor stresses are first expressed in natural isoparametric coordinates. They are then converted into physical components by using the mean values of the Jacobian [8]. Details of the assumed stress field for these elements are given in Ref. 7.

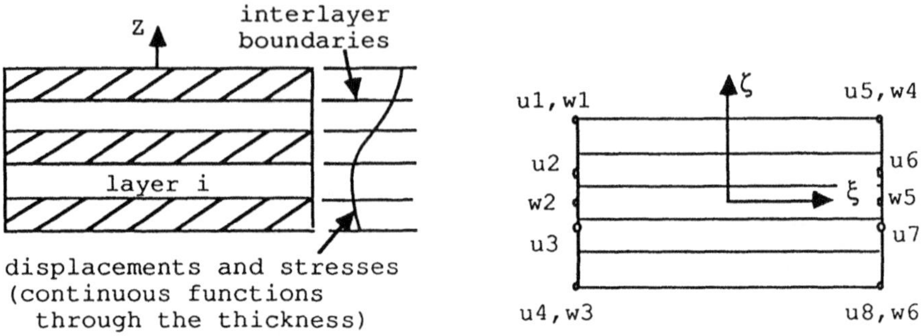

(a) type I element (b) 14 DOF type I element

Figure 1 Approximation for 2D type I element displacements and stresses

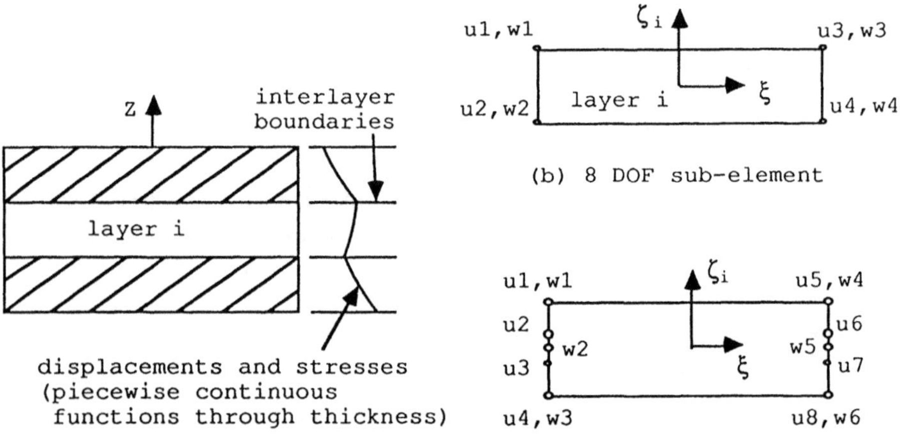

(a) type II element (c) 14 DOF sub-element

Figure 2 Approximation for 2D type II element displacements and stresses

Numerical Examples

The performances of these elements are verified by a large number of illustrative examples [7]. They include problems which do not include local stress concentration and those which do. The latter will require the use of very fine mesh in the region of stress concentration.

The former are for cylindrical bending of thick laminated plate consisting of layers of unidirectional fibrous composite material. An infinitely long laminate is simply supported along the two edges as shown in Fig. 3 and is subjected to sinusoidally distributed transverse loading $q_0\sin(\pi x/L)$. The plate is modeled as a plane strain problem in the xz plane. Finite element analysis is carried out over one half of the structure which is subdivided into 10 equal elements. Two different laminate configurations with span-to-thickness L/H ratio equal to 4 and with laminae of equal thickness are considered: (1) a 20-layer cross ply laminate $[90/0]_{10T}$ and (2) a 3-layer cross ply laminate $[0/90/0]_T$. The elasticity solutions for this plane strain problem is determined using the method given by Pagano [9]. The results based on classical lamination plate theory (CPT) are also presented in the 3-layer case. The material properties for each lamina are shown in Fig. 3. All results are presented in terms of stress values which are normalized by the applied pressure q_0.

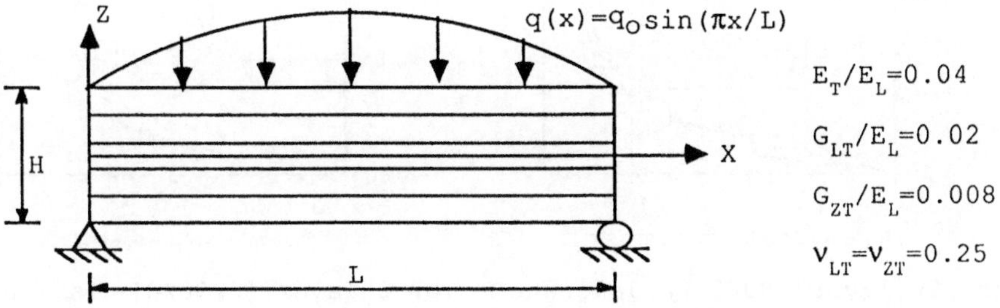

Figure 3 Thick laminated plate under cylindrical bending

The 20-layer laminate is modeled by a 14 DOF 2D type I mixed form hybrid stress element as shown in Fig. 1(b). The number of β used, in this case, is 7 which is the minimum number needed to suppress the kinematic deformation modes. The results of the transverse shear stress are in reasonable agreement with the elasticity solutions as shown in Fig. 4(a).

For the 3-layer case, each layer is modeled by a 8 DOF 2D type II mixed form element as shown in Fig. 2(b). Corresponding solution has also been obtained by using a conventional form of hybrid stress element which includes all six stress components but with the transverse stress continuity along the lamina interface maintained. Such scheme is similar to that used in Ref. 10. The numbers of stress parameters β used in the case of the mixed and conventional forms are respectively 13 and 19, while the minimum numbers required for suppressing the kinematic deformation modes are, respectively, 9 and 13. The resulting transverse shear stresses are shown in Fig. 4(b). Results obtained by the conventional and mixed form are indistinguishable. However, for constructing one element stiffness matrix, the mixed form element requires about 35% less computational effort than the conventional hybrid element. The results obtained by classical lamination theory are also plotted. Results compared with the elasticity solutions are similar to those in the 20-layer case.

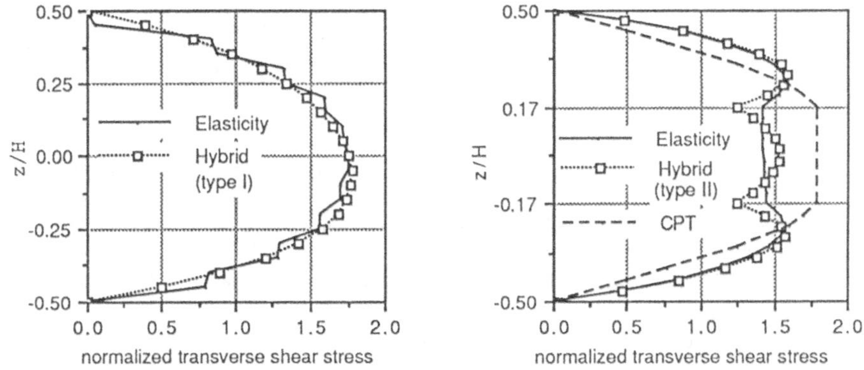

(a) 20-layer $[90/0]_{10T}$ laminate (b) 3-layer $[0/90/0]_T$ laminate

Figure 4 Through-the-thickness τ_{yz} distributions at the edge of the laminated plate under cylindrical bending

The example that involves local stress concentration is a 4-layer cross ply laminate with a circular hole of radius to laminate thickness ratio R/t equal to 5. It is subjected to uniform inplane stress σ_0 at $x=\pm L$ where L is half of the length of laminate. Because of symmetry, the analysis is carried out over one eighth of the structure. In order to reduce the cost of computation, the multi-step global local finite element method of Ref. 11 is used. The material properties are given by

$$E_T/E_L=0.074;\ G_{LT}/E_L=0.031;\ G_{ZT}/E_L=0.025;\ \nu_{LT}=0.31;\ \nu_{ZT}=0.49$$

For the global analysis, 48 elements are used as shown in Fig. 5(a). The first local region which is shown in Fig. 5(b) and designated as local region #1 has an outer radius equal to 2.15R. A subsequent local region designated as local region #2 is used as shown in Fig. 5(b) which has an outer radius of 1.25R. The finite element mesh is similar to that used in the local region #1. The numerical results obtained by the local regions #1 and #2 are labelled as Hybrid-1 and Hybrid-2, respectively and are presented in terms of stress values which are normalized by the far field tensile stress σ_0.

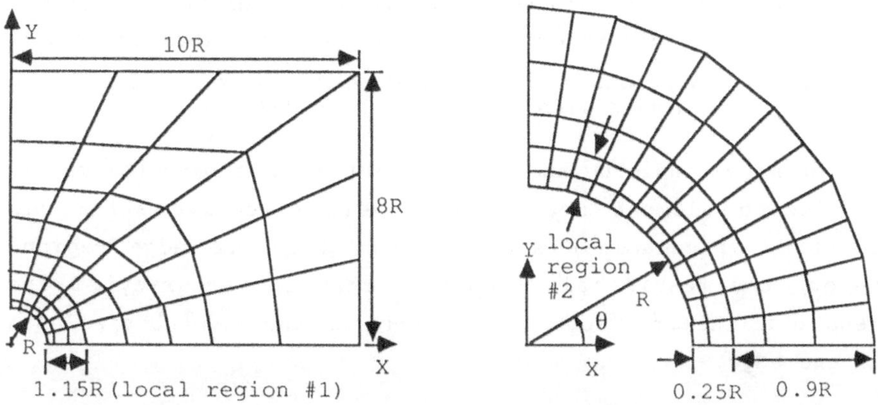

(a) model for global study (b) model for local study

Figure 5 Finite element model of a [0/90]s laminate with a hole

Each layer is modeled by one 44 DOF 3D type II mixed form hybrid stress element with nodal displacements similar to that shown in Fig. 2(c). The number of stress parameters for the two-layer model is 44 and the minimum number of β's required to suppress the kinematic modes is 29. A total of 1197 d.o.f. is used for the global model and 1482 d.o.f. for each local model. The results of the normalized $\tau_{\theta z}$ distribution around the hole at the ply interface and the normalized $\tau_{\theta z}$ and τ_{rz} distributions along the radial direction at the ply interface and $\theta=63^0$ are presented in Fig. 6. Results obtained by the three steps model with the present element are in very good agreement with the solutions obtained by the 20-node 3D isoparametric element with three steps global-local analysis [11], which used, respectively, 4149, 3507 and 3129 d.o.f.

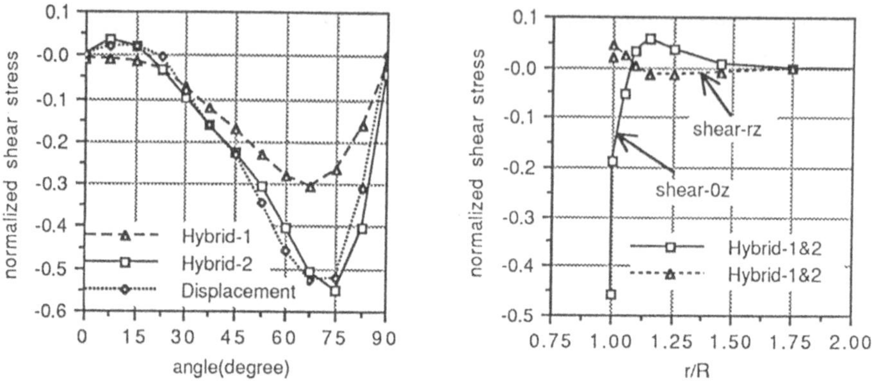

(a) $\tau_{\theta z}$ distributions around hole at the ply interface

(b) $\tau_{\theta z}$ and τ_{rz} distributions at the ply interface and $\theta=63^0$

Figure 6 Solutions of stress distributions for a [0/90]s laminate with a circular hole (R/t=5)

These examples indicate that the present element with the multi-step finite element method is a very effective scheme in predicting interlaminar stress around the free edge region. However, the results also indicate that the accuracy of the stresses at the free edge is dependent on the element size used near the edge.

Stress and Strain Smoothing

The continuous stress or strain field is determined by the condition that the integral of the difference between the continuous field and the discontinuous field obtained by either the conventional assumed displacement or hybrid stress element when weighted by the shape functions is zero. The scheme can be based on a consistent form or a lumped form [5]. However, since the computed smoothed stress or strain field, in general, does not satisfy the equilibrium condition, an iteration procedure is adopted in restoring such condition [6]. Both schemes applied to isotropic and orthotropic materials have been studied. Results have indicated that the lumped form is the better choice in the consideration of both accuracy and efficiency [7].

A 2-layer cross ply laminate subjected to cylindrical bending is used in this investigation. For the inplane portion, the smoothing scheme is applied to the strain and for the transverse portion, it is applied to the stress. Only a 5-element mesh is used. The results obtained by the mixed form 2D hybrid stress element are chosen as reference solutions to compare with the iterative results. Each layer is modeled by a 14 DOF 2D mixed form type II hybrid stress sub-element as shown in Fig. 2(c).

In order to maintain a similar DOF used in the thickness direction, each layer is modeled by a stack of three 4-node 5-β conventional hybrid stress elements and the lumped smoothing scheme is used. The computational efforts of the one with mixed form hybrid stress elements and the one with iteration scheme are almost the same. Figure 7 includes solutions with and without iteration. Obviously the former one trends to recover the shape of the elasticity solutions and the traction free conditions at the top and bottom surfaces of the laminate. The results show that the present mixed form of hybrid stress laminated plate element is still a better choice.

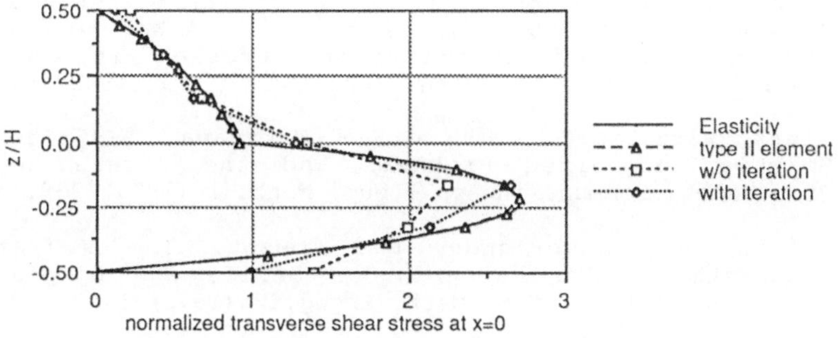

Figure 7 Thick $[0/90]_T$ laminated plate under cylindrical bending (L/H=4) by hybrid type II elements and by smoothing with and without iteration

Conclusions

1. For the case of laminated plates having large number of layers, the mixed form type I hybrid stress laminated plate element is computationally most efficient in evaluating the interlaminar stresses than other finite elements if there is no stress concentration.
2. The mixed form type II hybrid stress laminated plate elements are more computationally efficient than the type II hybrid stress elements with all six components of stress.

3. The present mixed form hybrid stress element with the multi-step global-local finite element method is a very cost effective scheme in predicting the interlaminar stress distributions around the free edge regions of laminates.
4. For laminated plates, the smoothing and iteration schemes are not computationally efficient in comparison with the present mixed form hybrid stress plate element.

References

1. Reissner, E.: On a Certain Mixed Variational Theorem and a Proposed Application. Int. J. Num. Meth. Engng. 20 (1984) 1366-1368.

2. Reissner, E.: On a Mixed Variational Theorem and on Shear Deformable Plate Theory. Int. J. Num. Meth. Engng. 23 (1986) 193-198.

3. Moriya, K.: Laminated Plate and Shell Elements for Finite Element Analysis of Advanced Fiber Reinforced Composite Structures (in Japanese). Nippon Kikai Gakkai Ronbunshu 52, No. 478, (1986) 1600-1607.

4. Huang, Q.: Variational Principle of Hybrid Energy and the Fundamentals of 3-D Laminate Theory -- A New Approach for the Analysis of Interlaminar Stresses in Composite Laminates. Appl. Math. Mech. (English ed.) 9 (1988) 649-666.

5. Zienkiewicz, O.C.; Li, X.K.; Nakazawa, S.: Iterative Solution of Mixed Problems and the Stress Recovery Procedures. Commun. in Appl. Num. Meth. 1 (1985) 3-9.

6. Cantin, G.; Loubrignac, G.; Touzot, G: An Iterative Algorithm to Build Continuous Stress and Displacement Solutions. Int. J. Num. Meth. Engng.12 (1978) 1493-1506.

7. Li, M.S.: Higher Order Laminated Composite Plate Analysis by Hybrid Finite Element methods. Ph.D. Thesis (1989) Mass. Inst. of Tech.

8. Pian, T.H.H.; Sumihara, K.: Rational Approach for Assumed Stress Finite Elements. Int. J. Num. Meth. Engng. 20 (1984) 1685-1694.

9. Pagano, N.J.: Exact Solutions for Composite Laminates in Cylindrical Bending. J. Comp. Mat. 3 (1969) 398-411.

10. Mau, S.T.; Tong, P.; T.H.H. Pian: Finite Element Solutions for Laminated Thick Plates. J. Comp. Mat. 6 (1972) 304-311.

11. Luckings, W.M.; Hoa, S.V.; Sankar, T.S.: The Effect of Geometry on Interlaminar Stresses of $[0/90]_s$ Composite Laminates with Circular Holes. J. Comp. Mat. 18 (1984) 188-198.

Liquid Loaded Membranes under Large Deformations

H. Bufler and H. Schneider

Institut for Mechanics (Civil Engineering)
University of Stuttgart, Stuttgart, FRG

Summary

Liquid loads, applied on a deformable body, are characterized by the fact, that during the loading the surface elements change their place, magnitude and direction, the pressure depending on the place and (in the most general case) on the deformation of the structure. These loads are therefore deformation dependent and belong to the follower forces. Nevertheless their potential property can be proved for all situations which make sense physically, and the corresponding potential allows a simple interpretation. Based on the principle of stationary potential energy or (equivalently, but much simpler in many cases) on the principle of virtual displacements the equilibrium of Mooney-Rivlin-membranes is studied for the following cases: Complete loading (i.e. the whole membrane being wet) and partly loading (the wet region being unknown a priori), both ones for a prescribed liquid volume (the final liquid level being unknown a priori) and a prescribed liquid level (demanding an influx from a reservoir) respectively. Gas pressure and dead loading is taken into account additionally. In certain situations the solution for a fixed liquid level turns out to be not unique and there exist stable and unstable equilibrium configurations. The discretization is performed by means of piecewise quadratic trial functions. Membrane strips, membranes of revolution and rectangular membranes are investigated numerically. Paper is a generalization of the authors' contribution [1] which is restricted to the symmetric membrane strip. Inflated isotropic membranes have been analyzed, among others, by Fried [2] and cord reinforced ones by Bolzon et al [3] who considered hydrostatic loading also.

1. The load operator and its potential

1.1 *General surface loading*

The position of a membrane point in the (load-free) reference configuration is denoted by $\underline{x} = \underline{e}_i x_i$ and in the actual one by $\underline{\xi} = \underline{e}_i \xi_i$ ($i = 1,2,3$), the surface coordinates are r_α ($\alpha = I, II$), and the deformation is described by $\underline{\xi} = \underline{\xi}(r_\alpha)$. For a configuration dependent surface load we write $\underline{q} = \underline{C}(\xi)$ which means that the load operator $\underline{C}(\xi)$ does depend on $\underline{\xi}$ and on its gradient. According to the Lagrangian description it is defined as the force in the actual configuration per surface element dS in the reference one. Its virtual work is therefore $\delta W = \int \underline{C}(\xi)\cdot\delta\underline{\xi}\, dS = \langle \underline{C}(\xi), \delta\underline{\xi} \rangle$. The load is said to be conservative (and

G. Kuhn · H. Mang (Eds.)
Discretization Methods in Structural Mechanis
IUTAM/IACM Symposium Vienna, Austria, 1989

the load operator to be potential), if the work $W = {}_{x}\!\int^{\xi} < \underline{C}(\xi), \delta\underline{\xi} >$ does not depend on the intermediate states ("path-independent"), or, equivalently if

$$\oint < \underline{C}(\xi), \delta\underline{\xi} > = 0 \quad \text{or} \quad < \delta\underline{C}(\xi; \delta\xi_1), \delta\underline{\xi}_2 > = < \delta\underline{C}(\xi; \delta\xi_2), \delta\underline{\xi}_1 >. \tag{1}$$

The first (second) equation in (1) is the potential condition in integral (differential) form, the second one being due to Volterra (1887) and Kerner (1933). In (1) $\delta\underline{C}(\xi; \delta\xi)$ means the Gâteaux differential (variation) of $\underline{C}(\xi)$ at $\underline{\xi}$ with the increment $\delta\underline{\xi}$:

$$\delta\underline{C}(\xi; \delta\xi) = \lim_{\alpha \to 0} \frac{\underline{C}(\xi + \alpha\delta\xi) - \underline{C}(\xi)}{\alpha} = \frac{d}{d\alpha} \underline{C}(\xi + \alpha\delta\xi)\Big|_{\alpha = 0} .$$

When (1) is satified, the potential P_c of the operator $\underline{C}(\xi)$ follows from

$$P_c(\xi) = {}_{x}\!\int^{\xi} < \underline{C}(\xi), \delta\underline{\xi} > = {}_{0}\!\int^{1} < \underline{C}(x + \lambda(\xi - x)), \xi - x > d\lambda = -\Pi^{(e)}(\xi) + \text{const} \tag{2}$$

where $\Pi^{(e)}$ is the physical potential. Vice versa from (2) the load operator can be calculated according to

$$\delta P_c = < \frac{\delta P_c}{\delta\underline{\xi}}, \delta\underline{\xi} > = < \underline{C}(\xi), \delta\underline{\xi} > = -\delta\Pi^{(e)}, \tag{3}$$

i.e., the load operator is the negative gradient of the external potential $\Pi^{(e)}$ and has the property (1), see [4].

1.2 *Pressure loading*

Denoting the actual pressure by $p(\xi)$, the (scalar) surface element in the reference and in the actual configuration by dS and ds respectively, the load operator follows either directly or from

$$\delta W = \int p(\xi)\delta\underline{\xi}\cdot d\underline{s} = \int \frac{1}{\sqrt{H}} p(\xi)(\underline{\xi}_{,I} \times \underline{\xi}_{,II})\cdot\delta\underline{\xi}\, dS = \int \underline{C}(\xi)\cdot\delta\underline{\xi}\, dS \tag{4}$$

($H = \det(x_{i,\alpha}\, x_{i,\beta})$, $x_{i,\alpha} = \partial x_i/\partial r_\alpha$, $dS = \sqrt{H}\, dr_I dr_{II}$, r_α such that $\underline{\xi}_{,I} \times \underline{\xi}_{,II}$ has the direction of the pressure force) and reads

$$\underline{C}(\xi) = \frac{1}{\sqrt{H}} p(\xi)(\underline{\xi}_{,I} \times \underline{\xi}_{,II}). \tag{5}$$

It has been shown in [4] that in this case the potential condition (1) is reduced to

$$\oint p(\xi)(\delta\underline{\xi}_1 \times \delta\underline{\xi}_2)\cdot d\underline{\xi} = 0 \tag{6}$$

where the integral has to be taken along the edge of the loaded structure and $\delta\underline{\xi}_1$ and $\delta\underline{\xi}_2$ are kinematically admissible variations. Condition (6) is met if (a) $p(\xi) = 0$ at the edge (**Fig.1a**) or if $p(\xi) \neq 0$ at the edge under the restriction $(\delta\underline{\xi}_1 \times \delta\underline{\xi}_2)\cdot d\underline{\xi} = 0$; this means that every edge point has either to be fixed or is allowed to move in a plane which contains this point and the edge tangent element $d\underline{\xi}$ (**Fig.1b**, where $d\underline{\xi}$ is perpendicular to the plane of the drawing paper).

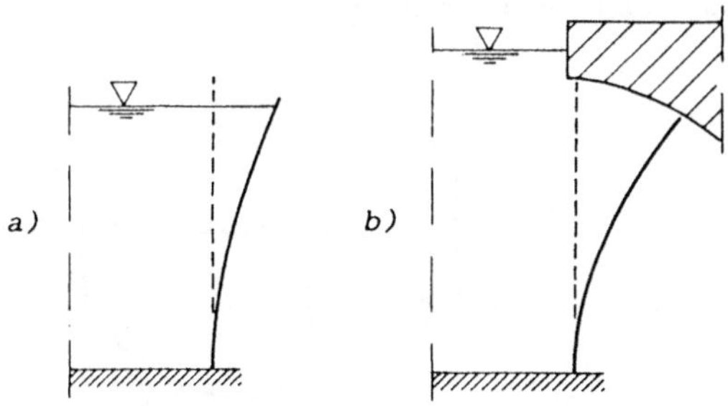

Fig.1 *Conservative pressure loading*

The physical interpretation shows that no gap is admissible along the edge of the pressure loaded structure, i.e. tightness has to be guaranteed. In this case a static equilibrium is possible and a physical potential exists. This potential can either be calculated formally via (2) with (5), or it can be written down directly by means of physical intuition as shown in [5] and applied in [1] for a hydrostatic pressure.

1.3 *Hydrostatic loading, gas pressure and dead loading*

Let us consider at first the following *hydrostatic loading cases* :
(IA) Prescribed liquid volume V and complete loading (**Fig.2**), the final liquid level h being unknown, (IIA) prescribed liquid level $\hat{h}$ and complete loading (**Fig.3**), (IB) prescribed liquid volume V and partial loading the final level h being unknown (**Fig.4**) and (IIB) prescribed liquid level $\hat{h}$ and partial loading (**Fig.5**). In the cases (IIA) and (IIB) a suitable reservoir is required in order to maintain a constant liquid level. Using (5) and physical arguments the load operator turns out to be

$$\underline{C}(\xi) = \frac{1}{\sqrt{H}}\, p(\xi)(\underline{\xi}_{,I} \times \underline{\xi}_{,II}) \quad \text{with} \quad p(\xi) = \gamma \begin{cases} \xi_3 - h & \text{(IA)} \\ \xi_3 - \hat{h} & \text{(IIA)} \\ \langle\!\langle \xi_3 - h \rangle\!\rangle & \text{(IB)} \\ \langle\!\langle \xi_3 - \hat{h} \rangle\!\rangle & \text{(IIB)} \end{cases} \tag{7}$$

$$\text{and } \langle\!\langle z \rangle\!\rangle = \begin{cases} z \\ 0 \end{cases} \text{for } z \begin{matrix} \geq 0 \\ < 0 \end{matrix} .$$

Additionally, the following subsidiary conditions have to be satisfied in the case of a presrcibed liquid volume:

$$\int (\xi_3 - h)\, ds_3 = V \quad \text{(for a cylindrical container)} \quad \text{(IA)}$$

$$\int \ll \xi_3 - h \gg ds_3 = V \quad \text{(IB)} \; . \qquad (8)$$

Here ds_3 means

$$ds_3 = \underline{e}_3 \cdot (\underline{\xi}_{,I} \times \underline{\xi}_{,II})/\sqrt{H}\, dS = (\xi_{1,I}\xi_{2,II} - \xi_{2,I}\xi_{1,II}) dr_I dr_{II}$$

the 3-projection of the deformed surface element $d\underline{s}$.

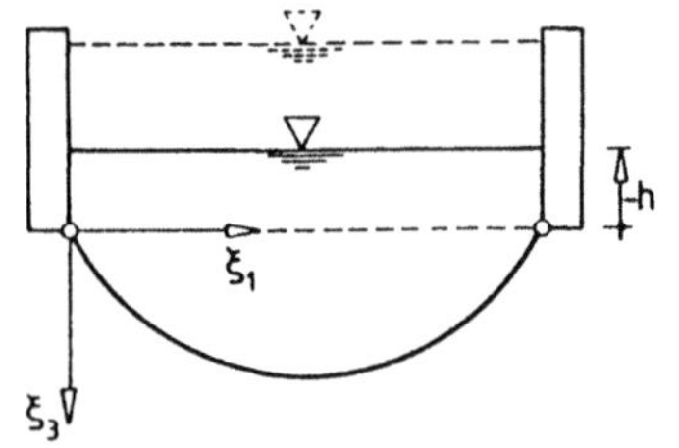

Fig.2 (IA): *Prescribed liquid volume, complete loading*

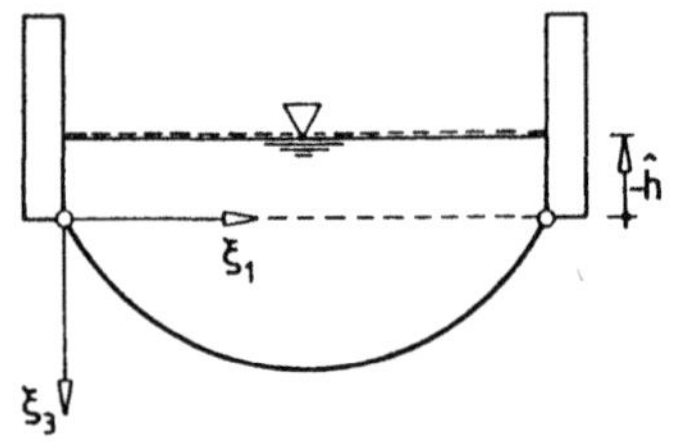

Fig.3 (IIA): *Prescribed liquid level, complete loading*

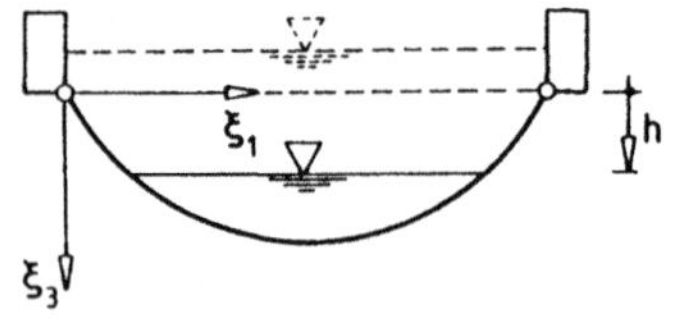

Fig.4 (IB): *Prescribed liquid volume, partial loading*

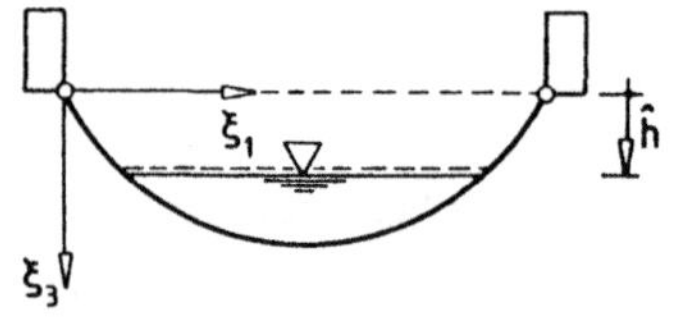

Fig.5 (IIB): *Prescribed liquid level, partial loading*

The external potential $\Pi^{(e)}$ follows from physical considerations:

$$-\Pi^{(e)} = \frac{\gamma}{2} \int \left\{ \begin{array}{ll} (\xi_3^2 - h^2) & \text{(cylindrical container)} \quad \text{(IA)} \\ (\xi_3 - \hat{h})^2 & \text{(cylindrical container)} \quad \text{(IIA)} \\ \ll \xi_3^2 - h^2 \gg & \text{(IB)} \\ \ll \xi_3 - \hat{h} \gg^2 & \text{(IIB)} \end{array} \right\} ds_3 \qquad (9)$$

(where (8) has to be taken into account). Note that the domain of the integrals in the cases (IB) and (IIB) is not known a priori. For this (open) domain the inequalities

$$\xi_3(r_\alpha) - h > 0 \quad \text{(IB)} \quad \text{and} \quad \xi_3(r_\alpha) - \hat{h} > 0 \quad \text{(IIB)} \qquad (10)$$

must hold, the edge line coordinates r^*_α following from

$$\xi_3(r^*_\alpha) - h = 0 \quad (IB) \quad \text{and} \quad \xi_3(r^*_\alpha) - \hat{h} = 0 \quad (IIB) \tag{10*}$$

respectively. In order to check the potential property of the load operator we evaluate the variation of $\Pi^{(e)}$. The result of the somewhat lenghty calculations confirms (3) with $\underline{C}(\xi)$ according to (7). It should be emphasized that the calculation of $\delta\Pi^{(e)}$ in the cases (IA) and (IB) requires the additional variation of $h = h(\xi)$ (a functional with respect to the deformation) which can, however, be eliminated with the help of the varied version of the corresponding subsidiary conditions in (8). For *gas pressure*, eq (7/1) holds with $p = \text{const} = p_o$ and the external potential is

$$-\Pi^{(e)} = p_0 \int \xi_3 ds_3. \tag{11}$$

Dead loading is characterized by

$$\underline{C}(\xi) = \underline{e}_3 q_3 \tag{12}$$

and

$$-\Pi^{(e)} = \int q_3 \xi_3 dS \tag{13}$$

where q_3 is the weight per area of the undeformed surface.

2. General variational formulation

The strain energy of a Mooney-Rivlin membrane with the original thickness t and the material parameters C_1 and C_2 is given by

$$\Pi^{(i)} = \int \pi^{(i)} dS \; ; \qquad \pi^{(i)} = t \{ C_1(I_1 - 3) + C_2(I_2 - 3) \} \tag{14}$$

with

$$I_1 = \text{tr}\,\underline{c} = \text{tr}\,\underline{\tilde{c}} = \lambda_1^2 + \lambda_2^2 + \lambda_3^2 \,,$$

$$I_2 = \frac{1}{2}\{ I_1^2 - \text{tr}(\underline{c}^2) \} = \frac{1}{2}\{ I_1^2 - \text{tr}(\underline{\tilde{c}}^2) \} = \lambda_1^2\lambda_2^2 + \lambda_2^2\lambda_3^2 + \lambda_3^2\lambda_1^2 \,,$$

$$I_3 = \det(\underline{c}) = \det(\underline{\tilde{c}}) = \lambda_1^2\lambda_2^2\lambda_3^2$$

and (using the summation convention)

$$\underline{c} = \begin{bmatrix} \begin{bmatrix} c_{\alpha\beta} \end{bmatrix} & \\ & \lambda_3^2 \end{bmatrix} \quad \text{with } c_{\alpha\beta} = \xi_{k,\alpha}\xi_{k,\beta} \quad \text{or} \quad \underline{\tilde{c}} = \begin{bmatrix} \lambda_1^2 & & \\ & \lambda_2^2 & \\ & & \lambda_3^2 \end{bmatrix}. \tag{15}$$

$\underline{c}$ being the Cauchy-Green deformation tensor and λ_i denoting the principal stretches ($\underline{\tilde{c}}$ is referred to the principal axes). Of course more general material laws could be used also, see Ogden [6]. The representation

(15/1) is valid for a membrane which is plane in the reference configuration with the cartesian surface coordinates r_α; it will be used later on for the approximation of an arbitrary initial membrane geometry by means of piecewise plane elements. The representation (15/2) is adequate whenever the principal axes are known a priori (axisymmetric problems for example). Alternatively, (15/2) can be obtained via the principal axes transformation of (15/1). Eliminating λ_3 by means of the incompressibility condition $I_3 = 1$ the strain energy density can be expressed either as a function of $c_{\alpha\beta}$ or as a function of λ_1 and λ_2

$$\pi^{(i)} = f(c_{\alpha\beta}) \quad , \quad \pi^{(i)} = \tilde{f}(\lambda_1, \lambda_2) \ . \tag{16}$$

From (16/2) the (Biot) stress resultants

$$N_k = \frac{\partial \pi^{(i)}}{\partial \lambda_k} \qquad (k = 1,2) \tag{17}$$

follow.

The equilibrium of the membrane is characterized by the principle of stationary potential energy

$$\Pi = \Pi^{(i)} + \Pi^{(e)} = \text{stat} \qquad (\forall \text{ kinematically admissible } \xi_k) \tag{18}$$

under the subsidiary conditions (8). It is, however, advantageous to replace (18) by the principle of virtual work

$$\delta\Pi^{(i)} - \int \underline{C}(\xi)\cdot\delta\underline{\xi} \, dS = 0 \tag{19}$$

which has to be supplemented by the conditions (8). The discretization by means of finite elements transforms (19) into a system of nonlinear equations for the unknowns ξ_1, ξ_2, ξ_3 at discrete points and the additional unknown h (cases (IA) and (IB)). For the wet region (10) has to be true and its edge line is obtained via (10*) (cases (IB) and (IIB)). This system is solved using a program based on Newton-Raphson's method. Alternatively, the linear Newton-Raphson equations for the increments $\Delta\xi_i$ can be produced by demanding the equilibrium in an adjacent state $\xi_i + \Delta\xi_i$. In this state the potential energy is (cases (IIA) and (IIB))

$$\Pi(\xi + \Delta\xi) = \Pi(\xi) + \int \left\{ \frac{\partial \pi^{(i)}}{\partial \xi_{i,\alpha}} \Delta\xi_{i,\alpha} + \frac{1}{2} \frac{\partial^2 \pi^{(i)}}{\partial \xi_{i,\alpha} \, \partial \xi_{j,\beta}} \Delta\xi_{i,\alpha} \Delta\xi_{j,\beta} + \ldots - C_i(\xi)\Delta\xi_i - \frac{1}{2} C_{\xi ij}(\xi)\Delta\xi_i\Delta\xi_j \ldots \right\} dS \tag{20}$$

where the index notation $\delta C_i(\xi;\Delta\xi) = C_{\xi ij}(\xi)\Delta\xi_j$ for the variation $\delta\underline{C}(\xi;\Delta\xi)$ has been used. Note that the coefficients of the nonlinear terms in (20) are symmetric operators due to their potential property. Carrying out the discretization the stationarity condition reads (with $\Delta\bar{\xi}_i$ as nodal quantities, i = 1,2,...n corresponding to the number of the degrees of freedom)

$$\frac{1}{2}(E_{ij} - C_{\xi ij})\Delta\bar{\xi}_i\Delta\bar{\xi}_j + (E_i - C_i)\Delta\bar{\xi}_i = \text{stat}$$

with $E_{ij} = E_{ji}$ and $C_{\xi ij} = C_{\xi ji}$, yielding finally

$$(E_{ij} - C_{\xi ij})\Delta\bar{\xi}_j = C_i - E_i \; . \tag{21}$$

The stability of an equilibrium state can be checked by means of the second variation of the potential energy. With respect to the discretized equations there is stability if

$$\delta^2\Pi(\xi;\Delta\bar{\xi}) = (E_{ij} - C_{\xi ij})\Delta\bar{\xi}_i\Delta\bar{\xi}_j > 0 \qquad \forall\ \Delta\bar{\xi}_i \neq 0 \; . \tag{22}$$

This means that all eigenvalues of the tangent matrix $[E_{ij} - C_{\xi ij}]$ must be positive.

3. Examples

According to the various examples which are discussed the general equations of section 2 have to be specialized and adapted. The discretization is based on piecewise quadratic shape functions. The following data are taken for all examples: t = 0.001m (membrane thickness), C_1 = 100kN/m^2 and C_2 = 10kN/m^2 (material parameters), q_3 = 0.01kN/m^2 (dead load), γ = 10kN/m^3 (specific weight of the fluid).

In **Fig.6** the equilibrium configurations of a membrane strip with the original length ℓ = 0.2m and breadth b = 1m are shown for the fluid volume V = 0.002m^3. The left edge is fixed and for the right one there is prescribed $\hat{\xi}_1$ = 0.1m and $\hat{\xi}_3$ = 0.1m, $\hat{\xi}_3$ = 0.0m and $\hat{\xi}_3$ = -0.1m.

Fig.7a illustrates the results for a circular cylindrical membrane (height H = 0.2m, radius R = 0.1m) with a moveable rigid bottom for the fluid volumes V_1 = 0.001π m^3, V_2 = 0.004π m^3, and V_3 = 0.011π m^3 respectively. To each volume V there belongs a certain fluid level h according to the diagram in **Fig.7b**. For certain fluid levels h, however, more

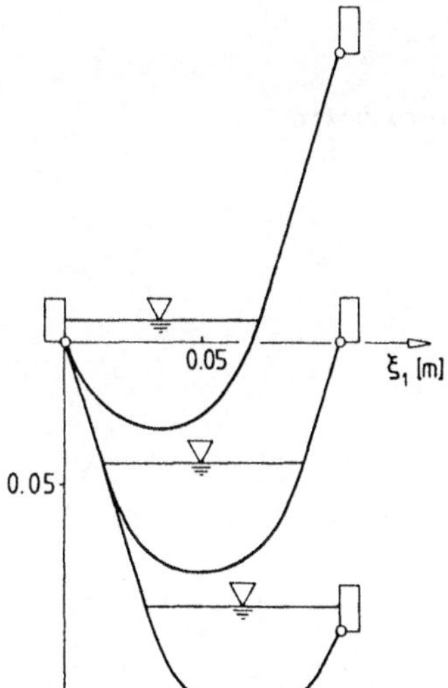

Fig.6 *Equilibrium configurations for a membrane strip*

than one equilibrium configuration are possible. The investigations based on (22) show that the equilibrium is stable whenever V is prescribed, and that it is unstable, when h is prescribed and the slope dh/dV > 0 (i.e., an increasing fluid volume results in a decreasing fluid level).

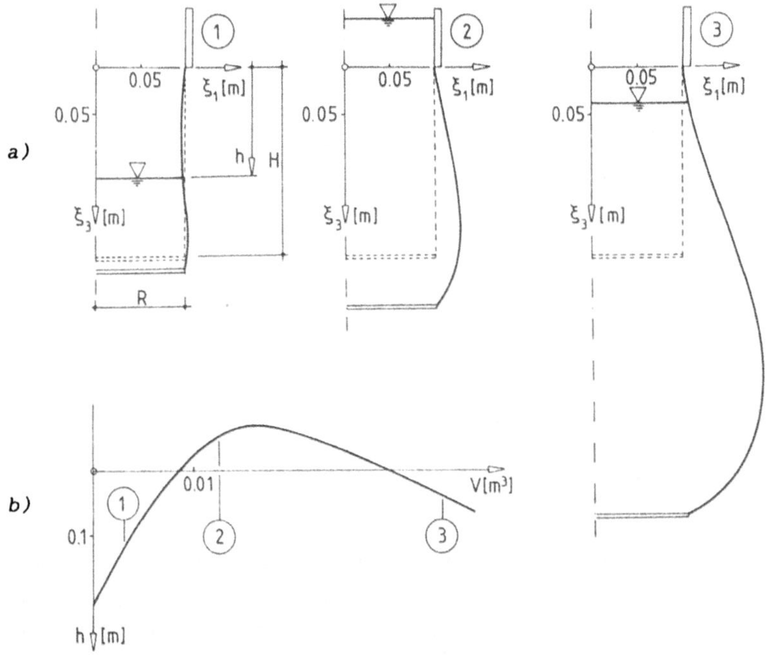

Fig.7 *Circular cylindrical membrane with a rigid bottom*
a) Equilibrium configurations,
b) Fluid level versus fluid volume

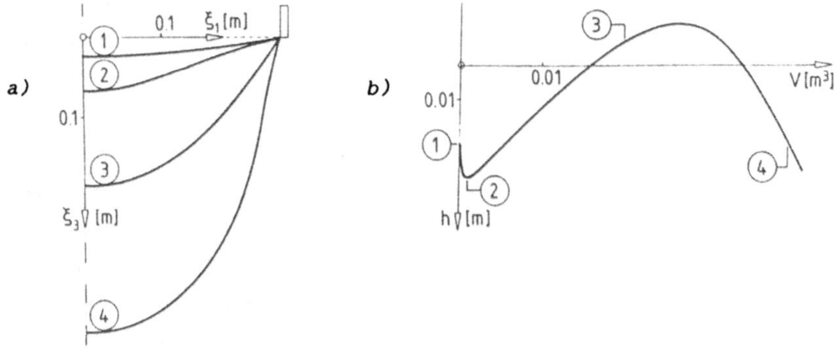

Fig.8 *Circular membrane*
a) Equilibrium configurations
b) Fluid level versus fluid volume

In **Fig.8a** the equilibrium configurations of an originally plane circular membrane (radius R = 0.25m) are plotted for the fluid volumes $V_1 = 0.0m^3$, $V_2 = 0.003125\pi\ m^3$, $V_3 = 0.006250\pi\ m^3$, and $V_4 = 0.012500\pi\ m^3$ respectively. The fluid level versus the fluid volume is documented in **Fig.8b**. The qualitative behaviour corresponds to that in **Fig.7b**.

The results for a quarter of an originally plane quadratic membrane (side length a = 0.2m) are listed in **Fig.9**. **Fig.9a** shows the equilibrium con-

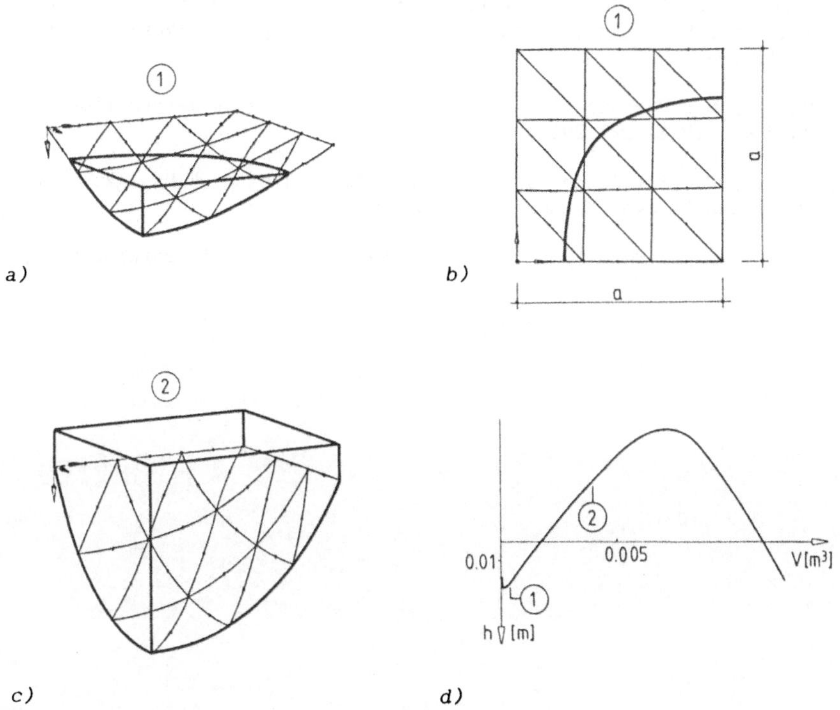

Fig.9 *Quadratic membrane*
a) and c) Equilibrium configurations for different fluid volumes,
b) Wet region corresponding to a),
d) Fluid level versus fluid volume

figuration for the fluid volume $V_1 = 0.0004m^3$ and **Fig.9b** the corresponding edge line of the wet region. In **Fig.9c** the case of a complete loading with $V_2 = 0.004m^3$ is realized. At last in **Fig.9d** the level-volume-diagram is reproduced.

Note that the initial value of h (for $V = 0.0m^3$) in **Fig.8b** and **Fig.9d** are due to the dead weight of the membrane which has been taken into account.

References

[1] Bufler, H. and Schneider, H.: Equilibrium of liquid loaded membranes taking into account the deformation dependent variation of the liquid level. In: Computational Mechanics '88, Proc. Intern. Conf. Eng. Sc., S. N. Atluri and G. Yagawa (eds.), Springer-Verlag 1988, 22 III 1-4

[2] Fried, A.: Finite element computation of large rubber membrane deformations. Int. J. Num. Meth. Eng. 18 (1982) 635-660

[3] Bolzon, G., Scheffler B. A. and Vitaliani, R.: A 3-D geometrically nonlinear analysis of inflated cord-reinforced membranes of rubber-like materials. In: Computational Mechanics '88, Proc. Intern. Conf. Eng. Sc., S. N. Atluri and G Yagawa (eds.), Springer-Verlag 1988, 27 II 1-4

[4] Bufler, H.: Pressure loaded structures under large deformations. ZAMM 64 (1984) 287-295

[5] Bufler, H.: Zur Potentialeigenschaft der von einer Flüssigkeit herrührenden Druckbelastung. ZAMM 65 (1985) 130-132

[6] Ogden, R. W.: Non-linear Elastic Deformations. Ellis Horwood Ltd., Chichester (1984)

Stabilizing and Destabilizing Effects of Small Damping for Structures with Finite Number of Degrees of Freedom

N.V.BANICHUK, A.S.BRATUS

Institute for Problems in Mechanics
USSR Academy of Sciences, Moscow

Summary

A finite element formulation ([1, 2]) is used to investigate the dynamical behaviour of elastic structures loaded by nonconservative forces ([3]). Stability analysis is reduced to study of vibration of nonconservative mechanical system finite number of degrees of freedom. Dynamic properties of the system are described by a set of homogeneous ordinary differential equations. Necessary conditions for the perturbations to be ideal and the asymptotic formulas for critical values are considered. The structure of the matrices which realize the ideal perturbations is discussed. The formula, evaluating the defect of the damping matrix, is presented for the case of defective perturbations.

Introductory Remarks

There is increasing activity in the development of general techniques for including the effects of various forms of damping at the initial design stage. The availability of accurate efficient analysis is crucial to the design of reliable structures that include damping materials. This paper describes the finite element approach which has been used to analyze dynamic stability of a structures compressed by a nonconservative forces. To obtain the stability and instability conditions we use spectral approach based on representation of state variables in the form of time exponentials. Arising eigenvalue problem for frequencies and amplitude vectors is nonselfadjoint. The multiplicity of eigenvalues and linear dependence of the corresponding eigenvectors characterize the instability of the considered system. The matrix of the eigenvalue problem has Jordan cell. This property results in singularities of dynamic behaviour

G. Kuhn · H. Mang (Eds.)
Discretization Methods in Structural Mechanis
IUTAM/IACM Symposium Vienna, Austria, 1989

typical for problems of stability of elastic structures, subjected to follow forces, and for some flutter problems of aeroelasticity.

An important consideration in the study of a class of problems for compressed structures, is the determination of the effects of viscosity in reducing dynamic stability. It was discovered before and illustrated with the help of variety of examples that the small viscous damping, when applied to nonconservative system, can induce instability [4-10]. Thus, when the viscosity decreases, the critical load of perturbed system tends to the limit, which is equal or less then the critical load of undisturbed system. Perturbation is called ideal or defective if the first or the second case is realized. In the second case the value of the loss of stability parameter is called the defect of the damping. It was the purpose of many papers to prove that the critical load of a system with finite number degrees of freedom without damping, subjected to nonconservative forces which are linear functions of generalized coordinates, is an upper bound for critical load of the same system perturbed by some sufficiently small damping.

Particular emphasis of the paper has been focused on the determination of the dependence of critical stability parameter on small damping. Asymptotic approach based on fractional power expansion (with respect to small damping parameter) is suggested for the multiple eigenvalue and respective vibration mode. This approach combined with finite element method leads to the regular procedure for analysis of nonconservative dynamical systems and to an effective technique for sensitivity analysis of structural behaviour with respect to small variations of external loadings and geometric parameters.

Basic Assumptions

Consider elastic thin-walled structure, loaded by nonconservative forces, and take into account small damping. To model the dynamic behaviour of the structure let us use finite element approach [1, 2]. Performing the partitioning of the structure we obtain discrete, time - invariant damped linear multiple degree-of-freedom mechanical system described by the set of homogeneous ordinary differential equations

$$a\ddot{w} + \varepsilon b\dot{w} + cw = 0 \qquad (1)$$

Here a, b and c are n by n real matrices representing inertia, damping and generalized stiffness, respectively; $w = (w_1(t), w_2(t), \ldots, w_n(t))$ - state vector-function; t - time; $\varepsilon > 0$ - small parameter. Matrices c(p) and b(p) depend analytically on real loading parameter p for $p \in D$, where D - given open interval. The superposed dot denotes the time derivative. For later convenience we replace (1) with "normalized" notation according to
$U = a^{1/2}w$, $B = a^{-1/2}ba^{-1/2}$, $C = a^{-1/2}ca^{-1/2}$ the system of equations (1) now becomes

$$\ddot{U} + \varepsilon B\dot{U} + CU = 0 \qquad (2)$$

Appearance of the term $\varepsilon B(p)\dot{U}(t)$ in (2) ($\varepsilon b(p)\dot{w}(t)$ in (1)) is stipulated by accounting of small damping. The perturbed system (2) turns into undamped system

$$\ddot{U} + CU = 0 \qquad (3)$$

for $\varepsilon = 0$. To investigate the stability and instability conditions of perturbed system we shall seek a special solution of (2) in the form $U(t) = u \exp(i\omega t)$, in which u - amplitude vector, ω - frequency, i - imaginary unit. Putting this vector expression into (2) and cancelling out $\exp(i\omega t)$ we get

$$C(p)u + i\varepsilon\omega B(p)u = \omega^2 u \qquad (4)$$

Characteristic exponents ω (frequencies) and amplitude vectors u are the eigenvalues and eigenvectors for the eigenvalue problem (4). Value p_ε is termed the critical value of loading parameter p if it is the lower bound of the values p for which there exists at least one eigenvalue of (4) with negative imaginary part. If there not exist such p we suppose $p_\varepsilon = \sup_{p\in D} p$. Introduced definition means that the system (2) is stable for $p<p_\varepsilon$ (in any case, if all $\omega_{j,\varepsilon}$ are simple) and become unstable for all sufficiently small values of the difference $p-p_\varepsilon>0$. The set of the points $(\varepsilon, p_\varepsilon)$ at plane ε, p defines certain curve which separates stability domain from unstability domain. It is called the critical curve. The value p_0 is taken as critical value of undisturbed system

$$C(p)u = \omega^2 u \tag{5}$$

obtained from (4) for $\varepsilon=0$.

Let us denote by p_d the lower limit of p_ε when $\varepsilon\to +0$. It follows from analyticity of the functions under consideration that this limit exists. Addition of dissipative forces to the conservative system (3) as it is well known does not violate the its stability [11]. In general it is not right for nonconservative systems. Here the small damping can reduce to destabilizing effects [4-10]. This effect is expressed mathematically by means of inequality $p_d<p_o$. The value $d_c(B)=p_o-p_d$ will be called defect of matrix B with respect to matrix C; it was shown in [7] that $d_c(B)\geq 0$. Let us call matrix B ideal with respect to matrix C, if $d_c(B)=0$. The corresponding perturbation of the system (3) will be called ideal. The operator B will be called defective with respect to matrix C if $d_c(B)>0$. The corresponding perturbations are termed destabilizing in this case. Ideal perturbations reduce to stabilizing effect if $p_\varepsilon\geq p_o$ for sufficiently small interval $0<\varepsilon<\varepsilon_o$.

Introduce the following two basic assumptions.

1. There exists the value $p_o\in D$ of parameter p, such that for $p<p_o$ all eigenvalues $\omega^2_j(p)$ $(j=1, 2, \ldots, n)$ of the problem (5) are real, simple and positive: $\omega_j^2=\tau_j^2, \tau_j\neq 0$.
2. For $p=p_o$ the matrix $C(p_o)$ is nonsymmetric, the eigenvalues of (5) ω_1^2 and ω_2^2 become double $\omega_1^2(p_o)=\omega_2^2(p_o)=\tau_1^2(\tau_1^2\neq 0)$ and the corresponding eigenvectors are linear dependent (the only one

eigenvector corresponds to these eigenvalues). Eigenvalues $\omega_1^2(p)$ and $\omega_2^2(p)$ become negative ore complex when $p>p_o$. The rest $\omega_j^2(j=3, 4, \ldots, n)$ remain real, simple and positive.
As it may be shown with the help of implicit function theorem the critical curve $p_\varepsilon=p_\varepsilon(\varepsilon)$ is given by the equation with analytical right-hand side

$$p_\varepsilon=p_d+\varepsilon^2 p_2+0(\varepsilon^2) \tag{6}$$

at the band $0<\varepsilon<\varepsilon_o$ for sufficiently small ε_o. The solution of the equation $\det\left(C(p)+i\varepsilon\omega B(p)-\omega^2 I\right)=o$, where I - unit matrix, satisfy condition $\operatorname{Im}\omega_i(\varepsilon, p)\geq 0$, $i=1, 2, \ldots, 2n$ along the critical curve.

Asymptotic Approach

Suppose $p=p_o$. It follows from assumptions 1, 2 and condition $p_d=p_o$ that along the critical curve (6) the values $\omega_{j,\varepsilon}^2(j=3, 4, \ldots, n)$ are analytical functions of ε and $\omega_{1,\varepsilon}^2$, $\omega_{2,\varepsilon}^2$ are the functions of $\varepsilon^{1/2}$ in the neighborhood of the point $\varepsilon=0$([12]). Thus, we have

$$\omega_{1,\varepsilon}^2=\tau_1^2+\varepsilon^{1/2}\mu_{11}+\varepsilon\mu_{12}+\ldots, \qquad \omega_{j,\varepsilon}^2=\tau_j^2+\varepsilon\mu_{j1}+\varepsilon^2\mu_{j2}+\ldots \tag{7}$$

The values μ_{ij} do not depend on ε. The following expressions are correct for frequencies

$$\omega_{1,\varepsilon}=\pm\tau_1\pm\frac{\varepsilon^{1/2}}{2\tau_1}\mu_{11}\pm\frac{\varepsilon}{2\tau_1}\left(\mu_{12}-\frac{\mu_{11}^2}{4\tau_1}\right)+\ldots \tag{8}$$

$$\omega_{j,\varepsilon}=\pm\tau_j\pm\frac{\varepsilon}{2\tau_j}\mu_{j1}\pm\frac{\varepsilon^2}{2\tau_j}\left(\mu_{j2}-\frac{\mu_{j1}^2}{4\tau_j}\right)+\ldots \tag{9}$$

Here the upper sign corresponds to one branch of the function, and the lower sign-to another branch. The values $\mu_{ik}(j=1, 2, \ldots, n;\ k=1, 2, \ldots)$ determine the corrections to unperturbed eigenvalues ω_j^2 and are to be found. Expansions in a series for $\omega_{2,\varepsilon}$ and $\omega_{2,\varepsilon}^2$ have the form similar to series expansions (8), (7) for $\omega_{1,\varepsilon}$ and $\omega_{1,\varepsilon}^2$. It was shown in [12] that the eigenvectors u_j^ε and $u_j^\varepsilon (j=3, 4, \ldots, n)$ of the problem (4) admit series expansions with respect to small parameters $\varepsilon^{1/2}$ and ε

$$u_1^\varepsilon = u_1^o + \varepsilon^{1/2} u_1^1 + \varepsilon u_1^2 + \ldots, \qquad u_j^\varepsilon = u_j^o + \varepsilon u_j^1 + \varepsilon^2 u_j^2 + \ldots \tag{10}$$

where u_1^o - the eigenvector corresponding to double eigenvalue $\omega_1^2(p_o)$ of the problem (5) for $p=p_o$, and $u_j^o (j=3, 4, \ldots, n)$ - eigenvectors corresponding to the eigenvalues $\omega_j^2(p_o)$. Vectors $u_j^k (j=1, 2, \ldots, n;\ k=1, 2, \ldots)$ are to be found.

Without loss of generality we consider that the eigenvectors u_j^ε are normalized in the following way: $(u_j^\varepsilon, u_j^o)=1$; $j=1, 2, \ldots, n$. Here and later the parenthesis mean the scalar product of the vectors. Let us expand the coefficients of the equation (4) with respect to p and use the expressions (6)-(8), (10) for p_ε, $\omega_{1,\varepsilon}^2$, $\omega_{1,\varepsilon}$, u_1^ε. The next step is to arrange into groups the coefficients for each power of ε. Since these equations must be satisfied for arbitrary values of ε and the sequence of powers of ε is linearly independent, then the coefficients at any power of ε are equal to zero. We obtain comparatively simple equations for the desired values μ_{1i} and vectors u_1^i

$$\begin{gathered} L_o u_1^o = 0, \quad L_o u_1^1 = \mu_{11} u_1^o \\ L_o u_1^2 = \mu_{11} u_1^1 + \mu_{12} u_1^o \mp i\tau_1 B_o u_1^o \end{gathered} \tag{11}$$

Here $L_o = C(p_o) - \tau_1^2 I$, $B_o = B(p_o)$. These equations must be successively solved.

For simple eigenvalues and eigenvectors $(j=3, 4, \ldots, n)$ we have the following equations

$$L_o u_j^o = 0, \quad L_o u_j^1 = \mu_{j1} u_j^o \mp i\tau_j B_o u_j^o \tag{12}$$

The equalities (11), (12) are obtained taking into account that the perturbation is ideal and $p=p_o$. If the perturbation is defective $(p_d < p_o,\ p_d \in D)$, then as it follows from assumptions 1, 2 only simple eigenvalues correspond to the value $p=p_d$. Therefore corresponding expansions in ε for $\omega^2_{j,\varepsilon}$, $\omega_{j,\varepsilon}$, u^{ε}_j have the form (7), (9), (10). The components of these expansions are determined with the help of the equations (12). Note that the value p_d must be substitute into equations (12)instead of the value p_o.

The Structure of the Ideal Perturbation

Theorem 1. Let the assumption 1, 2 are fulfilled and the matrix B realizes the ideal perturbation of the system (3) in the form (4). Then

$$(B_o u^o_1,\ v^o_1)=0 \tag{13}$$

where u^o_1 - eigenvector of the original problem (5), and v^o_1 - eigenvector of the problem adjoint to (5)

$$C^T(p_o)v^o_1=\tau^2_1 v^o_1 \tag{14}$$

Describe the structure of the matrices, which realize the ideal perturbations. It follows from the assumption 2 that $(u^o_1,\ v^o_1)=0$. Taking into account (13) we conclude that u^o_1 is the eigenvector of the matrix B_o and v^o_1 is the eigenvector of the matrix B^T_o. Let there exist real values ℓ_1 and $\ell_2(\ell_1+\ell_2\neq 0)$ such that

$$B_o u^o_1=\ell_1 u^o_1,\ B^T_o v^o_1=\ell_2 v^o_1$$

Construct an arbitrary orthogonal basis $W_1=u^o_1$, $W_2=v^o_1$, W_3, ..., W_n. Let basis vectors (columns) $W_i$put together the matrix W. Then the matrices realizing the ideal perturbation have the form

$$B=W^{-1}\Lambda W$$

Here Λ - diagonal matrix with diagonal elements ℓ_1 ℓ_2 ..., ℓ_n. In this case it can be shown that

$$p_\varepsilon=p_o - \frac{\tau^2_1\ell_1\ell_2(G_o u^o_1,v^o_1)}{(R_o u^o_1,v^o_1)}\ \varepsilon^2 + o(\varepsilon^2)$$

where $R_o=(dC(p)/dp)_{p=p_o}$ and G_o is the inverse matrix with respect to the matrix $L_o=C_o-\tau_1^2 I$ in the subspace of vectors w satisfying the condition $(w,u_1^o)=0$.

Example. If

$$C(p) = \begin{bmatrix} -p+3 & -2+p \\ -5+p & -p+4 \end{bmatrix}$$

then the eigenvalue problem (5) has double eigenvalue $\omega_1^2=\omega_2^2=\sqrt{2}$ and unique eigenvector $u_1^o=(3-2\sqrt{2},\ 1)$ for $p_o=(7-2\sqrt{2})/2$. The eigenvector of the adjoint problem (14) has the form $v_1^o=(-(3+2\sqrt{2}),\ 1)$. The matrix, realizing ideal symmetric perturbations, is reduced in our case to the form

$$B = \frac{1}{6}\begin{bmatrix} 3(\ell_1-\ell_2)-2\sqrt{2}(\ell_1-\ell_2) & \ell_1-\ell_2 \\ \ell_1-\ell_2 & 3(\ell_1+\ell_2)+2\sqrt{2}(\ell_1-\ell_2) \end{bmatrix}$$

where $\ell_1+\ell_2>0$.

Consider the case when the ideal perturbation is realized by nonsymmetric matrix. Let the matrix C(p) has the form

$$C(p) = \begin{bmatrix} p-2 & -1 \\ 1 & 0 \end{bmatrix}$$

The eigenvalues of the problem (5) are double and equal to -1 for $p_o=0$. The unique eigenvector of (5) and eigenvector of adjoint problem (14) are given by the formulas $u_1^o=(-1,1)$, $v_1^o=(1,1)$. The general form of the ideal perturbation is represented by the matrix

$$B = \begin{bmatrix} t & t-\ell_1 \\ \ell_2-1 & \ell_1+\ell_2-t \end{bmatrix} \qquad \ell_1+\ell_2 > 0,\ t \in R$$

Defective Perturbation

Theorem. Let the matrix B realizes the defective perturbation of the system (3) in the form (2). For stability of (2) it is necessary that

$$(u_i^o,\ v_i^o)(B_d u_i^o,\ v_i^o)\geq 0 \quad i=1,\ 2,\ \ldots,\ n \tag{15}$$

where u_i^o, v_i^o - the solutions of the eigenvalue problems

$$C_d u_i^o=\omega_i^2 u_i^o,\ C_d^T v_i^o=\omega_i^2 v_i^o \tag{16}$$

The subscript d is used in (15), (16) for the matrices C and B determined at $p=p_d$, i.e. $C_d=C(p_d)$, $B_d=B(p_d)$.

Corollary. Let $p_\varepsilon^1=p_d^1+\varepsilon^2 p_2^1+\ldots$ is certain curve in place ε, p of the form (6) outgoing from the point p_d^1 of the axis p. If (15) are satisfied as strict inequalities then for sufficiently small ε_o the system (2) will be asymptotic stable along the curve $p^1(\varepsilon)$ in the band $0<\varepsilon<\varepsilon_o$.

For given matrix B(p) let us consider the set $Q_B(p)$ of the values of parameter p such that

$$Q_B(p)=\{p:p<p_o,\ (u(p),\ v(p))(Bu(p),\ v(p))\geq 0\}$$

where u(p) are the eigenvectors of the problems

$$C(p)u(p)=\omega^2(p)u(p),\ C^T(p)v(p)=\omega^2(p)v(p)$$
$$(u(p),\ v(p))\neq 0$$

Let $p^*(B)$ - the upper bound of the values of $p \in Q_B(p)$. If the set $Q_B(p)$ is empty then we assume $p^*(B) = -\infty$.

Theorem. Defect of the matrix B(p) is determined by the formula

$$d_C(B)=p_o-p^*(B) \quad (p_d=p^*(B))$$

References

1. Bathe K.-J., Wilson E.L. Numerical methods in finite analysis. New York: Prentice-Hall, Englewood Cliffs, 1976.
2. Zienkiewicz O.C., Morgan K. Finite elements and approximation New York: Wiley-Interscience, 1983.
3. Bolotin V.V. Nonconservative problems of the theory of elastic stability. Moscow: Fizmatgiz, 1961. Englishtranslation published by Pergamon Press, New York, 1962.

4. Ziegler H.: Die Stabilitätskriterien der Elastomechanik, Ing.-Arch. 20 (1952), 49-56.

5. Bolotin V.V., Zinzher N.I.: Effects of damping on stability of elastic systems subjected to conconservative forces. Int. J. Sol. Struct. 5, 9 (1969) 965-989.

6. Herrmann G., Jong I.C.: On the destabilizing effect of damping in nonconservative elastic systems. J. Appl. Mech. 32 (1965) 592-597.

7. Nemal-Nasser S., Herrmann G.: Some general considerations concerning the destabilizing effect in nonconservative systems. ZAMP. 17 (1966) 305-313.

8. Andreichikov I.P., Yudovich V.I.: On stability of visco-elastic rods. Mechanics of Solids. 2 (1974) 78-87 (in Russian).

9. Denisov G.G., Novikov V.V.: On stability of elastic systems with small internal damping. Mechanics of Solids. 3 (1978) 41-47 (in Russian).

10. Miloslavsky A.I.: Stabilizing influence of small damping on abstract nonconservative system. Advances of Mathematical Sciences. 41, 1 (1986) 199-200 (in Russian).

11. Chetayev N.G.: The stability of motion. Moscow: Nauka, 1965. English translation published by Pergamon Press, Inc. Oxford, London, New York (1961).

12. Vishik M.I., Lyusternik L.A.: Solution of certain perturbation problems in the case of matrices and selfadjoint and nonselfadjoint differential equations.-I. Advances of Mathematical Sciences. 15. 3 (1960) 3-80 (in Russian).

Treatment of Micro-Mechanical Phenomena by Finite Elements

F.G. RAMMERSTORFER[1], F.D. FISCHER[2] and H.J. BÖHM[1]

[1] Institute of Lightweight Construction,
Vienna Technical University, Vienna, Austria
[2] Institute of Mechanics,
University of Mining and Metallurgy, Leoben, Austria

Abstract

The characteristics and parameter values of constitutive laws for the continuum mechanical description of solids are governed by micro-mechanical phenomena. It is a common - but often very cumbersome - practice to evaluate the parameters of theoretically derived constitutive equations by many experiments. In addition to cost and time consuming manipulations these experimental methods often do not provide a fundamental insight into the micro-mechanical mechanisms responsible for the typical material behavior.
An alternative approach is to use micro-mechanical methods for investigating the micro-structure of the material and to develop models for a computational derivation of the macro-scopic (continuum mechanical) material description.
This paper shows the micro-mechanical treatment by numerical methods of composite materials as well as metals under phase transformation.

1. Introduction

In the course of the last decade numerical methods in micro-mechanics have become very popular in materials science. They represent tools which - in combination with or in addition to - experimental investigations on the microscopic (or even submicro-scopic) scale allow improved understanding of the material behavior.

At the first look the use of continuum mechanical methods for considerations on a scale where the discontinuity of the material properties is an essential feature of the investigation appears to be contradictory. However, this approach has become very common in micro-mechanics [1] in the sense that the individual phases are each treated as a continuum taking into account the local material behavior and individual orientation of the local material axes.

This approach can be justified by the results achieved even for inhomogeneities at a very small scale, i.e. less that 1 micron [2]. This way a number of phenomena, which previously could only be described phenomenologically, such as anisotropic plasticity, damage etc., can be explained as the consequence of material behavior at the micro-mechanical level [3].

G. Kuhn · H. Mang (Eds.)
Discretization Methods in Structural Mechanis
IUTAM/IACM Symposium Vienna, Austria, 1989

2. Computation of overall composite material behavior

A fair number of papers deal with the numerical determination of the overall elasticity matrix of particle or fiber reinforced plastics or metals [4]. The behavior of such composites is strongly dependent on the material behavior of the constituents, the shape and size of the particles or fibers and the volume fraction and distribution of the reinforcement. An efficient finite element method taking into account the above parameters on the basis of a "transformation strain" approach has been published by Accorsi [5].

Despite the fact that such approaches lead to reasonably good results for the overall elastic properties of the composite, sufficient information cannot be obtained with respect to the local stress fields, stress concentrations etc., which are essential for strength considerations [6,7].

A worldwide interest in metal matrix composites (MMCs) can currently be recognized and, hence, worldwide theoretical and experimental investigations are concerned with the modelling of their material behavior, which is predominantly elastic-plastic even for low load levels.

An analytical micro-mechanical approach has been developed by Dvorak and his research group [8], which is partially based on a theory by Hill [9]. This approach starts with rather rough assumptions and simplifications:

$$d\bar{\sigma}_{ij} = d\sigma^f_{ij} = d\sigma^m_{ij} \qquad \text{for } ij \neq 33 \tag{1}$$

$$d\bar{\sigma}_{33} = c_f\, d\sigma^f_{33} + c_m\, d\sigma^m_{33} \tag{2}$$

$$d\bar{\varepsilon}_{ij} = c_f\, d\varepsilon^f_{ij} + c_m\, d\varepsilon^m_{ij} \qquad \text{for } ij \neq 33 \tag{3}$$

$$d\bar{\varepsilon}_{33} = d\varepsilon^f_{33} + d\varepsilon^m_{33} \tag{4}$$

where $d\bar{\underset{\sim}{\sigma}}$ and $d\bar{\underset{\sim}{\varepsilon}}$ are the overall quantities of the composite, c_f and c_m are the volume fractions of fiber and matrix, respectively, and the 3-axis corresponds to the fiber axis.

As long as the stress state in the matrix is purely elastic this theory assumes

$$d\underset{\sim}{\sigma}^r = \underset{\approx}{\mathbf{B}}^e_r\, d\bar{\underset{\sim}{\sigma}}\,, \qquad d\underset{\sim}{\varepsilon}^r = \underset{\approx}{\mathbf{A}}^e_r\, d\bar{\underset{\sim}{\varepsilon}} \qquad (r=f,m) \tag{5}$$

with the stress- and strain concentration matrices $\underset{\approx}{\mathbf{B}}^e_r$ and $\underset{\approx}{\mathbf{A}}^e_r$, respectively.

Assuming a von Mises type elastic plastic matrix with Ziegler's kinematic hardening and linear elastic reinforcing fibers, this theory leads an overall initial yield surface given by

$$\mathbf{F}(\bar{\underset{\sim}{\sigma}}) = \bar{\underset{\sim}{\sigma}}\, (\underset{\approx}{\mathbf{B}}^e_m)^T\, \underset{\approx}{\mathbf{C}}^T\, \underset{\approx}{\mathbf{B}}^e_m\, \bar{\underset{\sim}{\sigma}} - Y^2 = 0\,, \qquad \mathbf{F}(\bar{\underset{\sim}{\sigma}}) \triangleq \mathbf{F}_m(\underset{\sim}{\sigma}^m) \tag{6}$$

Here, $\underset{\approx}{C}$ is the von Mises projection matrix and Y is the initial yield stress of the matrix. Subsequent yield surfaces are described by

$$F(\underset{\sim}{\bar{\sigma}} - \underset{\sim}{\bar{\alpha}}) = 0 \tag{7}$$

$$d\underset{\sim}{\bar{\alpha}} = (\underset{\approx}{B}{}^{e}_{m})^{-1} \; d\underset{\sim}{\alpha}{}^{m} - [\,(\underset{\approx}{B}{}^{e}_{m})^{-1} \; \underset{\approx}{B}{}^{ep}_{m} \; - \; \underset{\approx}{I}\,] \, d\underset{\sim}{\bar{\sigma}} \tag{8}$$

$$d\underset{\sim}{\alpha}{}^{m} = \left(\frac{1}{Y^2}\right)(\underset{\sim}{\sigma}{}^{m} - \underset{\sim}{\alpha}{}^{m}) \; \underset{\approx}{C} \; (\underset{\sim}{\sigma}{}^{m} - \underset{\sim}{\alpha}{}^{m}) \, d\underset{\sim}{\sigma}{}^{m} \tag{9}$$

where $\underset{\sim}{\bar{\alpha}}$ is the translation tensor of the mid-point of the overall yield-surface, $\underset{\sim}{\alpha}{}^{m}$ is the matrix translation tensor and $\underset{\approx}{B}{}^{ep}_{m}$ is the state dependent elastic-plastic stress concentration matrix, compare [8].

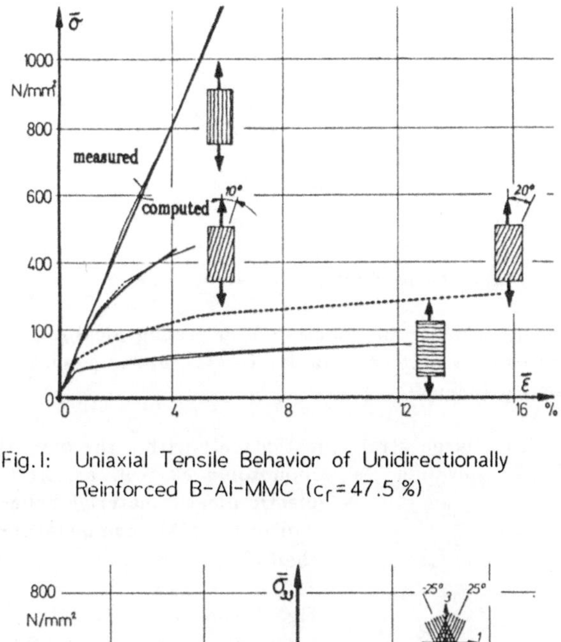

Fig. I: Uniaxial Tensile Behavior of Unidirectionally Reinforced B-Al-MMC (c_f = 47.5 %)

This approach has been implemented in a non-linear finite element program using first order Euler Forward Integration with subincrementation and stress correction [10]. Fig. 1 shows the efficiency of this approach. The numerical prediction of the uniaxial stress-strain behavior for different angles between applied load and fiber direction is in reasonably good agreement with experimental results published by Kenaga et al. [11].

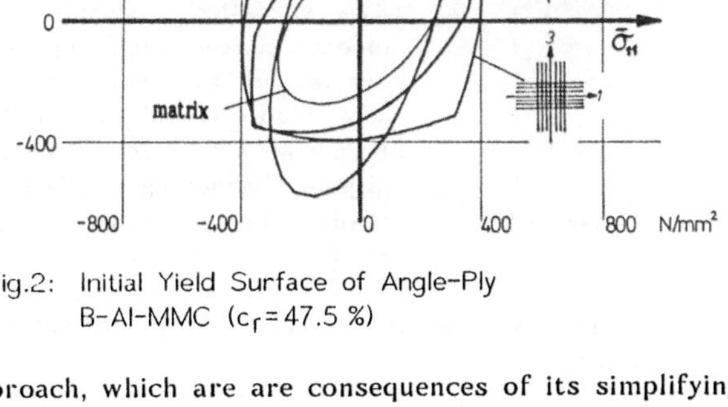

Fig.2: Initial Yield Surface of Angle-Ply B-Al-MMC (c_f = 47.5 %)

Furthermore, this finite element formulation of Dvorak's VFD theory in combination with overlay techniques [10] allows the pointwise computation of the initial yield surface of multilayered MMCs, as shown in Fig. 2 for some examples in the case of plane stress situations with the principal stress axes coinciding with the material axes 1 and 3.

There are, however, a number of shortcomings to this approach, which are are consequences of its simplifying assumptions:

* the actual local stress field cannot be investigated,
* possibly poor representation of the cross fiber behaviour,
* coupling effects between neighbouring fibers are not taken into account and, thus, no influence of different packings can recognized,
* the influence of initial micro-stresses (residual stresses resulting from the manufacturing process) cannot be taken into account.

Therefore, a more detailed micro-mechanical study seems to be justified.

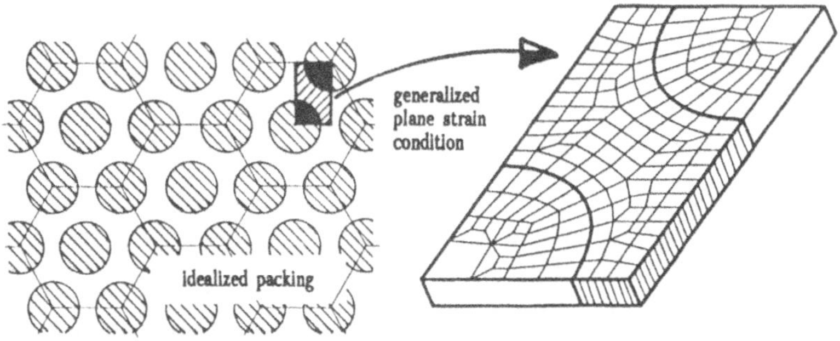

Fig.3: Representative Unit-Cell (B-Al-MMC; c_f = 47.5 %)

Assuming regular hexagonal packing, which - of course - is a strong idealization, numerical studies were performed using a unit cell similar to [12], under generalized plane strain conditions as shown in Fig.3.

Using this model in combination with a large strain plasticity algorithm the overall monotonic as well as cyclic elastic plastic material behavior of the MMC can be investigated.

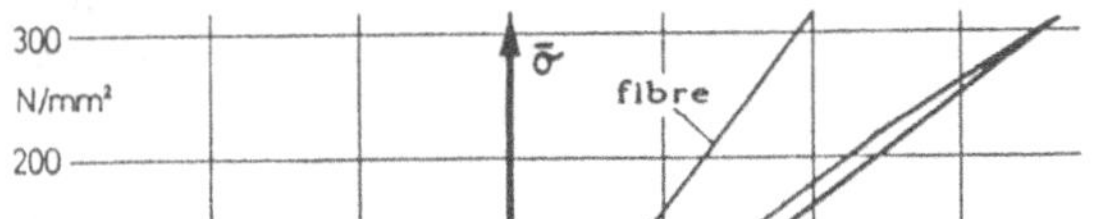

Fig.4: Uniaxial Stress-Strain Behavior Obtained from Unit Cell Approach, Virgin Material

Fig. 4 shows the results of some numerical tensile-compression-tests based on a completely stress-free virgin material.

The so-called "Bauschinger effect", i.e. the significant reduction of the yield stress after load reversal, observable in MMCs even for essentially non-hardening matrix behavior, can be explained by those micro-mechanical investigations as the result of the action of residual micro-stresses being developed during the initial plastification of the originally virgin material; see [12].

In the above analysis a completely stress free configuration was assumed as a starting point. This assumption probably is never met in real MMCs, which are typically produced either by diffusion bonding or by infiltration processes, both of which require elevated temperatures. Since the different components, i.e. matrix and fiber, have significantly different coefficients of thermal expansion, residual micro-stresses develop during the cooling down process. Of course, these micro-stresses influence the overall elastic plastic behavior of the MMC at room temperature.

The development of these micro-stresses can be simulated by micro-mechanical modelling using a thermoelastic-plastic analysis. Furthermore, subsequent numerical "tensile tests" can be performed immediately with the same model as a continuation of the thermal stress analysis. Some computational results showing the influence of the residual stresses on the overall stress-strain behavior are given in Fig. 5.

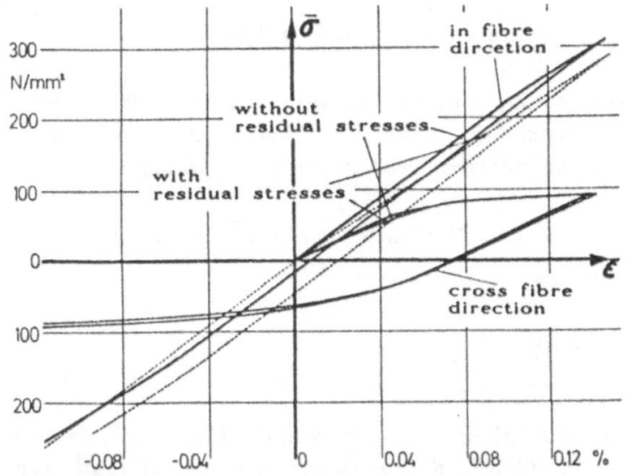

Fig.5: Influence of Residual Micro-Stresses due to Material Processing

Fig. 5, while neglecting viscosity effects and thus describing a highly idealized case, shows several interesting phenomena. The initial stiffness of the material with respect to loading in fiber direction depends on the load orientation: in compression it is stiffer than in tension. The reason for this rather strange behavior is the fact that the matrix has a residual stress state which locally lies at the matrix yield surface and continued plastification takes place in case of tensile loading. On the other hand, compression in fiber direction initially leads to elastic unloading and the same stiffness as in the virgin material is achieved. Furthermore, one can conclude from Fig. 5 that the overall stress-strain behavior for cross-fiber loading is less affected by the residual microstresses than for in-fiber loading.

3. Micromechanical Treatment of Phase Transforming Materials

Metals and alloys generally consist of regions with different metallurgical structure. The regions of each type may be distinguished by a common set of parameters (e.g. crystal structure) constituting a phase. Since the regions are usually portions of a grain a microscale must be introduced for the continuum mechanical descriptions. A new or product phase develops in an old or parent phase either by nucleation and growth transformation (diffusional transformation) or by displacive transformation (martensitic transformation), where nearly no time is necessary for the formation of the new phase, see [13]. The product phase differs in its volume and, in the case of martensite, its shape from the parent phase. The adaptation of this "misfit" may lead to a significant mircro-stress state varying inside each grain and from grain to grain.

The thermodynamic state function describing transformation kinetics is the Gibbs-Free-Energy $G(\underset{\sim}{\sigma}^M, T, V_i)$, where $\underset{\sim}{\sigma}^M$ is the actual stress state in a micro-region, T is the absolute temperature and V_i is a set of internal variables. Assuming a unique function G, transformation takes place if G reaches the same value for the parent and product phases; for a detailed account see [14]. G contains the chemical energy, the strain energy, the entropy and the potential energy of the load stress state and must - in principle - be calculated simultaneously with $\underset{\sim}{\sigma}^M$ and $\underset{\sim}{\varepsilon}^M$ for each micro-region. The numerical realization of this coupled problem, however, is generally beyond current capabilities and constitutes an exciting task for the future. In the case of ferrous materials the above coupling is "weak", and the transformation is usually described by a TTT (Time-Temperature-Transformation) diagram. The influence of the stress state on transformation kinetics is formulated e.g. in [15], [16].

Since it is hardly feasible to consider a metallic continuum grain by grain or even at the scale of a part of a grain, a macroscopic material law formulation must be elaborated which takes into account the micro-stress state in an average sense.

Following e.g. Leblond et al. [18], the overall plastic strain $\underset{\sim}{\bar{\varepsilon}}^P$ can be decomposed into three terms,

$$d\underset{\sim}{\bar{\varepsilon}}^P = \underset{\approx}{A}^P d\underset{\sim}{\bar{\sigma}} + \underset{\sim}{b}\, dT + \underset{\sim}{c}\, d\xi \tag{10}$$

where the first two terms are "classical" plasticity terms and the third term describes the transformation induced plasticity (TRIP); $\underset{\sim}{\bar{\sigma}}$ stands for the overall load stress. A micromechanical model for describing the TRIP term going back to Mitter [19] will shortly be explained.

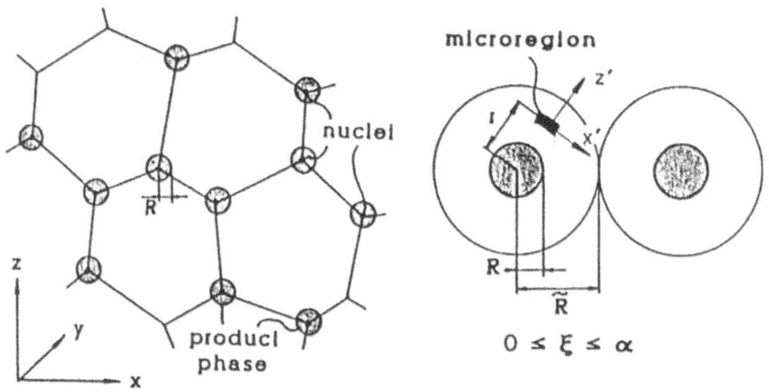

Fig.6: Micro-Mechanical Model

These considerations are restricted to diffusional transformations, such as the austenite-pearlite transformation, see e.g. [17]. The new pearlite phase with a volume fraction $\xi(t)$ grows in nodules nucleated at the junction of austenitic grains and can be represented by spheres with a radius R situated at an average distance $\tilde{D}$ and having a sphere packing factor α with $(0.524 \leq \alpha \leq 0.740)$, see Fig.6. The volume fraction $\xi(t)$ then is

$$\xi(t) = \alpha \left(\frac{R}{\tilde{R}}\right)^3 \tag{11}$$

Assuming a fully plastified material state (and thus incompressibility), the displacement $\rho(r)$ caused in a micro-region by the transformation volume strain δ can be described as

$$\rho(r) = \frac{\delta}{3} \frac{R^3}{r^2}, \qquad R \le r \le \tilde{R}, \tag{12}$$

the sphere with radius $\tilde{R}$ being isolated from the neighbouring spheres. The corresponding "free" strain rate in the global z-direction can then be described as

$$d\varepsilon_{c,z} = \frac{\delta}{3\alpha} \frac{R^3}{r^2} \, d\xi \, (1 - 3\cos^2\vartheta) \tag{13}$$

where $0 \le \vartheta \le \pi$ is one of the Eulerian angles.

Now a specimen is investigated under a constant load stress $\bar{\sigma}$ in z-direction. Following Taylor's Full Constraint Concept [20], one and the same global strain tensor is assumed in each micro-region, its components being

$$\varepsilon_{11} = \varepsilon_{22} = \varkappa, \quad \varepsilon_{33} = \varepsilon, \quad 2\varkappa + \varepsilon = 0. \tag{14}$$

The difference between the "free" strain rate $d\varepsilon_{c,z}$, and the "actual" strain rate, $d\varepsilon$, is compensated by plastification:

$$d\varepsilon - \frac{\delta}{3\alpha} \frac{\tilde{R}^3}{r^2} \, d\xi \, (1 - 3\cos^2\vartheta) = d\lambda \, s_z \tag{15}$$

where $d\lambda$ is a multiplier and s_z is the deviator component in z-direction. With the yield stresses of the old and new phases, σ_{F_0} and σ_{F_N}, and using the flow rule, the average of σ_z on a "shell" with radius r can be calculated as

$$\frac{\langle \sigma_z \rangle_r}{\sigma_{F_0}} = \int_0^{\pi} \frac{\left[\dot{\varepsilon} - \frac{\dot{\tilde{\delta}}}{3}(1 - 3\cos^2\vartheta)\right] \sin\vartheta \, d\vartheta}{\left\{\frac{2}{3}\left[\frac{3}{2}\dot{\varepsilon}^2 - \dot{\varepsilon}\dot{\tilde{\delta}}(1 - 3\cos^2\vartheta) + \frac{2}{3}\dot{\tilde{\delta}}^2\right]\right\}} =$$

$$= \frac{27\bar{\varepsilon}^2 - 6\bar{\varepsilon} - 4}{36\sqrt{2\bar{\varepsilon}^{3/2}}} \, arsinh\left(\sqrt{18\bar{\varepsilon}/(9\bar{\varepsilon}^2 - 6\bar{\varepsilon} + 4)}\right) + \frac{1}{4} + \frac{1}{6\bar{\varepsilon}} = f(\bar{\varepsilon})$$

$$\dot{\varepsilon} \ge 0, \quad \bar{\varepsilon} = \frac{\dot{\varepsilon}}{\dot{\tilde{\delta}}}, \quad \dot{\tilde{\delta}} = \frac{\delta}{\alpha} \frac{R^3}{r^3} \dot{\xi} \tag{16}$$

Since the product phase is under a high hydrostatic stress, an elastic state with σ being the average load stress component in z-direction can be assumed. This leads to the following average stress balance

$$(1-\xi)\int_{R}^{\tilde{R}} \frac{\langle\sigma_z\rangle_r \, 4r^2\pi}{4\pi/3\,(\tilde{R}^3-R^3)}\,dr + \xi\bar{\sigma} = \bar{\sigma}, \qquad \xi \le \alpha. \tag{17}$$

This is a nonlinear differential equation for ε. In the case of small values of σ/σ_{F_0} a linearization is possible and leads to

$$\dot{\varepsilon} = \frac{5}{6}\frac{\sigma}{\sigma_{F_0}}\frac{2\dot{\xi}}{\alpha+\xi} \qquad \text{for } \frac{\sigma}{\sigma_{F_0}} << 1. \tag{18}$$

The total strain rate of the specimen is now

$$\dot{\mathcal{E}} = (1-\xi)\dot{\varepsilon} + \frac{\delta}{3}\dot{\xi}, \qquad \xi \le \alpha ;$$

$$\mathcal{E} = \frac{5}{6}\delta\frac{\sigma}{\sigma_{F_0}}\left[2(1+\alpha)\, ln(1+\frac{\xi}{\alpha}) - 2\xi\right] + \frac{\delta}{3}\xi \qquad \text{for } \frac{\sigma}{\sigma_{F_0}} << 1. \tag{19}$$

The quasi "inverse" model for the wedge after the coalescence of the product phase spheres, $\alpha<\xi\le 1$, leads to a similar relation and is discussed elsewhere. For small values of σ/σ_{F_0} one can obtain

$$\mathcal{E} = \mathcal{E}|_{\xi=\alpha} + \frac{5}{6}\delta\frac{\sigma}{\sigma_{F_N}}\frac{(\xi^2-\alpha^2)}{(2-\alpha)} + \frac{\delta}{3}\xi \qquad \text{for } \frac{\sigma}{\sigma_{F_0}} << 1. \tag{20}$$

For $\alpha = 0.632$ and $\xi = 1$ the following relation holds:

$$\mathcal{E} = \frac{5}{6}\delta\frac{\sigma}{\sigma_{F_0}}\left[1+0.425\frac{\sigma_{F_0}}{\sigma_{F_N}}\right] + \frac{\delta}{3} \qquad \text{for } \frac{\sigma}{\sigma_{F_0}} << 1 \tag{21}$$

which for $\sigma_{F_0}/\sigma_{F_N} \to 0$ yields the well known Greenwood/Johnson relation, see [21].

This "micro-region" concept may be extended to a triaxial load stress state and a transformation shape change, see Fischer et al. [22]. In that paper the justification of a TRIP term in form of

$$d\underset{\sim}{\bar{\varepsilon}}^{TRIP} = \frac{3}{2}K\underset{\sim}{\bar{s}}\,\varphi(\xi)\,d\xi, \tag{22}$$

where $\underset{\sim}{\bar{s}}$ is the overall stress deviator and K is a constant (in the above case $K=5\delta/6\sigma_{F_0}$ for $\sigma_{F_0}/\sigma_{F_N} << 1$), is given, too.

This model implies some simplifications (neglection of elastic strain components, no hardening, Taylor's Full Constraint Concept). To understand their distinct influences a finite element study was performed.

A regular lattice of nuclei was assumed, the distance between them being a fraction of the grain size.

Now an "ordered" transformation was introduced with the product phase regions growing from each nucleus in the same way. Due to this periodicity a unit cell can be defined with the condition that its confining planes must remain parallel to their initial position, compare Fig.7. The use of symmetry conditions allows the use of a triangular prism as the final model geometry, which comprises 1/16 of the unit cell. A rather crude 7x7x7 grid with 3636 DOFs was selected, this being the finest mesh that could be handled on the SUN 4/110 work station used for the computations (about 30 iterations were required, each using approximately 50 minutes of CPU time).

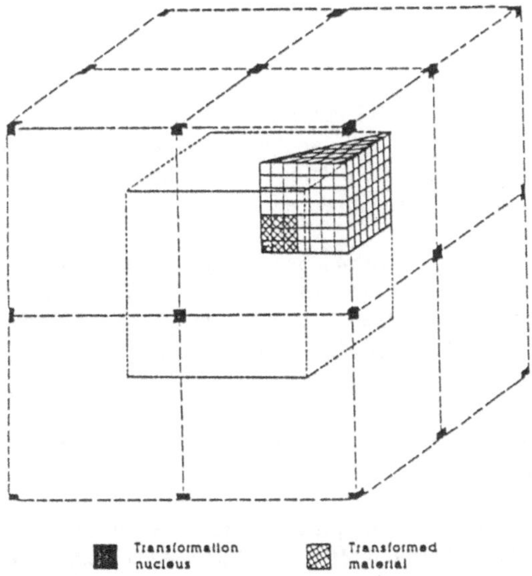

Fig.7: Representative Unit Cell for an Ordered Transformation

In a first step the load stress $\bar{\sigma}$ in z-direction was applied to the parent phase. The progress of the transformation was simulated by an affine inner prism of the product material growing from a 1x1x1 grid to a nxnxn grid, $n \leq 7$. The volume change was simulated by thermal expansion with a temperature-dependent thermal expansion coefficient $\alpha(T)$. When the transformation proceeded through the i-th element row the volume was changed linearly. After undergoing a full transformation in the i-th row the volume was kept constant by specifying $\alpha(T) = \delta/3T$. Generally, the transformation was modelled using linearly increasing temperatures.

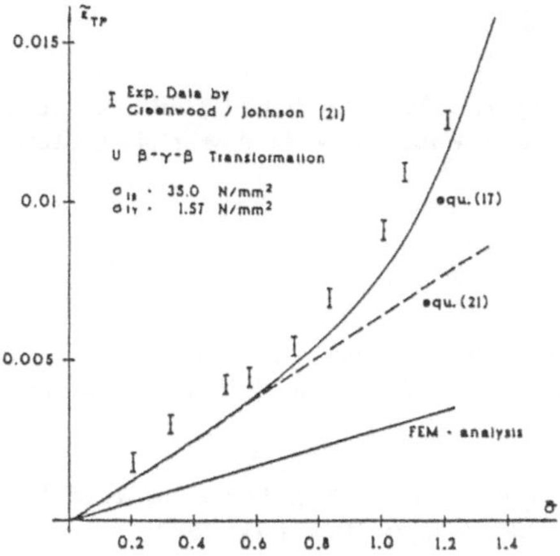

Fig.8: Calculated and Experimental Data for TRIP-Strains

To allow a comparison with the experimental data by Greenwood et al.[21], the Uranium β-γ-β transformation from the "hard" phase to the "weak" phase was simulated using $\sigma_{F\beta} = 35$ N/mm^2 and $\sigma_{F\gamma} = 1.57$ N/mm^2. Unfortunately, the results were somewhat disappointing. The computed final irreversible strain after transformation, $\tilde{\varepsilon}_{TP}$, was less than the half the measured value. The analytical model fits the experimental data considerably better. The reason for this probably lies in the excessive "regularity" of the model. Specifically, the confining planes of the prism are free of shear and, therefore, under too small a constraint. A similar observation is reported in the literature: Sjöström et al. [23] performed an analysis of an isothermal pearlite transformation with a 6x6x6 grid representing an octant of a unit cell with the faces of the cell moving uniformly in the respective coordinate directions. The calculated irreversible strain after deformation was about one third of the measured value for an ordered transformation and about two thirds for a random transformation.

In a further paper Leblond et al. [24] investigated a martensitic transformation by modelling volume changes only. They used a 5x5x5 octant and found the irreversible strain to be about one half to three quarters of the analytical value which was also reported to be smaller than the measured one. Surprisingly, the random transformation gave a smaller irreversible strain than the ordered transformation.

In the authors' opinion realistic results can only be achieved if a group of transformation nuclei is assumed in more or less random positions. This leads to meshes with so many degrees of freedom (e.g. Sjöström et al. suggest grids larger than 20x20x20) that the running time will be probably enormous even on supercomputers.

Simplified models with axisymmetric and plane cells have also been investigated and led to even worse results, e.g. loosing the spherical symmetry in the case of transformation without any load stress.

Finally it should be mentioned that the "classical" plasticity term $\underset{\approx}{A}^{P} d\underset{\sim}{\bar{\sigma}}$ can also be investigated by discretization methods. In his book Mitter published artificial stress-strain diagrams obtained in the course of simulating a dilatometer test. These diagrams show a deviation from linear behavior even at low stresses. The computed flow stress σ_f is significantly lower than the value obtained by the mixture rule:

$$\sigma_f^M = (1-\xi)\sigma_{f_0} + \xi\sigma_{f_N}, \qquad \sigma_f < \sigma_{f_N} \quad \text{for } 0 \le \xi \le 1. \tag{23}$$

For further information the reader is referred to chapter 2.2.1.2.1 of [19] or to [25].

4. Acknowledgements

The financial support of this investigation by the Austrian Ministry of Science and Technology within the project "Computer Aided Material Development" is gratefully acknowledged.

5. References

[1] Needleman A.: *Computational Mechanics.* In: Theoretical and Applied Mechanics (Eds.: P.Germain, M.Piau, D.Caillere), 217-240. North-Holland, Amsterdam, 1989.

[2] Tvergaard V.: *Analysis of Tensile Properties for a Whisker-Reinforced Metal Matrix Composite.* DCAMM Report 382, Lyngby, Denmark, 1989.

[3] Needleman A.: *A Continuum Model for Void Nucleation by Inclusion Debonding.* J.Appl.Meth. 109 (1987), 525-531.

[4] Hashin Z.: *Analysis of Composite Materials.* J.Appl.Mech. 50 (1983), 481-505.

[5] Accorsi M.L.: *A Method for Modelling Microstructural Material Discontinuities in a Finite Element Analysis.* Int.J.Num.Meth.Eng. 26 (1988), 2187-2197.

[6] Needleman A.; Nutt S.R.: *Void Nucleation at Fiber Ends in Al-SiC Composites.* Script.Met. 21 (1987), 705-710.

[7] Dorninger K.; Rammerstorfer F.G.: *Nonlinear Investigation of Composite Shells by Finite Element Method.* Proc. 12th Int. Conf. Energy-Sources Technology (Eds.: D.Hui, T.J.Kozik). ASME, New York, 1988.

[8] Dvorak G.J.; Bahei-El-Din Y.A.: *Plasticity Analysis of Fibrous Composites.* J.Appl.Mech. 49 (1982), 327-335.

[9] Hill R.: *Elastic Properties of Reinforced Solids: Some Theoretical Principles.* J.Mech.Phys.Sol. 11 (1963), 357-372.

[10] Svobodnik A.J.; Böhm H.J.; Rammerstorfer F.G.: *Elastic-Plastic Behavior of Metal Matrix Composites with Special Emphasis on Finite Element Analysis.* Proc. Plasticity '89. To be published.

[11] Kenaga D.; Doyle J.F.; Sun C.T.: *The Characterization of Boron/Aluminum Composite in the Nonlinear Range as an Orthotropic Elastic-Plastic Material.* J.Comp.Mat. 21 (1987), 516-531.

[12] Böhm H.J.; Rammerstorfer F.G.: *Computational Investigation of the Anisotropic Elastic-Plastic Material Behavior of Fiber-Reinforced Metals as Influenced by Residual Micro-Stresses.* Proc. Int. Conf. Resid. Stresses, Nancy, 1988. To be published.

[13] Christian J.W.: *The Theory of Transformations in Metals and Alloys.* Chapt.1. Pergamon Press, Oxford, 1975.

[14] Verhoeven J.D.: *Fundamentals of Physical Metallurgy.* Chapt.13. J.Wiley & Sons, New York, 1975.

[15] Inoue T.; Raniecki B.: *Determination of Thermal-Hardening Stress in Steels by Use of Thermoplasticity Theory.* J.Mech.Phys.Sol. 26 (1978), 187-212.

[16] Tanaka K.; Iwasaki R.; Nagaki S.: *On T-T-T and C-C-T Diagrams of Steels: A Phenomenological Approach to Transformation Kinetics.* Ingenieur-Archiv 54 (1984), 81-90.

[17] Marder A.R.: *Structure-Property Relationships in Ferrous Transformation Products.* In: Phase Transformations in Ferrous Alloys (Eds.: A.R.Marder, J.I.Goldstein). AIME, Warrendale, PA, 1984.

[18] Leblond J.B.; Mottet G.; Devaux J.C.: *A Theoretical and Numerical Approach to the Plastic Bevaiour of Steels during Phase Transformations - I. Derivation of General Relations.* J.Mech.Phys.Sol. 34 (1986), 395-409.

[19] Mitter W.: *Umwandlungsplastizität und ihre Berücksichtigung bei der Berechnung von Eigenspannungen.* Materialkundlich-Technische Reihe 7, Gebr. Borntäger, Berlin, 1987.

[20] Taylor G.I.: *Plastic Strain in Metals.* J.Inst.Metals 62 (1938), 307-324.

[21] Greenwood G.W.; Johnson R.M.: *The Deformation of Metals under Small Stresses During Phase Transformation.* Proc.Roy.Soc. A283 (1965), 403-422.

[22] Fischer F.D.; Pichler A.; Mitter W.: *A Micromechanical Explanation of Transformation Induced Plasticity (TRIP) Applied to Austenite-Martensite Transformations.* Proc. Int. Symp. Martensitic Transf., Bochum, 1989. To be published.

[23] Sjöström S.; Denis S.; Gautier E.; Simon A.: *FEM Calculation of the Micromechanics a Diffusional Transformation.* Proc. Int. Conf. Resid. Stresses, Nancy, 1988. To be published.

[24] Leblond J.B.; Devaux J.C.: *Mathematical Modelling of Transformation Plasticity in Steels - I: Case of Weak-Plastic Phases.* Submitted to Int.J.Plasticity, 1988.

[25] Rammerstorfer F.G.; Fischer F.D.; Mitter W.; Bathe K.J.; Snyder M.D.: *On Thermo-Elastic-Plastic Analysis of Heat-Treatment Processes Including Creep and Phase Changes.* Comp.Struct. 13 (1981), 771-779.

On the Coupling of Finite Elements and Boundary Elements

W.L. Wendland
Mathematics Institute A
University of Stuttgart, W.-Germany

Summary

The coupling of finite elements with boundary elements plays an increasingly important rôle in computational mechanics since corresponding advantages are combined in a "Marriage à la mode. The best of both worlds" (Bettess, Kelly and Zienkiewicz [4]). This marriage currently either combines both methods in two respective different subregions or in macro-elements within a hybrid finite element formulation. Here we present some asymptotic error analysis for couplings based on weak formulations
-for simple variational coupling with differently refined grids of the FEM and BEM
-the unsymmetric positive definite Galerkin coupling and
-first results for macro-elements modelled with boundary elements.

1 Introduction

The coupling of finite elements and boundary elements provides a very efficient combination of complementary advantages of both methods: The flexibility of the FEM in view of inhomogenities and non-linearities on one hand, with the natural involvement of boundary geometries and of unbounded domains in the BEM on the other hand. Moreover, the coupling is one case of domain decomposition.
In this lecture we give a brief survey on several different formulations for the coupling based on a weak formulation. For references and complete proofs see [27,28] and [16]. In practical computations, however, the coupling is often formulated by *pointwise transmission conditions* [4], here a rigorous error analysis could not be established yet. At present, there are basically two different concepts for the combination of FEM and BEM. One consists of subdividing the original domain $\Omega \subset \mathbb{R}^n$, $n = 2$ or 3, into a finite number of subregions and using in each of them either FEM or BEM. The second concept uses BEM for the modelling of special finite element functions within a hybrid FEM formulation [16,24]. The domain Ω of the boundary value problem will be divided into Ω_F and Ω_B with respective boundaries Γ_F, Γ_B, where $\Gamma_F \cap \Gamma_B$ and the remaining boundary components are either separated (see Figure 1) or the coupling boundary cuts Ω (see Figure 2) or we consider a quasi-regular family of triangulations where on some elements B_j we will define special macro-elements via individual BEM (see Figure 3).

G. Kuhn · H. Mang (Eds.)
Discretization Methods in Structural Mechanis
IUTAM/IACM Symposium Vienna, Austria, 1989

$D_y u = \sigma(u)(y)\cdot\vec{n}(y)$ is the boundary traction, ^{T} denotes the transposed. $u_{|\Gamma_B}$ and $Du_{|\Gamma_B}$ will be related to each other by boundary integral equations on Γ_B obtained from (1.7). In Ω_F we shall use a FEM approximation, here for brevity we take conforming elements. On Γ_B we use the boundary integral equations of the first kind and corresponding BEM as characterized in [10]. Asymptotic error estimates for the *coupled* FEM-BEM methods go back to Brezzi, Johnson and Nedelec [5,6,17] who considered the Laplacian. The commonly used method is based on a direct variational Ritz-Galerkin formulation yielding an unsymmetric non definite bilinear form. However, one needs the strong assumption that the double layer potential operator on Γ_B is a compact mapping in $H^{\frac{1}{2}}(\Gamma_B)$ which is violated in elasticity. Therefore, following Brezzi and Johnson [5], we require instead on $\Gamma_B \cap \Gamma_F$ a faster mesh refinement of the BEM than of the FEM. This also implies convergence of the FEM coupling with collocation BEM in two dimensions [27]. Moreover, this analysis can be applied to the BEM modelling of macro-elements as developed in [7,16,18,24]. The different rates of refinements can also be used the other way around: Convergence can be shown for faster FEM than BEM refinement. Recently, the variational coupling of FEM-BEM Galerkin was formulated as an antisymmetric definite variational bilinear equation, which in elasticity is the Hellinger-Reissner principle, see [11,12,13,15,20,21]. (See [27,28] for further references.) Here Galerkin formulations will provide methods converging of optimal order for all mesh refinements of the finite and the boundary elements.
Note that all results involving Galerkin BEM hold for $n = 2$ and for $n = 3$ as well.

2 Direct variational coupling

The simplest formulation of coupling is based on (1.5) and (1.7). From Ω_B we take $x \to \Gamma_B$ and D_x on Γ_B of (1.7) to obtain weak boundary integral equations on Γ_B,

$$\langle V\Lambda, \chi\rangle_{\Gamma_B} = \langle -\frac{1}{2}u + Ku, \chi\rangle_{\Gamma_B} \text{ for all } \chi \in W_1(\Gamma_B) \quad , \qquad (2.1)$$

$$\langle \frac{1}{2}\Lambda + K'\Lambda, \eta\rangle_{\Gamma_B} = -\langle Du, \eta\rangle_{\Gamma_B} \qquad \text{for all } \eta \in W_2(\Gamma_B) \quad , \qquad (2.2)$$

where $\langle\ ,\ \rangle$ is the L_2 scalar product (or duality pairing), W_1 and W_2 are the respective test spaces and the boundary integral operators are given by

$$Ku(x) = \int_{y\in\Gamma_B\setminus\{x\}} \left(D_y \underset{\sim}{E}(x,y)\right)^T \cdot u(y)\, ds_y \quad , \qquad (2.3)$$

$$K'\Lambda(x) = \int_{y\in\Gamma_B\setminus\{x\}} \left(D_x \underset{\sim}{E}(x,y)\right) \cdot \Lambda(y)\, ds_y \quad , \qquad (2.4)$$

$$Du(x) = -\int_{y\in\Gamma_B\setminus\{x\}} \left(D_x \left(D_y \underset{\sim}{E}(x,y)\right)^T\right) \cdot (u(y) - u(x))\, ds_y \quad , \qquad (2.5)$$

for $x \in \Gamma_B\setminus\gamma$ where γ denotes all corner (and edge) points for $n = 2$ (and 3). (2.1) and (2.2) form an overdetermined system of equations for

$$\Lambda(x) := \sigma(u)\cdot\vec{n}(x)_{|\Gamma_B} = D_x u(x)_{|\Gamma_B} \text{ on } \Gamma_B\setminus\Gamma_N \text{ and for } u_{|\Gamma_B} \text{ on } \Gamma_B\setminus\Gamma_D \quad . \qquad (2.6)$$

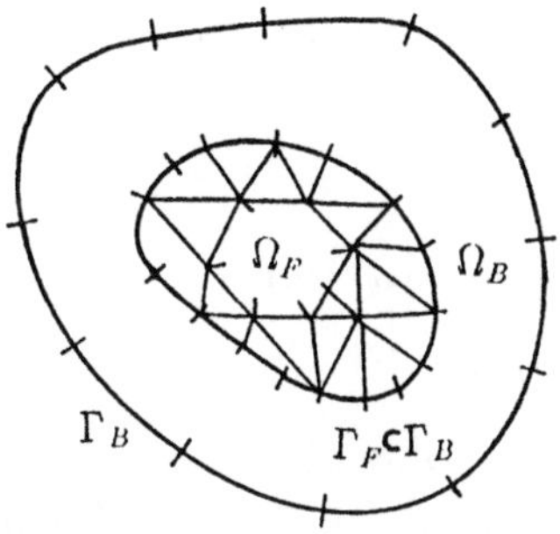

Figure 1

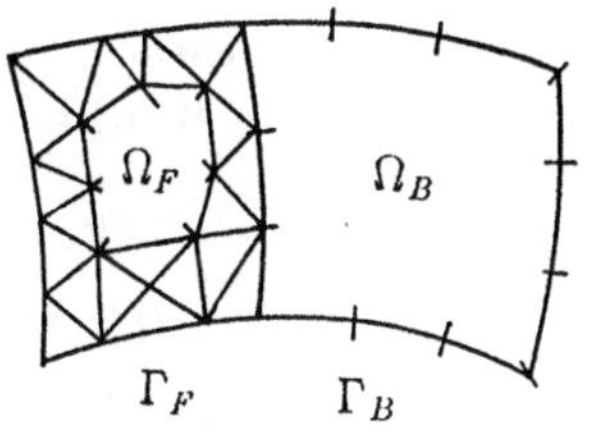

Figure 2

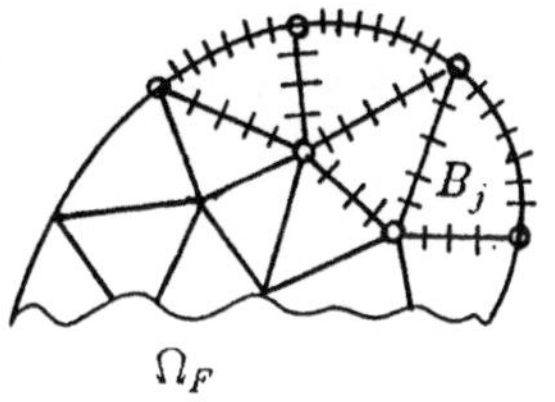

Figure 3

As a simple model problem, let us consider the displacement problem and the traction problem or the mixed problem of homogeneous isotropic ideal elasticity for the vector-valued displacement u satisfying the Navier system in $\Omega \subset \mathbf{R}^n$, $n = 2$ or 3,

$$\mu \Delta u + (\lambda + \mu) grad\ div\ u = f \tag{1.1}$$

where $\mu > 0$ and $\lambda > -\frac{2}{n}\mu$ are the Lamé constants. f is a given right hand side which we require to vanish in Ω_B. In addition, we require boundary conditions with

$$u_{|\Gamma_D} = \varphi \in H^{\frac{1}{2}}(\Gamma_D) \quad \text{on the Dirichlet part } \Gamma_D \subset \Gamma \ , \tag{1.2}$$

$$\sigma(u) \cdot \vec{n}_{|\Gamma_N} = \psi \in H^{-\frac{1}{2}}(\Gamma_N) \quad \text{on the Neumann part } \Gamma_N \subset \Gamma \tag{1.3}$$

with $\Gamma = \overline{\Gamma}_D \cup \overline{\Gamma}_N$ where $\overline{\Gamma}_D$, $\overline{\Gamma}_N$ are the closures. Here,

$$(\sigma(u))_{ij} = \lambda\ div\ u \delta_{ij} + \mu(\partial_{x_i} u_j + \partial_{x_j} u_i) \tag{1.4}$$

denotes the stress tensor, where δ_{ij} is the Kronecker symbol, $\vec{n}$ denotes the exterior normal to Γ and $\cdot$ denotes the scalar product in $\mathbf{R}^n$. The well known weak formulation corresponding to the principle of virtual work then reads as to *find* $u \in H^1(\Omega)$ *with* u *satisfying* (1.2) *such that for all* $v \in V(\Omega) := \left\{ v \in H^1(\Omega) \mid v_{|\Gamma_D} = 0 \right\}$,

$$\int_\Omega \sigma(u) \cdot e(v)\, dx = -\int_\Omega f \cdot v\, dx + \int_{\Gamma_N} \psi \cdot v\, ds \ . \tag{1.5}$$

$$(e(v))_{ij} = \frac{1}{2}(\partial_{x_i} u_j + \partial_{x_j} u_i) \quad , \quad \sigma \cdot e = \sum_{i,j=1}^{n} \sigma_{ij} e_{ij} \ . \tag{1.6}$$

In the coupling methods, we split the left hand side of (1.5) into two terms by integrating over Ω_F and Ω_B and use the Green identity and the Betti representation theorem,

$$u(x) = -\int_{\Gamma_B} \underset{\sim}{E}(x,y) \cdot (D_y u(y))\, ds_y + \int_{\Gamma_B} \left(D_y \underset{\sim}{E}(x,y) \right)^T \cdot u(y)\, ds_y \ , \ x \in \Omega_B \ . \tag{1.7}$$

Here $\underset{\sim}{E}(x,y)$ denotes the fundamental solution of the Navier system,

$$\left(\underset{\sim}{E}(x,y) \right)_{ij} = k_0 \left\{ \gamma(x,y)\delta_{ij} + \frac{\lambda+\mu}{\lambda+3\mu} \frac{(x_i - y_i)(x_j - y_j)}{(x-y)^n} \right\} \quad , \quad i,j = 1,.,n \ ;$$

$$k_0 = \frac{\lambda + 3\mu}{4\pi(n-1)(\lambda+2\mu)} \quad , \quad \gamma = \begin{cases} -\log|x-y| & \text{for} \quad n = 2, \\ |x-y|^{-1} & \text{for} \quad n = 3. \end{cases}$$

Since here the boundary conditions can change, for the correct function spaces we first *extend the given boundary data* $\varphi \in H^{r-\frac{1}{2}}(\Gamma_D)$ to some $\varphi^* \in H^r(\Omega)$ with

$$\varphi^*_{|\Gamma_D} = \varphi_{|\Gamma_D} \tag{2.7}$$

and, correspondingly, the given data $\psi \in H^{\tilde{r}}(\Gamma_N)$ to $\psi^* \in H^{\tilde{r}+\frac{1}{2}}(\Omega)$ with

$$\psi^*_{|\Gamma_N} = \psi_{|\Gamma_N} \text{ and } \psi^*_{|\Gamma_B} \in H^{\tilde{r}}(\Gamma_B) \ . \tag{2.8}$$

Then the unknown are $\tilde{u} = u - \varphi^*$ in Ω with associated $\tilde{\Lambda} = D(u) - \psi^*$.
For the finite element approximation in Ω_F we use a family $S_h^{k,r} \subset H^1(\Omega_F)$ of finite elements in the sense of Babuška and Aziz [1] providing the *approximation property*

$$\inf_{w \in S_h^{k,r}} \|u - w\|_{H^l(\Omega_F)} \leq ch^{k-l}\|u\|_{H^k(\Omega_F)} \text{ for } l \leq r < k,\ 1 \leq r \ . \tag{2.9}$$

From now on, c will denote a generic constant independent of h and u. h denotes the parameter of the maximum FEM mesh width in Ω_F. On the boundary element curve Γ_B we use a corresponding family $S_{\tilde{h}}^{\tilde{k},\tilde{r}}$ of $C^{\tilde{r}-1}$ - splines with respect to some parametric representation of Γ_B. $\tilde{h}$ denotes the maximum mesh width of the partition of $\Gamma_B, \tilde{k}-1$ the degree of piecewise polynomials. Again, we have the *approximation property*

$$\inf_{\chi \in S_{\tilde{h}}^{\tilde{k},\tilde{r}}} \|v - \chi\|_{H^l(\Gamma_B)} \leq c\tilde{h}^{\tilde{k}-l}\|v\|_{H^{\tilde{k}}(\Gamma_B)} \text{ for } l \leq \tilde{r} < \tilde{k} \ ,\ l < \tilde{k} - \frac{1}{2} \ . \tag{2.10}$$

For $\tilde{r} = 0,\ \tilde{k} = 1$, these are the piecewise constants and for $\tilde{r} = -1, \tilde{k} = 0$, these are the Diracs associated with collocation. For the direct variational coupling find

$$u_F \in H_h(\Omega_F) := \left\{v_F \in S_h^{k,r} \mid v_{F|\Gamma_D} = 0\right\} \ , \tag{2.11}$$

$$\Lambda_{\tilde{h}} \in H_{\tilde{h}}(\Gamma_{BD}) := \left\{\chi_{\tilde{h}} \in S_{\tilde{h}}^{\tilde{k},\tilde{r}}(\Gamma_B) \mid \chi_{\tilde{h}|\Gamma_N} = 0\right\} \ , \tag{2.12}$$

$$u_{\tilde{h}} \in H_{\tilde{h}}(\Gamma_{BN}) := \left\{w_{\tilde{h}} \in S_{\tilde{h}}^{\tilde{k},\tilde{r}}(\Gamma_B) \mid w_{\tilde{h}|\Gamma_D} = 0\right\} \ , \tag{2.13}$$

such that

$$\int_{\Omega_F} \sigma(u_F) \cdot e(v_F)\, dx + \int_{\Gamma_B \backslash \Gamma_N} \Lambda_{\tilde{h}} \cdot v_F\, ds = \mathcal{L}_F(v_F) \text{ for all } v_F \in H_h(\Omega_F) \tag{2.14}$$

$$\text{where } \mathcal{L}_F(v_F) = -\int_{\Omega_F} \{\sigma(\varphi^*) \cdot e(v_F) + f \cdot v_F\}\, dx - \int_{\Gamma_B \backslash \Gamma_N} \psi^* \cdot v_F\, ds + \int_{\Gamma_N \backslash \Gamma_B} \psi^* \cdot v_F\, ds \ ,$$

$$\langle V\Lambda_{\tilde{h}}, \eta_{\tilde{h}}\rangle_{\Gamma_B} + \langle \frac{1}{2}\tilde{u} - K\tilde{u}, \eta_{\tilde{h}}\rangle_{\Gamma_B} = \mathcal{L}_B(\eta_{\tilde{h}}) \quad \text{for all } \eta_{\tilde{h}} \in W_{\tilde{h}}(\Gamma_{BD}) \cup W_{\tilde{h}}(\Gamma_{BN}) \tag{2.15}$$

$$\text{where } \mathcal{L}_B(\eta_{\tilde{h}}) = -\langle v_{\psi^*}, \eta_{\tilde{h}}\rangle_{\Gamma_B} - \langle \frac{1}{2}\varphi^* - K\varphi^*, \eta_{\tilde{h}}\rangle_{\Gamma_B}$$

$$\text{and where } \tilde{u} = \begin{cases} u_F & \text{on } \Gamma_B \backslash \Gamma_N \ , \\ u_{\tilde{h}} & \text{on } \Gamma_B \cap \Gamma_N \text{ and} \end{cases} \tag{2.16}$$

$$\tilde{u}(p) = u_F(p) = u_{\tilde{h}}(p) \text{ for } p \in (\overline{\Gamma_N \backslash \Gamma_B}) \cap (\overline{\Gamma_N \backslash \Gamma_B}).$$

For boundary element Galerkin equations we choose in (2.15) $W_{1\tilde{h}}(\Gamma_{BD}) = H_{\tilde{h}}(\Gamma_{BD})$ and $W_{1\tilde{h}}(\Gamma_{BN}) = H_{\tilde{h}}(\Gamma_{BN})$. The last equations sometimes are replaced by Galerkin equations (2.2) on Γ_{BN} for $u_{\tilde{h}}$ on $\Gamma_B \cap \Gamma_N$, i.e.

$$\begin{aligned} \langle D\tilde{u}, \eta_{\tilde{h}} \rangle_{\Gamma_B} + \langle \frac{1}{2}\Lambda_{\tilde{h}} + K'\Lambda_{\tilde{h}}, \eta_{\tilde{h}} \rangle_{\Gamma_B} &= \mathcal{L}_{BN}(\eta_{\tilde{h}}) \\ \text{where } \mathcal{L}_{BN}(\eta_{\tilde{h}}) &= -\langle D\varphi^*, \eta_{\tilde{h}} \rangle_{\Gamma_B} - \langle \frac{1}{2}\psi^* + K'\psi^*, \eta_{\tilde{h}} \rangle_{\Gamma_B} \ . \end{aligned} \tag{2.17}$$

In general, one may also choose different trial and test spaces on Γ_{BN} and on Γ_{BD}. For *BEM collocation methods* we consider only the case $n = 2$ and choose naive collocation with "smoothest" splines, i.e. $W_{1\tilde{h}} = span\,\{\delta(\cdot - x_k),\ k = 1, ..., M\}$ and $H_{\tilde{h}} = S_{\tilde{h}}^{\tilde{k},\tilde{r}}$ with $\tilde{k} = \tilde{r} + 1$. The collocation points x_k are the break points for $\tilde{r}$ odd and the mid-points between neighboring break points for $\tilde{r}$ even. If (2.17) is used, we need $\tilde{r} \geq 2$ for collocation on Γ_{BN}. In case of macro-elements, Γ_B will be replaced by Γ_{B_j}.

3 Finer boundary element grid

For the case $\Gamma_{BN} = \emptyset$, Brezzi, Johnson and Nedelec prove in [6,17] convergence of the direct variational Galerkin coupling for $h \to 0$, $\tilde{h} \to 0$ by using strong ellipticity of the bilinear form ((2.14) +2× (2.15)) and compactness of K in $H^{\frac{1}{2}}(\Gamma_B)$. For our equations from elasticity, K is never compact and it is still open, whether (2.14), (2.15) define here a stable and convergent method if Γ_B has corners.
However, finer meshes for BEM guarantee asymptotic convergence. Let

$$\tilde{h} \leq c_0 h \quad \textit{with sufficiently small } c_0 > 0 \ . \tag{3.1}$$

Let the finite element family along the coupling boundary satisfy

$$\| v_F \|_{L_2(\Gamma_B \cap \Gamma_F)} \leq c h^{-\frac{1}{2}} \| v \|_{H^1(\Omega_F)} \ . \tag{3.2}$$

Then the result by Brezzi and Johnson [5] can be slightly generalized.

Theorem 1 : *Let (3.1), (3.2) be satisfied and $\Gamma_D \neq \emptyset$. Then there exist $c_0 > 0$, $h_0 > 0$ such that for all $0 < h \leq h_0$, the Galerkin equations (2.14), (2.15) and (2.17) are uniquely solvable. Moreover, we have for $0 \leq l,\ l' < \frac{1}{2}$,*

$$\begin{aligned} &\| u - \varphi^* - u_F \|_{H^{1-l}(\Omega_F)} + \| D(u) - \psi^* - \Lambda_{\tilde{h}} \|_{\tilde{H}^{-\frac{1}{2}-l}(\Gamma_B \setminus \Gamma_N)} \\ &+ \| u - \varphi^* - u_{\tilde{h}} \|_{H^{\frac{1}{2}-l}(\Gamma_B \cap \Gamma_N)} \leq c_1 h^{l'+l} \| u \|_{H^{1+l'}(\Omega)} + c_3 \tilde{h}^{l'+l} \| u \|_{H^{\frac{1}{2}+l'}(\Gamma_B \cap \Gamma_N)} \\ &+ c_2 \tilde{h}^{l'+l} \| D(u) \|_{H^{-\frac{1}{2}+l'}(\Gamma_B \setminus \Gamma_N)} \ . \end{aligned} \tag{3.3}$$

The proof can be given as in Wendland [27] with a slight modification due to (3.1). The proof rests on Gårding inequalities for each of the variational formulations (2.14), (2.15) and (2.17) and associated regularity of the solution, which are still valid for domains with Lipschitz boundaries (see Costabel [9]). In case $n = 2$, following Babuška and Rosenzweig [2] and Ruotsalainen [22], appropriate mesh grading in Ω_F and on Γ_B near

the corners, respectively, should improve (3.3) to larger l and l' in appropriate norms. Corresponding analysis and experiments with coupling, however, are yet to be done. If we use (2.15) also on Γ_{BN} instead of (2.17) then the convergence is not clear.
If Γ_F and Γ_B both are smooth -as for Figure 1- and if, in addition, Γ_D and Γ_N are separated smooth closed curves, then Theorem 1 also holds for (2.15) on Γ_{BN}. Moreover, (3.3) extends to $0 \le l,\ l' \le \min\{k, \tilde{k}\}$. In this case we also have convergence of the coupling with BEM collocation in (2.15) and (2.17), respectively.

Theorem 2 [27]: *For $n = 2$ let Γ_D and Γ_N be separated $C^\infty-$ smooth curves and let also Γ_F and Γ_B be $C^\infty-$ smooth. Let (3.1) and (3.2) be satisfied and $\Gamma_D \neq \emptyset$. Then there exist $c_0 > 0$, $h_0 > 0$ such that (2.14), (2.15) with 'smoothest' splines and naive boundary element collocation are uniquely solvable for all $0 < h \le h_0$ and*

$$\begin{aligned} \| u - \varphi^* & - u_F \|_{H^l(\Omega_F)} + \| D(u) - \psi^* - \Lambda_{\tilde{h}} \|_{H^{\tilde{l}}(\Gamma_B \setminus \Gamma_N)} + \| u - \varphi^* - u_{\tilde{h}} \|_{H^{\tilde{\tilde{l}}}(\Gamma_B \cap \Gamma_N)} \\ & \le c_1 h^{k-l} \| u \|_{H^k(\Omega)} + c_2 \tilde{h}^{\tilde{k}+\tilde{l}} \| u \|_{H^{\tilde{k}+\frac{3}{2}}(\Omega)} + c_3 \tilde{h}^{\tilde{k}-\tilde{\tilde{l}}} \| u \|_{H^{\tilde{k}+\frac{1}{2}}(\Omega)} \end{aligned} \tag{3.4}$$

provided $2 - k \le l \le 1 \le r,\ -1 \le \tilde{l} \le \frac{\tilde{k}-1}{2},\ 0 \le \tilde{\tilde{l}} \le \frac{\tilde{k}}{2}$.

Remark: If (2.17) with collocation on $\Gamma_B \cap \Gamma_N$ is used instead of (2.15) then Theorem 2 is still valid for $1 \le \tilde{\tilde{l}} \le \frac{\tilde{k}+1}{2}$ and $2 \le \tilde{r} = \tilde{k} - 1$.

4 Finer finite element grid [27]

Very often, as e.g. in the transonic flow problem [3], the FEM in Ω_F needs to be more accurate than the BEM of the coupling on Γ_B. Hence, we now consider

$$h \le c_0 \tilde{h} \text{ with sufficiently small } c_0 > 0 \ , \tag{4.1}$$

$$\| \chi_{\tilde{h}} \|_{L_2(\Gamma_B)} \le c \tilde{h}^{-\frac{1}{2}} \| \chi_{\tilde{h}} \|_{H^{-\frac{1}{2}}(\Gamma_B)} \text{ for all } \chi_{\tilde{h}} \in S_{\tilde{h}}^{\tilde{k},\tilde{r}}(\Gamma_B) \ . \tag{4.2}$$

(4.2) is an inverse property for the BEM which holds for quasi-uniform meshes on Γ_B.

Theorem 3 [27]: *Let (4.1), (4.2) be satisfied and $\Gamma_D \neq \emptyset$. Then there exist $c_0 > 0$, $\tilde{h}_0 > 0$ such that for all $0 < \tilde{h} \le \tilde{h}_0$, the Galerkin equations (2.14), (2.15) and (2.17) are uniquely solvable. Moreover, we have (3.3) with $0 \le l,\ l' < \frac{1}{2}$.*

The proof is based on Gårding inequalities for each of the variational formulations (2.14), (2.15), (2.17) on Lipschitz domains [9]. For nonsmooth Γ_B, we have the same situation as in Section 3. If Γ_F and Γ_B are smooth and Γ_D and Γ_N are separated, then Theorem 5 and (3.3) extend to $0 \le l,\ l' \le \min\{k, \tilde{k}\}$, also for (2.15) on Γ_{BN}.

5 Symmetric or positive definite Galerkin coupling

The variational formulation (2.14), (2.15) does not correspond to basic principles of mechanics for the *complete* fields in $\Omega_F \cup \Omega_B$. Costabel [8], Grannell [13], Mang et. al. [20], Polizzotto [21] independently proposed a variational formulation, which corresponds for the complete fields to the Hellinger-Reissner principle and which provides convergence analysis for the corresponding BEM-FEM treatment. Here one combines (2.14) with (2.17) which yields the following positive definite coupled FEM–BEM Galerkin system, *find* $u_F \in H_h(\Omega_F)$, $\Lambda_{\tilde{h}} \in H_{\tilde{h}}(\Gamma_{BD})$, $u_{\tilde{h}} \in H_{\tilde{h}}(\Gamma_{BN})$ *such that*

$$\int_{\Omega_F} \sigma(u_F) \cdot e(v_F)\, dx + \langle D\tilde{u}, v_F\rangle - \frac{1}{2}\langle \Lambda_{\tilde{h}}, v_F\rangle + \langle \Lambda_{\tilde{h}}, Kv_F\rangle = \mathcal{L}_F(v_F) + \mathcal{L}_{BN}(v_F) \quad (5.1)$$

for all $v_F \in H_h(\Omega_F)$ *and*

$$\langle V\Lambda_{\tilde{h}}, \eta_{\tilde{h}}\rangle_{\Gamma_B} + \frac{1}{2}\langle \tilde{u}, \eta_{\tilde{h}}\rangle_{\Gamma_B} - \langle K\tilde{u}, \eta_{\tilde{h}}\rangle_{\Gamma_B} = \mathcal{L}_B(\eta_{\tilde{h}}) \text{ for all } \eta_{\tilde{h}} \in H_{\tilde{h}}(\Gamma_{BD}) \cup H_{\tilde{h}}(\Gamma_{BN}) \ . \quad (5.2)$$

The sum of the two bilinear forms $\{(5.1)+(5.2)\}$ defines a bilinear form which corresponds to the work of $(u_F, \Lambda_{\tilde{h}}, u_{\tilde{h}})$ subject to $(v_F, \eta_{\tilde{h}|_{\Gamma_{BN}}}, \eta_{\tilde{h}|_{\Gamma_{BD}}})$ and which, consequently, is positive definite satisfying a Gårding inequality on Lipschitz domains [9].

Theorem 4 [8]: *Let* $\Gamma_D \neq \emptyset$. *Then there exists* $h_0 > 0$ *such that for all* $0 < h \leq h_0$, $0 < \tilde{h} \leq h_0$, *the coupled FEM-BEM Galerkin equations* (5.1), (5.2) *are uniquely solvable. We also have the asymptotic error estimate* (3.3).

Note that here different refinements are not necessary. Moreover, the method also works for nonlinear problems [12]. However, implementation of (5.1) is rather tedious since it involves a mixture of domain and hypersingular boundary integrals. (3.3) holds for larger l, l' if Γ_F and Γ_B are smooth and Γ_{BN} and Γ_{BD} are separated.

6 Macro-elements

The BEM modelling of finite macro-elements within a FEM computation turned out to be extremely efficient for complicated geometries and stress concentration. Corresponding formulations go back to Stein [25], Tong, Pian, Lasry [26] and Schnack [23,24]. The method in [19] is also closely related in. First stability and convergence results are based on the variational formulation by Hsiao [14] and the forthcoming paper by Hsiao, Schnack and Wendland [16].

Let us consider FEM triangulations of Ω with maximum meshwidth h. Some of the triangular domains B_j, $j = 1, ..., N_B$ will carry macro-elements whereas Ω_F will be the part with usual continuous piecewise linear finite elements v_F. The traces of v_F, however, will be piecewise linearly, continuously extended onto all edges or faces of the *complete* triangulation of Ω. Hence u_F, $v_F \in H_h(\Omega)$ will be piecewise linear functions in Ω_F and, moreover, also on every Γ_{B_j}, the boundary of the macro-element B_j of meshsize h_j. Every Γ_{B_j} will now be subdivided into an individual BEM partition with maximum meshsize $\tilde{h}_j$ (see Figure 3). In B_j we will approximate the elastic field to u_F on Γ_{B_j} by a BEM solution to the Navier system (1.1) via a hybrid formulation as follows: *Find* $u_F \in H_h(\Omega)$, $u_{\tilde{h}_j} \in H_{\tilde{h}_j}(\Gamma_{B_j}) := S_{\tilde{h}_j}^{\tilde{k},\tilde{r}}(\Gamma_{B_j})$, $\lambda_{\tilde{h}_j} \in H_{\tilde{h}_j}(\Gamma_{B_j})$ *such that for* $j = 1, ..., N_B$

$$\int_{\Omega_F} \sigma(u_F) \cdot e(v_F)\, dx + \sum_{j=1}^{N_B} \oint_{\Gamma_{B_j}} \lambda_{h_j} \cdot v_F\, ds \;=\; \mathcal{L}_F(v_F) \text{ for all } v_F \in H_h(\Omega) \ , \quad (6.1)$$

$$\langle \eta_{\tilde{h}_j}, u_{\tilde{h}_j} - u_F \rangle_{\Gamma_{B_j}} = 0 \ , \tag{6.2}$$

$$\langle \eta_{\tilde{h}_j}, V_{\Gamma_{B_j}} \lambda_{\tilde{h}_j} \rangle_{\Gamma_{B_j}} + \langle \eta_{\tilde{h}_j}, \frac{1}{2} u_{\tilde{h}_j} - K_{\Gamma_{B_j}} u_{\tilde{h}_j} \rangle_{\Gamma_{B_j}} = \mathcal{L}_{B_j}(\eta_{\tilde{h}_j}) \text{ for all } \eta_{\tilde{h}_j} \in H_{\tilde{h}_j}(\Gamma_{B_j}). \tag{6.3}$$

For fixed h_j but $\tilde{h}_j \to 0$, (6.2) will give $u_{\tilde{h}_j} \to u_F$ in $L_2(\Gamma_{B_j})$ and (6.3) will provide $\lambda_{\tilde{h}_j} \to Du_F|_{\Gamma_{B_j}} - \psi^*$ in $H^{-\frac{1}{2}}(\Gamma_{B_j})$ for every B_j where u_F is in B_j the solution of the Navier system (1.1) with boundary values $u_F - \varphi^*$. For $\tilde{h}_j$ satisfying

$$\tilde{h}_j \leq c_0 h_j \tag{6.4}$$

with sufficiently small $c_0 > 0$, the corresponding solution of (6.2), (6.3) via (1.7) in B_j will define a good approximation to u_F in B_j. Note that for given u_F, the system (6.2), (6.3) can be solved for $\lambda_{\tilde{h}_j}$ and $u_{\tilde{h}_j}$ separately which offers resolution for $j = 1, ..., N_B$ and the given degrees of freedom in u_F *in parallel*. The resulting $\lambda_{\tilde{h}_j}$ then can be inserted into (6.1). The terms $\oint_{\Gamma_{B_j}} \lambda_{h_j} \cdot v_F \, ds$ finally depend only on u_F and v_F defining the macro-element entries of the stiffness matrix.
In [16] will be shown that under appropriate local regularity assumptions for the B_j,

$$\sum_{j=1}^{N_B} \oint_{\Gamma_{B_j}} \lambda_{\tilde{h}_j} \cdot v_F \, ds = \sum_{j=1}^{N_B} \int_{B_j} \sigma(u_F) \cdot e(v_F) \, dx + O\left(c_0^{\frac{1}{2}} \parallel u_F \parallel_{H^1(\Omega)} \parallel v_F \parallel_{H^1(\Omega)}\right) \ . \tag{6.5}$$

Hence, for the analysis of system (6.1) we can use standard arguments yielding:

Theorem 5 [16]: *Let c_0 in (6.4) be sufficiently small and $\Gamma_D \neq \emptyset$. Then there exists $h_0 > 0$ such that for all $0 < h \leq h_0$ the system (6.1), (6.2), (6.3) will be uniquely solvable. Moreover, we have the asymptotic error estimate for $1 \leq l < \frac{3}{2}$,*

$$\begin{aligned} &\parallel u - \varphi^* - u_F \parallel_{H^1(\Omega_F)} + \sum_{j=1}^{N_B} \parallel D(u)|_{\Gamma_{B_j}} - \psi^* - \lambda_{\tilde{h}_j} \parallel_{H^{-\frac{1}{2}}(\Gamma_{B_j})} \\ &+ \sum_{j=1}^{N_B} \parallel u - \varphi^* - u_F \parallel_{H^{\frac{1}{2}}(\Gamma_{B_j})} \leq c_1 h^{l-1} \parallel u \parallel_{H^l(\Omega)} + c_2 \sum_{j=1}^{B_N} \tilde{h}_j^{l-1} \parallel u \parallel_{H^l(B_j)} \ . \end{aligned} \tag{6.6}$$

The low rate of convergence in (6.6) is due to the corner points of the macro-elements. Appropriate individual mesh grading on Γ_{B_j} could surely improve the method.
The computational results by Carmine [7] for two- and by Karaosmanoglu [18] for three-dimensional problems show the excellent performance of this method and its ability of high resolution of the displacement and stress fields in the macro-elements.

References

[1] Babuška, I.; Aziz, A.K.: Survey lectures on the mathematical foundations of the finite element method. In: The Mathematical Foundation of the Finite Element Method with Applications to Partial Differential Equations, Aziz A.K. (ed.). New York: Academic Press (1972) pp. 3-359.

[2] Babuška, I.; Rosenzweig, M.: A finite elemet scheme for domains with corners. Numer. Math. 20 (1973) pp. 1-21.

[3] Berger, H.; Warnecke, G.; Wendland, W.: Finite Element Berechnungen für transsonische Strömungen unter Berücksichtigung verschiedener Fernfeldrandbedingungen. In: Strömungen mit Ablösung. Bonn: DGLR-Bericht 88-5 (1988) pp. 233-242.

[4] Bettess, P.; Kelly, D.W.; Zienkiewicz, O.C.: Marriage à la mode. The best of both worlds (Finite elements and boundary integrals). Energy Methods in Finite Element Analysis. Glowinski R.; Rodin E.Y.; Zienkiewicz O.C. (eds.). Chichester: J. Wiley (1979).

[5] Brezzi, F.; Johnson, C.: On the coupling of boundary integral and finite element methods. Calcolo 16 (1979) pp. 189-201.

[6] Brezzi, F.; Johnson, C.; Nedelec, J.C.: On the coupling of boundary and finite element methods. In: Basic Problems of Numerical Analysis. Pilsen: IVth LIBLICE Conf. (1978).

[7] Carmine, R.: Ein Kopplungsverfahren von FEM und BEM zur Berechnung ebener Spannungskonzentrationen. Univ. Karlsruhe, W.-Germany: Doctoral Thesis (1989).

[8] Costabel, M.: Principles of boundary element methods. Computer Physics Reports 6 (1987) pp. 243-274.

[9] Costabel, M.: Boundary integral operators on Lipschitz domains: Elementary results. SIAM J. Math. Anal. 19 (1987/88) pp. 613-626.

[10] Costabel, M.; Wendland, W.L.: Strong ellipticity of boundary integral operators. J. Reine Angew. Math. 372 (1986) pp. 34-62.

[11] Feng, Kang: Finite element method and natural boundary reduction. In: Proc. Intern. Congress Math., Warsaw: Polish Acad. Sci. (1983) pp. 1439-1453.

[12] Gatica, G.N.; Hsiao, G.C.: The coupling of boundary element and finite element methods for a nonlinear exterior boundary value problem. Zeitschr. Anal. Anw., to appear.

[13] Grannell, J.J.: On simplified hybrid methods for coupling of finite elements and boundary elements. In: Boundary Elements IX Vol.1. Brebbia, C.A.; Wendland, W.L.; Kuhn, G. (eds.). Berlin: Springer-Verlag (1987) pp. 447-460.

[14] Hsiao, G.C.: The coupling of boundary element and finite element methods. ZAMM (1989), to appear.

[15] Hsiao, G.C.; Kleinman, R.E.; Schultz, L.S.: On variational formulations of boundary value problems for fluid-solid interactions. In: Elastic Wave Propagation. Mc Carthy, M.F.; Hayes, M.A. (eds.). Amsterdam: North-Holland (1989) pp. 321-326.

[16] Hsiao, G.C.; Schnack, E.; Wendland, W.: A hybrid coupled finite-boundary element method (1990), in preparation.

[17] Johnson, J.; Nedelec, J.C.: On coupling of boundary integral and finite element methods. Math. Comp. 35 (1980) pp. 1063-1079.

[18] Karaosmanoglu, N.: Kopplung von Randelement- und Finite-Element-Verfahren für dreidimensionale elastische Strukturen. Univ. Karlsruhe, W.-Germany: Doctoral Thesis. (1989).

[19] Khoromskiĭ, B.N.: The integral-grid method for the solution of the Dirichlet problem for Laplace's equation. J. Vyčislit. Mat. i Mat. Physics 24 (1984) pp. 53-64.

[20] Li, Hong-Bao; Han, Guo-Ming; Mang, H.A.; Torzicky, P.A.: A new method for the coupling of finite element and boundary element discretized subdomains of elastic bodies. Comput. Methods Appl. Mech. Engrg. 54 (1986) pp. 161-185.

[21] Polizzotto, C. (1987): A symmetric-definite BEM formulation for the elastoplastic rate problem. Boundary Elements IX Vol 2. Brebbia, C.A.; Wendland, W.L.; Kuhn, G. (eds.). Berlin: Springer-Verlag (1987) pp. 315-334.

[22] Ruotsalainen, K.: On the boundary element method with mesh refinement on curves with corners. J. Comp. Appl. Math. (1989), to appear.

[23] Schnack, E.: Ein Beitrag zu rotationssymmetrischen Spannungskonzentrationsproblemen mit der Methode der finiten Elemente. Univ. Munich, W.-Germany: Doctoral Thesis (1973).

[24] Schnack, E.: Stress analysis with a combination of HSM and BEM. In: The Mathematics of Finite Elements and Applications V. Whiteman, J. (ed.). London: Academic Press (1985) pp. 273-282.

[25] Stein, E.: Die Kombination des modifizierten Trefftzschen Verfahrens mit der Methode der finiten Elemente. In: Finite Elemente in der Statik. Berlin: Verlag Ernst u. Sohn (1973) pp. 172-185 and pp. 242-259 (with G. Ruoff).

[26] Tong, P.; Pian, T.H.; Lasry, S.J.: A hybrid element approach to crack problems in plane elasticity. J. Numer. Methods Engrg. 7 (1973) pp. 297-308.

[27] Wendland, W.L.: On asymptotic error estimates for combined BEM and FEM. In: CISM Lecture Notes No.301. Finite and Boundary Element Techniques from Mathematical and Engineering Point of View. Stein, E.; Wendland, W.L. (eds.). Vienna, New York: Springer-Verlag (1988) pp. 273-333.

[28] Wendland, W.L.: On the coupling of the finite and boundary elements. Proc. XIV Intern. School Varna, Bulgarian Academy of Sci. (1989), to appear.

Three-dimensional Coupling of FEM and BEM in Elasticity

E. SCHNACK, I. BECKER and N. KARAOSMANOGLU

Institute of Solid Mechanics
Karlsruhe University, Karlsruhe/FRG

Abstract

Working with the Finite Element Method for Lame systems, several disadvantages are well-known as for instance the violation of the divergence theorem. As another important disadvantage, the Neumann boundary condition cannot be fulfilled in each point. These are reasons for the low convergence order in high gradient fields. But a significant advantage of Finite Element Method exists in the symmetric positive definite band matrix. Working with BEM, the influence matrix is fully populated. The numerical effort for solving the linear system is large. The condition number of the influence matrix depends on the problem and especially is bad if one has elongated shapes in structures. Additionally, in BEM there are problems for convex corners. Consequently, one starts thinking about combining the advantages and avoiding disadvantages of FEM and BEM by a coupling procedure. Against to some other direct coupling methods, we are proposing a coupling procedure using the Lagrange-Multiplier-Technique. This leads to an extended variational formulation, which results in a symmetric and definite operator for two- and three-dimensional problems. Therefore, one can work with the optimized solvers from Finite-Element-Sofware.

Model Problem

The following formulation refers to linear elasticity, see Schnack [10,11,12] but can also be extended to elasto-plastic material behaviour, see Schnack [13]. We consider Dirichlet and Neumann boundary conditions only for interior problems, where the domain is bounded. We have homogeneous, isotropic materials but geometrical discontinuities in form of cracks are possible. In Fig. 1, you see the definition of domain Ω with at first smooth boundary Γ.

$$\Omega \subset \mathbb{R}^i \quad \text{with } i = 2,3 \tag{1}$$

The boundary Γ is defined by:

$$\Gamma := \Gamma_t \cup \Gamma_u \tag{2}$$

G. Kuhn · H. Mang (Eds.)
Discretization Methods in Structural Mechanis
IUTAM/IACM Symposium Vienna, Austria, 1989

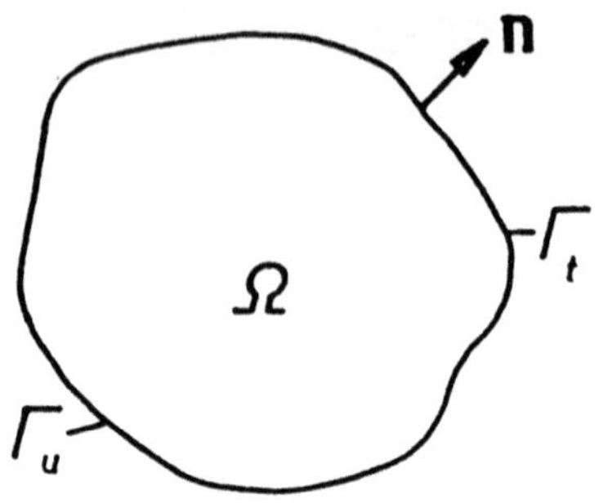

Fig. 1. Definition of the domain Ω

The Navier equation is given by the Δ^*-operator:

$$\Delta^* \mathbf{u} := \mu \, \Delta \mathbf{u} + (\lambda + \mu) \, \text{grad div } \mathbf{u} = \mathbf{0} \text{ in } \Omega \tag{3}$$

where λ, μ are the Lame-constants. The traction can be computed by the traction operator T:

$$T[\mathbf{u}]\Big|_{\Gamma_t} := \lambda (\text{div } \mathbf{u}) \, \mathbf{n} + 2 \, \mu \frac{\partial}{\partial n} \mathbf{u} + \\ + \mu \mathbf{n} \times \text{rot } \mathbf{u}\Big|_{\Gamma_t} = \psi \quad \text{on } \Gamma_t \tag{4}$$

In addition to the Neumann condition we have the Dirichlet one:

$$\mathbf{u}\Big|_{\Gamma_u} = \phi \quad \text{on } \Gamma_u \tag{5}$$

Here $\mathbf{n}$ denotes the outer normal to the boundary. In Ω we have the Betti-formula:

$$\mathbf{u}(x) := - \int_{\Gamma_u} \mathbf{T}(y,x) \, \phi(y) \, ds_y + \int_{\Gamma_u} \mathbf{U}(y,x) \, T[\mathbf{u}](y) \, ds_y \\ - \int_{\Gamma_t} \mathbf{T}(y,x) \, \mathbf{u}(y) \, ds_y + \int_{\Gamma_t} \mathbf{U}(y,x) \, \psi(y) \, ds_y \\ \text{with } x \in \Omega \tag{6}$$

$\mathbf{U}(y,x)$ defines the fundamental matrix:

$$\mathbf{U}(y,x) := \frac{\lambda + 3\mu}{4(i-1)\pi(\lambda+2\mu)\mu} \left\{ \gamma(y,x)\mathbf{I} + \right.$$
$$\left. + \frac{\lambda+\mu}{\lambda+3\mu} \frac{1}{|x-y|^i} (x-y)(x-y)^T \right\} \tag{7}$$

where $\gamma(y,x)$ is the fundamental solution of the Laplacian:

$$\gamma(y,x) := \begin{cases} -\log|x-y| & \\ & \text{for} \\ \dfrac{1}{|x-y|} & \end{cases} \quad \begin{array}{l} i = 2 \\ \\ i = 3 \end{array} \tag{8}$$

The fundamental traction matrix is given by:

$$\mathbf{T}(y,x) := \left[T_y \left(\mathbf{U}(y,x) \right) \right]^T \tag{9}$$

The index y indicates the differentiation to y. If the point x is going to the surface Γ, we get the Cauchy singular integral equation considered by Mikhlin [8]:

$$\frac{1}{2}\mathbf{u}(x) := -\int_{\Gamma_u} \mathbf{T}(y,x)\,\boldsymbol{\phi}(y)\,ds_y + \int_{\Gamma_u} \mathbf{U}(y,x)\,T[\mathbf{u}](y)\,ds_y$$
$$- \int_{\Gamma_t} \mathbf{T}(y,x)\,\mathbf{u}(y)\,ds_y + \int_{\Gamma_t} \mathbf{U}(y,x)\,\boldsymbol{\psi}(y)\,ds_y$$
$$\text{with } x \in \Gamma \tag{10}$$

In the next step we apply the traction operator to $\mathbf{u}$ on Γ:

$$\mathbf{t} := T[\mathbf{u}] \tag{11}$$

The result is:

$$\frac{1}{2}\mathbf{t}(x) := -T_x \int_{\Gamma_u} \mathbf{T}(y,x)\,\boldsymbol{\phi}(y)\,ds_y + \int_{\Gamma_u} \left(\mathbf{T}(x,y)\right)^T \mathbf{t}(y)\,ds_y$$
$$- T_x \int_{\Gamma_t} \mathbf{T}(y,x)\,\mathbf{u}(y)\,ds_y + \int_{\Gamma_t} \left(\mathbf{T}(x,y)\right)^T \boldsymbol{\psi}(y)\,ds_y$$
$$\text{where } x \in \Gamma \tag{12}$$

That equation is uniquely solvable with a kinematical operator M of x, taking into account the rigid body motions $\boldsymbol{\omega}$, see

Hsiao and Wendland [4], with the side condition :

$$\int_{\Gamma^-} M(y)^T\, u(y)\, ds_y = 0 \tag{13}$$

Additionally, we have the definition for T_x:

$$T_x\, U(y,x) = \left(T(x,y)\right)^T \tag{14}$$

In the next step, we define the boundary integral operators:

$$W\, u(x) := -\, T_x \int_{\Gamma} T(y,x)\, u(y)\, ds_y \tag{15}$$

$$K'\, t(x) := \int_{\Gamma} \left(T(x,y)\right)^T\, t(y)\, ds_y \tag{16}$$

$$V\, t(x) := \int_{\Gamma} U(y,x)\, t(y)\, ds_y \tag{17}$$

$$K\, u(x) := \int_{\Gamma} T(y,x)\, u(y)\, ds_y \tag{18}$$

While **V** is a regular operator, **K** and **K'** are singular operators and **W** is a hypersingular operator, which is very difficult to handle from numerical point of view. Besides of working with integral equations, we have the possibility to solve the problem by a variational formulation, starting with equation $\Delta^* u = 0$ and multiplying it with a test function **v** :

$$\int_{\Omega} (\Delta^* u)\, v\, d\Omega_x = 0 \qquad \forall v \quad \text{with } x \in \Omega \tag{19}$$

Integration by parts results in the bilinear form:

$$\int_{\Omega} a(u,v)\, d\Omega_x := \int_{\Gamma_t} v\, \psi\, ds_y + \int_{\Gamma_u} v\, T[u]\, ds_y \tag{20}$$

v is a test function with the homogeneous Dirichlet condition :

$$v\big|_{\Gamma_u} = 0 \tag{21}$$

From that follows:

$$\int_{\Omega} a(\mathbf{u},\mathbf{v})\ d\Omega_x := \int_{\Gamma_t} \mathbf{v}\ \psi\ ds_y \qquad \forall\ \mathbf{v} \tag{22}$$

Direct Coupling Procedure

The problem can be reformulated as a transmission problem. The domain Ω is divided into Ω^+ and Ω^- with the common boundary Γ_0. The boundaries with Dirichlet or Neumann boundary conditions are Γ_u and Γ_t, see Fig. 2.

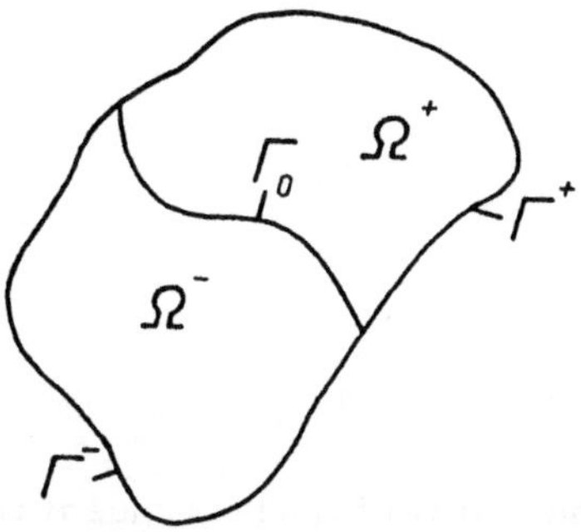

$$\Omega^- \subset \Omega$$

$$\Omega^+ := \Omega \setminus \Omega^-$$

$$\Gamma^+ := \Gamma_t^+ \cup \Gamma_u^+$$

$$\Gamma^- := \Gamma_t^- \cup \Gamma_u^-$$

Fig. 2. Coupling reformulation as a transmission problem

On Γ_0 we have the transmission condition:

$$\left.\begin{aligned} \mathbf{u}^- &= \mathbf{u}^+ = \mathbf{u}^0 \\ T[\mathbf{u}]^- &= -T[\mathbf{u}]^+ = \boldsymbol{\sigma} \end{aligned}\right\} \text{ on } \Gamma_0 \tag{23}$$

with the boundary conditions :

$$\mathbf{u}|_{\Gamma_u^+} = \boldsymbol{\phi}^+ \qquad \text{on } \Gamma_u^+ \tag{24}$$

$$T[\mathbf{u}]|_{\Gamma_t^+} = \psi^+ \qquad \text{on } \Gamma_t^+ \tag{25}$$

$$\mathbf{u}|_{\Gamma_u^-} = \boldsymbol{\phi}^- \qquad \text{on } \Gamma_u^- \tag{26}$$

$$T[\mathbf{u}]|_{\Gamma_t^-} = \psi^- \qquad \text{on } \Gamma_t^- \tag{27}$$

We have the equations:

$$\frac{1}{2}\,\mathbf{u}^{0}: = -\,\mathbf{K}\,\mathbf{u}^{0} - \mathbf{K}\,\boldsymbol{\phi}^{+} - \mathbf{K}\,\mathbf{u}^{+} - \mathbf{V}\,\sigma + \mathbf{V}\,\psi^{+} + \mathbf{V}\,\iota^{+} \tag{28}$$

and:

$$\frac{1}{2}\,\sigma: \;= +\,\mathbf{W}\,\mathbf{u}^{0} - \mathbf{W}\,\boldsymbol{\phi}^{+} - \mathbf{W}\,\mathbf{u}^{+} + \mathbf{K}'\sigma + \mathbf{K}'\psi^{+} + \mathbf{K}'\iota^{+} \tag{29}$$

with $x \in \Gamma_0$.

These equations are defined in the domain Ω^{+}. Now we can define the nonlocal problem:

Find $(\mathbf{u}^{0},\sigma)$ such that:

$$\Delta^{*}\mathbf{u}: = \mathbf{0} \text{ in } \Omega^{-} \tag{30}$$

$$\mathbf{u}\Big|_{\Gamma_u^-}: = \boldsymbol{\phi}^{-} \text{ on } \Gamma_u^{-} \tag{31}$$

$$\iota\;: = T[\mathbf{u}]^{-}\Big|_{\Gamma_\iota^-} = \psi^{-} \quad \text{on } \Gamma_\iota^{-} \tag{32}$$

For the domain Ω^{-}, we have the variational formulation:

$$\int_{\Omega^-} a(\mathbf{u},\mathbf{v})\; d\Omega_x - \int_{\Gamma_0} \mathbf{v}^{-}\;\underbrace{T[\mathbf{u}]^{-}}_{\sigma}\; ds_y = \int_{\Gamma_\iota^-} \psi^{-}\;\mathbf{v}\; ds_y \tag{33}$$

where we can insert σ from Equation (28) or Equation (29). For an exterior problem, existence, uniqueness and convergence for the solution can be proved, see Hsiao [6].

Hybrid Formulation

In the following we develop a mixed nonconforming formulation with a Super Boundary Element (SBE) in Ω^{-}:

$$\Omega_e^{-} \subset \Omega^{-} \tag{34}$$

We define:

$$\mathbf{U}: \;= \left\{ \mathbf{H}^{1/2}(\partial\Omega_e) \;\middle|\; \mathbf{u} = \boldsymbol{\phi}^{-} \text{ on } \Gamma_u^{-} \right\} \tag{35}$$

and an approximation:

$$\mathbf{U}_h: \;= \left\{ \mathbf{u}^{h} \in \mathbf{U} \;\middle|\; \mathbf{u}^{h} \in \mathbf{S}_d^{0}(\Delta_h) \right\} \tag{36}$$

with the condition:

$$\mathbf{u}^h\Big|_{\Gamma_0} = \mathbf{u}^H\Big|_{\Gamma_0} \tag{37}$$

h is the grid from BEM side and H the grid from FEM side. We can define:

$$\mathbf{u}^h\Big|_{\partial\Omega_e^-} = \sum_{k=1}^{N} \alpha_k \, \mathbf{N}_k \qquad \text{on } \partial\Omega_e^- \tag{38}$$

Additionally we have:

$$\mathbf{V}: = \Big\{ \mathbf{H}^{-1/2}(\partial\Omega_e^-) \,|\, \mathbf{t} = \psi^- \text{ on } \Gamma_t^-, \int_{\partial\Omega_e^-} \mathbf{t} \, ds_y = \mathbf{0}, \int_{\partial\Omega_e^-} \mathbf{r} \times \mathbf{t} \, ds_y = \mathbf{0} \Big\} \tag{39}$$

with the approximation:

$$\mathbf{V}_h: = \Big\{ \mathbf{t}^h \in \mathbf{V} \Big| \mathbf{t}^h \in \mathbf{S}_{d-1}^{-1}(\Delta_h) \Big\} \tag{40}$$

We can define:

$$\mathbf{t}^h\Big|_{\partial\Omega_e^-} = \sum_{k=1}^{M} \beta_k \, \mathbf{R}_k \tag{41}$$

From $\mathbf{t}^h$ the corresponding displacement $\mathbf{w}^h$ can be determined as the exact solution of:

$$\Delta^* \mathbf{w}^h = \mathbf{0} \quad \text{in } \Omega_e^- \tag{42}$$

and

$$\mathbf{T}[\mathbf{w}^h]\Big|_{\partial\Omega_e^-} = \mathbf{t}^h \tag{43}$$

and

$$\int_{\partial\Omega_e^-} \mathbf{M}(y) \, \mathbf{w}^h \, ds_y = \mathbf{0} \tag{44}$$

This Neumann problem can be solved either with the function theory, see Schnack [9], Drumm [3], Schnack, Wolf [14], Becker et al. [1], or by the following integral equation :

$$(C - K)w^h + V\, t^h = 0 \quad \text{on } \partial\Omega_e^- \tag{45}$$

The existence of a unique solution is clear. Therefore we get :

$$w^h = \sum_{k=1}^{M} \beta_k\, L_k \tag{46}$$

With the Lagrange-Multiplier-Technique the following condition can be obtained :

$$\int_{\partial\Omega_e^-} \overset{x}{t}{}^h \left(w^h - u^h\right) ds_y = 0 \qquad \forall\, \overset{x}{t}{}^h \in \overset{x}{V}_h \tag{47}$$

with $\overset{x}{t}{}^h$ as a Lagrange Multiplier and $\overset{x}{V}_h$ is the respective test space. From Galerkin's theorem in Ω^- follows:

$$\int_{\Omega_e^-} a\left(w^h,\, w^h\right) d\Omega_x = \int_{\partial\Omega_e^-} t^h\, w^h\, ds_y \tag{48}$$

So, we can see that $\langle t^h, w^h\rangle$ is positive definite. For having a weak equilibrium condition for the whole system $(\Omega^+ \cup \Omega^-)$ we need at first the definition of test space for homogeneous boundary conditions:

$$\overset{x}{U} := \left\{ H^{1/2}\left(\partial\Omega_e\right) \middle|\, \overset{x}{u} = 0 \text{ on } \Gamma_u^- \right\} \tag{49}$$

with the approximation:

$$\overset{x}{U}_H := \left\{ \overset{x}{u}{}^H \in \overset{x}{U} \,\middle|\, \overset{x}{u}{}^H \in S_d^0\left(\partial\Omega_e^-\right) \right\} \tag{50}$$

We define:

$$\overset{x}{u}{}^H := \sum_{k=1}^{N} \overset{x}{\alpha}_K\, N_K \qquad \text{on } \partial\Omega_e^- \tag{51}$$

Now we can formulate the weak equilibrium condition for the

whole system :

$$\underbrace{\int_{\Omega^+} a(u^H, \overset{x}{u}^H)\, d\Omega - \int_{\Gamma_t^+} \psi^+ \overset{x}{u}^H\, ds_y}_{FEM} + \underbrace{\int_{\partial\Omega_e^-} T[w^h]\overset{x}{u}^H\, ds_y - \int_{\Gamma_t^-} \psi^- \overset{x}{u}^H\, ds_y}_{BEM} = 0 \qquad (52)$$

One may construct the element-stiffness-matrix for the super boundary element by substituting the trial functions in the equations (47) and (52) and considering only the terms for Ω_e^-. This results in the following discretized equations :

$$\sum_{\substack{k=1\\ i=1}}^{M} \overset{x}{\beta}_K\, \beta_i \underbrace{\int_{\partial\Omega_e^-} R_k\, L_i\, ds_y}_{A_{ki}} - \sum_{k=1}^{M} \sum_{j=1}^{N} \overset{x}{\beta}_k\, \alpha_j \underbrace{\int_{\partial\Omega_e^-} R_k\, N_j\, ds_y}_{B_{kj}} = 0 \qquad (53)$$

$$\sum_{k=1}^{M} \sum_{j=1}^{N} \overset{x}{\alpha}_j\, \beta_k \underbrace{\int_{\partial\Omega_e^-} R_k N_j\, ds_y}_{B_{kj}} = \sum_{j=1}^{N} \overset{x}{\alpha}_j \underbrace{\int_{\Gamma_t^-} \psi^- N_j\, ds_y}_{F_j} \qquad (54)$$

β as a function of α can be inserted from equation (53) into equation (54). We obtain in the end:

$$\underbrace{B_{ik}\, A_{kp}^{-1}\, B_{pj}}_{k_{ij}}\ \alpha_i = F_j \qquad (55)$$

The stiffness matrix k_{ij} is symmetric and positive definite. It should be pointed out, that the collocation points (grid h) can be different from the positions of the Finite Element nodes. That is a very important advantage against direct coupling. It leads to a small band stiffness matrix and we have flexibility in constructing grid h independant of grid H. A theoretical justification of this coupling method concerning stability and convergence will be available in a forthcoming paper by Hsiao, Schnack, Wendland [5]. Some asymptotic error estimates for coupling methods were done by Wendland [18].

Numerical results for problems of plane and three-dimensional elasticity show the advantages of combining BEM and FEM especially for stress concentration problems in the field of fracture mechanics, see Schnack [12], Carmine [2], Karaosmanoglu [7], Schnack et al. [15,16,17].

References

[1] Becker, I; Karaosmanoglu, N.; Schnack, E.: Mixed methods with BEM for three dimensional fracture mechanics. In: Boundary Elements IX (eds.: Brebbia, C.A.; Wendland, W.L.; Kuhn, G.) pp. 227-241. Berlin, Heidelberg: Springer Verlag 1987.

[2] Carmine, R.: Ein Kopplungsverfahren von FEM und BEM zur Berechnung ebener Spannungskonzentrationsprobleme. Dissertation Universität Karlsruhe 1989 (to appear).

[3] Drumm, R.: Zur effektiven FEM-Analyse ebener Spannungskonzentrationsprobleme. Dissertation Universität Karlsruhe 1982.

[4] Hsiao, G.C.; Wendland, W.L.: On a boundary integral method for some exterior problems in elasticity. In: Proceedings of Tbilisi University, vol. 257, pp. 31-60, Tbilisi, University Press 1985.

[5] Hsiao, G.C.; Schnack, E.; Wendland, W.L.: A hybrid coupled finite-boundary element method (in preparation).

[6] Hsiao, G.C.: The coupling of BEM and FEM - A brief review. In: Boundary Elements X, Vol. 1 (ed.: Brebbia, C.A.) pp. 431-445. Berlin, Heidelberg: Springer Verlag 1988.

[7] Karaosmanoglu, N.: Kopplung von Randelement- und Finite-Element-Verfahren für dreidimensionale Strukturen. Dissertation Universität Karlsruhe 1989 (in preparation).

[8] Mikhlin, S.G.: The problem of the minimum of a quadratic functional. San Francisco: Holden Day Inc. 1965.

[9] Schnack, E.: Beitrag zur Berechnung rotationssymmetrischer Spannungskonzentrationsprobleme mit der Methode der finiten Elemente. Dissertation Techn. Universität München 1973.

[10] Schnack, E.: Singularities of cracks with generalized finite elements. In: Proceedings of singularities and constructive methods for their treatment (eds.: Grisvard, P.; Wendland, W.L.; Whiteman, J.R.) pp. 258-277. Berlin, Heidelberg, New York: Springer Verlag 1985.

[11] Schnack, E.: Stress analysis with a combination of HSM and BEM. In: Proceedings of the MAFELAP 1984 V (ed.: Whiteman, J.R.), pp. 273-281. London: Academic Press 1985.

[12] Schnack, E.: A hybrid BEM-model. Int. J. Num. Engng. 5 (1987) 1015-1025.

[13] Schnack, E.: Mixed methods in BEM for elastic-plastic problems. In: Boundary Elements VIII, Vol. II (eds.: Tanaka, M.; Brebbia, C.A.) pp. 641-651. Berlin, Heidelberg, New York, Tokyo: Springer Verlag 1986.

[14] Schnack, E.; Wolf, M: Application of displacement and hybrid-stress methods to plane notch and crack problems. Int. J. Num. Meth. Engng. 12, No. 6 (1978) 963-975.

[15] Schnack, E.; Becker, I.; Karaosmanoglu, N.: Modern coupling of FEM and BEM with mixed variational formulations. In: Boundary Elemnts X, Vol. 1 (ed.: Brebbia, C.A.) pp. 59-73. Southampton, Springer Verlag 1988.

[16] Schnack, E.; Carmine, R.; Becker, I.; Karaosmanoglu, N.: Two- and three-dimensional coupling of FEM and BEM in elasticity. In: Proceedings of the European Boundary Element Meeting (EBEM) (ed.: Migeot, J.L.) pp. 1.93-1.103.Université Libre de Bruxelles 1988.

[17] Schnack, E.; Carmine, R.; Becker, I.; Karaosmanoglu, N.: Mixed non-conforming technique for coupling FEM and BEM. In: Computational Mechanics 88, Vol. 1(eds.: Atluri, S.N.; Yagawa, G.) pp. 5.iii.1-5. Berlin, Heidelberg, New York, London, Paris, Tokyo: Springer Verlag 1988.

[18] Wendland, W.L.: On asymptotic error estimates for combined BEM and FEM. In Finite Element and Boundary Element Techniques from Mathematical and Engineering points of View (eds. Stein, E.; Wendland, W.L.) in CISM Courses Lectures No. 301, (1988) 273-333

Coupling of FE- and BE-Discretizations for 3D-Stress Analysis of Tunnels in Layered Anisotropic Rock

Z.-S. Chen, G. Hofstetter, Z.-K. Li, H.A. Mang, P. Torzicky

Institute for Strength of Materials, Technical University of Vienna, Austria

Summary

After introducing the method developed for the coupling of finite element (FE) and boundary element (BE) discretizations of subregions of solids, comments on the inadmissibility of symmetrizing the resulting coupling stiffness matrix and on the solution of the edge problem within the framework of the BEM are made. Somigliana's identity is used to obtain BE discretizations of subregions of solids. It contains results from the fundamental solution for a transversely isotropic material. After describing the mode of consideration of layers of rock in the context of the BEM, selected results from 3D stress analysis of a stretch of a tunnel in layered, anisotropic rock are presented, considering sequences of driving and securing of the tunnel. Finally, the BEM is applied to obtain a 3D solution of the problem of lowering the ground water table by means of compressed air which plays an important role as a dewatering measure in modern tunnelling.

Motivation for the Coupling of FE and BE Discretizations for 3D Stress Analysis of Tunnels

Fig.1 shows the FE discretization of a subdomain containing a stretch of a tunnel to be excavated and its vicinity, and BE discretizations of the ground surface, the interface of two neighboring layers of anisotropic rock and the interface of the FE and the BE discretization. The motivation for such a discretization is to exploit the complementary advantages of the FEM and the BEM [1].

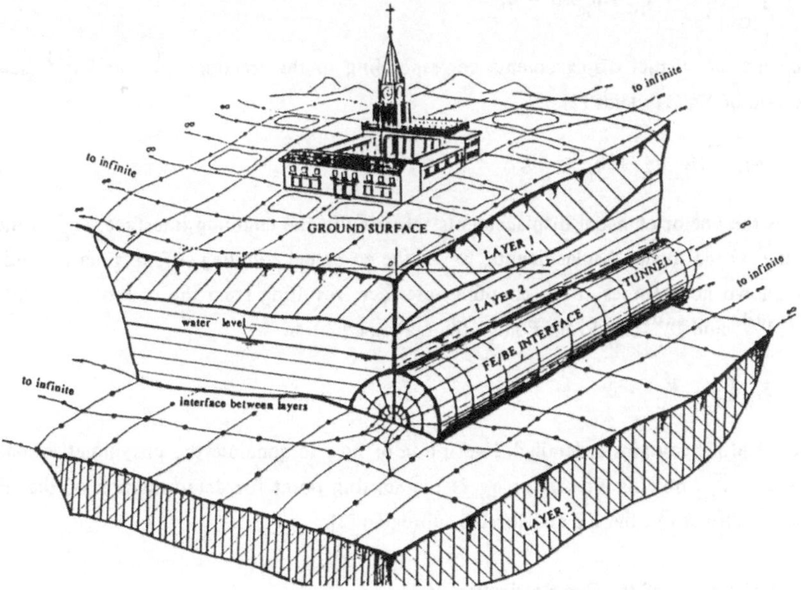

Fig.1. Coupled FE and BE discretizations for 3D stress analysis of a tunnel

G. Kuhn · H. Mang (Eds.)
Discretization Methods in Structural Mechanis
IUTAM/IACM Symposium Vienna, Austria, 1989

Coupling Method

Fig.2 shows the subdivision of a solid into a FE subdomain ($\Omega^{(F)}$) and a BE subregion ($\Omega^{(B)}$). The external surfaces of the two subregions are denoted as $\Gamma^{(F)}$ and $\Gamma^{(B)}$. The symbols "I" and "II" refer to the parts of $\Gamma^{(F)}$ and $\Gamma^{(B)}$, on which displacements $\bar{u}_i$ and tractions $\bar{t}_i$, respectively, are specified. $\Omega^{(e)}$ symbolizes a quadratic, hexahedral, isoparametric finite element. $\Gamma^{(e)}$ denotes a quadratic, quadrilateral, isoparametric boundary element. The designation $\Omega^{(g)}$ refers to a finite element whose side face $\Gamma^{(C^-,g)}$ is located on the coupling interface $\Gamma^{(C^-)}$. The designation $\Gamma^{(C^+,g)}$ pertains to the corresponding boundary element which is located on $\Gamma^{(C^+)}$ (see Fig.2).

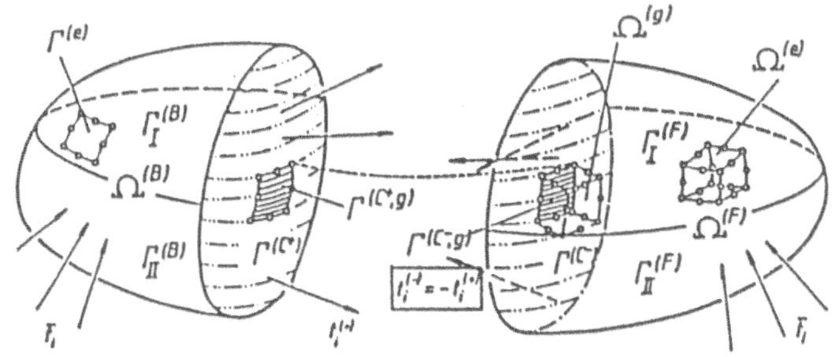

Fig.2. Subdivision of a solid into a FE subdomain and BE subregion

The theoretical background of the developed local (interface-related), FE-specific coupling method is the following weak form of the traction interface-continuity condition [2]:

$$\int_{\Gamma^{(C^+)}} (t_i^{(+)} + t_i^{(-)})\delta u_i \, dS = 0, \tag{1}$$

where δu_i are the virtual displacements corresponding to the tractions $t_i^{(+)}$ and $t_i^{(-)}$ (see Fig.2). Discretization of Eq.(1) yields [2]:

$$(\delta q_C)^T (K_C q_C + \bar{F}_C) = 0, \tag{2}$$

where q_C is the vector of nodal displacements referring to the coupling interface, δq_C is the vector of respective virtual nodal displacements, K_C is the co-called coupling stiffness matrix and $\bar{F}_C$ is a vector of known node forces at the coupling interface, resulting from the boundary conditions for $\Gamma_I^{(F)}$, $\Gamma_{II}^{(F)}$, $\Gamma_I^{(B)}$ and $\Gamma_{II}^{(B)}$ [2]. From Eq.(2), the matrix equation

$$K_C q_C + \bar{F}_C = 0 \tag{3}$$

is obtained. Ref.[3] contains a detailed description of how to generate the unsymmetric matrix K_C and the vector $\bar{F}_C$. Solving Eq.(3) for q_C is the starting point for determination of the displacements and the stresses in the FE and the BE subregion [3].

On the Inadmissibility of the Symmetrization of K_C

K_C is given as

$$K_C = K_C^{(+)} + K_C^{(-)}, \tag{4}$$

where $K_C^{(+)}$ and $K_C^{(-)}$ result from discretization of the BE subregion and the FE subdomain, respectively, followed by static condensation of nodal degrees of freedom. By contrast to $K_C^{(-)}$, $K_C^{(+)}$ is an unsymmetric matrix.

In Ref.[2] it is shown that

$$-\frac{1}{2}(q_C)^T\{\frac{1}{2}[K_C^{(+)} + (K_C^{(+)})^T]\}\, q_C \tag{5}$$

is not the potential of the vector of displacement-dependent node forces $K_C^{(+)}q_C$. It is demonstrated numerically [2] that disregard of the contribution of

$$\frac{1}{2}[K_C^{(+)} - (K_C^{(+)})^T]\, q_C \tag{6}$$

to these forces may result in a significant deterioration of the quality of the numerical results.

Edge Problem within the Framework of the BEM

Because of the discontinuity of the traction vector at the edges of solids, additional nodal degrees of freedom will appear at node points of BE discretizations, located on edges, if at least one traction component at more than one of the elements containing such a node point is unknown. Assuming that no component of anyone of the traction vectors is known at the considered node point, three additional algebraic equations each must be generated for the following pairs of neighboring elements containing this node point: elements 1 and 2, 1 and 3, etc. According to Chen, Hofstetter and Mang [4], for the elements 1 and 2, these equations are obtained as

$$[P^{(12)} + I]\, t_A^{(2)} - [P^{(21)} + I]\, t_A^{(1)} + Q^{(12)}F^{(2)}q^{(2)} - Q^{(21)}F^{(1)}q^{(1)} = 0, \tag{7}$$

where $t_A^{(1)}$ and $t_A^{(2)}$ are the traction vectors at node point A on the common edge of elements 1 and 2, and $q^{(1)}$ and $q^{(2)}$ are the two element node displacement vectors; I is a [3 x 3] unit matrix; the matrices $P^{(12)}$, $P^{(21)}$, $Q^{(12)}$, $Q^{(21)}$, $F^{(1)}$ and $F^{(2)}$ are given in [4].

BE Discretization of the Subregion of a Solid Consisting of a Transversely Isotropic Material

The starting point for such a discretization is Somigliana's identity. For the situation illustrated in Fig.2 it can be written as

$$C_{ji}(x)u_j(x) = \int_{\Gamma_I^{(B)}+\Gamma^{(C^+)}} U_{ji}(y,x)t_j(y)dS(y) - \int_{\Gamma_{II}^{(B)}+\Gamma^{(C^+)}} T_{ji}(y,x)u_j(y)dS(y) +$$

$$+ \int_{\Gamma_{II}^{(B)}} U_{ji}(y,x)\bar{t}_j(y)dS(y) - \int_{\Gamma_I^{(B)}} T_{ji}(y,x)\bar{u}_j(y)dS(y) + \int_{\Omega^{(B)}} U_{ji}(z,x)b_j(z)dV(z), \tag{8}$$

where x denotes the field point vector and y and z, respectively, denote the load point vector.

$U_{ji}(y,x)$ is the j-th component of the displacement vector at y, resulting from a unit point load acting in the direction i at x. $T_{ji}(y,x)$ is the corresponding component of the traction vector. $C_{ji}(x)$ depends on whether the field point is located in the interior of the solid or on its surface [5]. If the latter is true, then the value of $C_{ji}(x)$ depends on whether the tangent plane at this point is unique or not [5]. The repetition of an index in a term denotes a summation with respect to that index over its range (i,j = 1,2,3). By means of the Gauss integral theorem, the volume integral in Eq.(8), where $b_j(z)$ is the j-th component of the vector of volume forces at z, is transformed into a surface integral.

In general, it is sufficient to treat rock as a transversely isotropic material. In the isotropic plane (see Fig.3), all directions are principal directions of the material. The direction normal to this plane is also a principal direction of the material. For the case of linear elasticity, to which this work is restricted, the material stiffness matrix is obtained as (after correcting printing errors in Ref.[6])

$$C = \frac{E_2}{(1+\nu_1)(1-\nu_1-2\nu_2^2/n)} \begin{bmatrix} n(1-\nu_2^2/n) & n(\nu_1+\nu_2^2/n) & \nu_2(1+\nu_1) & 0 & 0 & 0 \\ & n(1-\nu_2^2/n) & \nu_2(1+\nu_1) & 0 & 0 & 0 \\ & & 1-\nu_1^2 & 0 & 0 & 0 \\ & & & (1/2)n(1-\nu_1-2\nu_2^2/n) & 0 & 0 \\ & \text{symmetric} & & & m(1+\nu_1)(1-\nu_1-2\nu_2^2/n) & 0 \\ & & & & & m(1+\nu_1)(1-\nu_1-2\nu_2^2/n) \end{bmatrix} \quad (9)$$

where $m = G_2/E_2$ and $n = E_1/E_2$. E_1 and ν_1 define the material properties in the isotropic plane, and E_2, ν_2 and G_2 in the direction normal to this plane. The ranges of m, n, ν_1 and ν_2 are restricted by the requirement of positive definiteness of C.

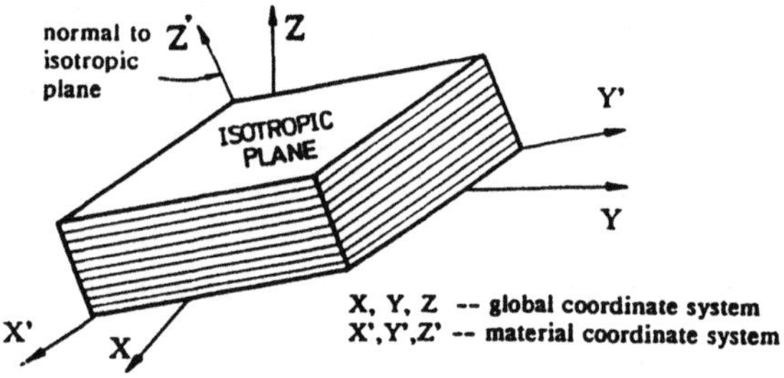

Fig.3. Transversely isotropic material

According to Pan and Chou [7], the displacement components $U'_{ji} = U'_{ij}(y',x'=0)$, j = 1,2,3, are referred to the principal directions of the material, given as

$$U'_{1i} = \frac{\partial^2 \Phi_i}{\partial x' \partial z'} + \frac{\partial \Psi_i}{\partial y'} , \qquad U'_{2i} = \frac{\partial^2 \Phi_i}{\partial y' \partial z'} + \frac{\partial \Psi_i}{\partial x'} , \qquad (10.1),(10.2)$$

$$U'_{3i} = - \frac{C_{11}}{C_{13} + C_{44}} \left(\frac{\partial^2}{\partial x'^2} + \frac{\partial^2}{\partial y'^2}\right)\Phi_i - \frac{C_{44}}{C_{13} + C_{44}} \frac{\partial^2 \Phi_i}{\partial z'^2} + \Theta_i , \qquad (10.3)$$

where Φ_i, Ψ_i and Θ_i are potential functions depending on the mode of application of the unit point load; C_{11}, C_{13} and C_{44} are coefficients depending on the elastic constants. For example, for the case of a unit point load acting normal to the isotropic plane (i=3) [7],

$$\Phi_3 = \sum_{k=1}^{2} [A_k z'_k \ln(R_k^*) + B_k R_k], \quad \Psi_3 = 0, \quad \Theta_3 = 0, \tag{11}$$

$$R_k^2 = x'^2 + y'^2 + z_k'^2, \quad R_k^* = R_k + z'_k \;, \quad z'_k = z'\nu_k, \quad k = 1,2 \;. \tag{12}$$

The coefficients A_k and B_k depend on the elastic constants. (The results in Ref.[7] contain a few printing errors.)

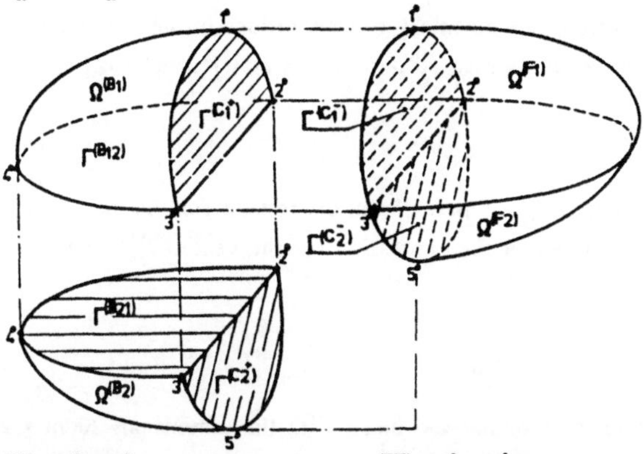

BE-subregion
divided into 2 subdomains

FE-subregion
consisting of 2 subdomains

Fig.4. Solid consisting of two layers of different materials

Consideration of Layers of Rock within the Framework of the BEM

Fig.4 refers to a solid consisting of two layers of different materials. Static condensation of those nodal degrees of freedom on the surface of $\Omega^{(B_1)}, \not\subset (\Gamma^{(B_{12})} \wedge \Gamma^{(C_1^+)})$, by means of Gaussian elimination yields

$$\left[\begin{array}{c|c|c|c|c} 0 & \multicolumn{4}{c}{\hat{A}^{(B_1)}} \\ \hline 0 & \hat{G}_{11}^{(B_1)} & \hat{H}_{11}^{(B_1)} & \hat{G}_{12}^{(B_1)} & \hat{H}_{12}^{(B_1)} \\ \hline 0 & \hat{G}_{21}^{(B_1)} & \hat{H}_{21}^{(B_1)} & \hat{G}_{22}^{(B_1)} & \hat{H}_{22}^{(B_1)} \end{array}\right] \left\{\begin{array}{c} q^{(B_1)} \\ \hline t^{(B_{12})} \\ \hline q^{(B_{12})} \\ \hline t^{(C_1^+)} \\ \hline q^{(C_1^+)} \end{array}\right\} = \left\{\begin{array}{c} \hat{R}^{(B_1)} \\ \hline \hat{R}^{(B_{12})} \\ \hline \hat{R}^{(C_1^+)} \end{array}\right\} \;. \tag{13}$$

An analogous set of simultaneous linear algebraic equations is obtained from static condensation of those nodal degrees of freedom on the surface of $\Omega^{(B_2)}, \not\subset (\Gamma^{(B_{21})} \wedge \Gamma^{(C_2^+)})$. Assemblage of this set of equations with Eq.(13) and use of the interface-continuity conditions $q^{(B_{12})} - q^{(B_{21})} = 0$, $t^{(B_{12})} + t^{(B_{21})} = 0$ gives

$$\left[\begin{array}{cc:cc:cc} \hat{G}_{11}^{(B_1)} & \hat{H}_{11}^{(B_1)} & \hat{G}_{12}^{(B_1)} & 0 & \hat{H}_{12}^{(B_1)} & 0 \\ -\hat{G}_{11}^{(B_2)} & \hat{H}_{11}^{(B_2)} & 0 & \hat{G}_{12}^{(B_2)} & 0 & \hat{H}_{12}^{(B_2)} \\ \hat{G}_{21}^{(B_1)} & \hat{H}_{21}^{(B_1)} & \hat{G}_{22}^{(B_1)} & 0 & \hat{H}_{22}^{(B_2)} & 0 \\ -\hat{G}_{21}^{(B_1)} & \hat{H}_{21}^{(B_2)} & 0 & \hat{G}_{22}^{(B_2)} & 0 & \hat{H}_{22}^{(B_2)} \end{array}\right] \left\{\begin{array}{c} t^{(B_{12})} \\ q^{(B_{12})} \\ t^{(C_1^+)} \\ t^{(C_2^+)} \\ q^{(C_1^+)} \\ q^{(C_2^+)} \end{array}\right\} = \left\{\begin{array}{c} \hat{R}^{(B_{12})} \\ \hat{R}^{(B_{21})} \\ \hat{R}^{(C_1^+)} \\ \hat{R}^{(C_2^+)} \end{array}\right\}. \tag{14}$$

Gaussian elimination, followed by elementary row operations in the coefficient matrix of Eq.(14) to obtain a unit matrix in the specially marked part of this matrix, yields [2]

$$t_C = Q_C q_C + \bar{t}_C, \tag{15}$$

where the unsymmetric matrix Q_C is part of the expression for the matrix $K_C^{(+)}$ (see Eq.(4)), and where the vector $\bar{t}_C$ is contained in the expression for the vector $\bar{F}_C^{(+)}$, which is part of the vector $\bar{F}_C$ (see Eq.(3)) [2].

3D Stress Analysis of a Stretch of a Tunnel

Fig.5(a) illustrates a quarter of the analyzed block with the dimensions 200m x 200m x 300m, containing a stretch of 28m length of a tunnel to be excavated. Numerical studies have shown [1] that at a distance of 20m from the excavation face the state of strain is practically a plane one. Fig.5a illustrates the BE discretization of the external boundaries of the block and of the interface between the two considered layers of rock. Material 1 is assumed to be anisotropic (transversely isotropic). The isotropic planes are assumed to be parallel to the x-y plane. Material 2 is assumed to be isotropic. Fig.5 contains the material data. Fig.5(b) shows the FE subregion containing the tunnel to be excavated and its vicinity. The BE discretization consists of 160 quadratic, quadrilateral, isoparametric boundary elements with altogether 492 node points. The FE discretization consists of 336 quadratic, hexahedral, isoparametric finite elements with altogether 1797 node points.

The first step of the analysis consists of determination of the primary state of stress by means of the BEM. The specific weight of the two materials, which is needed for this purpose, is chosen as $\gamma = 25$ kN/m^3. Thereafter, sequences of driving and securing of the tunnel are taken into account (Fig.6). Altogether, 28 analysis steps are considered. One such step consists of securing of the stope (by a shotcrete lining of 20 cm thickness) and driving of the calotte, or of securing of the calotte and driving of the stope. Instead of removing the finite elements in the part of the FE subregion representing the excavated tunnel, the original values of the elastic moduli of the rock are replaced by very small percentages of these values. Moreover, in this part the signs of γ and of the node forces from the primary state of stress must be reversed [1]. Since the shotcrete lining is built within a deformed system, stresses in the lining are produced only by post-construction deformations [1]. In the BE subdomain and in the remaining part of the FE subregion, γ must be set equal to zero [1].

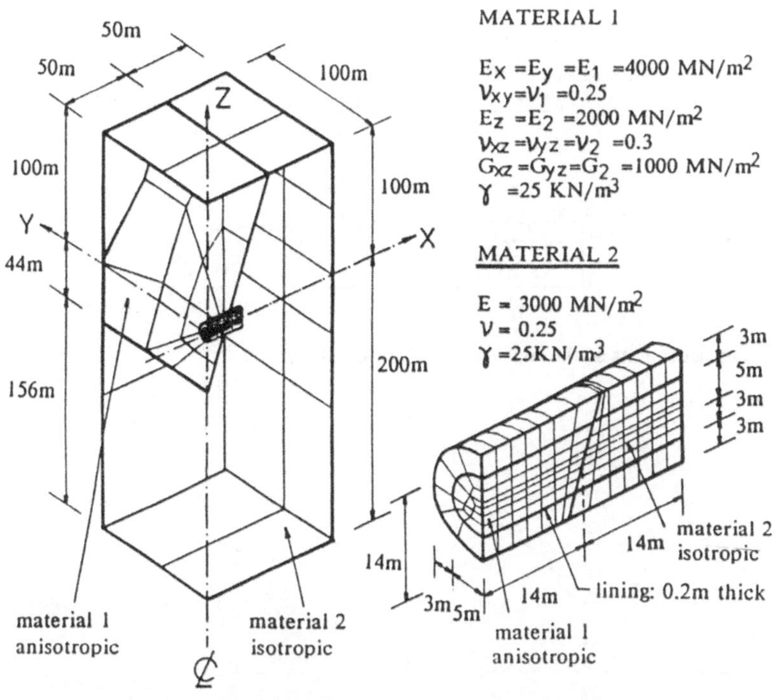

(a) quarter of the analysed block (b) FE subregion

Fig.5. BE-FE model for 3D stress analysis of a stretch of a tunnel

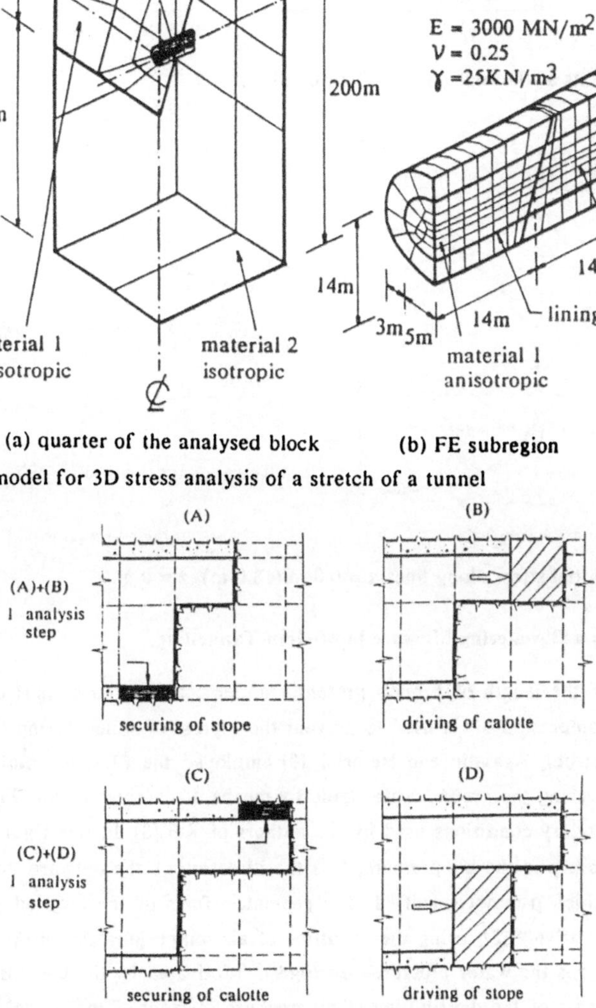

Fig.6. Typical analysis steps for the simulation of the excavation of the tunnel

Figs.7 and 8 contain selected results from the analysis. In Fig.8, $\sigma_z^{(p)}$ denotes the primary stress along lines characterized by x = 6.0 m and 18.0 m, respectively, and z = 0 m.

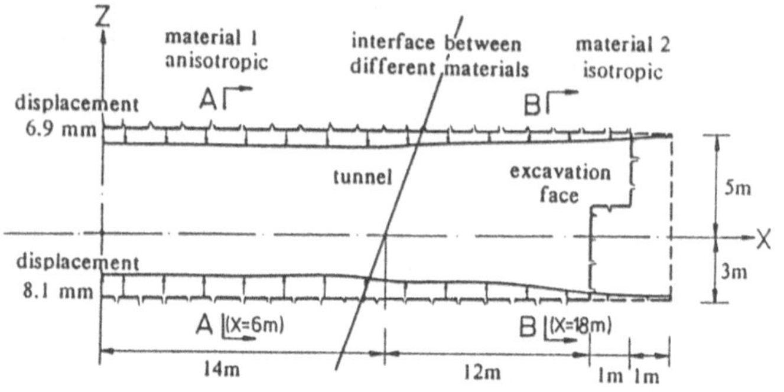

Fig.7. Displacements at the roof and the floor of the tunnel

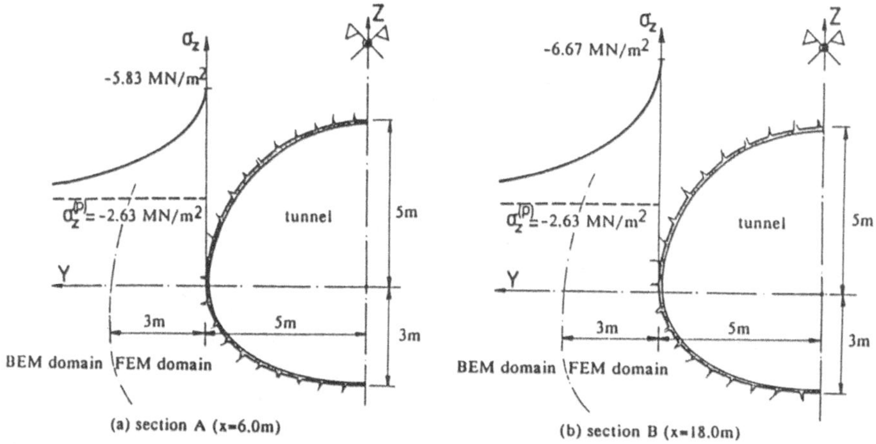

Fig.8. Circumferential stress along lines x = 6.0 m (18.0 m), z = 0 m

Compressed Air as a Dewatering Measure in Modern Tunnelling

For about 60 % of the length of tunnels presently under construction as part of Vienna's underground line U3, compressed air is used to prevent the ingress of water during the driving operation. Fuchsberger, Strobl, Ayaydin and Heinrich [8] employed the FEM to obtain a 2D solution of the problem of lowering the ground water table by means of compressed air. Fig.9 contains information on the boundary conditions used by the authors of Ref.[8]. In this figure, k is the air permeability of the soil, p is the air pressure, n is the direction of the outward normal at a point on the boundary at which p is not specified. The geometric form of the lowered ground water table was determined iteratively [8], using the condition of air imperviousness: $\partial p/\partial n = 0$ for $p = \gamma_w(z-z_0)$, where $\gamma_w(z-z_0)$ is the water pressure (see Fig.9). Motivated by the conclusion in Ref.[8] that "in-situ-measurements of the distribution of air pressure in front of the tunnel face showed lower values than the two-dimensional FEM calculation", a 3D solution of the addressed class of problems was sought. For this purpose, the BEM was applied. Determination of the lowered ground water table is a necessary prerequisite for the calculation of the demand of compressed air and of the pressure exerted on bottom slabs of buildings in order to assess the danger of uplift.

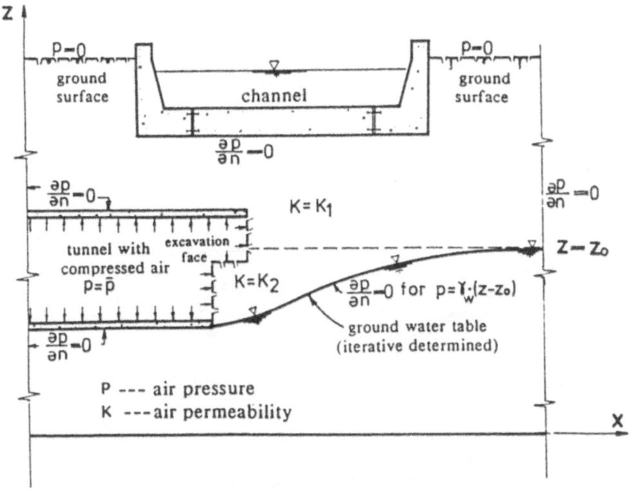

Fig.9. Boundary conditions for a 2D FE solution of the problem of lowering the ground water table using compressed air [8]

Somigliana's identity is the starting point of the analysis. It can be written as

$$q(x) = \int_{\Gamma_I} Q(y,x)\frac{\partial q}{\partial n}(y)dA(y) - \int_{\Gamma_{II}} \frac{\partial Q}{\partial n}(y,x)q(y)dA(y) + \int_{\Gamma_{II}} Q(y,x)(2p\bar{f}_n)dA(y) - \int_{\Gamma_I} \frac{\partial Q}{\partial n}(y,x)\bar{p}^2 dA(y), \quad (16)$$

where $q = p^2$, and Q is the fundamental solution, given as [9]

$$Q = \frac{(\mathrm{Det}[k^{ij}])^{\frac{1}{2}}}{4\pi R}, \quad [k^{ij}] = [k_{ij}]^{-1}, \quad R = [k^{ij}(x_i-y_i)(x_j-y_j)]^{\frac{1}{2}}, \quad i,j = 1,2,3, \quad (17)$$

with $[k_{ij}]$ representing the matrix of air permeabilities. On Γ_I, $q = \bar{q} = \bar{p}^2$, and on Γ_{II}, $\partial q/\partial n = -2pn_i k_{ij}(\partial p/\partial y_j) = 2p\bar{f}_n$, where n_i denotes the direction cosine of the outward normal n with respect to x_i, $i = 1,2,3$, and $f_n = -n_i k_{ij}(\partial p/\partial y_j) = \bar{f}_n$ is the flux in the direction of n. If Γ_{II} is an impervious boundary, then the respective integral in Eq.(16) will vanish. Eq.(16) is based on the assumption that the air obeys Darcy's law and that the density of the air is directly proportional to the pressure (Boyle's law) [9].

Fig.10 shows the iteratively determined final ground water table due to $p = \bar{p} = 60$ kN/m^2. The nonvanishing coefficients of the matrices of permeabilities, referred to the principal directions x', y', z' of the materials (assumed as $x'/\sqrt{2} = x = z$, $y' = y$, $z'/\sqrt{2} = -x = z$), were chosen as: (a) material 1 (above the original ground water table): $k'_{11} = 0.6\ 10^{-3}$ m/s, $k'_{22} = 1.5\ k'_{11}$, $k'_{33} = 2k'_{11}$, (b) material 2 (below the original ground water table): $(k'_{ij})_2 = 0.5(k'_{ij})_1$, $(k'_{ij})_1 \equiv k'_{ij}$, $i = j$. (In practice, the ratio $(k'_{ij})_2/(k'_{ij})_1$ is much smaller.) Although experiments by the authors of Ref.[8] showed that the assumption of air imperviousness of the shotcrete lining may be inadequate, the results in Fig.10 are based on this assumption. This figure also shows the distribution of the flux $f_z = -k_{zj}\ \partial p/\partial y_j$, $j = 1,2,3$, [kN/(m^{2}s)], of the air on the ground surface along lines parallel to the x- and the y-axis. $[k_{ij}]$ follows from a congruent transformation of the diagonal matrix $[k'_{ij}]$. It is noteworthy that the discretizations in Fig.10 contain infinite boundary elements.

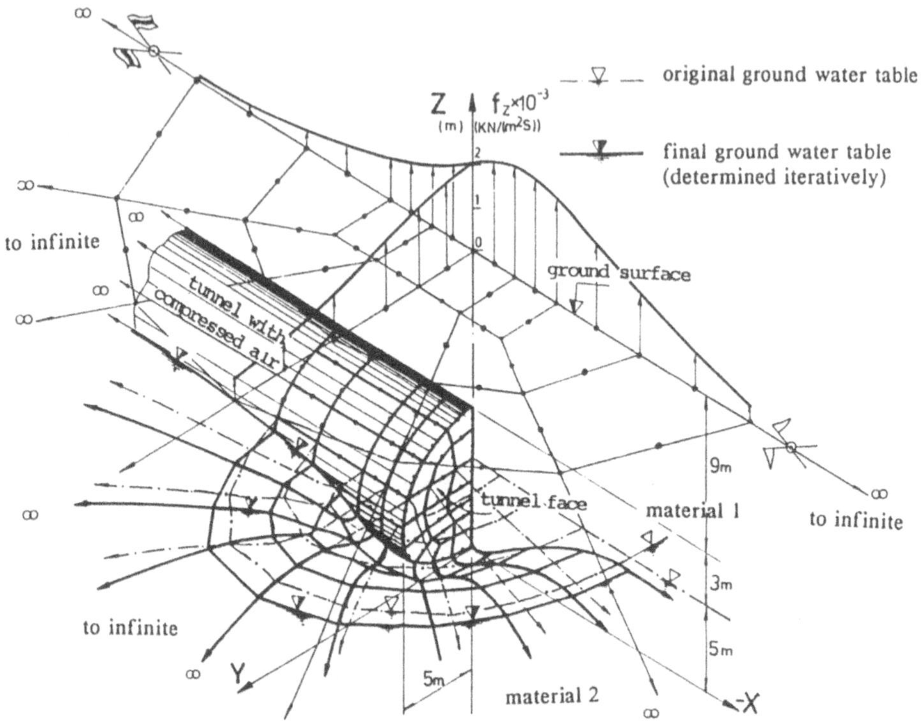

Fig.10. 3D BE solution of the problem of lowering the ground water table using compressed air

References

1. Haugeneder, E.; Mang, H.; Chen, Z.-S.; Heinrich, R.; Hofstetter, G.; Li, Z.-K.; Mehl, M.; Torzicky, P.: 3D Berechnungen von Tunnelschalen aus Stahlbeton. Final Report, Research Project Nr. 667, Straßenforschung, Bundesministerium für Wirtschaftliche Angelegenheiten, Vienna, 1988.
2. Mang, H.A.; Torzicky, P.; Chen, Z.-Y.: On the mechanical inconsistency of symmetrization of unsymmetric coupling matrices for BEFEM discretizations of solids. Computational Mechanics 4 (1989) 301-308.
3. Li, H.-B.; Han, G.-M.; Mang, H.A.; Torzicky, P.: A new method for the coupling of finite-element and boundary-element discretized subdomains of elastic bodies. Computational Methods in Applied Mechanics and Engineering 54 (1986) 161-185.
4. Chen, Z.-S.; Hofstetter, G.; Mang, H.A.: Ein neues Verfahren zur Lösung des Kantenproblems bei BEFEM-Diskretisierungen. Zeitschrift für Angewandte Mathematik und Mechanik 70 (1990) in print.
5. Hartmann, F.: Elastostatics. In: Progress in boundary element methods (ed.: Brebbia, C.A.). London: Pentech Press 1981.
6. Brady, B.H.G.; Brown, E.T.: Rock mechanics for underground mining. London: George Allen and Unwin 1985.
7. Pan, Y.-C.; Chou, T.-W.: Point force solution for an infinite transversely isotropic solid. Journal of Applied Mechanics 43 (1976) 608-612.
8. Fuchsberger, M.; Strobl, B.; Ayaydin, N.; Heinrich, R.: Tunnelling under compressed air in granular soils - research, design and site experience. In: Proceedings of the international congress on tunnels and water (ed.: Serrano, J.M.) Vol.1, 137-143. Rotterdam: A.A. Balkema 1988.
9. Chang, Y.-P.; Kang, C.-S.; Chen, D.-J.: The use of fundamental Green's functions for the solution of problems of heat conduction in anisotropic media. International Journal of Heat Mass Transfer 16 (1973) 1905-1918.

GPSR Compliance

The European Union's (EU) General Product Safety Regulation (GPSR) is a set of rules that requires consumer products to be safe and our obligations to ensure this.

If you have any concerns about our products, you can contact us on ProductSafety@springernature.com

In case Publisher is established outside the EU, the EU authorized representative is:

Springer Nature Customer Service Center GmbH
Europaplatz 3
69115 Heidelberg, Germany

Batch number: 10844669

Printed by Printforce, the Netherlands